Machine Learning and Medical Imaging

The Elsevier and MICCAI Society Book Series

Advisory board

Also available:

Zhou, Medical Image Recognition, Segmentation and Parsing, 9780128025819

Machine Learning and Medical Imaging

Guorong Wu

Dinggang Shen

Mert R. Sabuncu

AMSTERDAM • BOSTON • HEIDELBERG • LONDON
NEW YORK • OXFORD • PARIS • SAN DIEGO
SAN FRANCISCO • SINGAPORE • SYDNEY • TOKYO
Academic Press is an imprint of Elsevier

Academic Press is an imprint of Elsevier
125 London Wall, London EC2Y 5AS, United Kingdom
525 B Street, Suite 1800, San Diego, CA 92101-4495, United States
50 Hampshire Street, 5th Floor, Cambridge, MA 02139, United States
The Boulevard, Langford Lane, Kidlington, Oxford OX5 1GB, United Kingdom

Library of Congress Cataloging-in-Publication Data
A catalog record for this book is available from the Library of Congress

British Library Cataloguing-in-Publication Data
A catalogue record for this book is available from the British Library

ISBN: 978-0-12-804076-8

For information on all Academic Press publications
visit our website at https://www.elsevier.com/

Publisher: Todd Green
Acquisition Editor: Tim Pitts
Editorial Project Manager: Charlotte Kent
Production Project Manager: Nicky Carter
Designer: Greg Harris

Contents

PART 2 SUCCESSFUL APPLICATIONS IN MEDICAL IMAGING

Contributors

K. Bhatia
Imperial College London, London, United Kingdom

L. Bonilha
Medical University of South Carolina, Charleston, SC, United States

T. Brosch
The University of British Columbia, Vancouver, BC, Canada

J. Caballero
Imperial College London, London, United Kingdom

C. Davatzikos
University of Pennsylvania, Philadelphia, PA, United States

M. De Craene
Philips Medisys, Suresnes, France

M. Dewan
Flipkart, Palo Alto, CA, United States

A. Dong
University of Pennsylvania, Philadelphia, PA, United States

L. Du
Indiana University School of Medicine, Indianapolis, IN, United States

N. Duchateau
Inria Sophia Antipolis, Sophia Antipolis, France

H. Eavani
University of Pennsylvania, Philadelphia, PA, United States

A. Frangi
University of Sheffield, Sheffield, United Kingdom

J. Fridriksson
University of South Carolina, Columbia, SC, United States

Y. Gao
School of Software, Tsinghua University, Beijing, China

B. Gaonkar
University of Pennsylvania, Philadelphia, PA, United States

T. Ge
Massachusetts General Hospital/Harvard Medical School, Boston; Massachusetts General Hospital, Boston; Broad Institute of MIT and Harvard, Cambridge, MA, United States

T. He
Houston Methodist Research Institute, Houston, TX, United States

J. Huang
University of Texas at Arlington, Arlington, TX, United States

N. Honnorat
University of Pennsylvania, Philadelphia, PA, United States

B. Jian
Google, Mountain View, CA, United States

X. Jiang
The University of Georgia, Athens, GA, United States

S. Keller
University of Liverpool, Liverpool, United Kingdom

R. Kong
National University of Singapore, Singapore

Y. Li
University of Texas at Arlington, Arlington, TX, United States

M. Liu
University of North Carolina at Chapel Hill, Chapel Hill, NC, United States; Nanjing University of Aeronautics and Astronautics, Nanjing; Taishan University, Taian, China

T. Liu
The University of Georgia, Athens, GA, United States

R. Min
University of North Carolina at Chapel Hill, Chapel Hill, NC, United States

B.C. Munsell
College of Charleston, Charleston, SC, United States

Z. Peng
Siemens Healthcare, Malvern, PA, United States

G. Piella
Universitat Pompeu Fabra, Barcelona, Spain

D. Rueckert
Imperial College London, London, United Kingdom

M.R. Sabuncu
Massachusetts General Hospital/Harvard Medical School, Boston; Massachusetts Institute of Technology, Cambridge, MA, United States

A. Schaefer
National University of Singapore, Singapore

D. Shen
University of North Carolina at Chapel Hill, Chapel Hill, NC, United States

L. Shen
Indiana University School of Medicine, Indianapolis, IN, United States

J.W. Smoller
Massachusetts General Hospital/Harvard Medical School, Boston; Broad Institute of MIT and Harvard, Cambridge, MA, United States

A. Sotiras
University of Pennsylvania, Philadelphia, PA, United States

M. Stoner
University of North Carolina at Chapel Hill, Chapel Hill, NC, United States

R. Tam
The University of British Columbia, Vancouver, BC, Canada

L.Y.W. Tang
The University of British Columbia, Vancouver, BC, Canada

B.T. Thomas Yeo
National University of Singapore, Singapore

P. Thompson
University of Southern California, Los Angeles, CA, United States

T. Tong
Imperial College London, London, United Kingdom

E. Varol
University of Pennsylvania, Philadelphia, PA, United States

J. Wang
University of Michigan, Ann Arbor, MI, United States

B. Weber
University of Bonn, Bonn, Germany

G. Wu
University of North Carolina at Chapel Hill, Chapel Hill, NC, United States

F. Xing
University of Florida, Gainesville, FL, United States

Z. Xue
Houston Methodist Research Institute, Houston, TX, United States

J. Yan
Indiana University School of Medicine, Indianapolis, IN, United States

L. Yang
University of Florida, Gainesville, FL, United States

T. Yang
Arizona State University, Tempe, AZ, United States

X. Yao
Indiana University School of Medicine, Indianapolis, IN, United States

J. Ye
University of Michigan, Ann Arbor, MI, United States

Y. Yoo
The University of British Columbia, Vancouver, BC, Canada

Y. Zhan
Siemens Healthcare, Malvern, PA, United States

D. Zhang
Nanjing University of Aeronautics and Astronautics, Nanjing, China

S. Zhang
University of North Carolina at Charlotte, Charlotte, NC, United States

X. Zhang
University of North Carolina at Charlotte, Charlotte, NC, United States

X.S. Zhou
Siemens Healthcare, Malvern, PA, United States

D. Zhu
University of Southern California, Los Angeles, CA, United States

Editor Biographies

Guorong Wu is an Assistant Professor of Radiology at the Biomedical Research Imaging Center (BRIC) at the University of North Carolina at Chapel Hill. Dr. Wu received his PhD degree from the Department of Computer Science at Shanghai Jiao Tong University in 2007. After graduation, he worked for Pixelworks and joined the University of North Carolina at Chapel Hill in 2009. Dr. Wu's research aims to develop computational tools for biomedical imaging analysis and computer-assisted diagnosis. He is interested in medical image processing, machine learning, and pattern recognition. He has published more than 100 papers for international journals and conferences. Dr. Wu is active in the development of medical image processing software that facilitates scientific research on neuroscience and radiology therapy.

Dinggang Shen is a Professor of Radiology at the Biomedical Research Imaging Center (BRIC), and of Computer Science and Biomedical Engineering at the University of North Carolina at Chapel Hill (UNC-CH). He is currently directing the Center for Image Informatics and Analysis, the Image Display, Enhancement, and Analysis (IDEA) Lab in the Department of Radiology, and the medical image analysis core at the BRIC. He was a tenure-track assistant professor at the University of Pennsylvanian (UPenn), and a faculty member at Johns Hopkins University. Dr. Shen's research interests include medical image analysis, computer vision, and pattern recognition. He has published more than 700 papers for international journals and conference proceedings. He serves as an editorial board member for six international journals. Since January of 2012, he has served on the Board of Directors of The Medical Image Computing and Computer Assisted Intervention (MICCAI) Society.

Mert R. Sabuncu is an Assistant Professor of Radiology at the A.A. Martinos Center for Biomedical Imaging, Massachusetts General Hospital, Harvard Medical School, and a Research Affiliate at MIT's Computer Science and Artificial Intelligence Lab (CSAIL). Mert completed his postdoctoral training with the medical vision group at MIT's CSAIL. He received his PhD degree from the Department of Electrical Engineering at Princeton University in 2006. Mert's research aims to develop computational tools that allow us to draw inferences from biomedical image data, particularly in the context of neurology, psychiatry, and neuroscience, as well as from relationships with other data modalities, such as genetics and clinical variables. He is interested in the application of machine learning techniques for analyzing large-scale biomedical datasets and for deriving automated image processing pipelines that can automate and/or augment clinical workflows.

Preface

Machine learning is playing an increasingly essential role in the field of medical imaging, including problems as diverse as computer-assisted diagnosis, lesion detection, segmentation of anatomical regions of interest, brain mapping, spatial normalization or registration, multimodal data fusion, image-guided therapy, image annotation, and database retrieval.

Medical images offer an invaluable and rich source of information for a wide range of clinical and scientific applications. Furthermore, the technologies that allow us to capture medical images are advancing rapidly, providing higher quality images of previously unmeasured biological features at decreasing costs. Thus, in the dawning era of data-driven health sciences, image scans will make up a significant portion of the growing repositories of digital biomedical data. Machine learning, which to date has largely focused on nonmedical, artificial intelligence applications such as computer vision, promises to be an instrumental toolkit that will help unleash the potential of medical imaging.

The significant variation of anatomy across individuals and the complexities of the underlying biology often make it impossible to derive rule-based or predesigned algorithmic solutions for the problems we encounter in medical imaging applications. Therefore learning from prior data to build models about multivariate and dynamic relationships between variables and utilizing these models to make inferences is an indispensable strategy in tackling the challenges of medical image analysis. Machine learning provides an array of tools to execute this strategy.

The central aim of this book is to present the state-of-the-art of machine learning methods in medical image analysis. We created this edited volume for medical doctors, graduate and undergraduate students, academic researchers, and industry staff who are interested in translating these technologies from the academic setting to a commercial product. We anticipate that readers interested in this book will have diverse backgrounds, so we have organized the book into two parts. The first part summarizes cutting-edge machine learning algorithms in medical imaging. The topics covered in this part include not only classical probabilistic modeling and learning methods, but also recent breakthroughs in deep learning, sparse representation/coding, and big data hashing. Each chapter has a self-contained and thorough literature review, as well as an illustration of how to apply these ideas in medical imaging. In the second part, we have invited leading groups in the world to report their most recent progress, covering a wide spectrum of machine learning methods along with their application to different medical imaging modalities, clinical domains, and organs. The biomedical imaging modalities in this book include ultrasound, magnetic resonance imaging (MRI), computed tomography (CT), histology, and microscopy images. The targeted organs span the lung, liver, brain, and prostate, while there is also a treatment examining genetic associations. We roughly categorized the applications covered in this book into five topics, which include computer-assisted diagnosis, image-guided radiation therapy, landmark detection, imaging genomics, and brain connectomics. Since each application is within a specific clinical and biological context, we encouraged contributing authors to introduce the background, and provide the explanation and intuition behind each technique, aiming to make each chapter self-contained.

The topics covered in this book reflect an ongoing evolution of the field of medical imaging. As medical imaging datasets grow larger and become more complicated, more advanced machine learning methods to analyze and interpret the data will need to be developed. We hope that this book will aid research in medical image analysis by stimulating new ideas and opportunities for developing cutting-edge computational methods, which will in turn accelerate scientific discoveries and enable the translation of these ideas and discoveries into successful therapies for patients. All the contributing authors of this book share this vision, and herein are gratefully recognized for donating their knowledge, talent, and time.

Guorong Wu
Dinggang Shen
Mert R. Sabuncu

Acknowledgments

The editors are grateful for the support of the MICCAI society. Also, this work was supported in part by National Institutes of Health (NIH) grants HD081467, EB006733, EB008374, EB009634, MH100217, AG041721, AG049371, AG042599, CA140413, EB013649, and AG052246.

PART 1

Cutting-edge machine learning techniques in medical imaging

CHAPTER

Functional connectivity parcellation of the human brain

1

A. Schaefer[a], **R. Kong, B.T. Thomas Yeo**

National University of Singapore, Singapore

CHAPTER OUTLINE

1.1 INTRODUCTION

Brain disorders, comprising psychiatric and neurological disorders, are seen as one of the core health challenges of the 21st century (Wittchen et al., 2011). Their prevalence in developed countries surpasses those of cardiovascular diseases and cancer (Collins et al., 2011). Furthermore, while there have been strong advances in treatment of cardiovascular diseases, which translates to saving millions of lives each year, this progress has been absent in the treatment of psychiatric disorders (Insel, 2009). In order to develop new treatments, we need a deeper understanding of brain organization and function.

The largest structure of the brain is the cerebral cortex, which has the topology of a 2-D sheet, and is responsible for many higher order functions such as

[a]Supported by a fellowship within the Postdoc-Program of the German Academic Exchange Service (DAAD).

Machine Learning and Medical Imaging. http://dx.doi.org/10.1016/B978-0-12-804076-8.00001-3

consciousness, memory, attention, and language. The cerebral cortex can be subdivided into a larger number of different areas (Brodmann, 1909; Vogt and Vogt, 1919). Identifying these areas is important as it is believed that complex human behavior is mainly enabled by their interaction. For example, it has been shown that visual information is processed by distinct parallel pathways through the brain (Ungerleider, 1995). While each of the areas along these pathways are specialized, only their interplay allows a complex process such as visual recognition. To understand these complex interactions, identification of the distinct functional areas is needed. A map of all cortical areas is called a cortical parcellation.

Cortical areas are defined by their distinct microarchitecture, connectivity, topology, and function (Kaas, 1987; Felleman and Van Essen, 1991). Ideally we want to estimate all of these features in vivo as the location of cortical areas can vary between different subjects by more than 10 mm (Amunts et al., 1999; Fischl et al., 2008). As a consequence their results cannot be accurately translated to other (living) subjects. However, microarchitecture of cortical areas is commonly estimated by ex vivo methods which combine staining and chemicals to estimate myelo-, cyto-, or chemo-architectonics (Zilles and Amunts, 2010). While efforts have been made to develop noninvasive neuroimaging methods (Mackay et al., 1994; Glasser and Van Essen, 2011; Lodygensky et al., 2012), their resolution is much coarser compared to ex vivo methods. The function of different cortical locations can be approximated by in vivo task-based activation studies (Belliveau et al., 1991) or lesion studies (Rorden and Karnath, 2004; Preusser et al., 2015). These studies are often focused on a single brain location at a time and are therefore not ideally suited for identifying all cortical areas. Meta-analytic approaches allow the combination of hundreds or thousands of these studies (Eickhoff et al., 2009; Yarkoni et al., 2011; Fox et al., 2014; Gorgolewski et al., 2015), which can then be used to derive cortical maps (Eickhoff et al., 2011; Yeo et al., 2015). Topology can also be derived from functional activation studies, for example, for the visual cortex (Sereno et al., 1995; Swisher et al., 2007). Although it is not possible to solely use this feature to create a complete cortical map, it may be combined with other features.

In contrast, connectivity can be estimated for the whole brain within minutes via noninvasive imaging methods (Biswal et al., 2010; Craddock et al., 2013; Eickhoff et al., 2015). Functional connectivity is commonly defined as the synchronization of functional activation patterns (Friston, 1994; Biswal et al., 1995; Smith et al., 2011). In recent years, functional connectivity has been widely used as a feature for estimating brain parcellation (Shen et al., 2010; Yeo et al., 2011; Craddock et al., 2012; Varoquaux and Craddock, 2013). Two examples are shown in Fig. 1.1. Structural connections can be assessed by water diffusivity, which aims to identify white matter axons (Tuch et al., 2003; Hagmann et al., 2007; Johansen-Berg and Behrens, 2013). Connectivity poses some advantages compared to other features as it is not bounded to a specific task and can therefore be assessed in a single experiment.

Human brain parcellation has become an exciting field for the application of machine learning approaches as the dimensionality and diversity of the acquired data keeps increasing. With the rise of large-scale imaging initiatives (Biswal et al., 2010; Van Essen et al., 2012; Nooner et al., 2012; Zuo et al., 2014; Holmes et al., 2015),

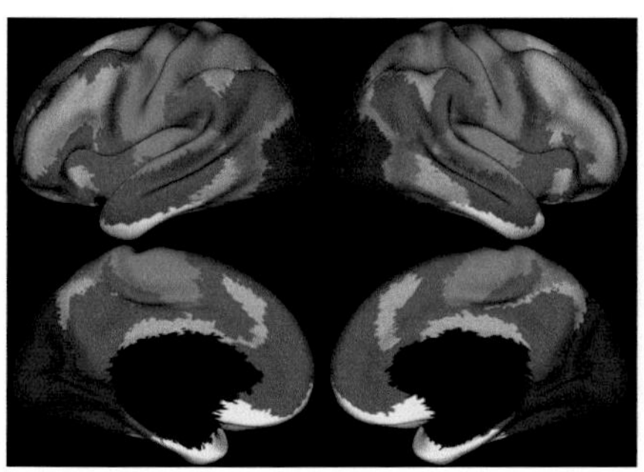
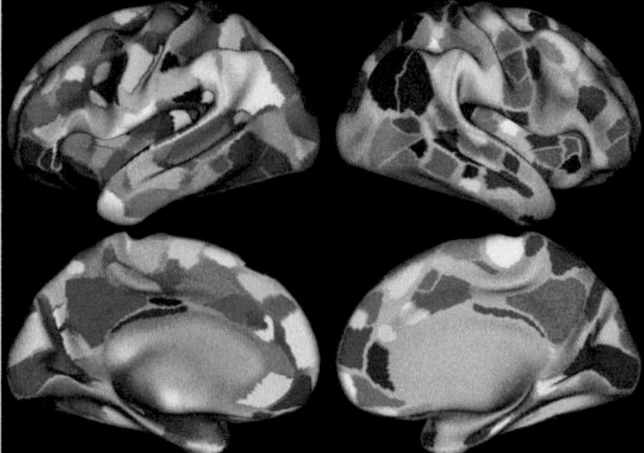

FIG. 1.1

Cortical labels based on functional connectivity. (Left) Average functional connectivity of 1000 subjects modeled with a mixture of von Mises-Fisher distributions and clustered into networks of areas with similar connectivity patterns (Yeo et al., 2011). (Right) Areas derived from estimating gradients of functional connectivity and applying a watershed transform (Gordon et al., 2014).

vast amounts of multimodal brain information have become publicly available, providing information about connectivity, function, and anatomy. These large datasets pose fascinating challenges for machine learning approaches.

In this chapter, we provide an overview of current approaches to connectivity-based parcellations with particular emphasis on mixture models and Markov random fields (MRFs) in Sections 1.3 and 1.4, respectively.

1.2 APPROACHES TO CONNECTIVITY-BASED BRAIN PARCELLATION

This section provides a general overview of different approaches to brain parcellation based on connectivity features. We will cover mixture and MRF models in more details in Sections 1.3 and 1.4.

In connectivity-based brain parcellations we want to model the relationship between connectivity features and cortical labels. Let us assume N brain locations denoted by $x_1, \ldots, x_N$. Let Y_n be the connectivity at brain location x_n. We assume that each connectivity feature Y_n is D-dimensional. We further assume L cortical labels of interest with $l_n \in \{1, \ldots, L\}$. Our goal is to estimate $l_1, \ldots, l_N$ for the N brain locations $x_1, \ldots, x_N$. Often we will just write $l_{1:N}$ instead of $l_1, \ldots, l_N$. Finding an assignment $l_{1:N}$ can be seen as a segmentation problem, which is a well-studied objective in pattern recognition and machine learning.

Common approaches to brain networks estimation include independent component analyses (ICAs) (Bell and Sejnowski, 1995; Calhoun et al., 2001; Beckmann and Smith, 2004), which in the context of brain parcellations is a form of temporal demixing of spatial factors. An ICA is commonly formulated as a matrix decomposition problem. More specifically, the decomposition is

$$S = EY, \tag{1.1}$$

where Y is a $D \times N$ matrix of D-dimensional observations for each of the N brain locations, S is an $L \times N$ matrix of L hidden factors, and E is an $L \times D$ demixing matrix. Here the matrix S represents a soft estimate of the labels. An ICA maximizes the independence of the signals in S. This independence assumption is sometimes criticized for its limited biological validity (Harrison et al., 2015). An ICA is a soft "clustering" algorithm which may result in overlapping areas and is therefore not directly a solution to the parcellation problem. Manual or automatic thresholding techniques (eg, random walks) can be applied to retrieve a parcellation (Abraham et al., 2014). ICAs belong to the class of linear signal decomposition models, which includes dictionary learning. This more general form can, for example, be used to build hierarchical models that model the variability between subjects (Varoquaux et al., 2011).

K-means clustering (Lloyd, 1982; Jain, 2010) is another widely used approach for connectivity-based parcellation (Mezer et al., 2009; Bellec et al., 2010; Kim et al., 2010; Cauda et al., 2011; Zhang and Li, 2012; Mars et al., 2012). The approach performs a mapping to a preselected number of K nonoverlapping clusters. In our notation K equals L. K-means clustering is performed in an iterative fashion by first hard-assigning the labels of the N brain regions to their respective closest centers. Then the L cluster centers are recomputed. The whole process is repeated until convergence. The initial cluster centers are usually assigned randomly and the final labeling is highly dependent on this initial assignment. Consequently, the algorithm is typically repeated many times and the solution with the best cost function value is selected.

Spectral clustering (Jianbo Shi and Malik, 2000; Ng et al., 2001) is also widely used for connectivity-based parcellation (Johansen-Berg et al., 2004; Thirion et al., 2006; van den Heuvel et al., 2008; Shen et al., 2010; Craddock et al., 2012). The approach is based on the affinity between each pair of locations x_i and x_j. Affinity is typically computed by some form of similarity between the corresponding observations y_i and y_j. In our case this similarity is often the similarity of structural or functional connectedness of x_i and x_j. Based on the affinity matrix A we can compute a Laplacian (von Luxburg, 2007), for example,

$$L = DA, \tag{1.2}$$

where D is the degree matrix $D(i,i) = \sum_j A(i,j)$. The idea of spectral clustering is to decompose this matrix into the eigenvectors of L. Then the assumption is that the data points are easier to separate in this eigenspace than in the original space. A related approach identifies areas that maximize modularity (Meunier et al., 2010; He et al., 2009). Modularity is maximized by selecting areas which have high within-module connections and low between-module connections. Modularity maximization is NP-hard but can be approximated by spectral methods (Newman, 2006).

Agglomerative hierarchical clustering (Eickhoff et al., 2011; Michel et al., 2012; Blumensath et al., 2013; Orban et al., 2015; Thirion et al., 2014; Moreno-Dominguez et al., 2014) is a bottom-up approach in which initially every brain location x_n has a separate label l_n. The agglomerative clustering builds a hierarchy by iteratively merging two labels based on a linkage criterion. This linkage criterion needs to be defined, for example, as the average similarity between all the connectivity features belonging to the two labels. The derived hierarchical organization allows a multiresolution parcellation by thresholding the hierarchy at different levels.

Gradient approaches identify rapid transitions in the connectivity pattern of adjacent regions (Cohen et al., 2008; Gordon et al., 2014; Wig et al., 2014). For example, transitions can be identified by applying a Canny edge detector (Canny, 1986) on the cortical surface. Gradient approaches have been applied on parts of the cortex (Cohen et al., 2008; Nelson et al., 2010a,b; Hirose et al., 2012), as well as the entire cerebral cortex (Wig et al., 2014; Gordon et al., 2014). A cortical parcellation (Fig. 1.1, right) can be derived by applying a watershed transform (Beucher and Lantuejoul, 1979) on the resulting gradient maps (Gordon et al., 2014). The first step of the watershed transform involves identifying local gradient minima as seed regions. The seed regions are then iteratively grown by including neighboring brain locations that have a gradient value lower than the current threshold. The threshold is iteratively increased until every location belongs to one of the seed regions corresponding to the local minima.

1.3 MIXTURE MODEL

In this section, we provide a more detailed introduction to the mixture model approach. We will show examples and describe a popular inference approach to learn the mixture parameters and labels.

Mixture models are a flexible way of clustering as they allow soft or weighted assignments. The value of the weight indicates the strength of the affinity to the corresponding cluster. In mixture models, this weight corresponds to the posterior probability that the data point belongs to a mixture component. A final parcellation or labeling can be gained by thresholding the posterior probabilities.

1.3.1 MODEL

A mixture model is a combination of component models which form a richer model:

$$p(Y) = \sum_{l=1}^{L} p(Y|l)p(l|\alpha_l). \tag{1.3}$$

The features Y are observed, whereas the labels l are hidden to us. The component models are given by $p(Y|l)$ and the prior probabilities of the different components are given by α_l, where α_l is non-negative and $\sum_{l=1}^{L} \alpha_l = 1$. We can interpret the above equation as a generative model to sample a data point Y. The probability of label l is α_l. We first draw a label l with $p(l|\alpha_l)$, and then draw an observation Y from $p(Y|l)$.

Let the probability distribution $p(Y|l)$ be parameterized by θ_l. Then the model becomes

$$p(Y|\Theta) = \sum_{l=1}^{L} p(Y|\theta_l)p(l|\alpha_l), \tag{1.4}$$

where $\Theta = \{\alpha_{1:L}, \theta_{1:L}\}$.

For our parcellation, we assume that each label l_n is independently drawn from a probability distribution $p(l_{1:N}) = \prod_n p(l_n)$. The features $Y_1, \ldots, Y_N$ are assumed to be independent conditioned on $\Theta = \{\alpha_{1:L}, \theta_{1:L}\}$, and so the mixture model is of the form:

$$p(Y_{1:N}|\Theta) = \prod_{n=1}^{N} p(Y_n|\Theta) = \prod_{n=1}^{N} \sum_{l_n=1}^{L} p(Y_n|\theta_{l_n})p(l_n|\alpha_{l_n}). \tag{1.5}$$

Conditioned on cortical label l_n at spatial location x_n, the observed features Y_n are assumed to be generated from the distribution $p(Y_n|\theta_{l_n})$.

In the case of a Gaussian mixture with parameter $\theta_{l_n} = \{\mu_{l_n}, \Sigma_{l_n}\}$ (Golland et al., 2007; Tucholka et al., 2008; Jbabdi et al., 2009), the distribution is set to

$$p(Y_n|\theta_{l_n}) = \mathcal{N}(Y_n|\mu_{l_n}, \Sigma_{l_n}), \tag{1.6}$$

where the Gaussian distribution $\mathcal{N}(Y_n|\mu_{l_n}, \Sigma_{l_n})$ is defined as

$$\mathcal{N}(Y_n|\mu_{l_n}, \Sigma_{l_n}) = \frac{1}{\sqrt{\det(2\pi\Sigma_{l_n})}} e^{-\frac{1}{2}(Y_n-\mu_{l_n})^{\mathrm{T}}\Sigma_{l_n}^{-1}(Y_n-\mu_{l_n})}, \tag{1.7}$$

where μ_{l_n} is the mean and Σ_{l_n} is the covariance matrix of cluster l_n.

Another example is a mixture of von Mises-Fisher distributions with parameter $\theta_{l_n} = \{\mu_{l_n}, \kappa\}$ (Yeo et al., 2011; Ryali et al., 2013; Liu et al., 2014), where

$$p(Y_n|\theta_{l_n}) = \mathrm{vmf}(Y_n|\mu_{l_n}, \kappa). \tag{1.8}$$

The von Mises-Fisher distribution $\mathrm{vmf}(Y_n|\mu_{l_n}, \kappa)$ is defined as

$$\mathrm{vmf}(Y_n|\mu_{l_n}, \kappa) = z_D(\kappa)e^{\kappa\mu_{l_n}^{\mathrm{T}} Y_n}, \tag{1.9}$$

with mean direction $||\mu_{l_n}|| = 1$, concentration parameter κ and dimensionality $D \geq 2$. The normalizing constant $z_D(\kappa)$ is given by

$$z_D(\kappa) = \frac{\kappa^{D/2-1}}{(2\pi)^{D/2} I_{D/2-1}(\kappa)}, \tag{1.10}$$

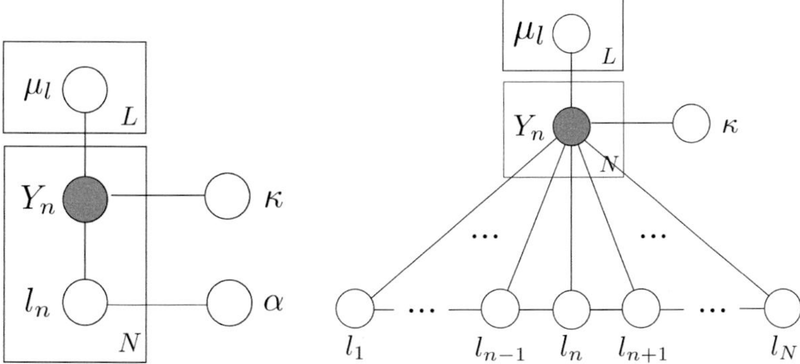

FIG. 1.2

Graphical representation of von Mises-Fisher mixture model with and without MRF. Shaded variables are observed. All other variables are hidden. Lines indicate dependencies between the variables. Any part of a graphical model within a plate with subscript X is replicated X times, where all dependencies between variables outside of the plate and variables on the same plate are preserved. There are no dependencies between elements on different replicates of a plate. (Left) Graphical model for a mixture of von Mises-Fisher distributions model (Yeo et al., 2011) (Section 1.3). Each label l_n is independently drawn from a probability distribution $p(l_{1:N}) = \prod_n p(l_n)$. The hidden label l_n indexes the mixture component. The probability of l_n is given by α. Given the label l_n and the corresponding von Mises-Fisher parameters κ and μ_{l_n}, the observations Y_n are generated via $p(Y_n|l_n, \mu_{l_n}, \kappa)$. (Right) Graphical model for a mixture of von Mises-Fisher distributions together with an MRF prior (Ryali et al., 2013) (Section 1.4). The MRF prior comprises dependencies between the labels of neighboring brain locations, for example, label l_n at brain location x_n is dependent on the labels at neighboring brain locations.

where $I_r(\cdot)$ represents the modified Bessel function of the first kind and order r. The graphical representation of the von Mises-Fisher mixture model is given in Fig. 1.2.

1.3.2 INFERENCE

The labels $l_{1:N}$ and parameters Θ are unknowns that need to be estimated. The optimal parameters Θ are typically estimated by maximum likelihood estimation:

$$\arg\max_{\Theta} p(Y_1, \ldots, Y_N|\Theta) = \arg\max_{\Theta} \prod_n p(Y_n|\Theta). \tag{1.11}$$

In other words we want to find the parameters that maximize the likelihood of our observations. A common approach to estimate hidden variables and parameters is expectation-maximization (EM) (Dempster et al., 1977). In EM we use an iterative two-step process. In the E-step we compute the posterior $p(l_{1:N}|Y_{1:N}, \Theta^t)$ of the labels $l_{1:N}$ given the current parameters Θ^t at iteration t. In the M-step we compute the parameters Θ^{t+1} based on the current estimate of the posterior probability $p(l_{1:N}|Y,\Theta^t)$. Notice that EM computes soft assignments instead of hard decisions.

As an example, we provide derivations for the mixture of von Mises-Fisher distributions (Lashkari et al., 2010). The likelihood of the mixture model is

$$p(Y_n|\Theta) = \prod_{n=1}^{N}\sum_{l_n=1}^{L} p(Y_n|\theta_{l_n})p(l_n|\alpha_{l_n}), \tag{1.12}$$

where $\theta_{l_n} = \{\mu_{l_n}, \kappa\}$ and $p(Y_n|\theta_{l_n}) = z_D(\kappa)e^{\kappa\mu_{l_n}^{\mathrm{T}}Y_n}$ as in Eq. (1.9). To derive EM, we apply log to the likelihood:

$$\log p(Y_{1:N}|\Theta) = \sum_{n=1}^{N}\log p(Y_n|\Theta) = \sum_{n=1}^{N}\log\sum_{l_n=1}^{L} p(Y_n|\theta_{l_n})p(l_n|\alpha_{l_n}). \tag{1.13}$$

Now we introduce a probability distribution $q_n(l_n)$ over the latent labels:

$$\log p(Y_{1:N}|\Theta) = \sum_{n=1}^{N}\log\sum_{l_n=1}^{L}\frac{p(Y_n|\theta_{l_n})p(l_n|\alpha_{l_n})q_n(l_n)}{q_n(l_n)}. \tag{1.14}$$

Using Jensen's inequality we can write

$$\log p(Y_{1:N}|\Theta) = \sum_{n=1}^{N}\log\sum_{l_n=1}^{L}\frac{p(Y_n|\theta_{l_n})p(l_n|\alpha_{l_n})q_n(l_n)}{q_n(l_n)} \tag{1.15}$$

$$\geq \sum_{n=1}^{N}\sum_{l_n=1}^{L} q_n(l_n)\log\frac{p(Y_n|\theta_{l_n})p(l_n|\alpha_{l_n})}{q_n(l_n)}. \tag{1.16}$$

If $q_n(l_n)$ is equal to $p(l_n|Y_n,\Theta)$, the inequality in Eq. (1.16) becomes an equality. Eq. (1.16) is sometimes referred to as the completed log likelihood.

In the expectation or E-step, we estimate the posterior probability of the labels given the current estimate of the parameters. Given the current estimate of $\Theta^t = \{\theta_{1:L}^t, \alpha_{1:L}^t\}$ with $\theta_l^t = \{\mu_l^t, \kappa^t\}$ at iteration t, we compute

$$q_n^{t+1}(l_n) = p(l_n|Y_n,\Theta^t) = \frac{p(Y_n|\theta_{l_n}^t)p(l_n|\alpha_{l_n}^t)}{\sum_{l'=1}^{L} p(Y_n|\theta_{l'}^t)p(l'|\alpha_{l'}^t)}, \tag{1.17}$$

where

$$p(Y_n|\theta_{l_n}^t) = z_D(\kappa^t)e^{\kappa^t\mu_{l_n}^{tT}Y_n}. \tag{1.18}$$

In the M-step, we estimate the parameters $\Theta^{t+1} = \{\theta_{1:L}^{t+1}, \alpha_{1:L}^{t+1}\}$ by maximizing the completed log likelihood (Eq. 1.16) using the current estimate of $q_n^{t+1}(l_n)$:

$$\arg\max_{\Theta} \sum_{n=1}^{N} \sum_{l_n=1}^{L} q_n^{t+1}(l_n) \log \frac{p(Y_n|\theta_{l_n})p(l_n|\alpha_{l_n})}{q_n^{t+1}(l_n)} \tag{1.19}$$

$$= \arg\max_{\Theta} \sum_{n=1}^{N} \sum_{l_n=1}^{L} q_n^{t+1}(l_n) \log p(Y_n|\theta_{l_n})p(l_n|\alpha_{l_n}). \tag{1.20}$$

Using Eq. (1.18), we can write the last equation as

$$\arg\max_{\Theta} \sum_{n=1}^{N} \sum_{l_n=1}^{L} q_n^{t+1}(l_n) \log p(l_n|\alpha_{l_n}) + \sum_{n=1}^{N} \sum_{l_n=1}^{L} q_n^{t+1}(l_n) \log p(Y_n|\theta_{l_n}) \tag{1.21}$$

$$= \arg\max_{\Theta} \sum_{n=1}^{N} \sum_{l_n=1}^{L} q_n^{t+1}(l_n) \log \alpha_{l_n} + \sum_{n=1}^{N} \sum_{l_n=1}^{L} q_n^{t+1}(l_n) \log z_D(\kappa) \tag{1.22}$$

$$+ \sum_{n=1}^{N} \sum_{l_n=1}^{L} q_n^{t+1}(l_n)(\kappa \mu_{l_n}^{\mathrm{T}} Y_n).$$

To estimate the parameters $\Theta = \{\mu_{1:L}, \alpha_{1:L}, \kappa\}$ we can write the lower bound (Eq. 1.22) with only the dependent terms. Together with the Lagrange multipliers η_l and β for constraints $\mu_l^{\mathrm{T}} \mu_l = 1$, $\sum_{l=1}^{L} \alpha_l = 1$, we have

$$\mathcal{L}_\mu = \sum_{n=1}^{N} \sum_{l=1}^{L} q_n^{t+1}(l_n = l)(\kappa \mu_l^{\mathrm{T}} Y_n) + \sum_{l=1}^{L} \eta_l (1 - \mu_l^{\mathrm{T}} \mu_l), \tag{1.23}$$

$$\mathcal{L}_\kappa = \sum_{n=1}^{N} \sum_{l=1}^{L} q_n^{t+1}(l_n = l) \log z_D(\kappa) + \sum_{n=1}^{N} \sum_{l=1}^{L} q_n^{t+1}(l_n = l)(\kappa \mu_l^{\mathrm{T}} Y_n), \tag{1.24}$$

$$\mathcal{L}_\alpha = \sum_{n=1}^{N} \sum_{l=1}^{L} q_n^{t+1}(l_n = l) \log \alpha_l + \beta \left(1 - \sum_{l=1}^{L} \alpha_l\right). \tag{1.25}$$

To estimate μ_l, we take the derivative of $\mathcal{L}_\mu$ and set it to zero:

$$\frac{\partial \mathcal{L}_\mu}{\partial \mu_l} = \sum_{n=1}^{N} q_n^{t+1}(l_n = l)\kappa Y_n - 2\eta_l \mu_l = 0, \tag{1.26}$$

$$\mu_l^{t+1} = \frac{\sum_{n=1}^{N} q_n^{t+1}(l_n = l)\kappa Y_n}{2\eta_l}. \tag{1.27}$$

The Lagrange multiplier η_l is determined by the fact that μ_l should be unit norm, and so we get

$$\mu_l^{t+1} = \frac{\sum_{n=1}^{N} q_n^{t+1}(l_n = l) Y_n}{\| \sum_{n=1}^{N} q_n^{t+1}(l_n = l) Y_n \|}. \tag{1.28}$$

The unknown concentration parameter κ was canceled out from the numerator and denominator in the last equation. To estimate κ, we take the derivative of $\mathcal{L}_\kappa$ from Eq. (1.24) and set it to zero:

$$\frac{\partial \mathcal{L}_\kappa}{\partial \kappa} = \sum_{n=1}^{N} \sum_{l=1}^{L} q_n^{t+1}(l_n = l) \frac{z_D'(\kappa)}{z_D(\kappa)} + \sum_{n=1}^{N} \sum_{l=1}^{L} q_n^{t+1}(l_n = l)(\mu_l^{\mathrm{T}} Y_n) \tag{1.29}$$

$$= N \frac{z_D'(\kappa)}{z_D(\kappa)} + \sum_{n=1}^{N} \sum_{l=1}^{L} q_n^{t+1}(l_n = l)(\mu_l^{\mathrm{T}} Y_n) = 0. \tag{1.30}$$

Solving the above equation for κ is hard (Banerjee et al., 2005), but there exist several approximation methods (Banerjee et al., 2005; Lashkari et al., 2010; Sra, 2011). Lashkari et al. (2010) suggested the following:

$$\kappa^{t+1} = \frac{(D-2)\Gamma}{1-\Gamma^2} + \frac{(D-1)\Gamma}{2(D-2)}, \tag{1.31}$$

where $\Gamma = \frac{1}{N} \sum_{n=1}^{N} \delta(l_n, l) Y_n^{\mathrm{T}} \mu_l$ and $\delta(l_n, l) = 1, \exists n : x_n = l$; 0 otherwise. Now we estimate α by taking the derivative of $\mathcal{L}_\alpha$ from Eq. (1.25):

$$\frac{\partial \mathcal{L}_\alpha}{\partial \alpha_l} = \frac{1}{\alpha_l} \sum_{n=1}^{N} q_n^{t+1}(l_n = l) - \beta = 0, \tag{1.32}$$

$$\alpha_l^{t+1} = \frac{1}{N} \sum_{n=1}^{N} q_n^{t+1}(l_n = l). \tag{1.33}$$

More details on inference in mixture models can also be found in the books by Bishop (2006), Barber (2012), and Koller and Friedman (2009).

1.4 MARKOV RANDOM FIELD MODEL

The previous independence assumption of $p(l_{1:N}) = \prod_n p(l_n)$ might be too strong as spatially neighboring locations often belong to the same brain area. Hence it is popular to make the weaker assumption that each $p(l_n)$ is independent given its neighborhood (Liu et al., 2011, 2012, 2014; Ryali et al., 2013; Honnorat et al., 2013, 2015). This is also called the Markov assumption.

1.4.1 MODEL

The Markov assumption can be modeled by MRFs, which are a form of probabilistic graphical models. For this we need an undirected graph which can be defined as: $G = (V, E)$, with vertex set V and edge set E. The graph structure is commonly given by the brain locations and their spatial proximity.

MRFs can then be defined by local joint probabilities which model a relationship of the assigned labels $\{l_1, \ldots, l_N\}$ using a neighborhood system $\mathcal{N}$. Here, the neighborhood system $\mathcal{N}$ constitutes for every vertex $v \in V : \mathcal{N}(v) = \{u \in V : (u, v) \in E\}$. Then the Markov assumption can be written as $p(l_n| - l_n) = p(l_n|l_x$ with $x \in \mathcal{N}(n))$, where $-l_n$ are all labels $l_{1:N}$ without l_n. However, this is only a local property. Based on the equivalence of MRFs and Gibbs fields (Besag, 1974; Geman and Geman, 1984), the Hammersly-Clifford theorem states that this local property can be transformed into a global property. A Gibbs field X takes the form of

$$p(X) = \frac{1}{\mathcal{Z}} \mathrm{e}^{-E(X)}, \tag{1.34}$$

where $\mathcal{Z}$ is a normalizing constant that guarantees that the function sums to one (ie, valid probability distribution). This involves a summation over all possible configurations of X:

$$\mathcal{Z} = \sum_{X} \mathrm{e}^{-E(X)}. \tag{1.35}$$

While $\mathcal{Z}$ is not practically tractable it might be approximated, for example, by a pseudo-likelihood (Ryali et al., 2013). The energy function $E(X)$ is the sum over all clique potentials Φ_c over all possible cliques C:

$$E(X) = \sum_{c \in C} \Phi_c(X_c), \tag{1.36}$$

where each clique potential Φ_c only depends on the variables in the corresponding cliques c. A clique is a subgraph in which every pair of vertices is connected. For example, every single vertex or every pair of connected vertices v_i, v_j with $(v_i, v_j) \in E$ are trivial examples of cliques. In the following we will focus on these single and pairwise clique potentials, although more complicated forms are possible. In our context the energy over X (Eq. 1.36) will be an energy over labelings $l_{1:N}$:

$$E(l_{1:N}) = \sum_{n=1}^{N} \Phi(l_n) + \sum_{n=1}^{N} \sum_{j \in N_n} \Phi_{\text{neigh}}(l_n, l_j). \tag{1.37}$$

One common approach is to combine a mixture model with a MRF prior (Jbabdi et al., 2009; Ryali et al., 2013; Liu et al., 2014). The singleton clique potentials $\Phi(l_n)$ often comprise a data cost term. This term gives a penalty for label assignments that fit the data poorly. For example, this can be formulated as the negative log-likelihood (Ryali et al., 2013):

$$\Phi(l_n) \triangleq \Phi_{\text{obs}}(l_n) = -\log p(Y_n|l_n), \tag{1.38}$$

where the likelihood can be given by a von Mises-Fisher distribution (Ryali et al., 2013) as in Eq. (1.9). For the pairwise potential there exist different options. A very common idea (Jbabdi et al., 2009; Ryali et al., 2013; Liu et al., 2014) is the Potts model (Potts, 1952), which includes the penalty that neighboring brain locations x_n and x_j are assigned different labels:

$$\Phi_{\text{neigh}}(l_n, l_j) = \begin{cases} 0 & \text{if } l_n = l_j \\ \gamma & \text{if } l_n \neq l_j. \end{cases}, \tag{1.39}$$

where γ is a positive parameter that is typically manually tuned. The Potts model enforces piecewise spatial consistency of the label assignment. When we combine the potentials from Eqs. (1.38) and (1.39), our energy term (Eq. 1.37) takes the following form:

$$E(l_{1:N}) = \sum_{n=1}^{N} \Phi_{\text{obs}}(l_n) + \sum_{n=1}^{N} \sum_{j \in N_n} \Phi_{\text{neigh}}(l_n, l_j). \tag{1.40}$$

This corresponds to the graph in Fig. 1.2 (right).

In the context of brain parcellations, we would often prefer each parcel to be topologically connected, rather than spatially distributed. Honnorat et al. (2015) provide an elegant MRF solution to enforce topologically connected parcels by first defining one cluster center i for each brain location x_i and comparing the data Y_n of the current location x_n with the data Y_i of all cluster centers i:

$$\Phi_{\text{obs}}(l_n = i) = -Y_i Y_n^{\text{T}}, \tag{1.41}$$

where we assume that each data vector Y_n has been normalized to zero mean and unit variance. We assume each brain location could potentially be a cluster center, that is, $L = N$. Honnorat et al. (2015) then utilized star shape priors to enforce connectedness of each cluster using a distance metric (Veksler, 2008). First, a distance metric between neighboring brain locations is defined as

$$d(n, m) = 1 - Y_n Y_m^{\text{T}}, \tag{1.42}$$

where $m \in \mathcal{N}(n)$. Based on this metric, the distance between any two brain locations x_i and x_j can then be computed by summing over the distances on the shortest path between x_i and x_j.

Recall that every brain location could be a potential cluster center. Accordingly, there exists as many potential cluster centers as brain locations. Then for each brain location x_j the neighboring brain location x_k along the shortest path towards the cluster center i is enforced to have the same label as the center by an infinite weight:

$$\Phi_{i,j,k}(l_j, l_k) = \begin{cases} \infty & \text{if } l_j = i \text{ and } l_k \neq i, \\ 0 & \text{otherwise.} \end{cases} \tag{1.43}$$

Here, in contrast to the Potts model, the pairwise potentials vary spatially.

Assigning a separate label l_n for each brain location x_n would maximize the data cost term. To prevent this overfit, Honnorat et al. (2015) introduce an additional cost (Delong et al., 2010) for the number of labels. The label cost balances a potential overfit by placing a penalty on the number of labels:

$$\Phi_{\text{label}}(l_{1:N}) = c \sum_{l=1}^{L} \delta_l(l), \tag{1.44}$$

where $\delta_l(l) = 1, \exists n : x_n = l$; 0 otherwise, and c controls the amount of overall label cost. Therefore, Eq. (1.44) semi-automatically estimates the final number of labels. Parcellations for varying label costs are displayed in Fig. 1.3.

Hence, the overall energy function (Honnorat et al., 2015) can be written as

$$E(l_{1:N}) = \sum_{n=1}^{N} \Phi_{\text{obs}}(l_n) + \Phi_{\text{label}}(l_{1:N}) + \sum_{n=1}^{N}\sum_{j=1}^{N} \Phi_{n,j,k}(l_j, l_k), \tag{1.45}$$

where k is the neighbor of j that is closest to n.

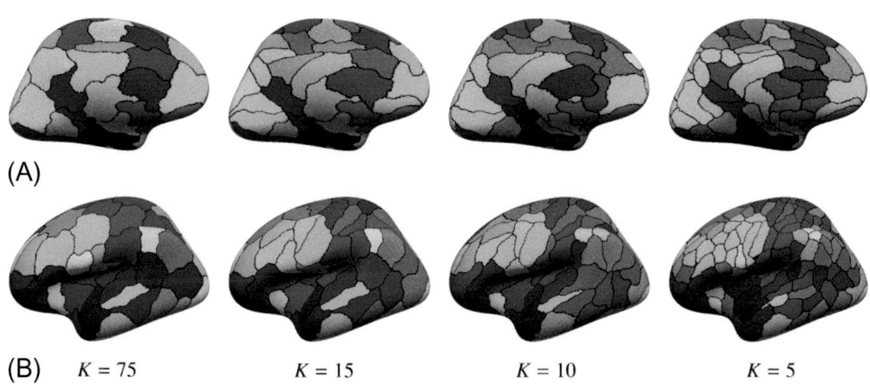

FIG. 1.3

Cortical labels based on functional connectivity from 859 fMRI scans modeled with a MRF model (Honnorat et al., 2015). Increasing number of labels for decreasing label cost K (Eq. 1.44). (A) Medial surface of the left hemisphere. (B) Lateral surface of the left hemisphere.

Source: Reprinted from Honnorat, N., Eavani, H., Satterthwaite, T., Gur, R., Gur, R., Davatzikos, C., 2015. GraSP: geodesic graph-based segmentation with shape priors for the functional parcellation of the cortex. NeuroImage 106, 207—221, with permission from Elsevier.

Most brain parcellation approaches assume that there is a single connectivity dataset (van den Heuvel et al., 2008; Yeo et al., 2011; Craddock et al., 2012) obtained by averaging connectivity over many subjects. However, it is known that there exists a large amount of variability betweens subjects (Amunts et al., 1999; Fischl et al., 2008). This variability can be modeled using an MRF approach. Liu et al. (2014) introduce a set of graphs $\{G_j = (V_j, E_j)\}$, one for each subject $j \in \{1, \ldots, J\}$, as well as one group graph $G_G = (V_G, E_G)$. Then the Potts model $\Phi_{\text{neigh}}(l_n, l_m)$ from Eq. (1.39) was used to specify a hierarchical MRF prior (Liu et al., 2014):

$$E_V(l_{1:N}) = \sum_{s,r \in V_G} \beta \Phi_{\text{neigh}}(l_s, l_r) \tag{1.46}$$

$$+ \sum_{j=1}^{J} \left(\sum_{s \in V_G, \hat{s} \in V_j} \alpha \Phi_{\text{neigh}}(l_s, l_{\hat{s}}) + \sum_{s,r \in V_j} \beta \Phi_{\text{neigh}}(l_s, l_r) \right), \tag{1.47}$$

where $\sum_{s,r \in V_G} \beta \Phi_{\text{neigh}}(l_s, l_r)$ and $\sum_{s,r \in V_j} \beta \Phi_{\text{neigh}}(l_s, l_r)$ enforce piecewise continuous labels on the group and subject level, respectively. $\alpha \Phi_{\text{neigh}}(l_s, l_{\hat{s}})$ penalizes different labels between parcellations at the subject and group levels. The parameters α and β control the strength of the respective potentials.

1.4.2 INFERENCE

Given our model and observed connectivity data $Y_{1:N}$, we want to perform inference of the unknown labels $l_{1:N}$. The complexity of the inference depends on the graph structure (Fig. 1.2), more precisely on the presence of loops in the graph. While exact inference in tree-like graphs is feasible, for general graphs we can often only approximate the inference (Bishop, 2006; Koller and Friedman, 2009; Barber, 2012).

To give an intuition of why exact inference is not possible in practice, the search space is huge as there are L^N possible solutions. Because of the interaction at each brain location x_n with its neighborhood the inference cannot be factorized.

Approximate inference can be categorized into two forms. The first form are stochastic approaches (Jbabdi et al., 2009) that, given an infinite amount of computational resources, produce exact results. The approximation arises from the natural limit of computational resources (Bishop, 2006), which can be problematic even for small instances. The second form are deterministic approximation schemes (Woolrich and Behrens, 2006; Tucholka et al., 2008; Ryali et al., 2013), which scale well, even on larger instances. Popular approximation schemes include graph cuts (Boykov et al., 2001; Delong et al., 2010), linear programming (Komodakis and Tziritas, 2007; Komodakis et al., 2011), and variational approaches, such as variational EM (Wainwright and Jordan, 2007). Here, we will give an example of deterministic approximation using variational EM. Variational EM analytically approximates the posterior probabilities by additional independence assumptions. As such they can never generate exact results even with infinite computational resources.

Given observed data $Y_{1:N}$, hidden labels $l_{1:N}$ and model parameters Θ, we aim to maximize the log likelihood. The log likelihood can be written as the marginal over the observed data $Y_{1:N}$ in terms of the sum over the joint distribution of hidden labels $l_{1:N}$ and observed data $Y_{1:N}$:

$$\log p(Y_{1:N}|\Theta) = \log \sum_{l_{1:N}} p(Y_{1:N}, l_{1:N}|\Theta). \tag{1.48}$$

Using Jensen's inequality, we can define a lower bound on the log likelihood:

$$\log p(Y_{1:N}|\Theta) = \log \sum_{l_{1:N}} \frac{p(Y_{1:N}, l_{1:N}|\Theta)q(l_{1:N})}{q(l_{1:N})} \tag{1.49}$$

$$\geq \sum_{l_{1:N}} q(l_{1:N}) \log \frac{p(Y_{1:N}, l_{1:N}|\Theta)}{q(l_{1:N})} \tag{1.50}$$

$$= \mathcal{L}(q, \Theta). \tag{1.51}$$

The difference between the log likelihood $\log p(Y_{1:N}|\Theta)$ and the lower bound $\mathcal{L}(q, \Theta)$ can be expressed by the Kullback-Leibler (KL) divergence:

$$\log p(Y_{1:N}|\Theta) - \mathcal{L}(q, \Theta) \tag{1.52}$$

$$= \log p(Y_{1:N}|\Theta) - \sum_{l_{1:N}} q(l_{1:N}) \log \frac{p(Y_{1:N}, l_{1:N}|\Theta)}{q(l_{1:N})} \tag{1.53}$$

$$= \log p(Y_{1:N}|\Theta) - \sum_{l_{1:N}} q(l_{1:N}) \log \frac{p(l_{1:N}|Y_{1:N}, \Theta)p(Y_{1:N}|\Theta)}{q(l_{1:N})} \tag{1.54}$$

$$= - \sum_{l_{1:N}} q(l_{1:N}) \log \frac{p(l_{1:N}|Y_{1:N}, \Theta)}{q(l_{1:N})} \tag{1.55}$$

$$= \mathrm{KL}(q(l_{1:N})||p(Y_{1:N}|l_{1:N}, \Theta)). \tag{1.56}$$

As a consequence, we can decompose the log likelihood as

$$\log p(Y_{1:N}|\Theta) = \mathcal{L}(q, \Theta) + \mathrm{KL}(q(l_{1:N})||p(Y_{1:N}|l_{1:N}, \Theta)). \tag{1.57}$$

As the KL divergence $\mathrm{KL}(q(l_{1:N})||p(Y_{1:N}|l_{1:N}, \Theta))$ is always non-negative, instead of directly maximizing $\log p(Y_{1:N}|\Theta)$, we can maximize its lower bound $\mathcal{L}(q, \Theta)$. For our model with an MRF prior, a common approximation (Bishop, 2006) is the mean-field approximation, where $q(l_{1:N})$ are assumed to be factorizable:

$$q(l_{1:N}) = \prod_{n=1}^{N} q_n(l_n), \tag{1.58}$$

where each l_n is a single variable in $l_{1:N}$. To maximize the lower bound $\mathcal{L}(q, \Theta)$, in the E-step we fix the model parameters Θ and make use of the factorization in Eq. (1.58) to optimize $q(l_{1:N})$. Therefore the E-step is equivalent to minimizing the

KL divergence $\mathrm{KL}(q(l_{1:N})||p(Y_{1:N}|l_{1:N},\Theta))$, which aims to find the $q(l_{1:N})$ closest to the exact posterior $p(l_{1:N}|Y_{1:N},\Theta)$:

$$q^{t+1}(l_{1:N}) = \underset{q(l_{1:N})}{\arg\max}\, \mathcal{L}(q,\Theta^t) = \underset{q(l_{1:N})}{\arg\min}\, \mathrm{KL}(q(l_{1:N})||p(Y_{1:N}|l_{1:N},\Theta^t)). \tag{1.59}$$

In the M-step, we optimize the model parameters Θ to maximize the lower bound $\mathcal{L}(q,\Theta)$ using $q_n(l_n)$ updated from the E-step:

$$\Theta^{t+1} = \underset{\Theta}{\arg\max}\, \mathcal{L}(q^{t+1},\Theta). \tag{1.60}$$

Here, we provide further detailed derivations in the context of a von Mises-Fisher MRF model (Fig. 1.2, right). The MRF prior corresponds to the example we have given in Eq. (1.39). Conditioned on the segmentation label l_n, we assume the connectivity features Y_n are generated from the von Mises-Fisher distribution with parameters $\theta_{l_n} = \{\mu_{l_n},\kappa\}$, and therefore

$$\Phi_{\mathrm{obs}}(l_n) = -\log p(Y_n|\theta_{l_n}) = -\log\left(z_D(\kappa)\mathrm{e}^{\kappa\mu_{l_n}^{\mathrm{T}}Y_n}\right). \tag{1.61}$$

Therefore the energy function over $l_{1:N}$ is

$$E(l_{1:N}) = \sum_{n=1}^{N}\Phi_{\mathrm{obs}}(l_n) + \sum_{n=1}^{N}\sum_{j\in\mathcal{N}_n}\Phi_{\mathrm{neigh}}(l_n,l_j) \tag{1.62}$$

$$= -\sum_{n=1}^{N}\log p(Y_n|\theta_{l_n}) + \sum_{n=1}^{N}\sum_{j\in\mathcal{N}_n}\Phi_{\mathrm{neigh}}(l_n,l_j). \tag{1.63}$$

Then the joint probability of hidden labels and observed data is

$$p(Y_{1:N},l_{1:N}|\Theta) = \frac{1}{\mathcal{Z}}\mathrm{e}^{-E(l_{1:N})} \tag{1.64}$$

$$= \frac{1}{\mathcal{Z}}\mathrm{e}^{\log\sum_{n=1}^{N}p(Y_n|\theta_{l_n})-\sum_{n=1}^{N}\sum_{j\in\mathcal{N}_n}\Phi_{\mathrm{neigh}}(l_n,l_j)}, \tag{1.65}$$

where model parameters $\Theta = \{\mu_{1:L},\kappa\}$, and Φ_{neigh} are defined as in Eq. (1.39). Given the observed data $Y_{1:N}$, hidden labels $l_{1:N}$ and model parameters $\Theta = \{\mu_{1:L},\kappa\}$, we maximize the log likelihood:

$$\underset{\Theta}{\arg\max}\,\log p(Y_{1:N}|\Theta) \tag{1.66}$$

$$= \underset{\Theta}{\arg\max}\,\log\sum_{l_{1:N}}p(Y_{1:N},l_{1:N}|\Theta) \tag{1.67}$$

$$= \underset{\Theta}{\arg\max}\,\log\sum_{l_{1:N}}\frac{1}{\mathcal{Z}}\mathrm{e}^{\sum_{n=1}^{N}p(Y_n|\theta_{l_n})-\sum_{n=1}^{N}\sum_{j\in\mathcal{N}_n}\Phi_{\mathrm{neigh}}(l_n,l_j)}. \tag{1.68}$$

Optimizing Eq. (1.68) is computationally intractable. We use variational EM to iteratively estimate Eq. (1.68), where we alternate between estimating the parameters from the von Mises-Fisher distributions $\Theta = \{\mu_{1:L}, \kappa\}$ and the hidden labels $l_{1:N}$.

To maximize Eq. (1.68), we maximize the lower bound:

$$\mathcal{L}(q, \Theta) = \sum_{l_{1:N}} q(l_{1:N}) \log \frac{p(Y_{1:N}, l_{1:N}|\Theta)}{q(l_{1:N})} \tag{1.69}$$

$$= \sum_{l_{1:N}} q(l_{1:N}) \log p(Y_{1:N}, l_{1:N}|\Theta) - \sum_{l_{1:N}} q(l_{1:N}) \log q(l_{1:N}), \tag{1.70}$$

where

$$\log p(Y_{1:N}, l_{1:N}|\Theta) = -\log \mathcal{Z} + \log \sum_{n=1}^{N} p(Y_n|\theta_{l_n}) - \sum_{n=1}^{N} \sum_{j \in \mathcal{N}_n} \Phi_{\text{neigh}}(l_n, l_j). \tag{1.71}$$

After expanding the terms, the lower bound $\mathcal{L}(q, \Theta)$ can be expressed as

$$\mathcal{L}(q, \Theta) = \sum_{n=1}^{N} \sum_{l_n=1}^{L} q_n(l_n) p(Y_n|\theta_{l_n}) - \sum_{n=1}^{N} \sum_{l_n=1}^{L} \sum_{j \in \mathcal{N}_n} \sum_{l_j=1}^{L} q_n(l_n) q_j(l_j) \Phi_{\text{neigh}}(l_n, l_j)$$
$$- \sum_{n=1}^{N} \sum_{l_n=1}^{L} q_n(l_n) \log q_n(l_n). \tag{1.72}$$

To simplify the notation in Eq. (1.72), let $\lambda_{n,l_n} = q_n(l_n)$, so the lower bound can be written as

$$\mathcal{L}(\lambda, \Theta) = \sum_{n=1}^{N} \sum_{l_n=1}^{L} \lambda_{n,l_n} \log p(Y_n|\theta_{l_n}) - \sum_{n=1}^{N} \sum_{l_n=1}^{L} \sum_{j \in \mathcal{N}_n} \sum_{l_j=1}^{L} \lambda_{n,l_n} \lambda_{j,l_j} \Phi_{\text{neigh}}(l_n, l_j)$$
$$- \sum_{n=1}^{N} \sum_{l_n=1}^{L} \lambda_{n,l_n} \log \lambda_{n,l_n}. \tag{1.73}$$

Variational EM proceeds as follows. We begin with the variational E-step where we estimate λ_{n,l_n} to maximize the lower bound $\mathcal{L}$. We add a Lagrange multiplier η_n to ensure that $\sum_{l_n} \lambda_{n,l_n} = 1$ for all n. We denote the current estimate of our parameters as Θ^t. The lower bound can be written as

$$\mathcal{L}(\lambda, \Theta^t) = \sum_{n=1}^{N} \sum_{l_n=1}^{L} \lambda_{n,l_n} \log p(Y_n|\theta^t_{l_n}) - \sum_{n=1}^{N} \sum_{l_n=1}^{L} \sum_{j \in N_n} \sum_{l_j=1}^{L} \lambda_{n,l_n} \lambda_{j,l_j} \Phi_{\text{neigh}}(l_n, l_j)$$
$$- \sum_{n=1}^{N} \sum_{l_n=1}^{L} \lambda_{n,l_n} \log \lambda_{n,l_n} + \sum_{n=1}^{N} \eta_n \left(\sum_{l_n=1}^{L} \lambda_{n,l_n} - 1 \right). \tag{1.74}$$

We will estimate $\lambda_{m,k}$ for each brain region $m \in \{1, \ldots, N\}$ and label $k \in \{1, \ldots, L\}$ by differentiating Eq. (1.74) and setting the derivative to 0:

$$\frac{\partial \mathcal{L}(\lambda, \Theta^t)}{\partial \lambda_{m,k}} = \log p(Y_m|\theta_k^t) - 2\sum_{j \in N_m}\sum_{l_j=1}^{L} \lambda_{j,l_j} \Phi_{\text{neigh}}(l_m = k, l_j)$$
$$- 1 - \log \lambda_{m,k} + \eta_m = 0. \tag{1.75}$$

Note the factor of 2 in the second term, which arises because we assume that if x_n is a neighbor of x_m, then x_m is also a neighbor of x_n and because of the symmetric nature of the Potts model. Rearranging the above equation, we get

$$\log \lambda_{m,k}^{t+1} \propto \log p(Y_m|\theta_k^t) - 2\sum_{j \in N_m}\sum_{l_j=1}^{L} \lambda_{j,l_j} \Phi_{\text{neigh}}(l_m = k, l_j), \tag{1.76}$$

$$\lambda_{m,k}^{t+1} \propto p(Y_m|\mu_k^t, \kappa^t) e^{-2\sum_{j \in N_m}\sum_{l_j=1}^{L} \lambda_{j,l_j} \Phi_{\text{neigh}}(l_m = k, l_j)}. \tag{1.77}$$

Since the update for $\lambda_{m,k}^{t+1}$ depends on λ_{j,l_j}, we estimate $\lambda_{m,k}^{t+1}$ via fixed point iterations using Eq. (1.77), and normalizing λs in each iteration so that $\sum_{l_n} \lambda_{n,l_n} = 1$ for all n.

In the variational M-step, we compute the parameters Θ^{t+1} with the lower bound (Eq. 1.73) based on the current estimate $\lambda_{1:N,1:L}^{t+1}$. By dropping the terms that do not contain the parameters Θ, we get

$$\Theta^{t+1} = \arg\max_{\Theta} \mathcal{L}(\lambda^{t+1}, \Theta) = \arg\max_{\mu_{1:L},\kappa} \sum_{n=1}^{N}\sum_{l_n} \lambda_{n,l_n}^{t+1} \log p(Y_n|\theta_{l_n}). \tag{1.78}$$

Using the Lagrange multiplier β_l for constraints $\mu_l^{\mathrm{T}}\mu_l = 1$, we can write the above lower bound as separate optimizations over $\mu_{1:L}$ and κ:

$$\mathcal{L}_\mu = \sum_{n=1}^{N}\sum_{l=1}^{L} \lambda_{n,l}^{t+1}(\kappa \mu_l^{\mathrm{T}} Y_n) + \sum_{l=1}^{L} \beta_l(1 - \mu_l^{\mathrm{T}}\mu_l), \tag{1.79}$$

$$\mathcal{L}_\kappa = \sum_{n=1}^{N}\sum_{l=1}^{L} \lambda_{n,l}^{t+1} \log z_D(\kappa) + \sum_{n=1}^{N}\sum_{l=1}^{L} \lambda_{n,l}^{t+1}(\kappa \mu_l^{\mathrm{T}} Y_n). \tag{1.80}$$

To compute μ_l^{t+1} we take the derivative of $\mathcal{L}_\mu$:

$$\frac{\partial \mathcal{L}_\mu}{\partial \mu_l} = \sum_{n=1}^{N} \lambda_{n,l}^{t+1} \kappa Y_n - 2\beta_l \mu_l = 0, \tag{1.81}$$

$$\mu_l^{t+1} = \frac{\sum_{n=1}^{N} \lambda_{n,l}^{t+1} \kappa Y_n}{2\beta_l}. \tag{1.82}$$

Just like in the case of the mixture model (Section 1.3), the Lagrange multiplier β_l is determined by the fact that μ_l should be unit norm, and so we get

$$\mu_l^{t+1} = \frac{\sum_{n=1}^{N} \lambda_{n,l}^{t+1} Y_n}{\| \sum_{n=1}^{N} \lambda_{n,l}^{t+1} Y_n \|}. \tag{1.83}$$

To estimate κ^{t+1}, we set the derivative of $\mathcal{L}_\kappa$ to 0:

$$\frac{\partial \mathcal{L}_\kappa}{\partial \kappa} = \sum_{n=1}^{N} \sum_{l=1}^{L} \lambda_{n,l}^{t+1} \frac{z_D'(\kappa)}{z_D(\kappa)} + \sum_{n=1}^{N} \sum_{l=1}^{L} \lambda_{n,l}^{t+1} \mu_l^{\mathrm{T}} Y_n \tag{1.84}$$

$$= N \frac{z_D'(\kappa)}{z_D(\kappa)} + \sum_{n=1}^{N} \sum_{l=1}^{L} \lambda_{n,l}^{t+1} \mu_l^{\mathrm{T}} Y_n = 0. \tag{1.85}$$

We can again use the approximation given by Lashkari et al. (2010):

$$\kappa^{t+1} \approx \frac{(D-2)\Gamma}{1-\Gamma^2} + \frac{(D-1)\Gamma}{2(D-2)}, \tag{1.86}$$

where $\Gamma = \frac{1}{N} \sum_{n=1}^{N} \delta(l_n, l) Y_n^{\mathrm{T}} \mu_l$ and $\delta(l_n, l) = 1, \exists n : x_n = l;\ 0$ otherwise.

We iterate between estimating the parameters $\Theta = \{\mu_{1:L}, \kappa\}$ (using Eqs. 1.83 and 1.86), and λ_{n,l_n} for each location x_n and label l_n (using Eq. 1.78). To generate a final parcellation, for each brain location n, the label l_n with highest λ_{n,l_n} can be chosen.

1.5 SUMMARY

Human brain parcellation is one of the major challenges in systems neuroscience and key for understanding complex human behavior. Machine learning has been and will continue to be a central element in deriving human brain parcellations as the underlying datasets become larger and more diverse. Here, we have focused on mixture and MRFs, which can be easily combined and extended to match a wide range of applications. These models can, for example, be used to create personalized brain parcellations while using population priors to increase stability (Jbabdi et al., 2009; Liu et al., 2011, 2014; Harrison et al., 2015). The resulting single subject parcellations can address the strong intersubject variability in brain organization and therefore improve sensitivity in clinical applications.

REFERENCES

Abraham, A., Dohmatob, E., Thirion, B., Samaras, D., Varoquaux, G., 2014. Region segmentation for sparse decompositions: better brain parcellations from rest fMRI, pp. 1–8, arXiv 1412.3925.

Amunts, K., Schleicher, A., Buergel, U., Mohlberg, H., Uylings, H.B., Zilles, K., 1999. Broca's region revisited: cytoarchitecture and intersubject variability. J. Comp. Neurol. 412 (2), 319–341. http://dx.doi.org/10.1002/(SICI)1096-9861(19990920)412:2<319::AID-CNE10>3.0.CO;2-7.

Banerjee, A., Dhillon, I.S., Ghosh, J., Sra, S., Ridgeway, G., 2005. Clustering on the unit hypersphere using von Mises-Fisher distributions. J. Mach. Learn. Res. 6 (9), 1345–1382.

Barber, D., 2012. Bayesian Reasoning and Machine Learning. Cambridge University Press, Cambridge, UK.

Beckmann, C.F., Smith, S.M., 2004. Probabilistic independent component analysis for functional magnetic resonance imaging. IEEE Trans. Med. Imaging 23 (2), 137–152. http://dx.doi.org/10.1109/TMI.2003.822821.

Bell, A.J., Sejnowski, T.J., 1995. An information-maximization approach to blind separation and blind deconvolution. Neural Comput. 7 (6), 1129–1159.

Bellec, P., Rosa-Neto, P., Lyttelton, O.C., Benali, H., Evans, A.C., 2010. Multi-level bootstrap analysis of stable clusters in resting-state fMRI. NeuroImage 51 (3), 1126–1139. http://dx.doi.org/10.1016/j.neuroimage.2010.02.082.

Belliveau, J.W., Kennedy, D.N., McKinstry, R.C., Buchbinder, B.R., Weisskoff, R.M., Cohen, M.S., Vevea, J.M., Brady, T.J., Rosen, B.R., 1991. Functional mapping of the human visual cortex by magnetic resonance imaging. Science 254 (5032), 716–719.

Besag, J., 1974. Spatial interaction and the statistical analysis of lattice systems. J. R. Stat. Soc. B 36 (2), 192–236. http://dx.doi.org/10.2307/2984812.

Beucher, S., Lantuejoul, C., 1979. Use of watersheds in contour detection. In: International Workshop on Image Processing: Real-Time Edge and Motion Detection/Estimation, Rennes, France.

Bishop, C.M., 2006. Pattern Recognition and Machine Learning. Springer, Berlin.

Biswal, B., Yetkin, F.Z., Haughton, V.M., Hyde, J.S., 1995. Functional connectivity in the motor cortex of resting human brain using echo-planar MRI. Magn. Reson. Med. 34 (4), 537–541.

Biswal, B.B., Mennes, M., Zuo, X.N., Gohel, S., Kelly, C., Smith, S.M., Beckmann, C.F., Adelstein, J.S., Buckner, R.L., Colcombe, S., Dogonowski, A.M., Ernst, M., Fair, D., Hampson, M., Hoptman, M.J., Hyde, J.S., Kiviniemi, V.J., Kötter, R., Li, S.J., Lin, C.P., Lowe, M.J., Mackay, C., Madden, D.J., Madsen, K.H., Margulies, D.S., Mayberg, H.S., McMahon, K., Monk, C.S., Mostofsky, S.H., Nagel, B.J., Pekar, J.J., Peltier, S.J., Petersen, S.E., Riedl, V., Rombouts, S.A.R.B., Rypma, B., Schlaggar, B.L., Schmidt, S., Seidler, R.D., Siegle, G.J., Sorg, C., Teng, G.J., Veijola, J., Villringer, A., Walter, M., Wang, L., Weng, X.C., Whitfield-Gabrieli, S., Williamson, P., Windischberger, C., Zang, Y.F., Zhang, H.Y., Castellanos, F.X., Milham, M.P., 2010. Toward discovery science of human brain function. Proc. Natl. Acad. Sci. USA 107 (10), 4734–4739. http://dx.doi.org/10.1073/pnas.0911855107.

Blumensath, T., Jbabdi, S., Glasser, M.F., Van Essen, D.C., Ugurbil, K., Behrens, T.E., Smith, S.M., 2013. Spatially constrained hierarchical parcellation of the brain with resting-state fMRI. NeuroImage 76, 313–324. http://dx.doi.org/10.1016/j.neuroimage.2013.03.024.

Boykov, Y., Veksler, O., Zabih, R., 2001. Fast approximate energy minimization via graph cuts. IEEE Trans. Pattern Anal. Mach. Intell. 23 (11), 1222–1239. http://dx.doi.org/10.1109/34.969114.

Brodmann, K., 1909. Vergleichende Lokalisationslehre der Grosshirnrinde: in ihren Prinzipien dargestellt auf Grund des Zellenbaues. JA Barth, Leipzig.

Calhoun, V., Adali, T., Pearlson, G., Pekar, J., 2001. A method for making group inferences from functional MRI data using independent component analysis. Hum. Brain Map. 14 (3), 140–151. http://dx.doi.org/10.1002/hbm.1048.

Canny, J., 1986. A computational approach to edge detection. IEEE Trans. Pattern Anal. Mach. Intell. 8 (6), 679–698. http://dx.doi.org/10.1109/TPAMI.1986.4767851.

Cauda, F., D'Agata, F., Sacco, K., Duca, S., Geminiani, G., Vercelli, A., 2011. Functional connectivity of the insula in the resting brain. NeuroImage 55 (1), 8–23. http://dx.doi.org/10.1016/j.neuroimage.2010.11.049.

Cohen, A.L., Fair, D.A., Dosenbach, N.U.F., Miezin, F.M., Dierker, D., Van Essen, D.C., Schlaggar, B.L., Petersen, S.E., 2008. Defining functional areas in individual human brains using resting functional connectivity MRI. NeuroImage 41 (1), 45–57. http://dx.doi.org/10.1016/j.neuroimage.2008.01.066.

Collins, P.Y., Patel, V., Joestl, S.S., March, D., Insel, T.R., Daar, A.S., Bordin, I.A., Costello, E.J., Durkin, M., Fairburn, C., Glass, R.I., Hall, W., Huang, Y., Hyman, S.E., Jamison, K., Kaaya, S., Kapur, S., Kleinman, A., Ogunniyi, A., Otero-Ojeda, A., Poo, M.M., Ravindranath, V., Sahakian, B.J., Saxena, S., Singer, P.A., Stein, D.J., Anderson, W., Dhansay, M.A., Ewart, W., Phillips, A., Shurin, S., Walport, M., 2011. Grand challenges in global mental health. Nature 475 (7354), 27–30. http://dx.doi.org/10.1038/475027a.

Craddock, R.C., James, G.A., Holtzheimer, P.E., Hu, X.P., Mayberg, H.S., 2012. A whole brain fMRI atlas generated via spatially constrained spectral clustering. Hum. Brain Map. 33 (8), 1914–1928. http://dx.doi.org/10.1002/hbm.21333.

Craddock, R.C., Jbabdi, S., Yan, C.g., Vogelstein, J.T., Castellanos, F.X., Di Martino, A., Kelly, C., Heberlein, K., Colcombe, S., Milham, M.P., 2013. Imaging human connectomes at the macroscale. Nat. Meth. 10 (6), 524–539. http://dx.doi.org/10.1038/nmeth.2482.

Delong, A., Osokin, A., Isack, H.N., Boykov, Y., 2010. Fast approximate energy minimization with label costs. In: 2010 IEEE Computer Society Conference on Computer Vision and Pattern Recognition. IEEE, Piscataway, NJ, pp. 2173–2180.

Dempster, A.P., Laird, N.M., Rubin, D.B., 1977. Maximum likelihood from incomplete data via the EM algorithm. J. R. Stat. Soc. B 39 (1), 1–38.

Eickhoff, S.B., Laird, A.R., Grefkes, C., Wang, L.E., Zilles, K., Fox, P.T., 2009. Coordinate-based activation likelihood estimation meta-analysis of neuroimaging data: a random-effects approach based on empirical estimates of spatial uncertainty. Hum. Brain Map. 30 (9), 2907–2926. http://dx.doi.org/10.1002/hbm.20718.

Eickhoff, S.B., Bzdok, D., Laird, A.R., Roski, C., Caspers, S., Zilles, K., Fox, P.T., United States, 2011. Co-activation patterns distinguish cortical modules, their connectivity and functional differentiation. NeuroImage 57 (3), 938–949. http://dx.doi.org/10.1016/j.neuroimage.2011.05.021.

Eickhoff, S.B., Thirion, B., Varoquaux, G., Bzdok, D., 2015. Connectivity-based parcellation: critique and implications. Hum. Brain Map. 36, 4771–4792. http://dx.doi.org/10.1002/hbm.22933.

Felleman, D.J., Van Essen, D.C., 1991. Distributed hierarchical processing in the primate cerebral cortex. Cerebral Cortex (NY) 1 (1), 1–47. http://dx.doi.org/10.1093/cercor/1.1.1.

Fischl, B., Rajendran, N., Busa, E., Augustinack, J., Hinds, O., Yeo, B.T.T., Mohlberg, H., Amunts, K., Zilles, K., 2008. Cortical folding patterns and predicting cytoarchitecture. Cerebral Cortex 18 (8), 1973–1980. http://dx.doi.org/10.1093/cercor/bhm225.

Fox, P.T., Lancaster, J.L., Laird, A.R., Eickhoff, S.B., 2014. Meta-analysis in human neuroimaging: computational modeling of large-scale databases. Ann. Rev. Neurosci. 37, 409–434. http://dx.doi.org/10.1146/annurev-neuro-062012-170320.

Friston, K.J., 1994. Functional and effective connectivity in neuroimaging: a synthesis. Hum. Brain Map. 2 (1–2), 56–78. http://dx.doi.org/10.1002/hbm.460020107.

Geman, S., Geman, D., 1984. Stochastic relaxation, Gibbs distributions, and the Bayesian restoration of images. IEEE Trans. Pattern Anal. Mach. Intell. 6 (6), 721–741. http://dx.doi.org/10.1109/TPAMI.1984.4767596.

Glasser, M.F., Van Essen, D.C., 2011. Mapping human cortical areas in vivo based on myelin content as revealed by T1- and T2-weighted MRI. J. Neurosci. 31 (32), 11597–11616. http://dx.doi.org/10.1523/JNEUROSCI.2180-11.2011.

Golland, P., Golland, Y., Malach, R., 2007. Detection of spatial activation patterns as unsupervised segmentation of fMRI data. In: Medical Image Computing and Computer-Assisted Intervention—MICCAI 2007. Springer, Berlin, pp. 110–118.

Gordon, E.M., Laumann, T.O., Adeyemo, B., Huckins, J.F., Kelley, W.M., Petersen, S.E., 2014. Generation and evaluation of a cortical area parcellation from resting-state correlations. Cerebral Cortex 26 (1), 288–303.

Gorgolewski, K.J., Varoquaux, G., Rivera, G., Schwarz, Y., Ghosh, S.S., Maumet, C., Sochat, V.V., Nichols, T.E., Poldrack, R.A., Poline, J.B., Yarkoni, T., Margulies, D.S., 2015. NeuroVault.org: a web-based repository for collecting and sharing unthresholded statistical maps of the human brain. Front. Neuroinformat. 9. http://dx.doi.org/10.3389/fninf.2015.00008.

Hagmann, P., Kurant, M., Gigandet, X., Thiran, P., Wedeen, V.J., Meuli, R., Thiran, J.P., 2007. Mapping human whole-brain structural networks with diffusion MRI. PLoS ONE 2 (7), e597. http://dx.doi.org/10.1371/journal.pone.0000597.

Harrison, S.J., Woolrich, M.W., Robinson, E.C., Glasser, M.F., Beckmann, C.F., Jenkinson, M., Smith, S.M., 2015. Large-scale probabilistic functional modes from resting state fMRI. NeuroImage 109, 217–231. http://dx.doi.org/10.1016/j.neuroimage.2015.01.013.

He, Y., Wang, J., Wang, L., Chen, Z.J., Yan, C., Yang, H., Tang, H., Zhu, C., Gong, Q., Zang, Y., Evans, A.C., 2009. Uncovering intrinsic modular organization of spontaneous brain activity in humans. PLoS ONE 4 (4), e5226. http://dx.doi.org/10.1371/journal.pone.0005226.

Hirose, S., Watanabe, T., Jimura, K., Katsura, M., Kunimatsu, A., Abe, O., Ohtomo, K., Miyashita, Y., Konishi, S., 2012. Local signal time-series during rest used for areal boundary mapping in individual human brains. PLoS ONE 7 (5), e36496. http://dx.doi.org/10.1371/journal.pone.0036496.

Holmes, A.J., Hollinshead, M.O., O'Keefe, T.M., Petrov, V.I., Fariello, G.R., Wald, L.L., Fischl, B., Rosen, B.R., Mair, R.W., Roffman, J.L., Smoller, J.W., Buckner, R.L., 2015. Brain Genomics Superstruct Project initial data release with structural, functional, and behavioral measures. Sci. Data 2, 150031. http://dx.doi.org/10.1038/sdata.2015.31.

Honnorat, N., Eavani, H., Satterthwaite, T.D., Davatzikos, C., 2013. A graph-based brain parcellation method extracting sparse networks. In: 2013 International Workshop on Pattern Recognition in Neuroimaging. IEEE, Piscataway, NJ, pp. 157–160.

Honnorat, N., Eavani, H., Satterthwaite, T., Gur, R., Gur, R., Davatzikos, C., 2015. GraSP: Geodesic graph-based segmentation with shape priors for the functional parcellation of the cortex. NeuroImage 106, 207–221. http://dx.doi.org/10.1016/j.neuroimage.2014.11.008.

Insel, T.R., 2009. Translating scientific opportunity into public health impact. Arch. Gen. Psychiat. 66 (2), 128. http://dx.doi.org/10.1001/archgenpsychiatry.2008.540.

Jain, A.K., 2010. Data clustering: 50 years beyond K-means. Pattern Recogn. Lett. 31 (8), 651–666. http://dx.doi.org/10.1016/j.patrec.2009.09.011. 0402594v3.

Jbabdi, S., Woolrich, M.W., Behrens, T.E.J., 2009. Multiple-subjects connectivity-based parcellation using hierarchical Dirichlet process mixture models. NeuroImage 44 (2), 373–384. http://dx.doi.org/10.1016/j.neuroimage.2008.08.044.

Jianbo Shi, Malik, J., 2000. Normalized cuts and image segmentation. IEEE Trans. Pattern Anal. Mach. Intell. 22 (8), 888–905. http://dx.doi.org/10.1109/34.868688.

Johansen-Berg, H., Behrens, T.E.J. (Eds.), 2013. Diffusion MRI: From Quantitative Measurement to In Vivo Neuroanatomy. Academic Press, London.

Johansen-Berg, H., Behrens, T.E.J., Robson, M.D., Drobnjak, I., Rushworth, M.F.S., Brady, J.M., Smith, S.M., Higham, D.J., Matthews, P.M., 2004. Changes in connectivity profiles define functionally distinct regions in human medial frontal cortex. Proc. Natl. Acad. Sci. USA 101 (36), 13335–13340. http://dx.doi.org/10.1073/pnas.0403743101.

Kaas, J., 1987. The organization of neocortex in mammals: implications for theories of brain function. Ann. Rev. Psychol. 38 (1), 129–151. http://dx.doi.org/10.1146/annurev.psych.38.1.129.

Kim, J.H., Lee, J.M., Jo, H.J., Kim, S.H., Lee, J.H., Kim, S.T., Seo, S.W., Cox, R.W., Na, D.L., Kim, S.I., Saad, Z.S., 2010. Defining functional SMA and pre-SMA subregions in human MFC using resting state fMRI: functional connectivity-based parcellation method. NeuroImage 49 (3), 2375–2386. http://dx.doi.org/10.1016/j.neuroimage.2009.10.016.

Koller, D., Friedman, N., 2009. Probabilistic Graphical Models: Principles and Techniques. MIT Press, Cambridge, MA.

Komodakis, N., Tziritas, G., 2007. Approximate labeling via graph cuts based on linear programming. IEEE Trans. Pattern Anal. Mach. Intell. 29 (8), 1436–1453. http://dx.doi.org/10.1109/TPAMI.2007.1061.

Komodakis, N., Paragios, N., Tziritas, G., 2011. MRF energy minimization and beyond via dual decomposition. IEEE Trans. Pattern Anal. Mach. Intell. 33 (3), 531–552. http://dx.doi.org/10.1109/TPAMI.2010.108.

Lashkari, D., Vul, E., Kanwisher, N., Golland, P., 2010. Discovering structure in the space of fMRI selectivity profiles. NeuroImage 50 (3), 1085–1098. http://dx.doi.org/10.1016/j.neuroimage.2009.12.106.

Liu, W., Awate, S.P., Anderson, J.S., Yurgelun-Todd, D., Fletcher, P.T., 2011. Monte Carlo expectation maximization with hidden Markov models to detect functional networks in resting-state fMRI. In: MICCAI Workshop on Machine Learning in Medical Imaging. Springer, New York, pp. 59–66.

Liu, W., Awate, S.P., Fletcher, P.T., 2012. Group analysis of resting-state fMRI by hierarchical Markov random fields. Med. Image Comput. Comput. Assis. Interven. 15 (3), 189–196. http://dx.doi.org/10.1126/science.3749875.

Liu, W., Awate, S.P., Anderson, J.S., Fletcher, P.T., 2014. A functional network estimation method of resting-state fMRI using a hierarchical Markov random field. NeuroImage 100, 520–534. http://dx.doi.org/10.1016/j.neuroimage.2014.06.001.

Lloyd, S.P., 1982. Least squares quantization in PCM. IEEE Trans. Inform. Theory 28 (2), 129–137.

Lodygensky, G.A., Marques, J.P., Maddage, R., Perroud, E., Sizonenko, S.V., Hüppi, P.S., Gruetter, R., 2012. In vivo assessment of myelination by phase imaging at high magnetic field. NeuroImage 59 (3), 1979–1987. http://dx.doi.org/10.1016/j.neuroimage.2011.09.057.

Mackay, A., Whittall, K., Adler, J., Li, D., Paty, D., Graeb, D., 1994. In vivo visualization of myelin water in brain by magnetic resonance. Magn. Reson. Med. 31 (6), 673–677. http://dx.doi.org/10.1002/mrm.1910310614.

Mars, R.B., Sallet, J., Schuffelgen, U., Jbabdi, S., Toni, I., Rushworth, M.F.S., 2012. Connectivity-based subdivisions of the human right "temporoparietal junction area": evidence for different areas participating in different cortical networks. Cerebral Cortex 22 (8), 1894–1903. http://dx.doi.org/10.1093/cercor/bhr268.

Meunier, D., Lambiotte, R., Bullmore, ET., 2010. Modular and hierarchically modular organization of brain networks. Front. Neurosci. 4, 200. http://dx.doi.org/10.3389/fnins.2010.00200.

Mezer, A., Yovel, Y., Pasternak, O., Gorfine, T., Assaf, Y., 2009. Cluster analysis of resting-state fMRI time series. NeuroImage 45 (4), 1117–1125. http://dx.doi.org/10.1016/j.neuroimage.2008.12.015.

Michel, V., Gramfort, A., Varoquaux, G., Eger, E., Keribin, C., Thirion, B., 2012. A supervised clustering approach for fMRI-based inference of brain states. Pattern Recogn. 45 (6), 2041–2049. http://dx.doi.org/10.1016/j.patcog.2011.04.006.

Moreno-Dominguez, D., Anwander, A., Knösche, T.R., 2014. A hierarchical method for whole-brain connectivity-based parcellation. Hum. Brain Map. 35 (10), 5000–5025. http://dx.doi.org/10.1002/hbm.22528.

Nelson, S.M., Cohen, A.L., Power, J.D., Wig, G.S., Miezin, F.M., Wheeler, M.E., Velanova, K., Donaldson, D.I., Phillips, J.S., Schlaggar, B.L., Petersen, S.E., 2010a, A parcellation scheme for human left lateral parietal cortex. Neuron 67 (1), 156–170. http://dx.doi.org/10.1016/j.neuron.2010.05.025.

Nelson, S.M., Dosenbach, N.U.F., Cohen, A.L., Wheeler, M.E., Schlaggar, B.L., Petersen, S.E., 2010b. Role of the anterior insula in task-level control and focal attention. Brain Struct. Funct. 214 (5–6), 669–680. http://dx.doi.org/10.1007/s00429-010-0260-2.

Newman, M.E.J., 2006. Modularity and community structure in networks. Proc. Natl. Acad. Sci. USA 103 (23), 8577–8582. http://dx.doi.org/10.1073/pnas.0601602103.

Ng, A.Y., Jordan, M.I., Weiss, Y., 2001. On spectral clustering: analysis and an algorithm. Adv. Neural Inform. Process. Syst. 2, 849–856.

Nooner, K.B., Colcombe, S.J., Tobe, R.H., Mennes, M., Benedict, M.M., Moreno, A.L., Panek, L.J., Brown, S., Zavitz, S.T., Li, Q., Sikka, S., Gutman, D., Bangaru, S., Schlachter, R.T., Kamiel, S.M., Anwar, A.R., Hinz, C.M., Kaplan, M.S., Rachlin, A.B., Adelsberg, S., Cheung, B., Khanuja, R., Yan, C., Craddock, C.C., Calhoun, V., Courtney, W., King, M., Wood, D., Cox, C.L., Kelly, A.M., Di Martino, A., Petkova, E., Reiss, P.T., Duan, N., Thomsen, D., Biswal, B., Coffey, B., Hoptman, M.J., Javitt, D.C., Pomara, N., Sidtis, J.J., Koplewicz, H.S., Castellanos, F.X., Leventhal, B.L., Milham, M.P., 2012. The NKI-Rockland sample: a model for accelerating the pace of discovery science in psychiatry. Front. Neurosci. 6, 152. http://dx.doi.org/10.3389/fnins.2012.00152.

Orban, P., Doyon, J., Petrides, M., Mennes, M., Hoge, R., Bellec, P., 2015. The richness of task-evoked hemodynamic responses defines a pseudohierarchy of functionally meaningful brain networks. Cerebral Cortex 25 (9), 2658–2669. http://dx.doi.org/10.1093/cercor/bhu064.

Potts, R.B., 1952. Some generalized order-disorder transformations. Math. Proc. Cambridge Philos. Soc., vol. 48 (01), pp. 106–109.

Preusser, S., Thiel, S.D., Rook, C., Roggenhofer, E., Kosatschek, A., Draganski, B., Blankenburg, F., Driver, J., Villringer, A., Pleger, B., 2015. The perception of touch and the ventral somatosensory pathway. Brain 138 (3), 540–548. http://dx.doi.org/10.1093/brain/awu370.

Rorden, C., Karnath, H.O., 2004. Opinion: using human brain lesions to infer function: a relic from a past era in the fMRI age? Nat. Rev. Neurosci. 5 (10), 812–819. http://dx.doi.org/10.1038/nrn1521.

Ryali, S., Chen, T., Supekar, K., Menon, V., 2013. A parcellation scheme based on von Mises-Fisher distributions and Markov random fields for segmenting brain regions using resting-state fMRI. NeuroImage 65, 83–96. http://dx.doi.org/10.1016/j.neuroimage.2012.09.067.

Sereno, M.I., Dale, A.M., Reppas, J.B., Kwong, K.K., Belliveau, J.W., Brady, T.J., Rosen, B.R., Tootell, R.B., 1995. Borders of multiple visual areas in humans revealed by functional magnetic resonance imaging. Science (New York) 268 (5212), 889–893.

Shen, X., Papademetris, X., Constable, R.T., United States, 2010. Graph-theory based parcellation of functional subunits in the brain from resting-state fMRI data. NeuroImage 50 (3), 1027–1035. http://dx.doi.org/10.1016/j.neuroimage.2009.12.119.

Smith, S.M., Miller, K.L., Salimi-Khorshidi, G., Webster, M., Beckmann, C.F., Nichols, T.E., Ramsey, J.D., Woolrich, M.W., 2011. Network modelling methods for fMRI. NeuroImage 54 (2), 875–891. http://dx.doi.org/10.1016/j.neuroimage.2010.08.063.

Sra, S., 2011. A short note on parameter approximation for von Mises-Fisher distributions and a fast implementation of $I_s(x)$. Comput. Stat. 27 (1), 177–190. http://dx.doi.org/10.1007/s00180-011-0232-x.

Swisher, J.D., Halko, M.A., Merabet, L.B., McMains, S.A., Somers, D.C., 2007. Visual topography of human intraparietal sulcus. J. Neurosci. 27 (20), 5326–5337. http://dx.doi.org/10.1523/JNEUROSCI.0991-07.2007.

Thirion, B., Flandin, G., Pinel, P., Roche, A., Ciuciu, P., Poline, J.B., 2006. Dealing with the shortcomings of spatial normalization: multi-subject parcellation of fMRI datasets. Hum. Brain Map. 27 (8), 678–693. http://dx.doi.org/10.1002/hbm.20210.

Thirion, B., Varoquaux, G., Dohmatob, E., Poline, J.B., 2014. Which fMRI clustering gives good brain parcellations? Front. Neurosci. 8 (July), 1–13. http://dx.doi.org/10.3389/fnins.2014.00167.

Tuch, D.S., Reese, T.G., Wiegell, M.R., Van J. Wedeen, 2003. Diffusion MRI of complex neural architecture. Neuron 40 (5), 885–895. http://dx.doi.org/10.1016/S0896-6273(03)00758-X.

Tucholka, A., Thirion, B., Perrot, M., Pinel, P., Mangin, J.F., Poline, J.B., 2008. Probabilistic anatomo-functional parcellation of the cortex: how many regions? In: Medical Image Computing and Computer-Assisted Intervention—MICCAI 2008. Springer, Berlin, Heidelberg, pp. 399–406.

Ungerleider, L.G., 1995. Functional brain imaging studies of cortical mechanisms for memory. Science (New York) 270 (5237), 769–775. http://dx.doi.org/10.1126/science.270.5237.769.

van den Heuvel, M., Mandl, R., Hulshoff Pol, H., 2008. Normalized cut group clustering of resting-state fMRI data. PLoS ONE 3 (4), e2001. http://dx.doi.org/10.1371/journal.pone.0002001.

Van Essen, D., Ugurbil, K., Auerbach, E., Barch, D., Behrens, T., Bucholz, R., Chang, A., Chen, L., Corbetta, M., Curtiss, S., Della Penna, S., Feinberg, D., Glasser, M., Harel, N., Heath, A., Larson-Prior, L., Marcus, D., Michalareas, G., Moeller, S., Oostenveld, R., Petersen, S., Prior, F., Schlaggar, B., Smith, S., Snyder, A., Xu, J., Yacoub, E., 2012. The Human Connectome Project: a data acquisition perspective. NeuroImage 62 (4), 2222–2231. http://dx.doi.org/10.1016/j.neuroimage.2012.02.018.

Varoquaux, G., Craddock, R.C., 2013. Learning and comparing functional connectomes across subjects. NeuroImage 80, 405–415. http://dx.doi.org/10.1016/j.neuroimage.2013.04.007. 1304.3880.

Varoquaux, G., Gramfort, A., Pedregosa, F., Michel, V., Thirion, B., 2011. Multi-subject dictionary learning to segment an atlas of brain spontaneous activity. In: Information Processing in Medical Imaging, pp. 562–573.

Veksler, O., 2008. Star shape prior for graph-cut image segmentation. Lecture Notes in Computer Science (including subseries Lecture Notes in Artificial Intelligence and Lecture Notes in Bioinformatics), LNCS vol. 5304 (PART 3), pp. 454–467. http://dx.doi.org/10.1007/978-3-540-88690-7-34.

Vogt, C., Vogt, O., 1919. Allgemeine Ergebnisse Unserer Hirnforschung, vol. 21. JA Barth, Leipzig.

von Luxburg, U., 2007. A tutorial on spectral clustering, pp. 1–32. arXiv 0711.0189.

Wainwright, M.J., Jordan, M.I., 2007. Graphical models, exponential families, and variational inference. Foundations and Trends in Machine Learning 1 (1–2), 1–305. http://dx.doi.org/10.1561/2200000001.

Wig, G.S., Laumann, T.O., Petersen, S.E., 2014. An approach for parcellating human cortical areas using resting-state correlations. NeuroImage 93 (2), 276–291. http://dx.doi.org/10.1016/j.neuroimage.2013.07.035.

Wittchen, H.U., Jacobi, F., Rehm, J., Gustavsson, A., Svensson, M., Jönsson, B., Olesen, J., Allgulander, C., Alonso, J., Faravelli, C., Fratiglioni, L., Jennum, P., Lieb, R., Maercker, A., van Os, J., Preisig, M., Salvador-Carulla, L., Simon, R., Steinhausen, H.C., 2011. The size and burden of mental disorders and other disorders of the brain in Europe 2010. Eur. Neuropsychopharmacol. 21 (9), 655–679. http://dx.doi.org/10.1016/j.euroneuro.2011.07.018.

Woolrich, M., Behrens, T., 2006. Variational Bayes inference of spatial mixture models for segmentation. IEEE Trans. Med. Imaging 25 (10), 1380–1391. http://dx.doi.org/10.1109/TMI.2006.880682.

Yarkoni, T., Poldrack, R.A., Nichols, T.E., Van Essen, D.C., Wager, T.D., 2011. Large-scale automated synthesis of human functional neuroimaging data. Nat. Meth. 8 (8), 665–670. http://dx.doi.org/10.1038/nmeth.1635.

Yeo, B.T.T., Krienen, F.M., Sepulcre, J., Sabuncu, M.R., Lashkari, D., Hollinshead, M., Roffman, J.L., Smoller, J.W., Zöllei, L., Polimeni, J.R., Fischl, B., Liu, H., Buckner, R.L., 2011. The organization of the human cerebral cortex estimated by intrinsic functional connectivity. J. Neurophysiol. 106 (3), 1125–1165. http://dx.doi.org/10.1152/jn.00338.2011.

Yeo, B.T.T., Krienen, F.M., Eickhoff, S.B., Yaakub, S.N., Fox, P.T., Buckner, R.L., Asplund, C.L., Chee, M.W., 2015. Functional specialization and flexibility in human association cortex. Cerebral Cortex 25 (10), 3654–3672. http://dx.doi.org/10.1093/cercor/bhu217.

Zhang, S., Li, C.s.R., 2012. Functional connectivity mapping of the human precuneus by resting state fMRI. NeuroImage 59 (4), 3548–3562. http://dx.doi.org/10.1016/j.neuroimage.2011.11.023.

Zilles, K., Amunts, K., 2010. Centenary of Brodmann's map—conception and fate. Nat. Rev. Neurosci. 11 (2), 139–145. http://dx.doi.org/10.1038/nrn2776.

Zuo, X.N., Anderson, J.S., Bellec, P., Birn, R.M., Biswal, B.B., Blautzik, J., Breitner, J.C.S., Buckner, R.L., Calhoun, V.D., Castellanos, F.X., Chen, A., Chen, B., Chen, J., Chen, X., Colcombe, S.J., Courtney, W., Craddock, R.C., Di Martino, A., Dong, H.M., Fu, X., Gong, Q., Gorgolewski, K.J., Han, Y., He, Y., He, Y., Ho, E., Holmes, A., Hou, X.H., Huckins, J., Jiang, T., Jiang, Y., Kelley, W., Kelly, C., King, M., LaConte, S.M., Lainhart, J.E., Lei, X., Li, H.J., Li, K., Li, K., Lin, Q., Liu, D., Liu, J., Liu, X., Liu, Y., Lu, G., Lu, J., Luna, B., Luo, J., Lurie, D., Mao, Y., Margulies, D.S., Mayer, A.R., Meindl, T., Meyerand, M.E., Nan, W., Nielsen, J.A., O'Connor, D., Paulsen, D., Prabhakaran, V., Qi, Z., Qiu, J., Shao, C., Shehzad, Z., Tang, W., Villringer, A., Wang, H., Wang, K., Wei, D., Wei, G.X., Weng, X.C., Wu, X., Xu, T., Yang, N., Yang, Z., Zang, Y.F., Zhang, L., Zhang, Q., Zhang, Z., Zhang, Z., Zhao, K., Zhen, Z., Zhou, Y., Zhu, X.T., Milham, M.P., 2014. An open science resource for establishing reliability and reproducibility in functional connectomics. Sci. Data 1, 140049. http://dx.doi.org/10.1038/sdata.2014.49.

CHAPTER

Kernel machine regression in neuroimaging genetics

T. Ge[1,2], J.W. Smoller[1,2], M.R. Sabuncu[1,3]

Massachusetts General Hospital/Harvard Medical School, Boston, MA, United States[1] Broad Institute of MIT and Harvard, Cambridge, MA, United States[2] Massachusetts Institute of Technology, Cambridge, MA, United States[3]

CHAPTER OUTLINE

Machine Learning and Medical Imaging. http://dx.doi.org/10.1016/B978-0-12-804076-8.00002-5

2.1 INTRODUCTION

The past few years have witnessed a tremendous growth of the amount of biomedical data, including the increasingly accessible medical images and genomic sequences. Techniques that can integrate different resources, extract reliable information from massive data, and reveal true relationships among biological variables have become essential and invaluable in biomedical research.

Kernel methods are a class of machine learning algorithms to study general types of relations in data sets, such as classifications, clusters and correlations, and are particularly powerful in high-dimensional and nonlinear settings (Vapnik, 1998; Cristianini and Shawe-Taylor, 2000; Schölkopf and Smola, 2002). The fundamental idea of kernel methods is built on the observation that, in many situations, relations between data points can be much more easily revealed or modeled if they are transformed from their original representations into a higher dimensional feature space via a user-specified feature map. For example, consider the simple classification problem as shown in Fig. 2.1. The gray and black dots cannot be linearly separated in one-dimensional space (Fig. 2.1, left), but when transformed into a two-dimensional space using the feature map $\varphi : x \mapsto (x, x^2)$, a linear separator can be easily found (eg, the dashed line in Fig. 2.1, right).

However, in practice, the explicit form of this feature map is often unknown, and the evaluation of a high-dimensional mapping can be computationally expensive. Kernel methods resolve this problem by employing a kernel function, which measures the resemblance between pairs of data points in the original space, and implicitly defines a (possibly infinite-dimensional) feature space and a feature map without actually accessing them. This makes kernel methods highly flexible since they can be easily applied to vectors, images, text, graphs, sequences, and any other data types, as long as a valid kernel function that converts any pair of data points into a scalar similarity measure can be defined on the particular data structure. Due to

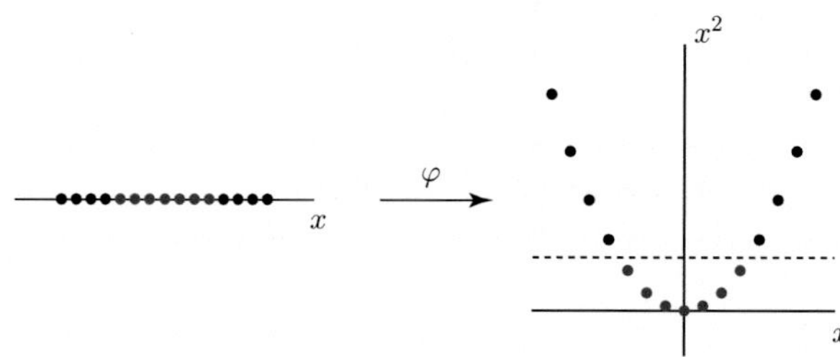

FIG. 2.1

An illustration of the power of feature mapping in a classification problem. Gray and black dots cannot be linearly separated in one-dimensional space (left), but when transformed into a two-dimensional space using the feature map $\varphi : x \mapsto (x, x^2)$, a linear separator can be found (right).

their generality, kernel methods have been applied to a wide range of data problems, such as classification, clustering, dimension reduction and correlation analysis, and have generalized support vector machine (SVM) (Schölkopf and Burges, 1999; Schölkopf and Smola, 2002), principal component analysis (PCA) (Schölkopf et al., 1998, 1997), Gaussian processes (Rasmussen and Williams, 2006), and many other techniques from linear to nonlinear settings.

Kernel machine regression (KMR) is a form of nonparametric regression and an application of the kernel methods to regression analysis. Data that exhibit complex nonlinear relationships in their original representations are implicitly operated in a higher dimensional feature space where a linear regression model is sufficient to describe the transformed data. In practice, KMR essentially regresses the traits (dependent variables) onto the similarity of attributes (independent variables) measured via the kernel function. Previously, KMR was fitted in a penalized regression framework, which can be computationally expensive and produce suboptimal model estimates. Recent theoretical advances have established the connection between KMR and mixed effects models in statistics, leading to elegant model fitting procedures and efficient statistical inferences about model parameters. This has spurred a rapidly expanding literature on applying KMR to biomedical research, especially in the field of genetics to identify cumulative effects of genetic variants on phenotypes, characterize the genetic architecture underlying complex traits, and test gene-by-gene or gene-by-environment interactions. Very recently the idea of KMR has been adapted to imaging genetics, an emerging field that identifies and characterizes genetic influences on brain structure, function and wiring, to dissect the genetic underpinning of features extracted from brain images and to understand the roles genetics plays in brain-related illnesses. In this chapter, we review both the mathematical basis and recent applications of KMR.

The remainder of this chapter is organized as follows. We first explain the intuitions and heuristics behind KMR from the perspective of linear regression analysis and give an illustrative example. We then formally introduce KMR, establish its connection to mixed effects models in statistics, and derive model fitting and statistical inference procedures. We also provide a framework for building and selecting kernel functions in a systematic and objective way. Recent theoretical extensions and applications of KMR are reviewed, with a focus on genetic association studies and imaging genetics. We close the chapter with a discussion of future directions. A rigorous mathematical treatment of the kernel methods and some technical aspects of the material presented in this chapter are included in the appendix.

2.2 MATHEMATICAL FOUNDATIONS

2.2.1 FROM REGRESSION ANALYSIS TO KERNEL METHODS

Kernel machine regression (KMR) is a form of nonparametric regression. We start with a simple model to help understand how it is related to linear regression analysis.

A rigorous mathematical treatment of the theory underlying kernel methods is provided in Appendix A.

Let y_i be a quantitative trait and z_i be a multidimensional attribute for the ith subject. Suppose that y_i is dependent on z_i through an unknown function f:

$$y_i = f(z_i) + \epsilon_i, \quad i = 1, 2, \ldots, n, \tag{2.1}$$

where ϵ_i is assumed to be independently Gaussian distributed with zero mean and homogeneous variance σ^2. The relationship between y_i and z_i can be highly nonlinear and thus the function f can be complex. The idea of kernel methods is to approximate f by a *feature map* φ, which transforms the attributes from their original input space to a higher dimensional *feature space*, $\varphi(z_i) = [\varphi_1(z_i), \varphi_2(z_i), \ldots]^{\mathrm{T}}$, such that the mappings $\{\varphi_k\}$, also known as the *basis functions* of the feature space, can be linearly combined to predict y_i. Expanding a function into the linear combination of a set of basis functions is called the *primal representation* of the function in kernel methods. Frequently used mappings include polynomial functions, spline-based functions (Wahba, 1990; Gu, 2013), and many others. Suppose we employ p different basis functions, model (2.1) then becomes

$$y_i = \sum_{k=1}^{p} \varphi_k(z_i)\omega_k + \epsilon_i, \quad i = 1, 2, \ldots, n, \tag{2.2}$$

where $\omega_k, k = 1, \ldots, p$, are scalar coefficients. If we define vectors $\boldsymbol{y} = [y_1, \ldots, y_n]^{\mathrm{T}}$, $\boldsymbol{\omega} = [\omega_1, \ldots, \omega_p]^{\mathrm{T}}$, $\boldsymbol{\epsilon} = [\epsilon_1, \ldots, \epsilon_n]^{\mathrm{T}}$, and the $n \times p$ *design matrix* $\mathbf{Z}_\varphi = [\varphi(z_1), \ldots, \varphi(z_n)]^{\mathrm{T}}$, Eq. (2.2) can be written in the matrix form:

$$\boldsymbol{y} = \mathbf{Z}_\varphi \boldsymbol{\omega} + \boldsymbol{\epsilon}, \tag{2.3}$$

which is a linear regression model. If $p \leqslant n$ and the design matrix is full rank, the *ordinary least squares* (OLS) estimator for the regression coefficients $\boldsymbol{\omega}$ is $\widehat{\boldsymbol{\omega}} = (\mathbf{Z}_\varphi^{\mathrm{T}}\mathbf{Z}_\varphi)^{-1}\mathbf{Z}_\varphi^{\mathrm{T}}\boldsymbol{y}$, which minimizes the loss function $\mathcal{L}(\boldsymbol{\omega}) = \|\boldsymbol{y} - \mathbf{Z}_\varphi\boldsymbol{\omega}\|^2 = \sum_{i=1}^{n}\left(y_i - \boldsymbol{\omega}^{\mathrm{T}}\varphi(z_i)\right)^2$, and is also the *best linear unbiased estimator* (BLUE) and the *maximum likelihood estimator* (MLE) because we have assumed the residual $\boldsymbol{\epsilon}$ to be normal. When $p > n$, that is, the number of basis functions is greater than the sample size, fitting linear regression model (2.3) becomes ill-posed because the problem is underdetermined, that is, there exists infinitely many $\boldsymbol{\omega}$ that can perfectly fit the model. To avoid the danger of overfitting, an effective and widely used approach is to penalize the norm of the regression coefficients $\boldsymbol{\omega}$ in the loss function. For example, consider the *Ridge regression* (also known as *Tikhonov regularization*), which penalizes the squared error loss by the squared Euclidean norm of $\boldsymbol{\omega}$:

$$\mathcal{J}(\boldsymbol{\omega}) = \frac{1}{2}\sum_{i=1}^{n}\left(y_i - \boldsymbol{\omega}^{\mathrm{T}}\varphi(z_i)\right)^2 + \frac{\lambda}{2}\|\boldsymbol{\omega}\|^2, \tag{2.4}$$

where λ is a tuning parameter, which balances the model fitting and model complexity. Taking the derivative of $\mathcal{J}$ with respect to $\boldsymbol{\omega}$ and equating it to zero gives

$$\widehat{\boldsymbol{\omega}} = (\mathbf{Z}_{\varphi}^{\mathrm{T}}\mathbf{Z}_{\varphi} + \lambda \boldsymbol{I}_{p\times p})^{-1}\mathbf{Z}_{\varphi}^{\mathrm{T}}\boldsymbol{y}, \tag{2.5}$$

where $\boldsymbol{I}_{p\times p}$ is a $p \times p$ identity matrix. A prediction of the trait y given the attribute z from a new subject is

$$\hat{y} = \hat{f}(z) = \widehat{\boldsymbol{\omega}}^{\mathrm{T}}\varphi(z) = \boldsymbol{y}^{\mathrm{T}}\mathbf{Z}_{\varphi}(\mathbf{Z}_{\varphi}^{\mathrm{T}}\mathbf{Z}_{\varphi} + \lambda \boldsymbol{I}_{p\times p})^{-1}\varphi(z). \tag{2.6}$$

It can be seen that for positive λ, the smallest eigenvalue of $\mathbf{Z}_{\varphi}^{\mathrm{T}}\mathbf{Z}_{\varphi} + \lambda \boldsymbol{I}_{p\times p}$ is bounded away from zero, and thus the matrix inverse in Eqs. (2.5) and (2.6) always exists. Ridge regression thus regularizes the linear regression model (2.3) by shrinking the solutions towards zero.

We now rewrite Eqs. (2.5) and (2.6) to provide insights into the connection between kernel-based regression and regularized linear regression. Specifically, we note that

$$(\mathbf{Z}_{\varphi}^{\mathrm{T}}\mathbf{Z}_{\varphi} + \lambda \boldsymbol{I}_{p\times p})\mathbf{Z}_{\varphi}^{\mathrm{T}} = \mathbf{Z}_{\varphi}^{\mathrm{T}}\mathbf{Z}_{\varphi}\mathbf{Z}_{\varphi}^{\mathrm{T}} + \lambda \mathbf{Z}_{\varphi}^{\mathrm{T}} = \mathbf{Z}_{\varphi}^{\mathrm{T}}(\mathbf{Z}_{\varphi}\mathbf{Z}_{\varphi}^{\mathrm{T}} + \lambda \boldsymbol{I}_{n\times n}). \tag{2.7}$$

Since both $\mathbf{Z}_{\varphi}^{\mathrm{T}}\mathbf{Z}_{\varphi} + \lambda \boldsymbol{I}_{p\times p}$ and $\mathbf{Z}_{\varphi}\mathbf{Z}_{\varphi}^{\mathrm{T}} + \lambda \boldsymbol{I}_{n\times n}$ are invertible, we have

$$\widehat{\boldsymbol{\omega}} = (\mathbf{Z}_{\varphi}^{\mathrm{T}}\mathbf{Z}_{\varphi} + \lambda \boldsymbol{I}_{p\times p})^{-1}\mathbf{Z}_{\varphi}^{\mathrm{T}}\boldsymbol{y} = \mathbf{Z}_{\varphi}^{\mathrm{T}}(\mathbf{Z}_{\varphi}\mathbf{Z}_{\varphi}^{\mathrm{T}} + \lambda \boldsymbol{I}_{n\times n})^{-1}\boldsymbol{y} := \mathbf{Z}_{\varphi}^{\mathrm{T}}\widehat{\boldsymbol{\alpha}} = \sum_{i=1}^{n}\hat{\alpha}_i\varphi(z_i), \tag{2.8}$$

where we have defined $\widehat{\boldsymbol{\alpha}} = [\hat{\alpha}_1, \ldots, \hat{\alpha}_n]^{\mathrm{T}} = (\mathbf{Z}_{\varphi}\mathbf{Z}_{\varphi}^{\mathrm{T}} + \lambda \boldsymbol{I}_{n\times n})^{-1}\boldsymbol{y}$. It can be seen that the p-dimensional regression coefficient $\widehat{\boldsymbol{\omega}}$ lies in the span of the n transformed observations $\{\varphi(z_i)\}_{i=1}^{n}$ even if $p \gg n$. This is expected since the model is linear in the feature space. A more important point can be made by noticing that the $n \times n$ matrix $\mathbf{Z}_{\varphi}\mathbf{Z}_{\varphi}^{\mathrm{T}}$ is *non-negative definite* and its ijth element is the dot product between $\varphi(z_i)$ and $\varphi(z_j)$, which we denote as $\langle\varphi(z_i), \varphi(z_j)\rangle$. Therefore we can define a *kernel function* for any pair of subjects i and j as follows:

$$k(z_i, z_j) = \langle\varphi(z_i), \varphi(z_j)\rangle. \tag{2.9}$$

$\boldsymbol{K} := \mathbf{Z}_{\varphi}\mathbf{Z}_{\varphi}^{\mathrm{T}} = \{k(z_i, z_j)\}_{i,j=1}^{n}$ is then the *Gram matrix* or *kernel matrix* associated with the kernel function k given the observed attributes $\{z_1, \ldots, z_n\}$. For a new subject with attribute z, the trait can be predicted as

$$\hat{y} = \hat{f}(z) = \widehat{\boldsymbol{\omega}}^{\mathrm{T}}\varphi(z) = \widehat{\boldsymbol{\alpha}}^{\mathrm{T}}\mathbf{Z}_{\varphi}\varphi(z) := \widehat{\boldsymbol{\alpha}}^{\mathrm{T}}\boldsymbol{\kappa} = \boldsymbol{y}^{\mathrm{T}}(\boldsymbol{K} + \lambda \boldsymbol{I}_{n\times n})^{-1}\boldsymbol{\kappa}, \tag{2.10}$$

where $\boldsymbol{\kappa} = \mathbf{Z}_{\varphi}\varphi(z) = [k(z_1, z), \ldots, k(z_n, z)]^{\mathrm{T}}$. Here the function f is represented by a linear combination of the kernel function centered at the observed data points, that is, $\hat{f}(z) = \widehat{\boldsymbol{\alpha}}^{\mathrm{T}}\boldsymbol{\kappa} = \sum_{i=1}^{n}\hat{\alpha}_i k(z_i, z)$. This is the *dual representation* of a function in kernel methods. Eq. (2.10) indicates that f can be estimated without accessing the feature map φ, which can be infinite-dimensional, expensive to evaluate, or difficult to explicitly specify in practice. Instead, we only need to define a kernel function which collapses the (possibly high-dimensional) attributes for each pair of individuals into a scalar similarity measure. Moreover, we can substitute any non-negative definite kernel function $\tilde{k}$ for k defined in Eq. (2.9) to measure the similarity between pairs of individuals in a different way. This technique is called the "*kernel trick*" in the machine learning literature. The theory of kernel methods ensures that any non-negative definite kernel function $\tilde{k}$ implicitly specifies a feature map $\tilde{\varphi}$ (which could be infinite-dimensional) such that $\tilde{k}$ can be expressed as a dot product in the feature space: $\tilde{k}(z_i, z_j) = \langle\tilde{\varphi}(z_i), \tilde{\varphi}(z_j)\rangle$. Therefore kernel methods greatly simplify the specification of a nonparametric model, especially for multidimensional attributes.

As an illustration, consider the nonlinear relationship, $y = \sin(z)$, $-\pi \leqslant z \leqslant \pi$, between the trait y and a scalar attribute z. We generated synthetic data for $n = 30$ samples using the model $y_i = \sin(z_i)+\epsilon_i$, $i = 1, 2, \ldots, n$, where each z_i was randomly selected between $-\pi$ and π, and ϵ_i was Gaussian distributed with zero mean and variance $\sigma^2 = 0.01$. The objective is to recover the unknown function $f(\cdot) = \sin(\cdot)$ from the observed trait y_i and attribute z_i for each sample. This can be achieved by evaluating the estimated function $\hat{f}$ on a dense and equally spaced grid between $-\pi$ and π, and plotting the predicted traits.

We first fitted the data using a linear feature map $\varphi_L(z) = [1, z]^{\mathrm{T}}$. This is equivalent to modeling the data using the linear regression $y_i = \omega_0 + z_i\omega_1 + \epsilon_i$, where ω_0 and ω_1 are scalar coefficients. By Eq. (2.9), a linear kernel can be defined as $k_L(z_i, z_j) = \langle\varphi_L(z_i), \varphi_L(z_j)\rangle = z_i z_j + 1$, for any pair of samples i and j. The kernel matrix associated with the observed attributes is $\boldsymbol{K}_L = \{z_i z_j + 1\}_{i,j=1}^{n}$. The predicted traits can then be computed, using either Eq. (2.6) or Eq. (2.10). The upper left panel of Fig. 2.2 shows $\hat{f}$ estimated at three different values of the tuning parameter λ, along with the ground truth, that is, $f(\cdot) = \sin(\cdot)$, and the observed data points. It can be seen that a linear feature map (and the corresponding linear kernel) can only capture linear relationships and is not sufficient to recover the nonlinear trigonometric function used to generate the data. Moreover, when λ increases, the estimated function tends to be "simpler," which in the linear case means a smaller slope.

Next, we fitted the data by employing a cubic feature map $\varphi_C(z) = [1, \sqrt{3}z, \sqrt{3}z^2, z^3]^{\mathrm{T}}$, which corresponds to the regression model $y_i = \omega_0+\sqrt{3}z_i\omega_1+\sqrt{3}z_i^2\omega_2+z_i^3\omega_3 + \epsilon_i$ in the feature space, where ω_k, $k = 0, 1, 2, 3$, are regression coefficients. This feature map defines the cubic kernel $k_C(z_i, z_j) = \langle\varphi_C(z_i), \varphi_C(z_j)\rangle = (z_i z_j + 1)^3$, for samples i and j. The upper right panel of Fig. 2.2 shows that this nonlinear feature map and the corresponding cubic kernel can represent nonlinear relationships. Also, with the increase of λ, the estimated function tends to be flatter. We note that polynomial kernel functions are monotonic for sufficiently large attributes and thus need to be used with caution when extrapolating values far beyond the range of the observed data.

Lastly, we fitted the data using a Gaussian kernel $k_G(z_i, z_j) = \exp\left\{(z_i - z_j)^2/\rho\right\}$, where ρ is a free parameter that determines the width of the kernel. The Gaussian kernel corresponds to an infinite-dimensional feature map because the exponential function can be viewed as an infinite sum of polynomials. Thus the feature map cannot be explicitly expressed and evaluated. However, as shown in Eq. (2.10) and discussed above, a major advantage of the kernel methods is that the unknown function can be estimated without assessing the feature map, as long as a kernel function is specified. The infinite-dimensional nature of the Gaussian kernel enables it to model any nonlinear relationship in the data. The lower left panel of Fig. 2.2 shows the estimated function using a Gaussian kernel with $\rho = 0.1$. We observe that when λ is small, the model tends to interpolate the observed data points. Actually, if λ is set to zero, the data will be perfectly fitted without error. However, this is a typical overfitting behavior, where the model describes the noise instead of the true relationship underlying the data, and has poor predictive performance and generalization power. As λ increases, complex models are more heavily penalized, and the estimated function becomes smoother. An alternative approach to alleviate

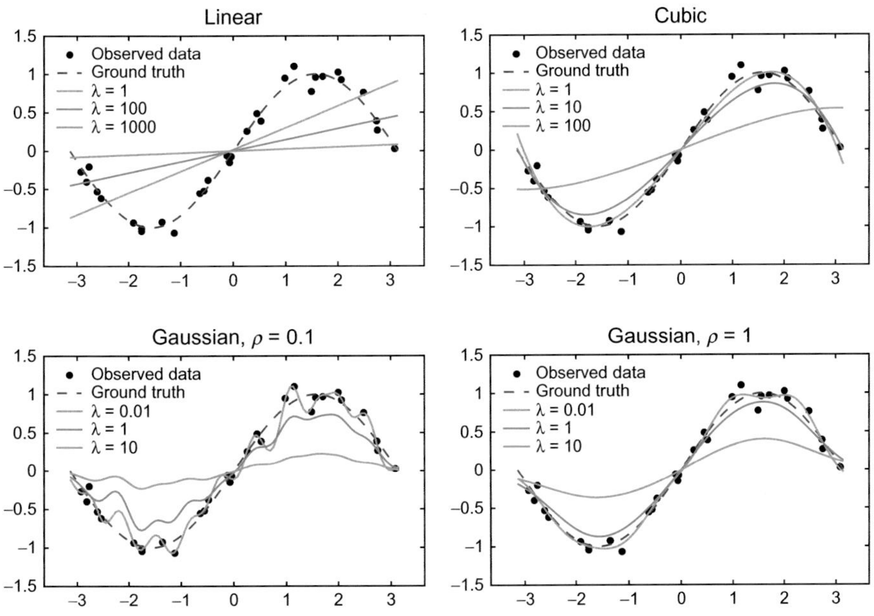

FIG. 2.2

A synthetic example to illustrate the penalized regression and kernel-based regression. Data were generated for 30 samples using the model $y_i = \sin(z_i) + \epsilon_i$, where each z_i was randomly selected between $-\pi$ and π, and ϵ_i was Gaussian distributed with zero mean and variance $\sigma^2 = 0.01$. The function $\sin(\cdot)$ was estimated from the observed traits y_i and attributes z_i using a linear kernel (upper left), a cubic kernel (upper right), a Gaussian kernel with the width parameter $\rho = 0.1$ (lower left), and a Gaussian kernel with $\rho = 1$ (lower right). In each panel, the estimated functions using different tuning parameters λ are shown, along with the ground truth of the function and the observed data points.

the overfitting issue is to increase the width of the Gaussian kernel. This leads to more data points being considered and averaged during model estimation and avoids overfitting the local trend. As shown in the lower right panel of Fig. 2.2, a Gaussian kernel with $\rho = 1$ produced a much better fitting than $\rho = 0.1$. However, a too large width may cause underfitting of the data. In practice, the tuning parameter λ and additional parameters in the kernel function, such as the width of a Gaussian kernel, need to be carefully selected. We will introduce techniques to determine these parameters and perform model selection in the following sections.

2.2.2 KERNEL MACHINE REGRESSION

We now introduce KMR in a more general and rigorous framework by extending model (2.1) as follows:

$$y_i = \boldsymbol{x}_i^{\mathrm{T}}\boldsymbol{\beta} + f(z_i) + \epsilon_i, \quad i = 1, 2, \ldots, n, \tag{2.11}$$

where $\boldsymbol{x}_i$ is a $q \times 1$ vector of covariates or nuisance variables, and $\boldsymbol{\beta}$ is a $q \times 1$ vector of regression coefficients. Here the covariates $\boldsymbol{x}_i$ are modeled parametrically (linearly), while the attributes z_i are modeled nonparametrically through the unknown function f. As shown in the previous section, we usually expand f into the linear combination of a set of basis functions (primal representation) or the linear combination of a kernel function centered at certain data points (dual representation) with the hope that these representations can be efficiently estimated and provide a good approximation to f. We now solidify this idea by assuming that f lies in a *reproducing kernel Hilbert space* (RKHS) $\mathcal{H}$ equipped with an inner product $\langle \cdot, \cdot \rangle_{\mathcal{H}}$ (Aronszajn, 1950; Saitoh, 1988). An RKHS is a function space defined on the input space $\mathcal{Z}$ where z_i resides, and is uniquely determined by a non-negative kernel function $k(\cdot, \cdot)$ on $\mathcal{Z} \times \mathcal{Z}$. It satisfies that, for any $f \in \mathcal{H}$ and an arbitrary attribute $z \in \mathcal{Z}$, $k(\cdot, z)$ as a function belongs to $\mathcal{H}$, and the inner product between f and $k(\cdot, z)$ is the evaluation of f at z: $\langle f, k(\cdot, z) \rangle_{\mathcal{H}} = f(z)$. The latter is known as the *reproducing property* of RKHSs. An RKHS also ensures the existence of the primal and dual representations of the functions belonging to it and implicitly regularizes their smoothness. See Appendix A for a mathematical characterization of RKHSs.

Model (2.11) can be fitted by minimizing the panelized likelihood function:

$$\mathcal{J}(\boldsymbol{\beta}, f) = \frac{1}{2}\sum_{i=1}^{n}\left(y_i - \boldsymbol{x}_i^{\mathrm{T}}\boldsymbol{\beta} - f(z_i)\right)^2 + \frac{\lambda}{2}\|f\|_{\mathcal{H}}^2, \tag{2.12}$$

where $\|\cdot\|_{\mathcal{H}}$ is the norm induced by the inner product on the RKHS. The first part of Eq. (2.12) is a loss function quantifying the goodness-of-fit of the model, and the second part of Eq. (2.12) is a regularization term controlling the smoothness of the optimizer. λ is a tuning parameter balancing the model fitting and model complexity. When $\lambda = 0$, the model interpolates the trait, whereas when $\lambda = +\infty$, model (2.11) degenerates to a linear regression model without f. Various other choices of the loss and penalty functions exist and can be used to handle a wide range of problems from regression to variable selection and to classification (see, eg, Schaid (2010a) for a review). Here we focus on the squared error loss and squared norm penalty because they are widely used for quantitative traits, lead to a closed form solution, and have a strong connection to linear mixed effects models, as shown in the next section.

Minimizing the functional (2.12) is a calculus of variations problem over an infinite-dimensional space of smooth curves, which can be difficult to resolve. However, the *Representer Theorem* (Kimeldorf and Wahba, 1971) shows that the solution of a very general class of optimization problems on RKHSs, which encompasses the problem under investigation here, has a finite-dimensional representation:

$$f(\cdot) = \sum_{j=1}^{n} \alpha_j k(\cdot, z_j), \tag{2.13}$$

where $\boldsymbol{\alpha} = [\alpha_1, \ldots, \alpha_n]^{\mathrm{T}}$ are unknown parameters. Substituting Eq. (2.13) into Eq. (2.12) gives

$$\begin{aligned} \mathcal{J}(\boldsymbol{\beta}, \boldsymbol{\alpha}) &= \frac{1}{2}\sum_{i=1}^{n}\left(y_i - \boldsymbol{x}_i^{\mathrm{T}}\boldsymbol{\beta} - \sum_{j=1}^{n}\alpha_j k(z_i, z_j)\right)^2 + \frac{\lambda}{2}\|f\|_{\mathcal{H}}^2 \\ &= \frac{1}{2}(\boldsymbol{y} - \boldsymbol{X}\boldsymbol{\beta} - \boldsymbol{K}\boldsymbol{\alpha})^{\mathrm{T}}(\boldsymbol{y} - \boldsymbol{X}\boldsymbol{\beta} - \boldsymbol{K}\boldsymbol{\alpha}) + \frac{\lambda}{2}\boldsymbol{\alpha}^{\mathrm{T}}\boldsymbol{K}\boldsymbol{\alpha}, \end{aligned} \tag{2.14}$$

where we have defined the *kernel matrix* $\boldsymbol{K} = \{k(z_i, z_j)\}_{n\times n}$ and used the matrix notation $\boldsymbol{X} = [\boldsymbol{x}_1, \ldots, \boldsymbol{x}_n]^{\mathrm{T}}$, and have made use of the reproducing property in the computation of the penalty term. Specifically,

$$\begin{aligned} \|f\|_{\mathcal{H}}^2 &= \left\|\sum\nolimits_{j=1}^{n}\alpha_j k(\cdot, z_j)\right\|_{\mathcal{H}}^2 = \sum\nolimits_{i=1}^{n}\sum\nolimits_{j=1}^{n}\alpha_i\alpha_j\langle k(\cdot, z_i), k(\cdot, z_j)\rangle_{\mathcal{H}}^2 \\ &= \sum\nolimits_{i=1}^{n}\sum\nolimits_{j=1}^{n}\alpha_i\alpha_j k(z_i, z_j) = \boldsymbol{\alpha}^{\mathrm{T}}\boldsymbol{K}\boldsymbol{\alpha}. \end{aligned} \tag{2.15}$$

It can been seen from Eq. (2.14) that the optimization of $\mathcal{J}(\boldsymbol{\beta}, \boldsymbol{\alpha})$ is now over finite-dimensional parameters $\boldsymbol{\alpha}$ and $\boldsymbol{\beta}$. Setting the derivatives of $\mathcal{J}(\boldsymbol{\beta}, \boldsymbol{\alpha})$ with respect to $\boldsymbol{\alpha}$ and $\boldsymbol{\beta}$ to zero yields the following first-order condition:

$$\begin{bmatrix} \boldsymbol{X}^{\mathrm{T}}\boldsymbol{X} & \boldsymbol{X}^{\mathrm{T}}\boldsymbol{K} \\ \boldsymbol{K}^{\mathrm{T}}\boldsymbol{X} & \boldsymbol{K}^{\mathrm{T}}\boldsymbol{K} + \lambda\boldsymbol{K} \end{bmatrix}\begin{bmatrix} \boldsymbol{\beta} \\ \boldsymbol{\alpha} \end{bmatrix} = \begin{bmatrix} \boldsymbol{X}^{\mathrm{T}}\boldsymbol{y} \\ \boldsymbol{K}^{\mathrm{T}}\boldsymbol{y} \end{bmatrix}. \tag{2.16}$$

It can be verified that the following pair of $\boldsymbol{\alpha}$ and $\boldsymbol{\beta}$ is a solution of Eq. (2.16):

$$\begin{aligned} \widehat{\boldsymbol{\beta}} &= \left[\boldsymbol{X}^{\mathrm{T}}(\boldsymbol{K} + \lambda\boldsymbol{I})^{-1}\boldsymbol{X}\right]^{-1}\boldsymbol{X}^{\mathrm{T}}(\boldsymbol{K} + \lambda\boldsymbol{I})^{-1}\boldsymbol{y}, \\ \widehat{\boldsymbol{\alpha}} &= (\boldsymbol{K} + \lambda\boldsymbol{I})^{-1}(\boldsymbol{y} - \boldsymbol{X}\widehat{\boldsymbol{\beta}}). \end{aligned} \tag{2.17}$$

Following Eq. (2.13), the optimizer $\hat{f}$, evaluated at an arbitrary attribute z, can be expressed as

$$\hat{f}(z) = \sum_{j=1}^{n} \hat{\alpha}_j k(z, z_j) = \widehat{\alpha}^{\mathrm{T}} \kappa = (y - X\widehat{\beta})^{\mathrm{T}} (K + \lambda I)^{-1} \kappa, \tag{2.18}$$

where $\kappa = [k(z_1, z), \ldots, k(z_n, z)]^{\mathrm{T}}$. Specifically, the vector $\widehat{f} = [\hat{f}(z_1), \ldots, \hat{f}(z_n)]^{\mathrm{T}}$, comprising $\hat{f}$ evaluated at the observed attributes $\{z_1, \ldots, z_n\}$, is

$$\widehat{f} = K\widehat{\alpha} = K(K + \lambda I)^{-1}(y - X\widehat{\beta}). \tag{2.19}$$

We emphasize that Eqs. (2.17) and (2.19) only depend on the kernel matrix K and the penalty parameter λ, while the explicit form of the function f does not need to be specified, making this nonparametric modeling approach highly flexible.

We note that the optimal value of λ is usually unknown and, moreover, the kernel function may rely on additional unknown parameters, such as the width in a Gaussian kernel. Few methods exist to jointly estimate these model parameters. In practice, they are usually fixed a priori, resulting in potentially suboptimal solutions. Alternatively, a sequence of tuning and/or kernel parameters can be applied, resulting in a collection of models, and the optimal parameters are determined by selecting the most appropriate model. Information criteria (eg, *Akaike information criterion*, AIC, and *Bayesian information criterion*, BIC) and cross-validation techniques are widely used in model selection. However, a fine search of potential parameter values can be computationally expensive. In a seminal paper, Liu et al. (2007) showed that there is a strong connection between KMR and linear mixed effects models, and thus model estimation and inferences can be conducted within the mixed model framework. In the next section, we review this connection as well as linear mixed effects models.

2.2.3 LINEAR MIXED EFFECTS MODELS

Linear mixed effects models (LMMs) (also known as variance component models) are widely used in statistics to model dependent data structures such as clustered data (McCulloch and Neuhaus, 2001; Snijders, 2011) and longitudinal data (Diggle et al., 2002; Verbeke and Molenberghs, 2009). Here we consider the following LMM:

$$y = X\beta + f + \epsilon, \tag{2.20}$$

where y is a vector of traits from n individuals, X is an $n \times q$ covariate matrix, β is a $q \times 1$ vector of regression coefficients, f is an $n \times 1$ random vector following the normal distribution $\mathrm{N}(\mathbf{0}, \tau^2 K)$, in which K is an $n \times n$ kernel matrix, $\tau^2 = \lambda^{-1}\sigma^2$, and ϵ is an $n \times 1$ vector of residuals following $\mathrm{N}(\mathbf{0}, \sigma^2 I)$. f is assumed to be independent of ϵ. Here, τ^2 is expressed as a function of the tuning parameter λ and the variance of the residual σ^2 in the previous section in order to build the connection between KMR and LMMs, which will soon be made clear. Also, note that although K can be an arbitrary non-negative definite matrix, as we will see below, it is connected to the kernel matrix in the previous section. Hence we use the somewhat confusing notation $\boldsymbol{K}$ to represent the covariance of the random effects in the LMM. In the context of mixed models, the regression coefficients $\boldsymbol{\beta}$ are called *fixed effects* since they model population-average effects, whereas $\boldsymbol{f}$ is a vector of *random effects* since it models subject-specific effects, which are assumed to be randomly sampled from a general population. The conditional distribution of $\boldsymbol{y}$ given the random effects $\boldsymbol{f}$ is normal:

$$\boldsymbol{y} \mid \boldsymbol{f} \sim \mathrm{N}(\boldsymbol{X\beta} + \boldsymbol{f}, \sigma^2 \boldsymbol{I}). \tag{2.21}$$

Marginally (averaged across individuals),

$$\boldsymbol{y} \sim \mathrm{N}(\boldsymbol{X\beta}, \tau^2 \boldsymbol{K} + \sigma^2 \boldsymbol{I}). \tag{2.22}$$

We denote the marginal covariance of $\boldsymbol{y}$ as $\boldsymbol{V} = \tau^2\boldsymbol{K} + \sigma^2\boldsymbol{I}$. It is clear from Eq. (2.22) that $\boldsymbol{\beta}$ can be estimated by *generalized least squares*: $\widehat{\boldsymbol{\beta}} = (\boldsymbol{X}^{\mathrm{T}}\boldsymbol{V}^{-1}\boldsymbol{X})^{-1}\boldsymbol{X}^{\mathrm{T}}\boldsymbol{V}^{-1}\boldsymbol{y}$, which coincides with the solution in Eq. (2.17) since $\boldsymbol{V} = \tau^2(\boldsymbol{K} + \lambda\boldsymbol{I})$. $\boldsymbol{f}$ can be estimated by noticing the fact that $\boldsymbol{y}$ and $\boldsymbol{f}$ are jointly normal and their covariance is $\tau^2\boldsymbol{K}$. Thus by making use of the conditional distribution of multivariate normal, the expectation of $\boldsymbol{f}$ given the observation $\boldsymbol{y}$ can be estimated as $\tau^2\boldsymbol{K}\boldsymbol{V}^{-1}(\boldsymbol{y} - \boldsymbol{X}\widehat{\boldsymbol{\beta}})$, which is the same as Eq. (2.19). Alternatively, $\boldsymbol{\beta}$ and $\boldsymbol{f}$ can be estimated by jointly maximizing the log likelihood of $[\boldsymbol{y}^{\mathrm{T}}, \boldsymbol{f}^{\mathrm{T}}]^{\mathrm{T}}$ with respect to $\boldsymbol{\beta}$ and $\boldsymbol{f}$. This gives the Henderson *mixed model equation* (MME):

$$\begin{bmatrix} \boldsymbol{X}^{\mathrm{T}}\boldsymbol{R}^{-1}\boldsymbol{X} & \boldsymbol{X}^{\mathrm{T}}\boldsymbol{R}^{-1} \\ \boldsymbol{R}^{-1}\boldsymbol{X} & \boldsymbol{R}^{-1} + \tau^{-2}\boldsymbol{K}^{-1} \end{bmatrix} \begin{bmatrix} \boldsymbol{\beta} \\ \boldsymbol{f} \end{bmatrix} = \begin{bmatrix} \boldsymbol{X}^{\mathrm{T}}\boldsymbol{R}^{-1}\boldsymbol{y} \\ \boldsymbol{R}^{-1}\boldsymbol{y} \end{bmatrix}, \tag{2.23}$$

where $\boldsymbol{R} = \sigma^2\boldsymbol{I}$. The solutions to the MME, $\widehat{\boldsymbol{\beta}}$ and $\widehat{\boldsymbol{f}}$, are the *best linear unbiased estimator* (BLUE) and the *best linear unbiased predictor* (BLUP) for $\boldsymbol{\beta}$ and $\boldsymbol{f}$, respectively. It is easy to verify that Eq. (2.23) and Eq. (2.16) are identical by using the identity $\boldsymbol{f} = \boldsymbol{K\alpha}$ (see Eq. 2.13). Therefore the estimates of $\boldsymbol{\beta}$ and $\boldsymbol{f}$ obtained by minimizing the penalized likelihood function in Eq. (2.12) are equivalent to the BLUE and BLUP of the LMM defined in Eq. (2.20). This connection bridges machine learning and statistics, specifically KMR and LMMs, and allows for a unified framework of model fitting and statistical inferences. In the mixed model framework, the tuning parameter λ can be interpreted as a ratio between the variance component parameters: $\lambda = \sigma^2/\tau^2$. When the kernel matrix explains a large portion of the trait variation, λ tends to be small and thus, in the context of KMR, the nonparametric function f is less penalized. When the kernel matrix captures little variation of the trait, λ tends to be large, and the KMR approaches a parametric linear regression.

The variance component parameters τ^2 and σ^2, and any unknown parameter in the kernel function, can now be estimated by maximizing the likelihood of the LMM.

Specifically, denoting $\boldsymbol{\theta}$ as a vector comprising all the unknown parameters in the marginal covariance structure $\boldsymbol{V}$, the log likelihood function of the LMM is

$$\ell(\boldsymbol{\beta},\boldsymbol{\theta}) = -\frac{1}{2}\log|\boldsymbol{V}(\boldsymbol{\theta})| - \frac{1}{2}(\boldsymbol{y}-\boldsymbol{X\beta})^{\mathrm{T}}\boldsymbol{V}(\boldsymbol{\theta})^{-1}(\boldsymbol{y}-\boldsymbol{X\beta}), \tag{2.24}$$

and the profile log likelihood is

$$\ell_{\mathrm{P}}(\boldsymbol{\theta}) = -\frac{1}{2}\log|\boldsymbol{V}(\boldsymbol{\theta})| - \frac{1}{2}(\boldsymbol{y}-\boldsymbol{X}\widehat{\boldsymbol{\beta}})^{\mathrm{T}}\boldsymbol{V}(\boldsymbol{\theta})^{-1}(\boldsymbol{y}-\boldsymbol{X}\widehat{\boldsymbol{\beta}}), \tag{2.25}$$

where we have replaced $\boldsymbol{\beta}$ in Eq. (2.24) by its generalized least squares estimator $\widehat{\boldsymbol{\beta}} = \widehat{\boldsymbol{\beta}}(\boldsymbol{\theta}) = \left(\boldsymbol{X}^{\mathrm{T}}\boldsymbol{V}(\boldsymbol{\theta})^{-1}\boldsymbol{X}\right)^{-1}\boldsymbol{X}\boldsymbol{V}(\boldsymbol{\theta})^{-1}\boldsymbol{y}$. Maximizing ℓ_{P} with respect to $\boldsymbol{\theta}$ gives the maximum likelihood estimate $\widehat{\boldsymbol{\theta}}_{\mathrm{MLE}}$. However, $\widehat{\boldsymbol{\theta}}_{\mathrm{MLE}}$ is biased since it does not account for the loss in degrees of freedom resulting from estimating the unknown fixed effects $\boldsymbol{\beta}$. In contrast, restricted maximum likelihood (ReML) estimation (Patterson and Thompson, 1971; Harville, 1977; Lindstrom and Bates, 1988) produces unbiased estimates of variance component parameters by applying a transformation to the LMM to remove the effect of covariates and calculating the log likelihood function based on the transformed data. The restricted log likelihood can be written as a correction to the profile log likelihood:

$$\ell_{\mathrm{R}}(\boldsymbol{\theta}) = \ell_{\mathrm{P}}(\boldsymbol{\theta}) - \frac{1}{2}\log|\boldsymbol{X}^{\mathrm{T}}\boldsymbol{V}(\boldsymbol{\theta})^{-1}\boldsymbol{X}|. \tag{2.26}$$

We review the Newton-Raphson method commonly employed to maximize ℓ_{R} with respect to $\boldsymbol{\theta}$ in Appendix B. Once the ReML estimate $\widehat{\boldsymbol{\theta}}_{\mathrm{ReML}}$ has been obtained, it can be plugged into Eqs. (2.17) and (2.19), producing *empirical* BLUE and BLUP of the fixed and random effects, respectively. The covariance matrices of $\widehat{\boldsymbol{\beta}}$ and $\widehat{\boldsymbol{f}}$ can be calculated as

$$\begin{aligned}\mathbf{cov}(\widehat{\boldsymbol{\beta}}) &= \left(\boldsymbol{X}^{\mathrm{T}}\boldsymbol{V}(\boldsymbol{\theta})^{-1}\boldsymbol{X}\right)^{-1},\\ \mathbf{cov}(\widehat{\boldsymbol{f}}) &= \mathbf{cov}(\widehat{\boldsymbol{f}}-\boldsymbol{f}) = \tau^2\boldsymbol{K} - (\tau^2\boldsymbol{K})\boldsymbol{P}(\tau^2\boldsymbol{K}),\end{aligned} \tag{2.27}$$

where $\boldsymbol{P} = \boldsymbol{P}(\boldsymbol{\theta}) = \boldsymbol{V}^{-1} - \boldsymbol{V}^{-1}\boldsymbol{X}(\boldsymbol{X}^{\mathrm{T}}\boldsymbol{V}^{-1}\boldsymbol{X})^{-1}\boldsymbol{X}^{\mathrm{T}}\boldsymbol{V}^{-1}$. Analogously, inserting $\widehat{\boldsymbol{\theta}}_{\mathrm{ReML}}$ into Eq. (2.27) gives empirical estimates of the covariance matrices. However, we note that these empirical covariance estimates are derived under the assumption that the covariance structure $\boldsymbol{V}$ is known. Therefore they are expected to underestimate the variation of the fixed and random effects as the variation in $\widehat{\boldsymbol{\theta}}_{\mathrm{ReML}}$ is not taken into account. A full Bayesian analysis (Gelman et al., 2013) based on sampling methods such as the Markov chain Monte Carlo (MCMC) can produce more accurate approximations. However, in this chapter, we will focus on making inferences about the kernel function and the corresponding variance component parameters. As we will see in the next section, efficient statistical tests have been devised, which can avoid expensive computation for fitting the full LMM.

2.2.4 STATISTICAL INFERENCE

Hypothesis testing on elements of the fixed effects $\widehat{\boldsymbol{\beta}}$ can be performed using the standard *likelihood ratio test* (LRT) or the *Wald test*, under the assumption that $\widehat{\boldsymbol{\beta}}$ is asymptotically normal with mean $\boldsymbol{\beta}$ and covariance calculated in Eq. (2.27). For example, to test the null hypothesis $\mathcal{H}_0 : \boldsymbol{C\beta} = \boldsymbol{d}$ versus the alternative $\mathcal{H}_1 : \boldsymbol{C\beta} \neq \boldsymbol{d}$, where $\boldsymbol{C}$ is a *contrast matrix* with rank r, the LRT statistic is

$$\mathcal{D} = -2\left[\ell(\widehat{\boldsymbol{\beta}}_{\mathrm{R}}, \widehat{\boldsymbol{\theta}}_{\mathrm{R}}) - \ell(\widehat{\boldsymbol{\beta}}, \widehat{\boldsymbol{\theta}})\right] \sim \chi^2_r, \tag{2.28}$$

where $\widehat{\boldsymbol{\beta}}$ and $\widehat{\boldsymbol{\theta}}$ are estimates in the full, unrestricted model, and $\widehat{\boldsymbol{\beta}}_{\mathrm{R}}$ and $\widehat{\boldsymbol{\theta}}_{\mathrm{R}}$ are estimates in the restricted model, that is, when $\boldsymbol{C\beta} = \boldsymbol{d}$ is satisfied. Under the null hypothesis, $\mathcal{D}$ approximately follows a chi-squared distribution with r degrees of freedom. Alternatively, the Wald test statistic is

$$\mathcal{W} = (\boldsymbol{C}\widehat{\boldsymbol{\beta}} - \boldsymbol{d})^{\mathrm{T}} \left(\boldsymbol{C}\boldsymbol{A}(\widehat{\boldsymbol{\theta}})\boldsymbol{C}^{\mathrm{T}}\right)^{-1} (\boldsymbol{C}\widehat{\boldsymbol{\beta}} - \boldsymbol{d}) \sim \chi^2_r, \tag{2.29}$$

where $\boldsymbol{A}(\widehat{\boldsymbol{\theta}}) = \left(\boldsymbol{X}^{\mathrm{T}}\boldsymbol{V}(\widehat{\boldsymbol{\theta}})^{-1}\boldsymbol{X}\right)^{-1}$. Under the null hypothesis, $\mathcal{W}$ also approximately follows a chi-squared distribution with r degrees of freedom.

However, in many applications, covariates or nuisance variables are not of primary interest, and people are more interested in testing whether the nonparametric function f significantly contributes to the trait in Eq. (2.11), that is, testing the null hypothesis $\mathcal{H}_0 : f(\cdot) = 0$. Using the LMM representation (2.20), it can be seen that testing this null hypothesis is equivalent to testing $\mathcal{H}_0 : \tau^2 = 0$ against the alternative $\mathcal{H}_1 : \tau^2 > 0$. Note that under the null hypothesis, τ^2 lies on the boundary of the parameter space (since it cannot be negative), and the kernel matrix $\boldsymbol{K}$ is not block diagonal, making standard LRT inapplicable (Self and Liang, 1987). Liu et al. (2007) proposed a *score test* to address this issue. The score test (also known as the *Lagrange multiplier test*) assesses whether a parameter of interest θ is equal to a particular value θ_0 under the null using a test statistic that in general takes the form:

$$\mathcal{S} = \left.\frac{\mathcal{U}(\theta)}{\mathcal{I}_{\mathrm{E}}(\theta)^{1/2}}\right|_{\theta=\theta_0}, \tag{2.30}$$

where $\mathcal{U}(\theta) = \partial\ell(\theta)/\partial\theta$ is the derivative of the log likelihood with respect to θ, known as the *score*, and $\mathcal{I}_{\mathrm{E}}(\theta)$ is the *Fisher information* (or expected information) of θ. Here, using the results derived in Eqs. (B.3) and (B.6) in Appendix B, an ReML version of the score and Fisher information can be calculated as

$$\begin{aligned}
\mathcal{U}(\tau^2)\Big|_{\tau^2=0} &= \left.\frac{\partial\ell_{\mathrm{R}}}{\partial\tau^2}\right|_{\tau^2=0} = -\frac{1}{2\sigma_0^2}\mathbf{tr}\{\boldsymbol{P}_0\boldsymbol{K}\} + \frac{1}{2\sigma_0^4}\boldsymbol{y}^{\mathrm{T}}\boldsymbol{P}_0\boldsymbol{K}\boldsymbol{P}_0\boldsymbol{y}, \\
\mathcal{I}_{\mathrm{E}}(\tau^2)\Big|_{\tau^2=0} &= \mathbf{E}\left[-\frac{\partial^2\ell_{\mathrm{R}}}{(\partial\tau^2)^2}\right]_{\tau^2=0} = \frac{1}{2\sigma_0^4}\mathbf{tr}\{\boldsymbol{P}_0\boldsymbol{K}\boldsymbol{P}_0\boldsymbol{K}\},
\end{aligned} \tag{2.31}$$

where $\mathbf{E}[\cdot]$ denotes expectation, σ_0^2 is the variance of the residual $\boldsymbol{\epsilon}_0$ under the null regression model $\boldsymbol{y} = \boldsymbol{X}\boldsymbol{\beta}_0 + \boldsymbol{\epsilon}_0$, and $\boldsymbol{P}_0 = \boldsymbol{I} - \boldsymbol{X}(\boldsymbol{X}^\mathrm{T}\boldsymbol{X})^{-1}\boldsymbol{X}^\mathrm{T}$ is the projection matrix under the null. Since $\mathbf{tr}\{\boldsymbol{P}_0\boldsymbol{K}\}$ and $\mathbf{tr}\{\boldsymbol{P}_0\boldsymbol{K}\boldsymbol{P}_0\boldsymbol{K}\}$ are constants independent of $\boldsymbol{y}$, the null hypothesis $\mathcal{H}_0 : \tau^2 = 0$ can be tested using the following score test statistic:

$$\mathcal{S}(\sigma_0^2) = \frac{1}{2\sigma_0^2}\boldsymbol{y}^\mathrm{T}\boldsymbol{P}_0\boldsymbol{K}\boldsymbol{P}_0\boldsymbol{y} = \frac{1}{2\sigma_0^2}\left(\boldsymbol{y} - \boldsymbol{X}\widehat{\boldsymbol{\beta}}_0\right)^\mathrm{T}\boldsymbol{K}\left(\boldsymbol{y} - \boldsymbol{X}\widehat{\boldsymbol{\beta}}_0\right), \tag{2.32}$$

which is a measure of the association between the residuals estimated from the null model and the kernel matrix. $\mathcal{S}(\sigma_0^2)$ is a quadratic function of $\boldsymbol{y}$ and thus follows a mixture of chi-squares under the null. Specifically,

$$\mathcal{S}(\sigma_0^2) = \sum_{i=1}^{n}\lambda_i\chi_{1,i}^2, \tag{2.33}$$

where $\{\lambda_i\}_{i=1}^n$ are eigenvalues of the matrix $\frac{1}{2}\boldsymbol{P}_0^{1/2}\boldsymbol{K}\boldsymbol{P}_0^{1/2}$, and $\chi_{1,i}^2$ are independent and identically distributed (i.i.d.) random variables following chi-squared distributions with 1 degree of freedom. The p-value of an observed score test statistic can be analytically computed by the Davies (1980) method or Kuonen's saddle point method Kuonen (1999) using Eq. (2.33). Liu et al. (2007) proposed to use the *Satterthwaite method* to approximate the distribution of $\mathcal{S}(\sigma_0^2)$ by a scaled chi-squared distribution $\delta\chi_\nu^2$, where δ is the scale parameter and ν denotes the degrees of freedom. The two parameters are estimated by matching the first two moments, mean and variance, of $\mathcal{S}(\sigma_0^2)$ with those of $\delta\chi_\nu^2$:

$$\begin{cases} \zeta := \mathbf{E}[\mathcal{S}(\sigma_0^2)] = \frac{1}{2}\mathbf{tr}\{\boldsymbol{P}_0\boldsymbol{K}\} = \mathbf{E}[\delta\chi_\nu^2] = \delta\nu, \\ \xi := \mathbf{var}[\mathcal{S}(\sigma_0^2)] = \frac{1}{2}\mathbf{tr}\{\boldsymbol{P}_0\boldsymbol{K}\boldsymbol{P}_0\boldsymbol{K}\} = \mathbf{var}[\delta\chi_\nu^2] = 2\delta^2\nu. \end{cases} \tag{2.34}$$

Solving the two equations yields $\delta = \xi/2\zeta$ and $\nu = 2\zeta^2/\xi$. In practice, σ_0^2 is often unknown and is replaced by its maximum likelihood estimate $\hat{\sigma}_0^2$ under the null model. To account for this substitution, ξ needs to be replaced by $\hat{\xi}$ based on efficient information (Zhang and Lin, 2003): $\hat{\xi} = \widehat{\mathcal{I}}_{\tau\tau} = \mathcal{I}_{\tau\tau} - \mathcal{I}_{\sigma\sigma}^{-1}\mathcal{I}_{\tau\sigma}^2$, where $\mathcal{I}_{\tau\tau} = \mathbf{tr}\{\boldsymbol{P}_0\boldsymbol{K}\boldsymbol{P}_0\boldsymbol{K}\}/2$, $\mathcal{I}_{\tau\sigma} = \mathbf{tr}\{\boldsymbol{P}_0\boldsymbol{K}\boldsymbol{P}_0\}/2$ and $\mathcal{I}_{\sigma\sigma} = \mathbf{tr}\{\boldsymbol{P}_0\boldsymbol{P}_0\}/2$. Here $\mathcal{I}_{\tau\tau}$, $\mathcal{I}_{\tau\sigma}$ and $\mathcal{I}_{\sigma\sigma}$ are proportional to elements in the Fisher information matrix of τ^2 and σ^2 (see Appendix B). With the adjusted parameters $\hat{\delta} = \hat{\xi}/2\zeta$ and $\hat{\nu} = 2\zeta^2/\hat{\xi}$, the p-value of an observed score statistic $\mathcal{S}(\hat{\sigma}_0^2)$ is then computed using the scaled chi-squared distribution $\hat{\delta}\chi_{\hat{\nu}}^2$. One advantage of this score test is that it only requires fitting of a linear fixed effects model under the null hypothesis, and thus can be highly computationally efficient and suitable for analyzing a large number of traits. Pan (2011) showed that when no covariate needs to be adjusted other than an intercept, the score test statistic is equivalent to the genomic distance-based regression (GDBR) (Wessel and Schork, 2006), which is based on the Gower distance (Gower, 1966) and the pseudo-F statistic (McArdle and Anderson, 2001). The F-statistic is also closely related to the sum of squared score (SSU) test (Pan, 2011) and the Goeman's test (Goeman et al., 2006; Pan, 2009).

2.2.5 CONSTRUCTING AND SELECTING KERNELS

Kernel methods are appealing for their flexibility and generality; any non-negative definite kernel function can be used to measure the similarity between attributes from pairs of individuals and explain the trait variation. However, this flexibility can sometimes make the selection and comparison of kernels challenging. For example, it can be difficult to select a kernel that best captures the characteristics of the data or most powerfully detects a specific mechanism from a collection of valid kernels. In most studies, a kernel is a priori selected from commonly used candidates, such as the linear kernel, polynomial kernel and Gaussian kernel, or specifically designed in order to address a unique scientific question. See Schaid (2010b) and Hofmann et al. (2008) for reviews of kernel functions proposed in genomic studies and machine learning, respectively. However, there are existing methods that can build, compare and select kernels in a more systematic and objective way.

First, new kernels can be created by using existing non-negative definite kernel functions as building blocks. Assuming that $\{k_t\}_{t=1}^{+\infty}$ is a sequence of kernels defined on $\mathcal{Z} \times \mathcal{Z}$ with the associated kernel matrices $\{\boldsymbol{K}_t\}_{t=1}^{+\infty}$ evaluated at a set of attributes $\{z_1, \ldots, z_n\}$, and $z, z' \in \mathcal{Z}$ are arbitrary attributes, then:

- For any $\gamma_1, \gamma_2 \geqslant 0$, the linear combination $(\gamma_1 k_1 + \gamma_2 k_2)(z, z') := \gamma_1 k_1(z, z') + \gamma_2 k_2(z, z')$ is a new kernel function with the associated kernel matrix $\gamma_1 \boldsymbol{K}_1 + \gamma_2 \boldsymbol{K}_2$. This property can be useful when jointly modeling data from different sources and/or at different scales using multiple kernels.
- The point-wise product $(k_1 k_2)(z, z') := k_1(z, z') \cdot k_2(z, z')$ is a new kernel function with the associated kernel matrix $\boldsymbol{K}_1 \circ \boldsymbol{K}_2$, where $\circ$ is the Hadamard product (element-wise product) of two matrices. This can be useful when modeling the interaction of two kernels defined on the same space.
- $k(z, z') := \lim_{t \to +\infty} k_t(z, z')$ is a new kernel function with the associated kernel matrix $\boldsymbol{K} = \lim_{t \to +\infty} \boldsymbol{K}_t$, if the limit exists for arbitrary z and z'.

Many more kernels can be created using a combination of these mathematical operations. For example, any polynomial of a kernel, $\tilde{k}(z, z') := \sum_l \gamma_l k^l(z, z')$ with $\gamma_l \geqslant 0$, gives a new kernel; the exponentiation of a kernel, $\tilde{k}(z, z') := \exp\{k(z, z')\}$, is also a kernel since it can be expanded into a convergent sequence of polynomials. Moreover, given two kernels k_1 and k_2 defined on $\mathcal{Z}_1 \times \mathcal{Z}_1$ and $\mathcal{Z}_2 \times \mathcal{Z}_2$, with the associated kernel matrices $\boldsymbol{K}_1$ and $\boldsymbol{K}_2$ evaluated at $\{z_{1,1}, \ldots, z_{1,n}\}$ and $\{z_{2,1}, \ldots, z_{2,n}\}$, respectively, and arbitrary attributes $z_1, z_1' \in \mathcal{Z}_1$, $z_2, z_2' \in \mathcal{Z}_2$, then:

- The tensor product $(k_1 \otimes k_2)\left((z_1,z_2),(z_1',z_2')\right) := k_1(z_1,z_1') \cdot k_2(z_2,z_2')$ is a new kernel on the product domain $(\mathcal{Z}_1 \times \mathcal{Z}_2) \times (\mathcal{Z}_1 \times \mathcal{Z}_2)$, with the associated kernel matrix $\boldsymbol{K}_1 \circ \boldsymbol{K}_2$. This can be useful to construct kernels on the tensor product of two RKHSs.

In practice, it can be difficult to directly specify a kernel to capture complex relationships between pairs of attributes. However, the properties presented above suggest that if the effect of interest can be decomposed into components that can be well characterized by primitive kernels, an advanced kernel can be constructed using a bottom-up approach.

Second, specifying a kernel function requires a similarity measure between pairs of attributes, but sometimes it is more natural to measure dissimilarity or distance. In this case, we note that any distance measure, $d = d(z,z')$, can be converted into a similarity measure $-\frac{1}{2}d^2$, which is known as the Gower distance (Gower, 1966). For arbitrary attributes $\{z_1,\ldots,z_n\}$, this gives a similarity matrix $-\frac{1}{2}\boldsymbol{D} \circ \boldsymbol{D}$, where $\boldsymbol{D} = [d_{ij}]_{n\times n}$ with $d_{ij} = d(z_i,z_j)$. We now center the Gower distance matrix and define $\boldsymbol{K} = \boldsymbol{H}\left[-\frac{1}{2}\boldsymbol{D} \circ \boldsymbol{D}\right]\boldsymbol{H}$, where $\boldsymbol{H}$ is a centering matrix with the ijth entry $\boldsymbol{H}_{ij} = \delta_{ij} - 1/n$, δ_{ij} being the Kronecker delta. Note that $\boldsymbol{K}$ is not guaranteed to be non-negative definite. However, when the Euclidean distance, $d_{ij} = \|z_i - z_j\| = \sqrt{\sum_l (z_{il} - z_{jl})^2}$, is used, we have

$$\boldsymbol{K} = \boldsymbol{H}\left[-\frac{1}{2}\boldsymbol{D} \circ \boldsymbol{D}\right]\boldsymbol{H} = \boldsymbol{H}\boldsymbol{Z}\boldsymbol{Z}^{\mathrm{T}}\boldsymbol{H}, \tag{2.35}$$

where $\boldsymbol{Z} = (z_1,\ldots,z_n)^{\mathrm{T}}$. Clearly, $\boldsymbol{K}$ is now non-negative definite, and more specifically, $\boldsymbol{K}$ is a centered linear kernel of the attributes. This offers a way to induce a valid kernel from Euclidean distance, perhaps the most widely used distance measure in practice.

Finally, Liu et al. (2007) pointed out that both model selection and variable selection are special cases of kernel selection within the KMR framework. They proposed AIC and BIC in the context of KMR, which can evaluate candidate kernels in a systematic and objective way. The calculation of AIC/BIC normally requires the number of estimated parameters in the model. However, KMR models are nonparametric and thus the number of model parameters is not explicitly defined. To address this issue, Liu et al. (2007) noted that, using Eqs. (2.17) and (2.19), the estimated trait, $\widehat{\boldsymbol{y}}$, can be expressed as

$$\begin{aligned}\widehat{\boldsymbol{y}} = \boldsymbol{X}\widehat{\boldsymbol{\beta}} + \widehat{\boldsymbol{f}} &= \left\{\hat{\tau}^2\boldsymbol{K}\widehat{\boldsymbol{V}}^{-1} + \hat{\sigma}^2\widehat{\boldsymbol{V}}^{-1}\boldsymbol{X}(\boldsymbol{X}^{\mathrm{T}}\widehat{\boldsymbol{V}}^{-1}\boldsymbol{X})^{-1}\boldsymbol{X}^{\mathrm{T}}\widehat{\boldsymbol{V}}^{-1}\right\}\boldsymbol{y} \\ &= \left(\boldsymbol{I} - \hat{\sigma}^2\widehat{\boldsymbol{P}}\right)\boldsymbol{y} := \boldsymbol{S}\boldsymbol{y},\end{aligned} \tag{2.36}$$

where $\widehat{V} = V(\widehat{\theta}_{\text{ReML}})$ and $\widehat{P} = P(\widehat{\theta}_{\text{ReML}}) = \widehat{V}^{-1} - \widehat{V}^{-1}X(X^{\text{T}}\widehat{V}^{-1}X)^{-1}X^{\text{T}}\widehat{V}^{-1}$. The matrix S smoothes the observed data y to produce the predicted trait $\widehat{y}$, and its trace, $\mathbf{tr}\{S\}$, can be interpreted as a measure of model complexity. To see this, we notice that Eq. (B.3) gives

$$\frac{\partial \ell_{\text{R}}}{\partial \sigma^2} = -\frac{1}{2}\mathbf{tr}\{P\} + \frac{1}{2}y^{\text{T}}PPy. \tag{2.37}$$

Thus at the ReML estimate $\hat{\sigma}^2$ that maximizes the restricted likelihood ℓ_{R}, the derivative is zero and we have $\mathbf{tr}\{\widehat{P}\} = y^{\text{T}}\widehat{P}\widehat{P}y$. Also, by the definition of S, we have $\mathbf{tr}\{S\} = \mathbf{tr}\left\{I - \hat{\sigma}^2\widehat{P}\right\} = n - \hat{\sigma}^2\mathbf{tr}\{\widehat{P}\}$. The residual sum of squares (RSS) can then be computed as

$$\text{RSS} = (y - \widehat{y})^{\text{T}}(y - \widehat{y}) = \hat{\sigma}^4 y^{\text{T}}\widehat{P}\widehat{P}y = \hat{\sigma}^4\mathbf{tr}\{\widehat{P}\} = \hat{\sigma}^2(n - \mathbf{tr}\{S\}), \tag{2.38}$$

and thus $\hat{\sigma}^2 = \text{RSS}/(n - \mathbf{tr}\{S\})$, indicating that $\mathbf{tr}\{S\}$ is the loss in degrees of freedom resulting from estimating the fixed effects $\boldsymbol{\beta}$ and the random effects $\boldsymbol{f}$ when estimating σ^2. Therefore Liu et al. (2007) proposed the KMR-based AIC and BIC as

$$\begin{aligned} \text{AIC} &= n\log(\text{RSS}) + 2\mathbf{tr}\{S\}, \\ \text{BIC} &= n\log(\text{RSS}) + \mathbf{tr}\{S\} \cdot \log(n). \end{aligned} \tag{2.39}$$

Both measures reward goodness-of-fit (the first term) and penalize complex models (the second term) in order to avoid overfitting. BIC has a larger penalty term than AIC for large n, and thus favors simpler models. Models with smaller AIC/BIC values are selected and believed to be better descriptions of the data.

2.2.6 THEORETICAL EXTENSIONS

In previous sections, we have introduced the basics of KMR, focusing on univariate (scalar) and quantitative traits collected from unrelated individuals whose attributes are modeled by a single nonparametric function. In this section, we review recent theoretical developments that generalized the classical KMR model to handle more complex data structures.

2.2.6.1 Generalized kernel machine regression

KMR can be extended to handle a much wider class of data types, such as binary and count data, whose distribution lies in the *exponential family* (McCullagh and Nelder, 1989; Liu et al., 2008; Wu et al., 2010). Specifically, suppose that the density function of the trait for the ith subject, y_i, takes the *canonical form* (or *natural form*):

$$p(y_i|\eta_i, \phi) = h(y_i, \phi)\exp\left\{\frac{1}{\phi}\left[\eta_i y_i - a(\eta_i)\right]\right\}, \tag{2.40}$$

where η_i is the *natural parameter*, ϕ is a scale or dispersion parameter, and $h(\cdot,\cdot)$ and $a(\cdot)$ are known functions. $a(\cdot)$ is a normalization factor (known as the *log-partition function*) that ensures the distribution sums or integrates to one, and is connected to the moments of y_i:

$$\mu_i = \mathbf{E}[y_i|\eta_i] = a'(\eta_i), \quad \mathbf{var}[y_i|\eta_i,\phi] = \phi a''(\eta_i) := \phi v(\mu_i), \tag{2.41}$$

where we have defined the variance function $v(\cdot)$ that associates the first and second moments of y_i. The generalized kernel machine regression (GKMR) then connects the mean of the distribution function, μ_i, with the covariates $\boldsymbol{x}_i$ and attributes z_i, using a monotonic *link function* g:

$$g(\mu_i) = \boldsymbol{x}_i^{\mathrm{T}}\boldsymbol{\beta} + f(z_i), \quad i = 1, 2, \ldots, n, \tag{2.42}$$

where $\boldsymbol{\beta}$ is a vector of regression coefficients, and f is an unknown function that lies in an RKHS $\mathcal{H}$ defined by a kernel function k. When $g^{-1}(\cdot) = a'(\cdot)$, g is the *canonical link* and we have $\eta_i = \boldsymbol{x}_i^{\mathrm{T}}\boldsymbol{\beta} + f(z_i)$. Table 2.1 lists the natural parameter, scale parameter, mean of the trait, variance function and canonical link for linear, logistic and Poisson regressions, which can handle normal, binary and count data, respectively. In the derivation below, we always assume that the canonical link is used.

Model (2.42) can be fitted by minimizing the panelized likelihood function:

$$\mathcal{J}(\boldsymbol{\beta}, f) = -\sum_{i=1}^{n}[\eta_i y_i - a(\eta_i)] + \frac{\lambda}{2}\|f\|_{\mathcal{H}}^2, \tag{2.43}$$

where the first part is proportional to the minus log likelihood of the model, ignoring a constant independent of $\boldsymbol{\beta}$ and f, while the second part is a regularization term penalizing rough functions. By the Representer theorem, the minimizer takes the form $f(\cdot) = \sum_{j=1}^{n}\alpha_j k(\cdot, z_j)$, and $\mathcal{J}$ has a finite-dimensional representation:

$$\mathcal{J}(\boldsymbol{\beta}, \boldsymbol{\alpha}) = -\sum_{i=1}^{n}[\eta_i y_i - a(\eta_i)] + \frac{\lambda}{2}\boldsymbol{\alpha}^{\mathrm{T}}\boldsymbol{K}\boldsymbol{\alpha}, \quad \eta_i = \boldsymbol{x}_i^{\mathrm{T}}\boldsymbol{\beta} + \boldsymbol{\kappa}_i^{\mathrm{T}}\boldsymbol{\alpha}, \tag{2.44}$$

Table 2.1 The Natural Parameter, Scale Parameter, Mean of the Trait, Variance Function and Canonical Link for Linear, Logistic and Poisson Regressions

Regression	η	ϕ	μ	$v(\mu)$	$g(\cdot)$
Linear	μ	σ^2	μ	1	id(·)
Logistic	$\log\left(\frac{\pi}{1-\pi}\right)$	1	π	$\mu(1-\mu)$	logit(·)
Poisson	$\log(\lambda)$	1	λ	μ	log(·)

id(·) is the identity function. π is the probability of an outcome of interest or event. logit(·) is the logit function. λ is the rate of occurrence of an event.

where $\boldsymbol{\alpha} = [\alpha_1, \ldots, \alpha_n]^{\mathrm{T}}$ and $\boldsymbol{\kappa}_i = [k(z_1, z_i), \ldots, k(z_n, z_i)]^{\mathrm{T}}$. Setting the derivatives of $\mathcal{J}(\boldsymbol{\beta}, \boldsymbol{\alpha})$ with respect to $\boldsymbol{\alpha}$ and $\boldsymbol{\beta}$ to zero yields the following equations:

$$\sum_{i=1}^{n} \frac{(y_i - \mu_i)\boldsymbol{x}_i}{v(\mu_i)g'(\mu_i)} = \mathbf{0}, \qquad \sum_{i=1}^{n} \frac{(y_i - \mu_i)\boldsymbol{\kappa}_i}{v(\mu_i)g'(\mu_i)} = \lambda \boldsymbol{K}\boldsymbol{\alpha}. \tag{2.45}$$

Using the canonical link implies that $g'(\mu_i) = 1/v(\mu_i)$, and the denominators in Eq. (2.45) thus vanish. We note that μ_i depends on $\boldsymbol{\alpha}$ and $\boldsymbol{\beta}$ through the link g, which is in general nonlinear. Thus the solution of Eq. (2.45) does not have a closed form and need to be estimated numerically. Breslow and Clayton (1993) showed that the solution to Eq. (2.45) via Fisher scoring is the iterative solution to the following linear system:

$$\begin{bmatrix} \boldsymbol{X}^{\mathrm{T}}\boldsymbol{W}\boldsymbol{X} & \boldsymbol{X}^{\mathrm{T}}\boldsymbol{W}\boldsymbol{K} \\ \boldsymbol{W}\boldsymbol{X} & \lambda\boldsymbol{I} + \boldsymbol{W}\boldsymbol{K} \end{bmatrix} \begin{bmatrix} \boldsymbol{\beta} \\ \boldsymbol{\alpha} \end{bmatrix} = \begin{bmatrix} \boldsymbol{X}^{\mathrm{T}}\boldsymbol{W}\tilde{\boldsymbol{y}} \\ \boldsymbol{W}\tilde{\boldsymbol{y}} \end{bmatrix}, \tag{2.46}$$

where $\boldsymbol{W}$ is an $n \times n$ diagonal matrix with the ith diagonal element $v(\mu_i)$, and $\tilde{\boldsymbol{y}}$ is an $n \times 1$ working vector with the ith element $\tilde{y}_i = \eta_i + (y_i - \mu_i)g'(\mu_i) = \boldsymbol{x}_i^{\mathrm{T}}\boldsymbol{\beta} + \boldsymbol{\kappa}_i^{\mathrm{T}}\boldsymbol{\alpha} + (y_i - \mu_i)/v(\mu_i)$. Note that both $\boldsymbol{W}$ and $\tilde{\boldsymbol{y}}$ depend on $\boldsymbol{\alpha}$ and $\boldsymbol{\beta}$, and thus Eq. (2.46) needs to be solved iteratively. More specifically, we first fix $\boldsymbol{W}$ and $\tilde{\boldsymbol{y}}$ to solve Eq. (2.46), and then update $\boldsymbol{W}$ and $\tilde{\boldsymbol{y}}$ using the new estimates of $\boldsymbol{\alpha}$ and $\boldsymbol{\beta}$. This is repeated until convergence.

It can be seen that Eq. (2.46) also depends on the tuning parameter λ, which is usually unknown. This can be resolved by establishing the connection between GKMR and the following generalized linear mixed effects model (GLMM):

$$g(\mu_i) = g\left(\mathbf{E}[y_i|\eta_i]\right) = \eta_i, \quad \eta_i = \boldsymbol{x}_i^{\mathrm{T}}\boldsymbol{\beta} + f_i, \quad i = 1, 2, \ldots, n, \tag{2.47}$$

where $\boldsymbol{\beta}$ is a vector of fixed effects and $\boldsymbol{f} = [f_1, \ldots, f_n]^{\mathrm{T}}$ is a vector of random effects following $\mathrm{N}(\mathbf{0}, \tau^2\boldsymbol{K})$ with $\tau^2 = \lambda^{-1}\phi$. The full likelihood-based estimation in GLMM is difficult due to the integral over the random effects. For LMMs, this integral can be analytically computed but in general the problem is intractable and numerical approximations have to be employed. A number of different approaches exist, which can be broadly categorized into integrand approximation (eg, the Laplace approximation), integral approximation (eg, the Gaussian quadrature approximation), and data approximation (eg, the penalized quasi-likelihood methods). Here we show that the penalized quasi-likelihood (PQL) approach is tightly connected to GKMR.

PQL does not specify a full probability distribution but approximates the data by the mean and variance: $y_i \approx \mu_i + \epsilon_i$, where ϵ_i is an error term with $\mathbf{var}[\epsilon_i] = \mathbf{var}[y_i|\eta_i, \phi] = \phi v(\mu_i)$. Then using a Taylor expansion, we have

$$\begin{aligned} y_i &\approx \mu_i + \epsilon_i = g^{-1}\left(\boldsymbol{x}_i^{\mathrm{T}}\boldsymbol{\beta} + f_i\right) + \epsilon_i \\ &\approx g^{-1}\left(\boldsymbol{x}_i^{\mathrm{T}}\widehat{\boldsymbol{\beta}} + \hat{f}_i\right) + a''\left(\boldsymbol{x}_i^{\mathrm{T}}\widehat{\boldsymbol{\beta}} + \hat{f}_i\right)\boldsymbol{x}_i^{\mathrm{T}}(\boldsymbol{\beta} - \widehat{\boldsymbol{\beta}}) + a''\left(\boldsymbol{x}_i^{\mathrm{T}}\widehat{\boldsymbol{\beta}} + \hat{f}_i\right)(f_i - \hat{f}_i) + \epsilon_i \\ &\approx \hat{\mu}_i + v(\hat{\mu}_i)\boldsymbol{x}_i^{\mathrm{T}}(\boldsymbol{\beta} - \widehat{\boldsymbol{\beta}}) + v(\hat{\mu}_i)(f_i - \hat{f}_i) + \epsilon_i, \end{aligned} \tag{2.48}$$

where we have used the identity $[g^{-1}(\eta_i)]' = a''(\eta_i) = v(\mu_i)$ for canonical links. We note that this expansion is exact for LMMs, in which g is an identity function. Reorganizing Eq. (2.48) gives

$$\tilde{y}_i := \hat{\eta}_i + (y_i - \hat{\mu}_i)/v(\hat{\mu}_i) = \boldsymbol{x}_i^{\mathrm{T}}\boldsymbol{\beta} + f_i + \epsilon_i/v(\hat{\mu}_i) := \boldsymbol{x}_i^{\mathrm{T}}\boldsymbol{\beta} + f_i + \tilde{\epsilon}_i, \tag{2.49}$$

where $\mathbf{var}[\tilde{\epsilon}_i] = \phi/v(\hat{\mu}_i)$ if we evaluate $\mathbf{var}[\epsilon_i]$ at $\hat{\mu}_i$. Therefore we have approximated the GLMM (2.47) by a working LMM for the pseudo-data $\tilde{y}_i$:

$$\tilde{\boldsymbol{y}} = \boldsymbol{X}\boldsymbol{\beta} + \boldsymbol{f} + \tilde{\boldsymbol{\epsilon}}, \quad \boldsymbol{f} \sim \mathrm{N}(\boldsymbol{0}, \tau^2\boldsymbol{K}), \quad \tilde{\boldsymbol{\epsilon}} \sim \mathrm{N}(\boldsymbol{0}, \phi\widehat{\boldsymbol{W}}^{-1}), \tag{2.50}$$

in which $\tilde{\boldsymbol{y}} = [\tilde{y}_1, \ldots, \tilde{y}_n]^{\mathrm{T}}$, $\boldsymbol{X} = [\boldsymbol{x}_1, \ldots, \boldsymbol{x}_n]^{\mathrm{T}}$, $\tilde{\boldsymbol{\epsilon}} = [\tilde{\epsilon}_1, \ldots, \tilde{\epsilon}_n]^{\mathrm{T}}$, and $\widehat{\boldsymbol{W}}$ is an $n \times n$ diagonal matrix with the ith diagonal element $v(\hat{\mu}_i)$. It is easy to verify that Henderson's MME of the LMM (2.50) is exactly the same as the linear system (2.46), by using the fact that $\boldsymbol{f} = \boldsymbol{K}\boldsymbol{\alpha}$. We have thus bridged GKMR and GLMMs by showing that the minimizer of the panelized likelihood function (2.43) is equivalent to the PQL estimate of the GLMM (2.47). The vector $\boldsymbol{\theta}$, containing ϕ, τ^2, and any unknown parameter in the kernel function, can be estimated by maximizing the restricted likelihood of the working LMM:

$$\ell_{\mathrm{R}}(\boldsymbol{\theta}) = -\frac{1}{2}\log|\boldsymbol{V}(\boldsymbol{\theta})| - \frac{1}{2}\log|\boldsymbol{X}^{\mathrm{T}}\boldsymbol{V}(\boldsymbol{\theta})^{-1}\boldsymbol{X}| - \frac{1}{2}(\tilde{\boldsymbol{y}} - \boldsymbol{X}\widehat{\boldsymbol{\beta}})^{\mathrm{T}}\boldsymbol{V}(\boldsymbol{\theta})^{-1}(\tilde{\boldsymbol{y}} - \boldsymbol{X}\widehat{\boldsymbol{\beta}}), \tag{2.51}$$

where $\boldsymbol{V}(\boldsymbol{\theta}) = \tau^2\boldsymbol{K} + \phi\widehat{\boldsymbol{W}}^{-1}$ and $\widehat{\boldsymbol{\beta}} = \widehat{\boldsymbol{\beta}}(\boldsymbol{\theta}) = \left(\boldsymbol{X}^{\mathrm{T}}\boldsymbol{V}(\boldsymbol{\theta})^{-1}\boldsymbol{X}\right)^{-1}\boldsymbol{X}\boldsymbol{V}(\boldsymbol{\theta})^{-1}\tilde{\boldsymbol{y}}$. Note that the pseudo-data $\tilde{\boldsymbol{y}}$ and the matrix $\widehat{\boldsymbol{W}}$ depend on the estimates of $\boldsymbol{\beta}$ and $\boldsymbol{f}$, and thus the GKMR model needs to be fitted by iteratively solving the working linear system (2.46) and maximizing the working restricted likelihood (2.51) until convergence.

Statistical inferences of the GKMR can also be conducted within the framework of GLMMs. Specifically, once the GLMM has been fitted, standard likelihood ratio test (LRT) and Wald test can be applied to the fixed effects $\widehat{\boldsymbol{\beta}}$. To test the null hypothesis $\mathcal{H}_0 : f(\cdot) = 0$, or equivalently $\mathcal{H}_0 : \tau^2 = 0$ against the alternative $\mathcal{H}_1 : \tau^2 > 0$ in the GLMM, a score statistic can be derived by using the linear approximation (2.50) and the null model $g(\mu_{0,i}) = \boldsymbol{x}_i^{\mathrm{T}}\boldsymbol{\beta}_0$:

$$\mathcal{S}(\phi_0) = \frac{1}{2\phi_0^2}\tilde{\boldsymbol{y}}^{\mathrm{T}}\boldsymbol{P}_0\boldsymbol{K}\boldsymbol{P}_0\tilde{\boldsymbol{y}} = \frac{1}{2\phi_0^2}(\boldsymbol{y} - \widehat{\boldsymbol{\mu}}_0)^{\mathrm{T}}\boldsymbol{K}(\boldsymbol{y} - \widehat{\boldsymbol{\mu}}_0), \tag{2.52}$$

where ϕ_0 is the scale parameter of the null GLMM, $\boldsymbol{P}_0 = \widehat{\boldsymbol{W}}_0 - \widehat{\boldsymbol{W}}_0\boldsymbol{X}(\boldsymbol{X}^{\mathrm{T}}\widehat{\boldsymbol{W}}_0\boldsymbol{X})^{-1}\boldsymbol{X}^{\mathrm{T}}\widehat{\boldsymbol{W}}_0$ is the null projection matrix, $\widehat{\boldsymbol{W}}_0$ is an $n \times n$ diagonal matrix with the ith element $v(\hat{\mu}_{0,i})$, $\hat{\mu}_{0,i} = g^{-1}\left(\boldsymbol{x}_i^{\mathrm{T}}\widehat{\boldsymbol{\beta}}_0\right)$ is the estimated mean under the null, $\widehat{\boldsymbol{\mu}}_0 = [\hat{\mu}_{0,1}, \ldots, \hat{\mu}_{0,n}]^{\mathrm{T}}$. $\mathcal{S}(\phi_0)$ follows a mixture of chi-squares under the null and can be approximated by a scaled chi-squared distribution using the Satterthwaite method. The estimated scale parameter and degrees of freedom of the chi-squared distribution literally take the same form as in the linear case if we assume that $\tilde{\boldsymbol{y}}$ is approximately normal. For very skewed data, taking into account the influences of high-order moments such as the *kurtosis* can provide more accurate approximations to the distribution of the score test statistic (Lin, 1997).

2.2.6.2 Multiple kernel functions

The KMR can also be extended to include multiple kernel functions (Gianola and van Kaam, 2008). These kernel functions can either be defined on the same input space, capturing different aspects of the same attribute or jointly modeling multiple attributes, or be defined on different input spaces, integrating data from different domains. In general, assuming quantitative traits, a KMR model with multiple kernel functions can be written as

$$y_i = \boldsymbol{x}_i^{\mathrm{T}} \boldsymbol{\beta} + \sum_{l=1}^{\varsigma} f_l(z_{il}) + \epsilon_i, \tag{2.53}$$

where ς is the number of kernel functions in the model, each attribute z_{il} belongs to an input space $\mathcal{Z}_l$, and f_l is an unknown function lying in an RKHS $\mathcal{H}_l$ defined by the kernel function k_l. A parallel proof of the classical KMR theory shows that fitting model (2.53) by minimizing its penalized likelihood function is equivalent to fitting the LMM:

$$\boldsymbol{y} = \boldsymbol{X\beta} + \sum_{l=1}^{\varsigma} \boldsymbol{f}_l + \boldsymbol{\epsilon}, \tag{2.54}$$

where $\boldsymbol{\epsilon} \sim \mathrm{N}(\boldsymbol{0}, \sigma^2 \boldsymbol{I})$, $\boldsymbol{f}_l \sim \mathrm{N}(\boldsymbol{0}, \tau_l^2 \boldsymbol{K}_l)$, and $\lambda_l = \sigma^2/\tau_l^2$ is the tuning parameter controlling the penalization on the lth nonparametric function f_l. Variance component parameters σ^2, τ_l^2, $l = 1, 2, \ldots, \varsigma$, and other unknown parameters in the kernel functions can be estimated by maximizing the restricted likelihood (B.1), in which the marginal covariance structure is $\boldsymbol{V} = \sum_{l=1}^{\varsigma} \tau_l^2 \boldsymbol{K}_l + \sigma^2 \boldsymbol{I}$.

A score test statistic can be computed to assess whether at least one function in a subset of the nonparametric functions $\{f_l\}_{l=1}^{\varsigma}$ is significantly different from zero. More specifically, let ϖ be any subset of the indexes: $\varpi \subset \{1, 2, \ldots, \varsigma\}$. Suppose we test the null hypothesis that all functions whose indexes in ϖ are zero, that is, $\mathcal{H}_0 : f_l(\cdot) = 0$, for all $l \in \varpi$, against the alternative that at least one of these functions is significantly different from zero, the null model is

$$\boldsymbol{y} = \boldsymbol{X\beta}_0 + \sum_{l \notin \varpi} \boldsymbol{f}_l + \boldsymbol{\epsilon}_0, \quad \boldsymbol{f}_l \sim \mathrm{N}(\boldsymbol{0}, \tau_{0,l}^2 \boldsymbol{K}_l), \quad \boldsymbol{\epsilon}_0 \sim \mathrm{N}(\boldsymbol{0}, \sigma_0^2 \boldsymbol{I}), \tag{2.55}$$

and the score test statistic can be constructed as

$$\mathcal{S}(\sigma_0^2; \tau_{0,l}^2, l \notin \varpi) = \frac{1}{2} \boldsymbol{y}^{\mathrm{T}} \boldsymbol{P}_0 \boldsymbol{K} \boldsymbol{P}_0 \boldsymbol{y}, \tag{2.56}$$

where $\boldsymbol{P}_0 = \boldsymbol{V}_0^{-1} - \boldsymbol{V}_0^{-1}\boldsymbol{X}(\boldsymbol{X}^{\mathrm{T}}\boldsymbol{V}_0^{-1}\boldsymbol{X})^{-1}\boldsymbol{X}^{\mathrm{T}}\boldsymbol{V}_0^{-1}$ is the projection matrix under the null, and $\boldsymbol{V}_0 = \sum_{l \notin \varpi} \tau_{0,l}^2 \boldsymbol{K}_l + \sigma_0^2 \boldsymbol{I}$ is the marginal covariance matrix under the null. The Satterthwaite method can be used to approximate the distribution of $\mathcal{S}(\sigma_0^2; \tau_{0,l}^2, l \notin$

ϖ) by a scaled chi-squared distribution. The unknown model parameters σ_0^2 and $\tau_{0,l}^2, l \notin \varpi$, can be replaced by their ReML estimates under the null model in practice, with the scale parameter and the degrees of freedom of the chi-squared distribution being adjusted using efficient information to account for this substitution.

2.2.6.3 Correlated phenotypes

Previous sections have focused on phenotypes collected from individuals in the general population, and thus can be reasonably assumed to be independent. However, correlated phenotypes are frequently seen in practice. For example, family or pedigree-based designs have been widely used in many fields, and are particularly valuable in health-related research such as the study of rare diseases. Phenotypes of closely related individuals are often highly correlated, due to shared genetics and common environmental factors. Direct application of classical KMR to family/pedigree data by ignoring the familial structure leads to misspecified models and may result in inflated type I error in statistical inferences. The KMR model (2.11) can be easily extended, by incorporating a random effect, to appropriately account for familial correlation (Schifano et al., 2012; Chen et al., 2013):

$$y_i = \boldsymbol{x}_i^{\mathrm{T}}\boldsymbol{\beta} + \pi_i + f(z_i) + \epsilon_i, \quad i = 1, 2, \ldots, n, \tag{2.57}$$

where $\boldsymbol{\pi} = [\pi_1, \ldots, \pi_n]^{\mathrm{T}} \sim \mathrm{N}(\boldsymbol{0}, \sigma_g^2 \boldsymbol{\Pi})$ is a vector of random effects modeling the familial correlation, σ_g^2 is the total additive genetic variance, and $\boldsymbol{\Pi} = 2\boldsymbol{\Phi}$ is twice the *kinship matrix* and indicates expected genetic covariance among individuals. The *ij*th entry of the kinship matrix, ϕ_{ij}, known as the kinship coefficient, defines genetic relatedness for subjects *i* and *j*, and can in general be derived from pedigree information. For example, for identical (monozygotic) twins $\phi_{ij} = 1/2$, for full siblings and parent-offspring $\phi_{ij} = 1/4$, for half siblings and grandparent-grandchild $\phi_{ij} = 1/8$. Model (2.57) can then be converted into an LMM, which has been used for decades in the field of quantitative genetics for human pedigree analysis:

$$\boldsymbol{y} = \boldsymbol{X}\boldsymbol{\beta} + \boldsymbol{\pi} + \boldsymbol{f} + \boldsymbol{\epsilon}, \tag{2.58}$$

where $\boldsymbol{\pi} \sim \mathrm{N}(\boldsymbol{0}, \sigma_g^2 \boldsymbol{\Pi})$, $\boldsymbol{f} \sim \mathrm{N}(\boldsymbol{0}, \tau^2 \boldsymbol{K})$, and $\boldsymbol{\epsilon} \sim \mathrm{N}(\boldsymbol{0}, \sigma^2 \boldsymbol{I})$. Testing the null hypothesis $\mathcal{H}_0 : f(\cdot) = 0$, or equivalently $\mathcal{H}_0 : \tau^2 = 0$ using a score test falls in the framework set in the above section by noticing that the null covariance matrix is $\boldsymbol{V}_0 = \sigma_{0,g}^2 \boldsymbol{\Pi} + \sigma_0^2 \boldsymbol{I}$, where $\sigma_{0,g}^2$ and σ_0^2 are parameters in the null model and can be estimated using the ReML approach.

2.2.6.4 Multidimensional traits

With the rapid advances in phenotyping technology, it is now not uncommon to collect multiple secondary phenotypes that characterize a health-related outcome from different angles. These phenotypes can be highly related and describe a common mechanism underlying biological processes, and thus a joint analysis that accounts for their correlation structure may improve statistical power relative to conducting analysis on individual phenotypes. The classical KMR can be extended to model multiple traits (Maity et al., 2012). In particular, consider a total of ϱ traits collected on each individual, each trait modeled by the following KMR:

$$y_{il} = \boldsymbol{x}_{il}^{\mathrm{T}}\boldsymbol{\beta}_l + f_l(z_i) + \epsilon_{il}, \quad l = 1, 2, \ldots, \varrho, \tag{2.59}$$

where f_l is an unknown function lying in an RKHS $\mathcal{H}_l$ defined by the kernel function k_l. The residuals are independent across individuals but are correlated across trait dimensions: $[\epsilon_{i1}, \ldots, \epsilon_{i\varrho}]^{\mathrm{T}} \sim \mathrm{N}(\mathbf{0}, \boldsymbol{\Sigma})$, reflecting the covariance among phenotypes. It is easy to see, using the classical KMR theory, that each trait has a corresponding LMM:

$$\boldsymbol{y}_l = \boldsymbol{X}_l\boldsymbol{\beta}_l + \boldsymbol{f}_l + \boldsymbol{\epsilon}_l, \quad l = 1, 2, \ldots, \varrho, \tag{2.60}$$

where $\boldsymbol{y}_l = [y_{1l}, \ldots, y_{nl}]^{\mathrm{T}}$, $\boldsymbol{X}_l = [\boldsymbol{x}_{1l}, \ldots, \boldsymbol{x}_{nl}]^{\mathrm{T}}$, $\boldsymbol{f}_l \sim \mathrm{N}(\mathbf{0}, \tau_l^2\boldsymbol{K}_l)$, and $\boldsymbol{\epsilon}_l = [\epsilon_{1l}, \ldots, \epsilon_{nl}]^{\mathrm{T}}$. By stacking these individual LMMs, we have the following joint LMM:

$$\boldsymbol{y} = \boldsymbol{X}\boldsymbol{\beta} + \boldsymbol{f} + \boldsymbol{\epsilon}, \tag{2.61}$$

where $\boldsymbol{y} = [\boldsymbol{y}_1^{\mathrm{T}}, \ldots, \boldsymbol{y}_\varrho^{\mathrm{T}}]^{\mathrm{T}}$, $\boldsymbol{X} = \mathbf{diag}\{\boldsymbol{X}_1, \ldots, \boldsymbol{X}_\varrho\}$ is a block diagonal matrix, $\boldsymbol{\beta} = [\boldsymbol{\beta}_1^{\mathrm{T}}, \ldots, \boldsymbol{\beta}_\varrho^{\mathrm{T}}]^{\mathrm{T}}$, $\boldsymbol{f} = [\boldsymbol{f}_1^{\mathrm{T}}, \ldots, \boldsymbol{f}_\varrho^{\mathrm{T}}]^{\mathrm{T}} \sim \mathrm{N}(\mathbf{0}, \boldsymbol{\Lambda}\boldsymbol{K})$, $\boldsymbol{\Lambda} = \mathbf{diag}\{\tau_1^2\boldsymbol{I}, \ldots, \tau_\varrho^2\boldsymbol{I}\}$, $\boldsymbol{K} = \mathbf{diag}\{\boldsymbol{K}_1, \ldots, \boldsymbol{K}_\varrho\}$, and $\boldsymbol{\epsilon} = [\boldsymbol{\epsilon}_1^{\mathrm{T}}, \ldots, \boldsymbol{\epsilon}_\varrho^{\mathrm{T}}]^{\mathrm{T}} \sim \mathrm{N}(\mathbf{0}, \boldsymbol{\Sigma} \otimes \boldsymbol{I})$, $\otimes$ being the Kronecker product between matrices.

To test the null hypothesis $\mathcal{H}_0 : f_l(\cdot) = 0, l = 1, 2, \ldots, \varrho$, or equivalently $\mathcal{H}_0 : \tau_l^2 = 0, l = 1, 2, \ldots, \varrho$, the score test statistic is

$$\mathcal{S}(\boldsymbol{\Sigma}_0) = \frac{1}{2}\boldsymbol{y}^{\mathrm{T}}\boldsymbol{P}_0\boldsymbol{K}\boldsymbol{P}_0\boldsymbol{y}, \tag{2.62}$$

where $\boldsymbol{P}_0 = \boldsymbol{V}_0^{-1} - \boldsymbol{V}_0^{-1}\boldsymbol{X}(\boldsymbol{X}^{\mathrm{T}}\boldsymbol{V}_0^{-1}\boldsymbol{X})^{-1}\boldsymbol{X}^{\mathrm{T}}\boldsymbol{V}_0^{-1}$ is the projection matrix under the null, and $\boldsymbol{V}_0 = \boldsymbol{\Sigma}_0 \otimes \boldsymbol{I}$ is the marginal covariance matrix under the null. When all the traits adjust for a common set of covariates and all the unknown functions f_l lie in an RKHS defined by the same kernel function, that is, $\boldsymbol{X}_1 = \boldsymbol{X}_2 = \cdots = \boldsymbol{X}_\varrho := \boldsymbol{X}_{\mathrm{c}}$ and $\boldsymbol{K}_1 = \boldsymbol{K}_2 = \cdots = \boldsymbol{K}_\varrho := \boldsymbol{K}_{\mathrm{c}}$, we have $\boldsymbol{X} = \boldsymbol{I} \otimes \boldsymbol{X}_{\mathrm{c}}$, $\boldsymbol{K} = \boldsymbol{I} \otimes \boldsymbol{K}_{\mathrm{c}}$, $\boldsymbol{P}_0 = \boldsymbol{\Sigma}_0^{-1} \otimes \boldsymbol{P}_{0,\mathrm{c}}$, where $\boldsymbol{P}_{0,\mathrm{c}} = \boldsymbol{I} - \boldsymbol{X}_{\mathrm{c}}(\boldsymbol{X}_{\mathrm{c}}^{\mathrm{T}}\boldsymbol{X}_{\mathrm{c}})^{-1}\boldsymbol{X}_{\mathrm{c}}^{\mathrm{T}}$, and the score test statistic can be simplified as

$$\begin{aligned}
\mathcal{S}(\boldsymbol{\Sigma}_0) &= \frac{1}{2}\boldsymbol{y}^{\mathrm{T}}\left[\boldsymbol{\Sigma}_0^{-2} \otimes (\boldsymbol{P}_{0,\mathrm{c}}\boldsymbol{K}_{\mathrm{c}}\boldsymbol{P}_{0,\mathrm{c}})\right]\boldsymbol{y} \\
&= \frac{1}{2}\mathbf{vec}(\boldsymbol{Y})^{\mathrm{T}}\left[\boldsymbol{\Sigma}_0^{-2} \otimes (\boldsymbol{P}_{0,\mathrm{c}}\boldsymbol{K}_{\mathrm{c}}\boldsymbol{P}_{0,\mathrm{c}})\right]\mathbf{vec}(\boldsymbol{Y}) \\
&= \frac{1}{2}\mathbf{tr}\left\{\boldsymbol{P}_{0,\mathrm{c}}\boldsymbol{K}_{\mathrm{c}}\boldsymbol{P}_{0,\mathrm{c}}\boldsymbol{Y}\boldsymbol{\Sigma}_0^{-2}\boldsymbol{Y}^{\mathrm{T}}\right\},
\end{aligned} \tag{2.63}$$

where $\boldsymbol{Y} = [\boldsymbol{y}_1, \ldots, \boldsymbol{y}_\varrho]$. Here $\mathcal{S}(\boldsymbol{\Sigma}_0)$ essentially quantifies the association between the attribute similarity matrix $\boldsymbol{K}_{\mathrm{c}}$ and the phenotypic similarity measured by the matrix $\boldsymbol{Y}\boldsymbol{\Sigma}_0^{-2}\boldsymbol{Y}^{\mathrm{T}}$. When the trait is a scalar, $\mathcal{S}(\boldsymbol{\Sigma}_0)$ degenerates to the score test statistic for the classical KMR. The distribution of $\mathcal{S}(\boldsymbol{\Sigma}_0)$ can again be approximated using the Satterthwaite method. In practice, we fit individual linear regression models under the null to obtain estimates of the residuals, $\widehat{\boldsymbol{\epsilon}}_{0,l} = \boldsymbol{y}_l - \boldsymbol{X}_l\widehat{\boldsymbol{\beta}}_{0,l}$, and then estimate the residual covariance matrix by $\widehat{\boldsymbol{\Sigma}}_0 = \frac{1}{n}\widehat{\mathcal{E}}^{\mathrm{T}}\widehat{\mathcal{E}}$, where $\widehat{\mathcal{E}} = [\widehat{\boldsymbol{\epsilon}}_{0,1}, \ldots, \widehat{\boldsymbol{\epsilon}}_{0,\varrho}]$.

2.3 APPLICATIONS

In this section, we review recent applications of KMR with a focus on biomedical research. Most of the work has a genetic component since kernel methods are particularly useful to flexibly model the joint effect of a collection of genetic variants on a phenotype of interest.

2.3.1 GENETIC ASSOCIATION STUDIES

The past decade has witnessed tremendous scientific and biological discoveries made through genome-wide association studies (GWASs), an experimental design that detects the association between millions of individual single nucleotide polymorphisms (SNPs, a DNA sequence variation occurring at a single nucleotide in the genome) and a wide range of clinical conditions (Visscher et al., 2012; Gratten et al., 2014; Psychiatric Genomics Consortium, 2014). However, GWASs require large sample size (hundreds of thousands) to achieve stringent statistical significance and identify robust and replicable associations, and can only be applied to common genetic variants (occurring more than 1% in a population). To date, a majority of the SNPs identified by GWASs are not in protein-coding regions, and thus do not have direct implications for disease diagnosis or treatment.

SNP set-based analyses offer a complementary method to GWASs by modeling a collection of genetic variants, which open opportunities to test the cumulative effect of rare genetic variants and dissect complex interactions in the genome. By grouping SNPs based on a priori biological knowledge such as genes, pathways, functional annotations and previous GWAS findings, set-based association studies can alleviate the burden of multiple testing correction, improve robustness, reproducibility and statistical power relative to univariate methods, and provide more interpretable results and biological insights.

Conventional methods model a collection of SNPs in a genomic region within the multiple regression framework, and use regularization techniques, such as the ridge regression, LASSO and elastic net (Kohannim et al., 2011, 2012b,a), or PCA (Hibar et al., 2011), to handle collinearity of SNP regressors due to linkage disequilibrium (LD, statistical associations between co-segregated SNPs). Recently, a large body of the literature has been devoted to the development of *burden tests*, which collapse a set of genetic variants in a genomic region into a single burden variable. The burden variable can be dichotomous, indicating the presence of any rare variant within a region (eg, cohort allelic sum test, CAST) (Morgenthaler and Thilly, 2007), the count of rare variants in the SNP set (Morris and Zeggini, 2010), a weighted average of the number of minor alleles (eg, weighted sum test, WST) (Madsen and Browning, 2009), or a hybrid of these methods (Li and Leal, 2008). Pan et al. (2014) showed that both the sum test and the sum of squared score (SSU) test are special cases of a family of tests, termed sum of powered score (SPU) tests, which also has an adaptive version (aSPU) (Pan et al., 2014, 2015). A major criticism of burden tests is that they rely on the strong assumptions that all variants being modeled are causal and the effects are in the same direction, and may suffer from dramatic power loss when these assumptions are violated.

KMR offers an alternative way to conduct SNP set-based association studies by employing a *sequence kernel* that defines the similarity between a pair of strings. Kernel-base association tests belong to the class of *nonburden tests* that is robust to the direction of SNPs and the proportion of causal SNPs in the set. A variety of kernels can be used to characterize different genetic contributions to the trait. For example, a linear kernel models the additive effects of SNPs (Yang et al., 2010, 2011), while a nonparametric identity-by-state (IBS) kernel provides a biologically informed way to capture the epistasis (interactions) in the SNP set. As a concrete illustration, consider two genetic variants A and B, with the corresponding minor and major alleles (the less and more common alleles in the population) represented with lower and upper case, respectively. Now, suppose that two subjects have the following genotypes at the two loci: [AA; bb] and [aa; Bb]. For the linear kernel, assuming that the minor allele is considered as the reference allele, the above genotypes can be coded as [0; 2] and [2; 1]. The linear similarity between the two subjects is then computed as an inner product between these two vectors, normalized by the number of loci, giving $(0 * 2 + 2 * 1)/2 = 1$. Note that, with this definition, the choice of the reference allele impacts the similarity measure and thus there is an implicit directionality. However, in practice, the genotypes are often standardized (subtracted by the mean and divided by the standard deviation across subjects) before the linear kernel is computed, and the linear similarity for the standardized genotype is independent of the choice of the reference allele. For the IBS kernel, the similarity between a pair of subjects is calculated as the number of identical alleles, normalized by the total number of alleles. Therefore the IBS similarity between the above two subjects is $(0 + 1)/4 = 1/4$. Note that here the reference allele does not need to be specified and no directionality is assumed. Different weighting strategies can also be used when building kernel functions to up-weight or down-weight SNPs based on their allele frequencies or a priori biological knowledge. When testing the aggregated genetic effect on the trait, the degrees of freedom of the fitted chi-squared distribution is adaptive to the correlation structure of the SNPs, and thus the KMR approach allows for modeling and testing highly correlated variants.

Liu et al. (2007) laid the theoretical foundation for the kernel-based association tests and applied the technique to testing the pathway effect of multiple gene expressions on prostate cancer. They termed the method sequence kernel association test (SKAT), which has become the basis of many follow-up theoretical extensions and biomedical applications. Specifically, SKAT has been generalized to handle binary data (case-control studies) (Liu et al., 2008), multidimensional traits (Maity et al., 2012), family data (Schifano et al., 2012; Chen et al., 2013; Ionita-Laza et al., 2013a; Jiang et al., 2014), and survival outcomes (Cai et al., 2011; Lin et al., 2011a; Chen et al., 2014). Kwee et al. (2008) and Wu et al. (2010) introduced SKAT to the genetic community and inspired a series of papers on testing the cumulative effects of rare variants (Wu et al., 2011, 2015; Lee et al., 2013; Ionita-Laza et al., 2013b; Lee et al., 2014), and modeling and detecting gene-by-gene and gene-by-environment (G×E) interactions (Maity and Lin, 2011; Li and Cui, 2012; Lin et al., 2013, 2015; Broadaway et al., 2015; Marceau et al., 2015). To maximize the statistical power of kernel-based association tests, Cai et al. (2012) developed an adaptive score test that up-weights or down-weights the contributions from individual SNPs based on their marginal effects. Lee et al. (2012b,a) proposed an optimal association test that combines SKAT with conventional burden tests, known as SKAT-O.

2.3.2 IMAGING GENETICS

Imaging genetics is an emerging field that identifies and characterizes the genetic basis of brain structure, function and wiring, which play important roles in fundamental cognitive, emotional and behavioral processes, and may be altered in brain-related illnesses (Meyer-Lindenberg and Weinberger, 2006; Thompson et al., 2013). Revealing the true relationship between neuroimaging and genetic variables is challenging because both data types are extremely high-dimensional (millions of genetic variants spanning the genome and hundreds of thousands of voxels/vertices across the image) and have complex covariance structures (genetic variants that are physically close in the genome are often correlated due to LD, and imaging data that are spatially close are also correlated), and sample sizes are typically limited (hundreds or thousands of subjects). Kernel-based methods, due to their modeling flexibility and computational efficiency, have shown promising potential to dissect the genetic underpinnings of the human brain.

Ge et al. (2012) were the first to introduce KMR to the context of imaging genetics. They modeled the aggregated effects and potential interactions of SNPs located in each gene using an IBS kernel, and conducted a voxel-wise, whole-genome, gene-based association study. The kernel method was also combined with a suite of other approaches including a fast implementation of the random field theory that takes use of the spatial information in images, and an efficient permutation procedure. The authors demonstrated that this multivariate analysis framework has boosted statistical power relative to previous massive univariate approaches (Stein et al., 2010). For the first time, some genes were identified to be significantly associated with local volumetric changes in the brain.

Recently, Ge et al. (2015a) proposed a flexible KMR-based method for detecting the interactive effects between multidimensional variable sets, which is particularly useful to identify G×E interactions. Specifically, they introduced three kernels in the KMR framework: one for modeling the joint and epistatic effect of a set of SNPs, one for accommodating multiple factors that potentially moderate genetic influences, and a third one, which is the Hadamard product of the first two kernels, for capturing the overall interactions between two sets of variables. An initial application of this method to imaging genetics has identified interactive effects between candidate late-onset Alzheimer's disease (AD) risk genes and a collection of cardiovascular disease (CVD) risk factors on hippocampal volume derived from structural brain magnetic resonance imaging (MRI) scans, an imaging biomarker associated with AD risk and future AD progression.

Lastly, Ge et al. (2015b) noticed that using a linear kernel function to combine all the SNPs spanning the genome assesses the total additive genetic effects on the trait and essentially gives a narrow-sense heritability estimate. Leveraging the efficient score test of the KMR, Ge et al. (2015b) proposed a statistical method, termed massively expedited genome-wide heritability analysis (MEGHA), for high-dimensional heritability analysis using genome-wide SNP data from unrelated individuals. This method is thousands of times faster than existing tools and makes heritability-based prioritization of millions of phenotypes tractable for the first time. The authors also developed a permutation-based nonparametric sampling technique within the KMR framework that enables flexible and accurate inferences for arbitrary statistics of interest. As a demonstration of application, Ge et al. (2015b) investigated the genetic basis of morphometric measurements derived from structural MRI, and have created and distributed high-resolution surface maps (containing approximately 300,000 vertices across the two hemispheres) for the heritability estimates and their significance of cortical thickness, sulcal depth, curvature and surface area (https://surfer.nmr.mgh.harvard.edu/fswiki/HeritabilityAnalysis_Ge2015). These maps can be useful to define regions of interest (ROIs) that are under substantial genetic influences. As an example, Fig. 2.3 shows the vertex-wise surface map for the heritability estimates of cortical thickness measurements constructed by MEGHA. The method can also be applied to qualify the heritability of other imaging modalities, or any other types of big data in a variety of settings.

2.4 CONCLUSION AND FUTURE DIRECTIONS

Kernel machine regression (KMR) is a powerful machine learning method, which allows for flexible modeling of multidimensional and heterogeneous data by implicitly specifying the complex relationship between traits and attributes via a knowledge-based similarity measure that characterizes the resemblance between pairs of attributes. Recent technical advances have bridged KMR with mixed effects models in statistics, enabling unified model fitting procedures, and accurate and efficient statistical inferences about model parameters. In this chapter, we have introduced the theoretical basis of KMR and highlighted some of its key extensions. Although we have focused the review on genetic research, which constitutes a large body of the expanding literature on the application of KMR, the method

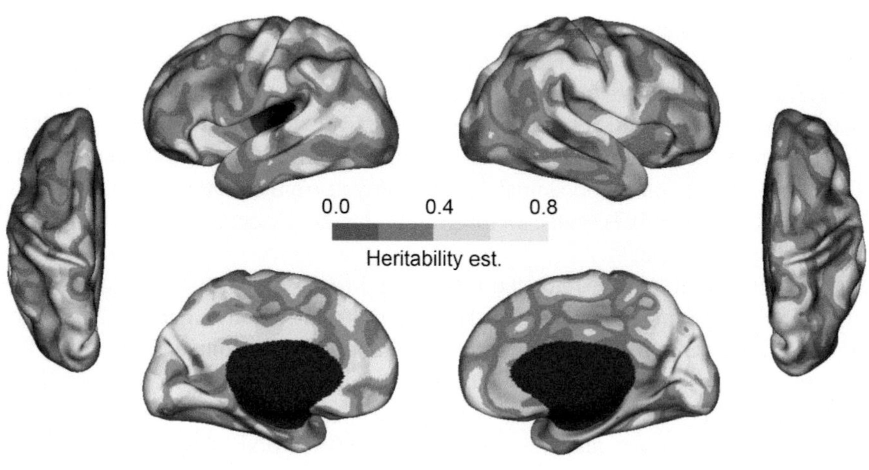

FIG. 2.3

Vertex-wise surface map for the heritability estimates of cortical thickness measurements constructed by MEGHA.

is general enough to explore the relationship between other data types, such as the association between neuroimaging measurements and cognitive, behavioral or diagnostic variables. In fact, the exponential progress in biological technologies is generating massive amounts of data spanning multiple levels of a biological system, from genomic sequences, to intermediate phenotypes such as medical images, and to high-level symptomatic variables. The KMR framework can potentially be used to integrate and jointly analyze different data sources, or be extended to respect the hierarchical structure of these data (Lin et al., 2011b; Huang et al., 2014). With the increasing availability of longitudinal imaging scans (Bernal-Rusiel et al., 2013a,b), KMR seems promising to exploit the high-dimensional imaging space and identify biomarkers that are related to the progression of a brain-related illness and the timing of a clinical event of interest. Last but not least, the research and application of kernel-based methods are not restricted to association detection, but can also encompass prediction, classification, clustering, learning, dimension reduction and variable selection problems, opening vast opportunities for both theoretical advancement and biological discoveries.

ACKNOWLEDGMENTS

This research was carried out at the Athinoula A. Martinos Center for Biomedical Imaging at the Massachusetts General Hospital (MGH), using resources provided by the Center for Functional Neuroimaging Technologies, P41EB015896, a P41 Biotechnology Resource Grant supported by the National Institute of Biomedical Imaging and Bioengineering (NIBIB), National Institutes of Health (NIH).

This research was also funded in part by an MGH Executive Committee on Research (ECOR) Tosteson Postdoctoral Fellowship Award (to TG); NIH grants R01 NS083534, R01 NS070963, and NIBIB 1K25EB013649-01 (to MRS); K24 MH094614 and R01 MH101486 (to JWS); and a BrightFocus Foundation grant AHAF-A2012333 (to MRS). JWS is a Tepper Family MGH Research Scholar.

Appendix A REPRODUCING KERNEL HILBERT SPACES

In this appendix, we briefly review the mathematical foundations of the kernel methods. We restrict our discussion to real vector spaces and kernel functions. For a more detailed introduction to kernel methods, see, for example, Cristianini and Shawe-Taylor (2000) and Schölkopf and Smola (2002).

Appendix A.1 INNER PRODUCT AND HILBERT SPACE

A function $\langle \cdot, \cdot \rangle_{\mathcal{H}} : \mathcal{H} \times \mathcal{H} \to \mathbb{R}$ is an *inner product* on the *vector space* (or *linear space*) $\mathcal{H}$ if the following conditions are satisfied:

- Symmetry: $\langle f, g \rangle_{\mathcal{H}} = \langle g, f \rangle_{\mathcal{H}}$, for any $f, g \in \mathcal{H}$.
- Bilinearity: $\langle \alpha_1 f_1 + \alpha_2 f_2, g \rangle_{\mathcal{H}} = \alpha_1 \langle f_1, g \rangle_{\mathcal{H}} + \alpha_2 \langle f_2, g \rangle_{\mathcal{H}}$, for any $f_1, f_2, g \in \mathcal{H}$ and $\alpha_1, \alpha_2 \in \mathbb{R}$.
- Positive definiteness: $\langle f, f \rangle_{\mathcal{H}} \geqslant 0$, for any $f \in \mathcal{H}$, with equality if and only if $f = 0$.

A vector space equipped with an inner product is called an *inner product space* or *pre-Hilbert space*. An inner product induces a *metric* or a *norm* by $\|f\|_{\mathcal{H}} = \sqrt{\langle f, f \rangle_{\mathcal{H}}}, f \in \mathcal{H}$, and a *distance* between f and g in $\mathcal{H}$ by $d(f, g) = \|f - g\|_{\mathcal{H}} = \sqrt{\langle f - g, f - g \rangle_{\mathcal{H}}}$.

In a vector space $\mathcal{H}$ with a metric $\| \cdot \|_{\mathcal{H}}$, a sequence $\{f_i\}_{i=1}^{+\infty}$ in $\mathcal{H}$ is said to be a *Cauchy sequence* if for any $\epsilon > 0$, there exists a positive integer $N(\epsilon)$ such that for all positive integers $m, n > N$, $\|f_m - f_n\|_{\mathcal{H}} < \epsilon$. A metric space $\mathcal{H}$ in which every Cauchy sequence converges to an element in $\mathcal{H}$ is *complete*. Intuitively a complete metric space suggests that when the terms of the sequence are getting closer and closer, a limit always exists and it never escapes from the space, that is, the space has no "holes." A *Hilbert space* is a complete inner product space with respect to the norm induced by the inner product.

Appendix A.2 KERNEL FUNCTION AND KERNEL MATRIX

Let $\mathcal{Z}$ be a nonempty set. A function $k : \mathcal{Z} \times \mathcal{Z} \to \mathbb{R}$ is called a *kernel function* if there exists a Hilbert space $\mathcal{H}$ with an inner product $\langle \cdot, \cdot \rangle_{\mathcal{H}}$ and a map $\varphi : \mathcal{Z} \to \mathcal{H}$ such that for any z and z' in the space $\mathcal{Z}$,

$$k(z, z') = \langle \varphi(z), \varphi(z') \rangle_{\mathcal{H}}. \tag{A.1}$$

Here φ is called a *feature map*, which transforms the data from the input space $\mathcal{Z}$ to a *feature space* $\mathcal{H}$, and can be highly complex and even infinite-dimensional. Kernel methods capture nonlinear patterns in the data by mapping the input to higher dimensions where linear models can be applied.

A function $k : \mathcal{Z} \times \mathcal{Z} \to \mathbb{R}$ is *non-negative definite* (or *positive semidefinite*) if for any finite subset $\{z_1, \ldots, z_n\}$ chosen from $\mathcal{Z}$, the *Gram matrix* (or *kernel matrix*) $\boldsymbol{K} = \{k(z_i, z_j)\}_{i,j=1}^n$ is symmetric and non-negative definite, ie, for any real numbers $a_1, \ldots, a_n$,

$$\sum_{i=1}^n \sum_{j=1}^n a_i a_j k(z_i, z_j) \geqslant 0. \tag{A.2}$$

Any kernel function k is clearly symmetric and we have

$$\sum_{i=1}^n \sum_{j=1}^n a_i a_j k(z_i, z_j) = \sum_{i=1}^n \sum_{j=1}^n \langle a_i \varphi(z_i), a_j \varphi(z_j) \rangle_{\mathcal{H}} = \left\| \sum_{i=1}^n a_i \varphi(z_i) \right\|_{\mathcal{H}}^2 \geqslant 0. \tag{A.3}$$

Therefore all kernel functions are non-negative definite. The reverse direction of the statement is also true, that is, for any non-negative definite function k, there exists a Hilbert space $\mathcal{H}$ and a feature map φ, such that Eq. (A.1) is satisfied. This is remarkable because the feature map is often expensive to compute or difficult to explicitly specify, while the kernel function may be easily evaluated and arbitrarily selected as long as it is non-negative definite. We note that a kernel function may rely on additional parameters, such as the width in a Gaussian kernel, as long as it is non-negative definite once the parameters are fixed. A widely used technique in the machine learning community is to substitute a kernel function $k(z, z')$ for a dot product between $\varphi(z)$ and $\varphi(z')$, which implicitly defines a feature map. This is known as the *kernel trick*.

Appendix A.3 REPRODUCING KERNEL HILBERT SPACE

Let $\mathcal{H}$ be a Hilbert space of real-valued functions defined on a nonempty set $\mathcal{Z}$. A function $k : \mathcal{Z} \times \mathcal{Z} \to \mathbb{R}$ is called a *reproducing kernel* of $\mathcal{H}$, and $\mathcal{H}$ is a *reproducing kernel Hilbert space* (RKHS) on $\mathcal{Z}$ (Aronszajn, 1950; Saitoh, 1988), if the followings are satisfied:

- For any $z \in \mathcal{Z}$, $k_z(\cdot) = k(\cdot, z)$ as a function on $\mathcal{Z}$ belongs to $\mathcal{H}$.
- The *reproducing property*: For any $z \in \mathcal{Z}$ and any $f \in \mathcal{H}$, $\langle f(\cdot), k(\cdot, z) \rangle_{\mathcal{H}} = f(z)$.

The reproducing property states that the evaluation of f at z can be expressed as an inner product in the feature space. By applying this property, we have, for any $z, z' \in \mathcal{Z}$,

$$k(z, z') = \langle k(\cdot, z), k(\cdot, z') \rangle_{\mathcal{H}}. \tag{A.4}$$

Since $\varphi(z) = k(\cdot, z)$ is a valid feature map of k, it can be seen from Eq. (A.4) that every reproducing kernel is indeed a kernel as defined in Eq. (A.1), and is thus non-negative definite.

It can be shown that if a Hilbert space $\mathcal{H}$ of functions on $\mathcal{Z}$ admits a reproducing kernel, then the reproducing kernel is uniquely determined by $\mathcal{H}$. Conversely, given any non-negative definite kernel $k(\cdot,\cdot)$ on $\mathcal{Z}$, there exists a uniquely determined Hilbert space $\mathcal{H}$ of functions on $\mathcal{Z}$, which admits the reproducing kernel k. In fact, the Hilbert space $\mathcal{H}$ can be constructed by completing the function space $\mathcal{H}_0$ spanned by $\{k(\cdot, z) \mid z \in \mathcal{Z}\}$, that is,

$$\mathcal{H}_0 = \left\{ f(\cdot) = \sum_{i=1}^{n} \alpha_i k(\cdot, z_i) \;\middle|\; n \in \mathbb{N}, \alpha_i \in \mathbb{R}, z_i \in \mathcal{Z} \right\}, \tag{A.5}$$

with the inner product of the functions f and $g(\cdot) = \sum_{j=1}^{m} \beta_j k(\cdot, y_j)$ from $\mathcal{H}_0$ defined as

$$\langle f, g \rangle_{\mathcal{H}_0} = \sum_{i=1}^{n}\sum_{j=1}^{m} \alpha_i \beta_j \langle k(\cdot, z_i), k(\cdot, y_j)\rangle_{\mathcal{H}_0} = \sum_{i=1}^{n}\sum_{j=1}^{m} \alpha_i \beta_j k(z_i, y_j), \tag{A.6}$$

where $m \in \mathbb{N}$, $\beta_j \in \mathbb{R}$, and $y_j \in \mathcal{Z}$. Therefore there is a one-to-one correspondence between non-negative definite kernels and RKHSs. Since $\mathcal{H}_0$ is dense in $\mathcal{H}$, any function in $\mathcal{H}$ can be represented as $f(z) = \sum_{i=1}^{+\infty} \alpha_i k(z, z_i)$, with $\alpha_i \in \mathbb{R}$ and $z_i \in \mathcal{Z}$. The infinite summation is due to the fact that a function in $\mathcal{H}$ may be a limit point of a sequence of functions in $\mathcal{H}_0$. This is the *dual representation* of functions in RKHSs.

Appendix A.4 MERCER'S THEOREM

Mercer's theorem (Mercer, 1909; Schölkopf and Smola, 2002; Cristianini and Shawe-Taylor, 2000) is essentially an analog of the singular value decomposition (SVD) of a matrix in an infinite-dimensional space. It states that under certain regulatory conditions, a kernel function k can be expanded in terms of eigenvalues and orthonormal eigenfunctions of an operator induced by k. More formally, suppose k is a continuous non-negative definite kernel function on a compact set $\mathcal{Z}$. Let $L^2(\mathcal{Z})$ be the space of square-integrable real-valued functions on $\mathcal{Z}$. Define the integral operator $T_k : L^2(\mathcal{Z}) \to L^2(\mathcal{Z})$ by

$$(T_k f)(\cdot) = \int_{\mathcal{Z}} k(\cdot, z) f(z) \mathrm{d}z. \tag{A.7}$$

Then $k(z, z')$ can be expanded into a set of orthonormal basis $\{\psi_i\}$ of $L^2(\mathcal{Z})$ consisting of the eigenfunctions of T_k, and the corresponding sequence of non-negative eigenvalues $\{\lambda_i\}$:

$$k(z, z') = \sum_{i=1}^{\infty} \lambda_i \psi_i(z) \psi_i(z'), \tag{A.8}$$

where the convergence is absolute and uniform. Any squared-integrable function on $\mathcal{Z}$ can thus be represented as $f(z) = \sum_{i=1}^{\infty} \omega_i \sqrt{\lambda_i}\psi_i(z)$, with $\omega_i \in \mathbb{R}$ and $\sum_{i=1}^{\infty} \omega_i^2 < +\infty$. This is the *primal representation* of functions in RKHSs. The inner product of the functions f and $g(\cdot) = \sum_{j=1}^{\infty} \upsilon_j \sqrt{\lambda_j}\psi_j(\cdot)$ from $\mathcal{H} = L^2(\mathcal{Z})$ is defined as $\langle f, g\rangle = \sum_{i=1}^{\infty} \omega_i \upsilon_i$. We thus have

$$\langle f(\cdot), k(\cdot, z)\rangle_{\mathcal{H}} = \sum_{i=1}^{\infty} \frac{\omega_i \lambda_i \sqrt{\lambda_i}\psi_i(z)}{\lambda_i} = \sum_{i=1}^{\infty} \omega_i \sqrt{\lambda_i}\psi_i(z) = f(z), \tag{A.9}$$

that is, the reproducing property of the kernel function. Moreover, a feature map can be explicitly written as $\varphi(z) = [\cdots, \lambda_\ell \psi_\ell(z), \cdots]^{\mathrm{T}}$.

Appendix A.5 REPRESENTER THEOREM

The representer theorem (Kimeldorf and Wahba, 1971) shows that solutions of a large class of optimization problems can be expressed as linear combinations of kernel functions centered on the observed data. Specifically, let $\mathcal{L} : \{\mathcal{Z} \times \mathbb{R}^2\}^n \to \mathbb{R}$ be an arbitrary loss function, and $\Omega : [0, \infty) \to \mathbb{R}$ be a strictly monotonic increasing function. Then each solution of the optimization problem

$$\operatorname{argmin}_{f\in\mathcal{H}} \mathcal{L}\left((z_1, f(z_1), y_1), \ldots, (z_n, f(z_n), y_n)\right) + \Omega\left(\|f\|_{\mathcal{H}}^2\right) \tag{A.10}$$

can be represented in the form $f = \sum_{i=1}^{n} \alpha_i k(\cdot, z_i)$. Moreover, if the loss function $\mathcal{L}$ is convex, a global minimum is ensured. Here $\Omega\left(\|f\|_{\mathcal{H}}^2\right)$ is a regularization term and controls smoothness of the minimizer. To see this, suppose that f can be expanded as $f(z) = \sum_{i=1}^{\infty} \omega_i \sqrt{\lambda_i}\psi_i(z)$ using its primal representation. We note that $\|f\|_{\mathcal{H}}^2 = \langle f, f\rangle_{\mathcal{H}} = \sum_{i=1}^{+\infty} \omega_i^2 < +\infty$, which indicates that ω_i decays with increasing i and thus de-emphasizes nonsmooth functions. The representer theorem is significant because it states that the solution of an optimization problem in an infinite-dimensional space $\mathcal{H}$, containing linear combinations of kernels centered on arbitrary points of $\mathcal{Z}$ (see Eq. A.5), has a finite-dimensional representation, which is the span of n kernel functions centered on the observed data points.

Appendix B RESTRICTED MAXIMUM LIKELIHOOD ESTIMATION

In this appendix, we review the Newton-Raphson method (Lindstrom and Bates, 1988; Kenward and Roger, 1997) that is often used to maximize the log restricted likelihood, ℓ_{R}, of linear mixed effects models (LMMs), producing unbiased estimates of the variance component parameters. Recall that ℓ_{R} can be written as a correction to the profile log likelihood:

$$
\begin{aligned}
\ell_{\mathrm{R}}(\boldsymbol{\theta}) &= \ell_{\mathrm{P}}(\boldsymbol{\theta}) - \frac{1}{2}\log|\boldsymbol{X}^{\mathrm{T}}\boldsymbol{V}^{-1}\boldsymbol{X}| \\
&= -\frac{1}{2}\log|\boldsymbol{V}| - \frac{1}{2}\log|\boldsymbol{X}^{\mathrm{T}}\boldsymbol{V}^{-1}\boldsymbol{X}| - \frac{1}{2}(\boldsymbol{y}-\boldsymbol{X}\widehat{\boldsymbol{\beta}})^{\mathrm{T}}\boldsymbol{V}^{-1}(\boldsymbol{y}-\boldsymbol{X}\widehat{\boldsymbol{\beta}}) \\
&= -\frac{1}{2}\log|\boldsymbol{V}| - \frac{1}{2}\log|\boldsymbol{X}^{\mathrm{T}}\boldsymbol{V}^{-1}\boldsymbol{X}| - \frac{1}{2}\boldsymbol{y}^{\mathrm{T}}\boldsymbol{P}\boldsymbol{y},
\end{aligned} \tag{B.1}
$$

where $\boldsymbol{V} = \boldsymbol{V}(\boldsymbol{\theta})$ is the marginal covariance of $\boldsymbol{y}$, which is dependent on an s-dimensional vector of unknown parameters $\boldsymbol{\theta}$, $\widehat{\boldsymbol{\beta}} = \widehat{\boldsymbol{\beta}}(\boldsymbol{\theta}) = (\boldsymbol{X}^{\mathrm{T}}\boldsymbol{V}^{-1}\boldsymbol{X})^{-1}\boldsymbol{X}\boldsymbol{V}^{-1}\boldsymbol{y}$, and we have defined $\boldsymbol{P} = \boldsymbol{P}(\boldsymbol{\theta}) = \boldsymbol{V}^{-1} - \boldsymbol{V}^{-1}\boldsymbol{X}(\boldsymbol{X}^{\mathrm{T}}\boldsymbol{V}^{-1}\boldsymbol{X})^{-1}\boldsymbol{X}^{\mathrm{T}}\boldsymbol{V}^{-1}$. The last equality in Eq. (B.1) is based on the identities $\boldsymbol{V}^{-1}(\boldsymbol{y}-\boldsymbol{X}\widehat{\boldsymbol{\beta}}) = \boldsymbol{P}\boldsymbol{y}$ and $\boldsymbol{P}\boldsymbol{V}\boldsymbol{P} = \boldsymbol{P}$. By making use of the following results on matrix derivatives:

$$
\frac{\partial \log|\boldsymbol{V}|}{\partial\theta_i} = \mathbf{tr}\left\{\boldsymbol{V}^{-1}\frac{\partial\boldsymbol{V}}{\partial\theta_i}\right\}, \quad \frac{\partial\boldsymbol{V}^{-1}}{\partial\theta_i} = -\boldsymbol{V}^{-1}\frac{\partial\boldsymbol{V}}{\partial\theta_i}\boldsymbol{V}^{-1}, \quad \frac{\partial\boldsymbol{P}}{\partial\theta_i} = -\boldsymbol{P}\frac{\partial\boldsymbol{V}}{\partial\theta_i}\boldsymbol{P}, \tag{B.2}
$$

where $\mathbf{tr}\{\cdot\}$ is the trace of a matrix, we have the gradient or the score:

$$
\frac{\partial\ell_{\mathrm{R}}}{\partial\theta_i} = -\frac{1}{2}\mathbf{tr}\left\{\boldsymbol{P}\frac{\partial\boldsymbol{V}}{\partial\theta_i}\right\} + \frac{1}{2}\boldsymbol{y}^{\mathrm{T}}\boldsymbol{P}\frac{\partial\boldsymbol{V}}{\partial\theta_i}\boldsymbol{P}\boldsymbol{y}, \quad i = 1, 2, \ldots, s, \tag{B.3}
$$

and the ijth element of the $s \times s$ observed information matrix $\mathcal{I}_{\mathrm{O}}$:

$$
\begin{aligned}
[\mathcal{I}_{\mathrm{O}}]_{ij} = -\frac{\partial^2\ell_{\mathrm{R}}}{\partial\theta_i\partial\theta_j} &= -\frac{1}{2}\mathbf{tr}\left\{\boldsymbol{P}\frac{\partial\boldsymbol{V}}{\partial\theta_i}\boldsymbol{P}\frac{\partial\boldsymbol{V}}{\partial\theta_j}\right\} + \boldsymbol{y}^{\mathrm{T}}\boldsymbol{P}\frac{\partial\boldsymbol{V}}{\partial\theta_i}\boldsymbol{P}\frac{\partial\boldsymbol{V}}{\partial\theta_j}\boldsymbol{P}\boldsymbol{y} \\
&+ \frac{1}{2}\mathbf{tr}\left\{\boldsymbol{P}\frac{\partial^2\boldsymbol{V}}{\partial\theta_i\partial\theta_j}\right\} - \frac{1}{2}\boldsymbol{y}^{\mathrm{T}}\boldsymbol{P}\frac{\partial^2\boldsymbol{V}}{\partial\theta_i\partial\theta_j}\boldsymbol{P}\boldsymbol{y}, \quad i,j = 1, 2, \ldots, s.
\end{aligned} \tag{B.4}
$$

We notice that

$$
\begin{aligned}
\mathbf{E}\left\{\boldsymbol{y}^{\mathrm{T}}\boldsymbol{P}\frac{\partial\boldsymbol{V}}{\partial\theta_i}\boldsymbol{P}\frac{\partial\boldsymbol{V}}{\partial\theta_j}\boldsymbol{P}\boldsymbol{y}\right\} &= \mathbf{tr}\left\{\boldsymbol{P}\frac{\partial\boldsymbol{V}}{\partial\theta_i}\boldsymbol{P}\frac{\partial\boldsymbol{V}}{\partial\theta_j}\right\}, \\
\mathbf{E}\left\{\boldsymbol{y}^{\mathrm{T}}\boldsymbol{P}\frac{\partial^2\boldsymbol{V}}{\partial\theta_i\partial\theta_j}\boldsymbol{P}\boldsymbol{y}\right\} &= \mathbf{tr}\left\{\boldsymbol{P}\frac{\partial^2\boldsymbol{V}}{\partial\theta_i\partial\theta_j}\right\},
\end{aligned} \tag{B.5}
$$

and thus the ijth element of the $s \times s$ Fisher information (expected information) matrix $\mathcal{I}_{\mathrm{E}}$ is

$$
[\mathcal{I}_{\mathrm{E}}]_{ij} = \mathbf{E}\left[-\frac{\partial^2\ell_{\mathrm{R}}}{\partial\theta_i\partial\theta_j}\right] = \frac{1}{2}\mathbf{tr}\left\{\boldsymbol{P}\frac{\partial\boldsymbol{V}}{\partial\theta_i}\boldsymbol{P}\frac{\partial\boldsymbol{V}}{\partial\theta_j}\right\}. \tag{B.6}
$$

When the covariance structure $\boldsymbol{V}(\boldsymbol{\theta})$ is a linear function of the unknown parameters $\boldsymbol{\theta}$, the second-order derivatives in Eq. (B.4) vanish, and the ijth element of the average information matrix, $\mathcal{I}_{\mathrm{A}} = (\mathcal{I}_{\mathrm{O}} + \mathcal{I}_{\mathrm{E}})/2$, has a simple form:

$$
[\mathcal{I}_{\mathrm{A}}]_{ij} = \frac{1}{2}[\mathcal{I}_{\mathrm{O}}]_{ij} + \frac{1}{2}[\mathcal{I}_{\mathrm{E}}]_{ij} = \frac{1}{2}\boldsymbol{y}^{\mathrm{T}}\boldsymbol{P}\frac{\partial\boldsymbol{V}}{\partial\theta_i}\boldsymbol{P}\frac{\partial\boldsymbol{V}}{\partial\theta_j}\boldsymbol{P}\boldsymbol{y}. \tag{B.7}
$$

Then given estimates of the unknown variance component parameters at the kth iteration $\boldsymbol{\theta}^{(k)}$, the parameters are iteratively updated by

$$\boldsymbol{\theta}^{(k+1)} = \boldsymbol{\theta}^{(k)} + \left[\mathcal{I}_{\bullet}^{(k)}\right]^{-1} \left.\frac{\partial \ell_{\mathrm{R}}}{\partial \boldsymbol{\theta}}\right|_{\boldsymbol{\theta}^{(k)}}, \quad k = 1, 2, \ldots, \tag{B.8}$$

where $\mathcal{I}_{\bullet}$ is either the expected information matrix $\mathcal{I}_{\mathrm{E}}$, leading to the Fisher scoring ReML, or the average information matrix $\mathcal{I}_{\mathrm{A}}$, leading to the average information ReML (Gilmour et al., 1995). At the beginning of the iteration process, all the variance component parameters need to be initialized to reasonable values $\theta_i^{(0)}$. An initial step of the expectation maximization (EM) algorithm (Laird et al., 1987), which is robust to poor starting values, may be used to determine the direction of the updates:

$$\theta_i^{(1)} = \frac{1}{n}\left[\left[\theta_i^{(0)}\right]^2 \boldsymbol{y}^{\mathrm{T}}\boldsymbol{P}\frac{\partial \boldsymbol{V}}{\partial \theta_i}\boldsymbol{P}\boldsymbol{y} + \mathbf{tr}\left\{\theta_i^{(0)}\boldsymbol{I} - \left[\theta_i^{(0)}\right]^2 \boldsymbol{P}\frac{\partial \boldsymbol{V}}{\partial \theta_i}\right\}\right], \quad i = 1, 2, \ldots, s. \tag{B.9}$$

The Newton-Raphson algorithm is terminated until the difference between successive log restricted likelihoods, or the gradient of the log restricted likelihood, is smaller than a predefined tolerance, such as 10^{-4}. In the iteration process, parameter estimates may escape from the parameter space (eg, negative estimates of variance parameters), in which case they should be reset to a feasible value close to the boundary of the parameter space.

REFERENCES

Aronszajn, N., 1950. Theory of reproducing kernels. Trans. Am. Math. Soc. 68 (3), 337–404.

Bernal-Rusiel, J.L., Greve, D.N., Reuter, M., Fischl, B., Sabuncu, M.R., et al., 2013a. Statistical analysis of longitudinal neuroimage data with linear mixed effects models. NeuroImage 66, 249–260.

Bernal-Rusiel, J.L., Reuter, M., Greve, D., Fischl, B., Sabuncu, M.R., et al., 2013b. Spatiotemporal linear mixed effects modeling for the mass-univariate analysis of longitudinal neuroimage data. NeuroImage 81, 358–370.

Breslow, N.E., Clayton, D.G., 1993. Approximate inference in generalized linear mixed models. J. Am. Stat. Assoc. 88 (421), 9–25.

Broadaway, K.A., Duncan, R., Conneely, K.N., Almli, L.M., Bradley, B., et al., 2015. Kernel approach for modeling interaction effects in genetic association studies of complex quantitative traits. Genet. Epidemiol. 39 (5), 366–375.

Cai, T., Tonini, G., Lin, X., 2011. Kernel machine approach to testing the significance of multiple genetic markers for risk prediction. Biometrics 67 (3), 975–986.

Cai, T., Lin, X., Carroll, R.J., 2012. Identifying genetic marker sets associated with phenotypes via an efficient adaptive score test. Biostatistics 13 (4), 776–790.

Chen, H., Meigs, J.B., Dupuis, J., 2013. Sequence kernel association test for quantitative traits in family samples. Genet. Epidemiol. 37 (2), 196–204.

Chen, H., Lumley, T., Brody, J., Heard-Costa, N.L., Fox, C.S., et al., 2014. Sequence kernel association test for survival traits. Genet. Epidemiol. 38 (3), 191–197.

Cristianini, N., Shawe-Taylor, J., 2000. An Introduction to Support Vector Machines and Other Kernel-Based Learning Methods. Cambridge University Press, Cambridge, MA.

Davies, R.B., 1980. The distribution of a linear combination of χ^2 random variables. J. R. Stat. Soc. C 29, 323–333.

Diggle, P., Heagerty, P., Liang, K.Y., Zeger, S., 2002. Analysis of Longitudinal Data. Oxford University Press, Oxford, UK.

Ge, T., Feng, J., Hibar, D.P., Thompson, P.M., Nichols, T.E., 2012. Increasing power for voxel-wise genome-wide association studies: the random field theory, least square kernel machines and fast permutation procedures. NeuroImage 63 (2), 858–873.

Ge, T., Nichols, T.E., Ghosh, D., Mormino, E.C., Smoller, J.W., et al., 2015a. A kernel machine method for detecting effects of interaction between multidimensional variable sets: an imaging genetics application. NeuroImage 109, 505–514.

Ge, T., Nichols, T.E., Lee, P.H., Holmes, A.J., Roffman, J.L., et al., 2015b. Massively expedited genome-wide heritability analysis (MEGHA). Proc. Natl. Acad. Sci. USA 112 (8), 2479–2484.

Gelman, A., Carlin, J.B., Stern, H.S., Rubin, D.B., 2013. Bayesian Data Analysis. Chapman & Hall, New York.

Gianola, D., van Kaam, J.B.C.H.M., 2008. Reproducing kernel Hilbert spaces regression methods for genomic assisted prediction of quantitative traits. Genetics 178 (4), 2289–2303.

Gilmour, A.R., Thompson, R., Cullis, B.R., 1995. Average information ReML: an efficient algorithm for variance parameter estimation in linear mixed models. Biometrics 51 (4), 1440–1450.

Goeman, J.J., Van De Geer, S.A., Van H.C., 2006. Testing against a high dimensional alternative. J. R. Stat. Soc. B 68 (3), 477–493.

Gower, J.C., 1966. Some distance properties of latent root and vector methods used in multivariate analysis. Biometrika 53 (3–4), 325–338.

Gratten, J., Wray, N.R., Keller, M.C., Visscher, P.M., 2014. Large-scale genomics unveils the genetic architecture of psychiatric disorders. Nat. Neurosci. 17 (6), 782–790.

Gu, C., 2013. Smoothing Spline ANOVA Models. Springer Science & Business Media, New York.

Harville, D.A., 1977. Maximum likelihood approaches to variance component estimation and to related problems. J. Am. Stat. Assoc. 72 (358), 320–338.

Hibar, D.P., Stein, J.L., Kohannim, O., Jahanshad, N., Saykin, A.J., et al., 2011. Voxelwise gene-wide association study (vGeneWAS): multivariate gene-based association testing in 731 elderly subjects. NeuroImage 56 (4), 1875–1891.

Hofmann, T., Schölkopf, B., Smola, A.J., 2008. Kernel methods in machine learning. Ann. Stat. 36 (3), 1171–1220.

Huang, Y.T., VanderWeele, T.J., Lin, X., 2014. Joint analysis of SNP and gene expression data in genetic association studies of complex diseases. Ann. Appl. Stat. 8 (1), 352.

Ionita-Laza, I., Lee, S., Makarov, V., Buxbaum, J.D., Lin, X., 2013a. Family-based association tests for sequence data, and comparisons with population-based association tests. Eur. J. Hum. Genet. 21 (10), 1158–1162.

Ionita-Laza, I., Lee, S., Makarov, V., Buxbaum, J.D., Lin, X., 2013b. Sequence kernel association tests for the combined effect of rare and common variants. Am. J. Hum. Genet. 92 (6), 841–853.

Jiang, Y., Conneely, K.N., Epstein, M.P., 2014. Flexible and robust methods for rare-variant testing of quantitative traits in trios and nuclear families. Genet. Epidemiol. 38 (6), 542–551.

Kenward, M.G., Roger, J.H., 1997. Small sample inference for fixed effects from restricted maximum likelihood. Biometrics 53 (3), 983–997.

Kimeldorf, G., Wahba, G., 1971. Some results on Tchebycheffian spline functions. J. Math. Anal. Appl. 33 (1), 82–95.

Kohannim, O., Hibar, D.P., Stein, J.L., Jahanshad, N., Jack Jr, C.R., et al., 2011. Boosting power to detect genetic associations in imaging using multi-locus, genome-wide scans and ridge regression. In: 2011 IEEE International Symposium on Biomedical Imaging: From Nano to Macro. IEEE, Piscataway, NJ, pp. 1855–1859.

Kohannim, O., Hibar, D.P., Jahanshad, N., Stein, J.L., Hua, X., et al., 2012a. Predicting temporal lobe volume on MRI from genotypes using l1-l2 regularized regression. In: 2012 IEEE International Symposium on Biomedical Imaging: From Nano to Macro. IEEE, Piscataway, NJ, pp. 1160–1163.

Kohannim, O., Hibar, D.P., Stein, J.L., Jahanshad, N., Hua, X., et al., 2012b. Discovery and replication of gene influences on brain structure using LASSO regression. Front. Neurosci. 6, Article 115.

Kuonen, D., 1999. Saddlepoint approximations for distributions of quadratic forms in normal variables. Biometrika 86 (4), 929–935.

Kwee, L.C., Liu, D., Lin, X., Ghosh, D., Epstein, M.P., 2008. A powerful and flexible multilocus association test for quantitative traits. Am. J. Hum. Genet. 82 (2), 386–397.

Laird, N., Lange, N., Stram, D., 1987. Maximum likelihood computations with repeated measures: application of the EM algorithm. J. Am. Stat. Assoc. 82 (397), 97–105.

Lee, S., Emond, M.J., Bamshad, M.J., Barnes, K.C., Rieder, M.J., et al., 2012a. Optimal unified approach for rare-variant association testing with application to small-sample case-control whole-exome sequencing studies. Am. J. Hum. Genet. 91 (2), 224–237.

Lee, S., Wu, M.C., Lin, X., 2012b. Optimal tests for rare variant effects in sequencing association studies. Biostatistics 13 (4), 762–775.

Lee, S., Teslovich, T.M., Boehnke, M., Lin, X., 2013. General framework for meta-analysis of rare variants in sequencing association studies. Am. J. Hum. Genet. 93 (1), 42–53.

Lee, S., Abecasis, G.R., Boehnke, M., Lin, X., 2014. Rare-variant association analysis: study designs and statistical tests. Am. J. Hum. Genet. 95 (1), 5–23.

Li, B., Leal, S.M., 2008. Methods for detecting associations with rare variants for common diseases: application to analysis of sequence data. Am. J. Hum. Genet. 83 (3), 311–321.

Li, S., Cui, Y., 2012. Gene-centric gene-gene interaction: a model-based kernel machine method. Ann. Appl. Stat. 6 (3), 1134–1161.

Liang, K.Y., Zeger, S.L., 1986. Longitudinal data analysis using generalized linear models. Biometrika 73 (1), 13–22.

Lin, X., 1997. Variance component testing in generalized linear models with random effects. Biometrika 84 (2), 309–326.

Lin, X., Cai, T., Wu, M.C., Zhou, Q., Liu, G., et al., 2011a. Kernel machine SNP-set analysis for censored survival outcomes in genome-wide association studies. Genet. Epidemiol. 35 (7), 620–631.

Lin, Y.Y., Liu, T.L., Fuh, C.S., 2011b. Multiple kernel learning for dimensionality reduction. IEEE Trans. Pattern Anal. Mach. Intell. 33 (6), 1147–1160.

Lin, X., Lee, S., Christiani, D.C., Lin, X., 2013. Test for interactions between a genetic marker set and environment in generalized linear models. Biostatistics 14 (4), 667–681.

Lin, X., Lee, S., Wu, M.C., Wang, C., Chen, H., et al., 2015. Test for rare variants by environment interactions in sequencing association studies. Biometrics 72 (1), 156–164. doi: 10.1111/biom. 12368.

Lindstrom, M.J., Bates, D.M., 1988. Newton-Raphson and EM algorithms for linear mixed-effects models for repeated-measures data. J. Am. Stat. Assoc. 83 (404), 1014–1022.

Liu, D., Lin, X., Ghosh, D., 2007. Semiparametric regression of multidimensional genetic pathway data: least-squares kernel machines and linear mixed models. Biometrics 63 (4), 1079–1088.

Liu, D., Ghosh, D., Lin, X., 2008. Estimation and testing for the effect of a genetic pathway on a disease outcome using logistic kernel machine regression via logistic mixed models. BMC Bioinformatics 9 (1), 292.

Madsen, B.E., Browning, S.R., 2009. A groupwise association test for rare mutations using a weighted sum statistic. PLoS Genet. 5 (2), e1000384.

Maity, A., Lin, X., 2011. Powerful tests for detecting a gene effect in the presence of possible gene-gene interactions using garrote kernel machines. Biometrics 67 (4), 1271–1284.

Maity, A., Sullivan, P.E., Tzeng, J., 2012. Multivariate phenotype association analysis by marker-set kernel machine regression. Genet. Epidemiol. 36 (7), 686–695.

Marceau, R., Lu, W., Holloway, S., Sale, M.M., Worrall, B.B., et al., 2015. A fast multiple-kernel method with applications to detect gene-environment interaction. Genet. Epidemiol. 39 (6), 456–468.

McArdle, B.H., Anderson, M.J., 2001. Fitting multivariate models to community data: a comment on distance-based redundancy analysis. Ecology 82 (1), 290–297.

McCullagh, P., Nelder, J.A., 1989. Generalized Linear Models. CRC Press, Boca Raton, FL.

McCulloch, C.E., Neuhaus, J.M., 2001. Generalized Linear Mixed Models. Wiley Online Library.

Mercer, J., 1909. Functions of positive and negative type, and their connection with the theory of integral equations. Philos. Trans. R. Soc. Lond. A 209, 415–446.

Meyer-Lindenberg, A., Weinberger, D.R., 2006. Intermediate phenotypes and genetic mechanisms of psychiatric disorders. Nat. Rev. Neurosci. 7 (10), 818–827.

Morgenthaler, S., Thilly, W.G., 2007. A strategy to discover genes that carry multi-allelic or mono-allelic risk for common diseases: a cohort allelic sums test (CAST). Mutat. Res./Fund. Mole. Mech. Mutagen. 615 (1), 28–56.

Morris, A.P., Zeggini, E., 2010. An evaluation of statistical approaches to rare variant analysis in genetic association studies. Genet. Epidemiol. 34 (2), 188.

Pan, W., 2009. Asymptotic tests of association with multiple SNPs in linkage disequilibrium. Genet. Epidemiol. 33 (6), 497.

Pan, W., 2011. Relationship between genomic distance-based regression and kernel machine regression for multi-marker association testing. Genet. Epidemiol. 35 (4), 211–216.

Pan, W., Kim, J., Zhang, Y., Shen, X., Wei, P., 2014. A powerful and adaptive association test for rare variants. Genetics 197 (4), 1081–1095.

Pan, W., Kwak, I.Y., Wei, P., 2015. A powerful pathway-based adaptive test for genetic association with common or rare variants. Am. J. Hum. Genet. 97 (1), 86–98.

Patterson, H.D., Thompson, R., 1971. Recovery of inter-block information when block sizes are unequal. Biometrika 58 (3), 545–554.

Psychiatric Genomics Consortium, 2014. Biological insights from 108 schizophrenia-associated genetic loci. Nature 511 (7510), 421–427.

Rasmussen, C.E., Williams, C.K.I., 2006. Gaussian Processes for Machine Learning. MIT Press, Cambridge, MA.

Saitoh, S., 1988. Theory of Reproducing Kernels and its Applications. Longman, Harlow.

Schaid, D.J., 2010a. Genomic similarity and kernel methods I: advancements by building on mathematical and statistical foundations. Hum. Hered. 70 (2), 109–131.

Schaid, D.J., 2010b. Genomic similarity and kernel methods II: methods for genomic information. Hum. Hered. 70 (2), 132–140.

Schifano, E.D., Epstein, M.P., Bielak, L.F., Jhun, M.A., Kardia, S.L.R., et al., 2012. SNP set association analysis for familial data. Genet. Epidemiol. 36 (8), 797–810.

Schölkopf, B., Burges, C.J.C., 1999. Advances in Kernel Methods: Support Vector Learning. MIT Press, Cambridge, MA.

Schölkopf, B., Smola, A.J., 2002. Learning With Kernels: Support Vector Machines, Regularization, Optimization, and Beyond. MIT Press, Cambridge, MA.

Schölkopf, B., Smola, A., Müller, K.R., 1997. Kernel principal component analysis. In: Artificial Neural Networks—ICANN'97. Springer, New York, pp. 583–588.

Schölkopf, B., Smola, A., Müller, K.R., 1998. Nonlinear component analysis as a kernel eigenvalue problem. Neural Comput. 10 (5), 1299–1319.

Self, S.G., Liang, K.Y., 1987. Asymptotic properties of maximum likelihood estimators and likelihood ratio tests under nonstandard conditions. J. Am. Stat. Assoc. 82 (398), 605–610.

Snijders, T.A.B., 2011. Multilevel Analysis. Springer, Berlin.

Stein, J.L., Hua, X., Lee, S., Ho, A.J., Leow, A.D., et al., 2010. Voxel-wise genome-wide association study (vGWAS). NeuroImage 53 (3), 1160–1174.

Thompson, P.M., Ge, T., Glahn, D.C., Jahanshad, N., Nichols, T.E., 2013. Genetics of the connectome. NeuroImage 80, 475–488.

Vapnik, V., 1998. Statistical Learning Theory. Wiley, New York.

Verbeke, G., Molenberghs, G., 2009. Linear Mixed Models for Longitudinal Data. Springer Science & Business Media, New York.

Visscher, P.M., Brown, M.A., McCarthy, M.I., Yang, J., 2012. Five years of GWAS discovery. Am. J. Hum. Genet. 90 (1), 7–24.

Wahba, G., 1990. Spline Models for Observational Data. SIAM Press, Philadelphia, PA.

Wessel, J., Schork, N.J., 2006. Generalized genomic distance-based regression methodology for multilocus association analysis. Am. J. Hum. Genet. 79 (5), 792–806.

Wu, M.C., Kraft, P., Epstein, M.P., Taylor, D.M., Chanock, S.J., et al., 2010. Powerful SNP-set analysis for case-control genome-wide association studies. Am. J. Hum. Genet. 86 (6), 929–942.

Wu, M.C., Lee, S., Cai, T., Li, Y., Boehnke, M., et al., 2011. Rare-variant association testing for sequencing data with the sequence kernel association test. Am. J. Hum. Genet. 89 (1), 82–93.

Wu, B., Pankow, J.S., Guan, W., 2015. Sequence kernel association analysis of rare variant set based on the marginal regression model for binary traits. Genet. Epidemiol. 39 (6), 399–405.

Yang, J., Benyamin, B., McEvoy, B.P., Gordon, S., Henders, A.K., et al., 2010. Common SNPs explain a large proportion of the heritability for human height. Nat. Genet. 42 (7), 565–569.

Yang, J., Lee, S.H., Goddard, M.E., Visscher, P.M., 2011. GCTA: a tool for genome-wide complex trait analysis. Am. J. Hum. Genet. 88 (1), 76–82.

Zhang, D., Lin, X., 2003. Hypothesis testing in semiparametric additive mixed models. Biostatistics 4 (1), 57–74.

CHAPTER

Deep learning of brain images and its application to multiple sclerosis

3

T. Brosch, Y. Yoo, L.Y.W. Tang, R. Tam

The University of British Columbia, Vancouver, BC, Canada

CHAPTER OUTLINE

3.1 INTRODUCTION

Deep learning is a field within machine learning that has been studied since the early 1980s (Fukushima, 1980). However, deep learning methods did not gain in popularity until the late 2000s with the advent of fast general-purpose graphics processors (Raina et al., 2009), layerwise pretraining methods (Hinton et al., 2006; Hinton and Salakhutdinov, 2006), and large datasets (Deng et al., 2009; Krizhevsky et al., 2012). Since then, deep learning methods have become the state-of-the-art in many nonmedical (Krizhevsky et al., 2012; Sainath et al., 2013) and medical (Ciresan et al., 2012; Kamnitsas et al., 2015) applications. There are many different algorithms and models that are commonly referred to as deep learning methods, all of which have two properties in common: (1) the use of multiple layers of nonlinear

Machine Learning and Medical Imaging. http://dx.doi.org/10.1016/B978-0-12-804076-8.00003-7

processing units for extracting features, and (2) the layers are organized to form a hierarchy of low-level to high-level features. Representing data in a feature hierarchy has many advantages for classification and other applications. To give an example of a feature hierarchy, let us consider the domain of face images. The lowest layer of the feature hierarchy is composed of the raw pixel intensities, which are the most basic features of an image. Multiple pixels can be grouped to form general image features like edges and corners, which can be further combined to form face parts such as different variations of noses, eyes, mouths, and ears. Finally, multiple face parts can be combined to form a variety of face images. Learning a feature hierarchy facilitates the parameterization of a large feature space with a small number of values by capturing complex relationships between feature layers. For example, a feature hierarchy consisting of three prototypical shapes for mouths, eyes, ears, and noses is able to represent $3 \times 3 \times 3 \times 3 = 81$ different prototypical faces with only $3 + 3 + 3 + 3 = 12$ features. Without a hierarchical representation of the data, a model would require 81 prototypical face features to span the same face manifold.

In this section, we will introduce the most commonly used deep learning methods for medical image analysis. We start with a description of unsupervised models like restricted Boltzmann machines (RBMs) (Freund and Haussler, 1992; Hinton, 2010), which are the building blocks of deep belief networks (DBNs) (Hinton et al., 2006), a model that can be used for learning a hierarchical set of features from input images without the need for labels. In the second part of the introduction, we will give a brief overview of dense neural networks (DNNs) (Farley and Clark, 1954; Werbos, 1974; Rumelhart et al., 1986) and convolutional neural networks (CNNs) (Fukushima, 1980; LeCun et al., 1989, 1998), which are the most commonly used supervised deep learning methods.

3.1.1 LEARNING FROM UNLABELED INPUT IMAGES

One of the most important applications of deep learning is to learn a feature hierarchy from unlabeled images. The key to learning such a hierarchy is the ability of deep models to be trained layer by layer, where each layer acts as a nonlinear feature extractor. Various methods have been proposed for feature extraction from unlabeled images. In this section, we will first introduce the RBMs (Freund and Haussler, 1992; Hinton, 2010), which are the building blocks of DBNs (Hinton et al., 2006), followed by a short introduction to alternative feature extractors such as stacked denoising autoencoders (SDAEs) (Vincent et al., 2010).

3.1.1.1 From restricted Boltzmann machines to deep belief networks

An RBM is a probabilistic graphical model defined by a bipartite graph as shown in Fig. 3.1. The units of the RBM are divided into two layers, one of visible units $\mathbf{v}$ and the other of hidden units $\mathbf{h}$. There are no direct connections between units within

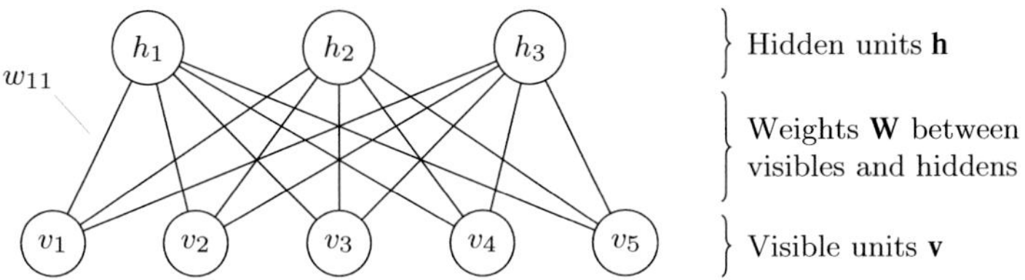

FIG. 3.1

Graphical representation of an RBM with three hidden and five visible units. An RBM models the joint probability of visible and hidden units. Edges between vertices denote conditional dependence between the corresponding random variables.

either layer. An RBM defines the joint probability of visible and hidden units in terms of the energy E:

$$p(\mathbf{v}, \mathbf{h} \mid \boldsymbol{\theta}) = \frac{1}{Z(\boldsymbol{\theta})} \mathrm{e}^{-E(\mathbf{v}, \mathbf{h} \mid \boldsymbol{\theta})}; \tag{3.1}$$

when the visible and hidden units are binary, the energy is defined as

$$-E(\mathbf{v}, \mathbf{h} \mid \boldsymbol{\theta}) = \sum_{i,j} v_i w_{ij} h_j + \sum_i b_i v_i + \sum_j c_j h_j \tag{3.2}$$

$$= \mathbf{v}^{\mathrm{T}} \mathbf{W} \mathbf{h} + \mathbf{b}^{\mathrm{T}} \mathbf{v} + \mathbf{c}^{\mathrm{T}} \mathbf{h}, \tag{3.3}$$

where $Z(\boldsymbol{\theta})$ is a normalization constant, $\mathbf{W}$ denotes the weight matrix that connects the visible units with the hidden units, $\mathbf{b}$ is a vector containing the visible bias terms, $\mathbf{c}$ is a vector containing the hidden bias terms, and $\boldsymbol{\theta} = \{\mathbf{W}, \mathbf{b}, \mathbf{c}\}$ are the trainable parameters of the RBM.

Inference

The hidden units represent patterns of similarity that can be observed in groups of images. Given a set of model parameters $\boldsymbol{\theta}$, the features of an image can be extracted by calculating the expectation of the hidden units. The posterior distribution of the hidden units given the visible units can be calculated by

$$p(h_j = 1 \mid \mathbf{v}, \boldsymbol{\theta}) = \mathrm{sigm}(\mathbf{w}_{\cdot,j}^{\mathrm{T}} \mathbf{v} + c_j), \tag{3.4}$$

where $\mathbf{w}_{\cdot,j}$ denotes the jth column vector of $\mathbf{W}$ and $\mathrm{sigm}(x)$ is the sigmoid function defined as $\mathrm{sigm}(x) = (1+\exp(-x))^{-1}, x \in \mathbb{R}$. An RBM is a generative model, which allows for the reconstruction of an input signal given its features. This is achieved by calculating the expectation of the visible units given the hidden units. The posterior distribution $p(v_i = 1 \mid \mathbf{h}, \boldsymbol{\theta})$ can be calculated by

$$p(v_i = 1 \mid \mathbf{h}, \boldsymbol{\theta}) = \mathrm{sigm}(\mathbf{w}_{i,\cdot}^{\mathrm{T}} \mathbf{h} + b_i), \tag{3.5}$$

where $\mathbf{w}_{i,\cdot}$ denotes the ith row vector of $\mathbf{W}$. Reconstructing the visible units can be used to visualize the learned features. To visualize the features associated with a particular hidden unit, all other hidden units are set to zero and the expectation of the visible units is calculated, which represents the pattern that causes a particular hidden unit to be activated.

Training

RBMs can be trained by maximizing the likelihood or, more commonly, the log-likelihood of the training data, $\mathcal{D} = \{\mathbf{v}_n \mid n \in [1, N]\}$, which is called maximum likelihood estimation (MLE). The gradient of the log-likelihood function with respect to the weights, $\mathbf{W}$, is given by the mean difference of two expectations:

$$\nabla_{\mathbf{W}} \log p(\mathcal{D} \mid \boldsymbol{\theta}) = \frac{1}{N} \sum_{n=1}^{N} \mathbb{E}[\mathbf{v}\mathbf{h}^{\mathrm{T}} \mid \mathbf{v}_n, \boldsymbol{\theta}] - \mathbb{E}[\mathbf{v}\mathbf{h}^{\mathrm{T}} \mid \boldsymbol{\theta}]. \tag{3.6}$$

The first expectation can be estimated using a mean field approximation:

$$\mathbb{E}[\mathbf{v}\mathbf{h}^{\mathrm{T}} \mid \mathbf{v}_n, \boldsymbol{\theta}] \approx \mathbb{E}[\mathbf{v} \mid \mathbf{v}_n, \boldsymbol{\theta}]\mathbb{E}[\mathbf{h}^{\mathrm{T}} \mid \mathbf{v}_n, \boldsymbol{\theta}] \tag{3.7}$$

$$= \mathbf{v}_n \mathbb{E}[\mathbf{h}^{\mathrm{T}} \mid \mathbf{v}_n, \boldsymbol{\theta}]. \tag{3.8}$$

The second expectation is typically estimated using a Monte Carlo approximation:

$$\mathbb{E}[\mathbf{v}\mathbf{h}^{\mathrm{T}} \mid \boldsymbol{\theta}] \approx \frac{1}{S} \sum_{s=1}^{S} \mathbf{v}_s \mathbf{h}_s^{\mathrm{T}}, \tag{3.9}$$

where S is the number of generated samples, and $\mathbf{v}_s$ and $\mathbf{h}_s$ are samples drawn from $p(\mathbf{v} \mid \boldsymbol{\theta})$ and $p(\mathbf{h} \mid \boldsymbol{\theta})$, respectively. Samples from an RBM can be generated efficiently using block Gibbs sampling, in which the visible and hidden units are initialized with random values and alternately sampled given the previous state using

$$h_j = \mathbb{I}(y_j < p(h_j = 1 \mid \mathbf{v}, \boldsymbol{\theta})) \qquad \text{with } y_j \sim \mathrm{U}(0, 1) \tag{3.10}$$

$$v_i = \mathbb{I}(x_i < p(v_i \mid \mathbf{h}, \boldsymbol{\theta})) \qquad \text{with } x_i \sim \mathrm{U}(0, 1), \tag{3.11}$$

where $z \sim \mathrm{U}(0, 1)$ denotes a sample drawn from the uniform distribution in the interval $[0, 1]$ and $\mathbb{I}$ is the indicator function, which is defined as 1 if the argument is true and 0 otherwise. After several iterations, a sample generated by the Gibbs chain is distributed according to $p(\mathbf{vh} \mid \boldsymbol{\theta})$.

If the Gibbs sampler is initialized at a data point from the training set and only one Monte Carlo sample is used to approximate the second expectation in (3.6), the learning algorithm is called contrastive divergence (CD) (Hinton, 2002). Alternatively, persistent contrastive divergence (PCD) (Tieleman, 2008) uses several separate Gibbs chains to generate data independent samples from the model, which results in a better approximation of the gradient of the log-likelihood than CD. To speed up the training, the dataset is usually divided into small subsets called mini-batches and a gradient step is performed for each mini-batch. To avoid confusion with a gradient step, the term "iteration" is generally avoided and the term "epoch" is used instead to indicate a sweep through the entire dataset. Additional tricks to monitor and speed up the training of an RBM can be found in Hinton's RBM training guide (Hinton, 2010).

Deep belief networks

A single RBM can be regarded as a nonlinear feature extractor. To learn a hierarchical set of features, multiple RBMs are stacked and trained layer by layer, where the first RBM is trained on the input data and subsequent RBMs are trained on the hidden unit activations computed from the previous RBM. The stacking of RBMs can be repeated to initialize DBNs of any depth.

3.1.1.2 Variants of restricted Boltzmann machines and deep belief networks

Convolutional DBNs

A potential drawback of DBNs is that the learned features are location dependent. Hence, features that can occur at many different locations in an image, such as edges and corners, must be relearned for every possible location, which dramatically increases the number of features required to capture the content of large images. To increase the translational invariance of the learned features, Lee et al. (2009, 2011) introduced the convolutional deep belief network (convDBN). In a convDBN, the units of each layer are organized in a multidimensional array that reflects the arrangement of pixels in the input image. The units of one layer are only connected to the units of a subregion of the previous layer, and share the same weights with all other units of the same layer. This greatly reduces the number of trainable weights, which reduces the risk of overfitting, reduces the memory required to store the model parameters, speeds up the training, and thereby facilitates the application to high-resolution images.

A convDBN consists of alternating convolutional and pooling layers, which are followed by one or more dense layers. Each convolutional layer of the model can be trained in a greedy layerwise fashion by treating it as a convolutional restricted Boltzmann machine (convRBM). The energy of a convRBM is defined as

$$E(\mathbf{v},\mathbf{h}) = -\sum_{i=1}^{N_c}\sum_{j=1}^{N_k}\mathbf{h}^{(j)} \bullet (\tilde{\mathbf{w}}^{(ij)} * \mathbf{v}^{(i)}) - \sum_{i=1}^{N_c} b_i \sum_{x,y=1}^{N_v} v_{xy}^{(i)} - \sum_{j=1}^{N_k} c_j \sum_{x,y=1}^{N_h} h_{xy}^{(j)}. \tag{3.12}$$

The key terms and notation are defined in Table 3.1. At the first layer, the number of channels N_c is one when trained on unimodal images, or equal to the number of input modalities when trained on multimodal images. For subsequent layers, N_c is equal to the number of filters of the previous layer.

Table 3.1 Key Variables and Notation (for Notational Simplicity, We Assume the Input Images to be Square 2D Images)

Symbol	Description
$\mathbf{v}^{(i)}$	A 2D array containing the units of the ith input channel
$\mathbf{h}^{(j)}$	A 2D array containing the units of the jth output channel or feature map
$\mathbf{w}^{(ij)}$	A 2D array containing the weights of filter kernels connecting visible units $\mathbf{v}^{(i)}$ to hidden units $\mathbf{h}^{(j)}$
b_i	Bias terms of the visible units
c_j	Bias terms of the hidden units
N_{c}	Number of channels of the visible units
N_{v}	Width and height of the image representing the visible units
N_{k}	Number of filters and feature maps
N_{h}	Width and height of a feature map
$\bullet$	Element-wise product followed by summation
$*$	Valid convolution
$\circledast$	Full convolution
$\tilde{\mathbf{w}}^{(ij)}$	Horizontally and vertically flipped version of $\mathbf{w}^{(ij)}$, that is, $\tilde{w}_{uv}^{(ij)} = w_{N_{\mathrm{w}}-u+1,N_{\mathrm{w}}-v+1}^{(ij)}$, where N_{w} denotes the width and height of a filter kernel

The posterior distributions $p(\mathbf{h} \mid \mathbf{v})$ and $p(\mathbf{v} \mid \mathbf{h})$ can be derived from the energy equation and are given by

$$p(h_{xy}^{(j)} = 1 \mid \mathbf{v}) = \mathrm{sigm}\left(\sum_{i=0}^{N_{\mathrm{c}}-1} (\tilde{\mathbf{w}}^{(ij)} * \mathbf{v}^{(i)})_{xy} + c_j\right), \tag{3.13}$$

$$p(v_{xy}^{(i)} = 1 \mid \mathbf{h}) = \mathrm{sigm}\left(\sum_{j=0}^{N_{\mathrm{k}}-1} (\mathbf{w}^{(ij)} \circledast \mathbf{h}^{(j)})_{xy} + b_i\right). \tag{3.14}$$

To train a convRBM on a set of images $\mathcal{D} = \{\mathbf{v}_n \mid n \in [1, N]\}$, the weights and bias terms can be learned by CD. During each iteration of the algorithm, the gradient of each parameter is estimated and a gradient step with a fixed learning rate is applied. The gradient of the filter weights can be approximated by

$$\Delta\mathbf{w}^{(ij)} \approx \frac{1}{N}(\mathbf{v}_n^{(i)} * \tilde{\mathbf{h}}_n^{(j)} - \mathbf{v}'^{(i)}_n * \tilde{\mathbf{h}}'^{(j)}_n), \tag{3.15}$$

where $\mathbf{h}_n^{(j)}$ and $\mathbf{h}'^{(j)}_n$ are samples drawn from $p(\mathbf{h}^{(j)} \mid \mathbf{v}_n)$ and $p(\mathbf{h}^{(j)} \mid \mathbf{v}'_n)$, and $\mathbf{v}'^{(i)}_n = \mathbb{E}[\mathbf{v}^{(i)} \mid \mathbf{h}_n]$.

Different types of operations (Scherer et al., 2010) have been proposed for the pooling layers, with the common goal of creating a more compact representation of the input data. The most commonly used type of pooling is max-pooling, in which the input to the pooling layer is divided into small blocks and only the maximum value of each block as passed on to the next layer, which makes the representation of the input invariant to small translations in addition to reducing its dimensionality.

Alternative unit types

To model real-valued inputs like the intensities of some medical images, the binary visible units of an RBM can be replaced with Gaussian visible units, which leads to the following energy function:

$$- E(\mathbf{v}, \mathbf{h} \mid \boldsymbol{\theta}) = \sum_{i,j} \frac{v_i}{\sigma_i} w_{ij} h_j + \sum_i \frac{(v_i - b_i)^2}{2\sigma_i^2} + \sum_j c_j h_j, \tag{3.16}$$

where the mean of the ith visible unit is encoded in the bias term b_i, and its standard deviation is given by σ_i. Although approaches have been proposed to learn the standard deviation (Cho et al., 2011), the training data is often simply standardized to have zero mean and unit variance, which yields the following simplification for the inference of the visible and hidden units:

$$\mathbb{E}[h_j \mid \mathbf{v}, \boldsymbol{\theta}] = \text{sigm}(\mathbf{w}_{\cdot,j}^{\mathrm{T}} \mathbf{v} + c_j), \tag{3.17}$$

$$\mathbb{E}[v_i \mid \mathbf{h}, \boldsymbol{\theta}] = \mathbf{w}_{i,\cdot}^{\mathrm{T}} \mathbf{h} + b_i. \tag{3.18}$$

A binary hidden unit can only encode two states. In order to increase the expressive power of the hidden units, Nair and Hinton (2010) proposed using noisy rectified linear units (NReLUs) as the hidden units, and showed that this can improve the learning performance of RBMs. The signal of an NReLU is the sum of an infinite number of binary units, all of which have the same weights but different bias terms. In the special case where the offsets of their bias terms are set to $-0.5, -1.5, \ldots$, the sum of their probabilities and therefore the expectation of an NReLU is extremely close to having a closed form:

$$\mathbb{E}[h_j \mid \mathbf{v}, \boldsymbol{\theta}] = \sum_{i=1}^{\infty} \text{sigm}(\mathbf{w}_{\cdot,j}^{\mathrm{T}} \mathbf{v} + c_j - i + 0.5) \tag{3.19}$$

$$\approx \log(1 + \exp(\mathbf{w}_{\cdot,j}^{\mathrm{T}} \mathbf{v} + c_j)). \tag{3.20}$$

However, sampling of this type of unit involves the repeated calculation of the sigmoid function, which can be time-consuming. If a sample is not constrained to being an integer, a fast approximation can be calculated with

$$h_j \sim \max(0, \mu_j + \mathcal{N}(0, \text{sigm}(\mu_j))), \tag{3.21}$$

$$\mu_j = \mathbf{w}_{\cdot,j}^{\mathrm{T}} \mathbf{v} + c_j, \tag{3.22}$$

where $\mathcal{N}(0, \sigma^2)$ denotes Gaussian noise.

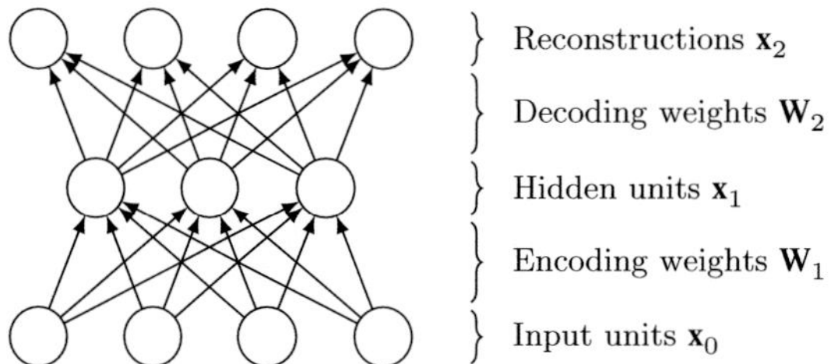

FIG. 3.2

Autoencoder with input units $\mathbf{x}_0$, hidden units $\mathbf{x}_1$, and reconstructions $\mathbf{x}_2$. If the network is trained on corrupted versions of the inputs with the goal of improving the robustness to noise, it is called a denoising autoencoder.

3.1.1.3 Stacked denoising autoencoders

Popular alternatives to DBNs for unsupervised feature learning are stacked autoencoders (SAEs) and SDAEs (Vincent et al., 2010) due to their ability to be trained without the need to generate samples, which speeds up the training compared to RBMs. A minimal autoencoder is a three-layer neural network (see Fig. 3.2) consisting of an input layer $\mathbf{x}_0$, a hidden layer $\mathbf{x}_1$, and an output layer $\mathbf{x}_2$. Similar to RBMs, there are many variants of autoencoders. In the following, we will only consider dense autoencoders with real-valued input units and binary hidden units. Alternative unit types are discussed by Vincent et al. (2010).

The input layer of an autoencoder is a vector containing the intensities of an input image. In the encoding step, features are extracted from the inputs as follows:

$$\mathbf{x}_1 = \text{sigm}(\mathbf{W}_1\mathbf{x}_0 + \mathbf{b}_1), \tag{3.23}$$

where $\mathbf{W}_1$ denotes a matrix containing the encoding weights and $\mathbf{b}_1$ denotes a vector containing the bias terms. In the decoding step, an approximation of the original input signal is reconstructed based on the extracted features:

$$\mathbf{x}_2 = \mathbf{W}_2\mathbf{x}_1 + \mathbf{b}_2, \tag{3.24}$$

where $\mathbf{W}_2$ denotes a matrix containing the decoding weights and $\mathbf{b}_2$ denotes a vector containing the bias terms. An autoencoder is trained by minimizing an error measure (eg, the sum of squared differences or cross-entropy) between the original inputs and their reconstructions. Given a training set $\mathcal{D} = \{\mathbf{x}^{(i)} \mid i \in [1, N]\}$, the optimization problem can be formalized as

$$\hat{\boldsymbol{\theta}} = \arg\min_{\boldsymbol{\theta}} \sum_{i=1}^{N} (\mathbf{x}_2^{(i)} - \mathbf{x}^{(i)})^{\mathrm{T}} (\mathbf{x}_2^{(i)} - \mathbf{x}^{(i)}), \tag{3.25}$$

where $\mathbf{x}_0^{(i)} = \mathbf{x}^{(i)}$ and $\boldsymbol{\theta} = \{\mathbf{W}_1, \mathbf{W}_2, \mathbf{b}_1, \mathbf{b}_2\}$ are the parameters of the autoencoder. The optimization problem can be solved using stochastic gradient descent (SGD)

(Rumelhart et al., 1986) (see Section 3.1.2.1). If the hidden layer contains fewer units than the input layer, the autoencoder learns a lower-dimensional representation of the input data, which allows the model to be used for dimensionality reduction.

The learning of the features can be improved by altering the input signal with random perturbations such as adding Gaussian noise or randomly setting a fraction of the input units to zero. This forces the model to learn features that are robust to noise and capture structures that are useful for reconstructing the original signal. An autoencoder trained on the corrupted versions of the input images is called a denoising autoencoder. Similar to DBNs, a stack of autoencoders can learn a hierarchical set of features, where subsequent autoencoders are trained on the extracted features of the previous autoencoder.

3.1.2 LEARNING FROM LABELED INPUT IMAGES

Features extracted by unsupervised feature learning methods are often fed into a separate supervised learning model, such as a random forest (Breiman, 2001a) or support vector machine (Cortes and Vapnik, 1995), to perform classification or prediction. Alternatively, classification and prediction can be performed with a single model that takes the raw input data and produces the desired output, such as class probabilities. This type of learning is called end-to-end learning and has shown great potential for medical image analysis (Ciresan et al., 2012). The most popular models for end-to-end learning are neural networks due to their ability to learn a hierarchical set of features from raw input data. The supervised learning framework allows for the learning of features that are tuned for a given combination of input modalities and classification tasks, but is more prone to overfitting than unsupervised feature learning, especially when the amount of labeled data is limited. In this section, we will start with an introduction to DNN, followed by a concise overview of CNNs.

3.1.2.1 Dense neural networks

A DNN is a deterministic function that maps input data to the desired outputs through the successive application of multiple nonlinear mappings of the following form:

$$\mathbf{z}_l = \mathbf{W}_l \mathbf{x}_{l-1} + \mathbf{b}_l, \tag{3.26}$$

$$\mathbf{x}_l = f_l(\mathbf{z}_l), \tag{3.27}$$

where l indexes a unit layer, $\mathbf{x}_0$ denotes a vector containing the input of the neural network, $\mathbf{x}_L$ denotes a vector containing the output, L is the number of computational layers, f_l are transfer functions, $\mathbf{W}_l$ are weight matrices, and $\mathbf{b}_l$ are bias terms. Popular choices for the transfer function are the sigmoid function $f(x) = \text{sigm}(x)$ and the rectified linear function $f(x) = \max(0, x)$. The same transfer function is typically used for all layers except for the output layer. The choice of the output transfer function depends on the learning task. For classification, a 1-of-n encoding of the output class is usually used in combination with the softmax transfer function defined as

$$\text{softmax}(\mathbf{a})_i = \frac{\exp(a_i)}{\sum_{j=1}^{n} \exp(a_j)}, \tag{3.28}$$

where $\mathbf{a}$ denotes an n-dimensional output vector.

Given a training set $\mathcal{D} = \{(\mathbf{x}_0^{(i)}, \mathbf{y}^{(i)}) \mid i \in [1, N]\}$, a neural network is trained by minimizing the error between the predicted outputs $\mathbf{x}_L^{(i)}$ and the given labels $\mathbf{y}^{(i)}$:

$$\hat{\boldsymbol{\theta}} = \arg\min_{\boldsymbol{\theta}} \sum_{i=1}^{N} E\left(\mathbf{x}_L^{(i)}, \mathbf{y}^{(i)}\right), \tag{3.29}$$

where $\boldsymbol{\theta}$ denotes the trainable parameters of the neural network. Typical choices for the error function are the sum of squared differences (SSD) and cross-entropy. The minimization problem can be solved using SGD (Rumelhart et al., 1986; Polyak and Juditsky, 1992), which requires the calculation of the gradient of the error function with respect to the model parameters. The gradient can be calculated by backpropagation (Werbos, 1974) as follows:

$$\boldsymbol{\delta}_L = \nabla_{\mathbf{x}_L} E \cdot f_L'(\mathbf{z}_L), \tag{3.30}$$

$$\boldsymbol{\delta}_l = (\mathbf{W}_{l+1}^{\mathrm{T}} \boldsymbol{\delta}_{l+1}) \cdot f_l'(\mathbf{z}_l) \quad \text{for } l < \mathrm{L}, \tag{3.31}$$

$$\nabla_{\mathbf{W}_l} E = \boldsymbol{\delta}_l \mathbf{x}_{l-1}^{\mathrm{T}}, \tag{3.32}$$

$$\nabla_{\mathbf{b}_l} E = \boldsymbol{\delta}_l, \tag{3.33}$$

where $\nabla_{\mathbf{x}_L} E$ denotes the gradient of the error function with respect to the predicted output and $\cdot$ denotes element-wise multiplication.

3.1.2.2 Convolutional neural networks

The structure of CNNs is inspired by the complex arrangement of simple and complex cells found in the visual cortex (Hubel and Wiesel, 1962, 1968). Simple cells are only connected to a small subregion of the previous layer and need to be tiled to cover the entire visual field. In a CNN (see Fig. 3.3), simple cells are represented by convolutional layers, which exhibit a similar mechanism of local connectivity and weight sharing. Complex cells combine the activation of simple cells to add robustness to small translations. These cells are represented in the form of pooling layers similar to the pooling layers found in convDBNs. After several alternating convolutional and pooling layers, the activations of the last convolutional layer are fed into one or more dense layers to carry out the final classification.

For multimodal 3D volumes, the neurons of convolutional and pooling layers are arranged in a 4D array, where the first three dimensions correspond to the dimensions of the input volume, and the fourth dimension indexes the input modality or channel. The activations of the output of a convolutional layer are calculated by

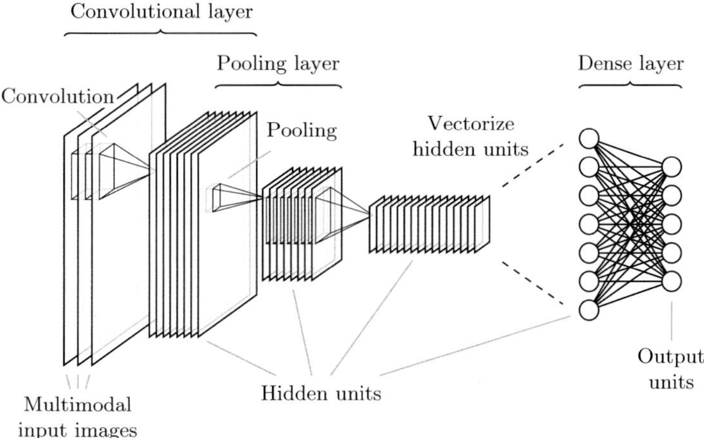

FIG. 3.3

Convolutional neural network with two convolutional layers, one pooling layer and one dense layer. The activations of the last layer are the output of the network.

$$x_j^{(l)} = f\left(\sum_{i=1}^{C} \tilde{w}_{ij}^{(l)} * x_i^{(l-1)} + b_j^{(l)}\right), \tag{3.34}$$

where l is the index of a convolutional layer, $x_j^{(l)}$ denotes the jth channel of the output volume, $w_{ij}^{(l)}$ is a 3D filter kernel connecting the ith channel of the input volume to the jth channel of the output volume, $b_j^{(l)}$ denotes the bias term of the jth output channel, and $\tilde{w}$ denotes a flipped version of w, that is, $\tilde{w}(a) = w(-a)$. CNNs can be trained using SGD, where the gradient can be derived analogously to DNNs and calculated using backpropagation (LeCun et al., 1989, 1998).

A major challenge for gradient-based optimization methods is the choice of an appropriate learning rate. Classic SGD (LeCun et al., 1998) uses a fixed or decaying learning rate, which is the same for all parameters of the model. However, the partial derivatives of parameters of different layers can vary substantially in magnitude, which can require different learning rates. In recent years, there has been an increasing interest in developing methods for automatically choosing independent learning rates. Most methods (eg, AdaGrad by Duchi et al. (2011); AdaDelta by Zeiler (2012); RMSprop by Dauphin et al. (2015); and Adam by Kingma and Ba (2014)) collect different statistics of the partial derivatives over multiple iterations and use this information to set an adaptive learning rate for each parameter. This is especially important for the training of deep networks, where the optimal learning rates often differ greatly for each layer.

3.2 OVERVIEW OF DEEP LEARNING IN NEUROIMAGING

Deep learning methods have been applied with great success to a number of image understanding tasks, such as the detection, segmentation, and classification of objects and regions in images. They have demonstrated impressive improvements over the traditional machine learning techniques and have approached human-level performance in visual and sound recognition applications (LeCun et al., 2015). Due to these successes, deep learning has attracted the attention of neuroimaging researchers to investigate the potential of deep learning for analyzing medical images such as computed tomography (CT), magnetic resonance imaging (MRI), positron emission tomography (PET), and so on.

More specifically, there are two main reasons why deep learning methods are seen as highly promising for neuroimage analysis. First, neuroimaging data is generally high-dimensional. For example, MR images are typically 3D volumes and contain several million voxels each. In addition, due to the continuing development of advanced MRI techniques such as diffusion tensor imaging (DTI) (Le Bihan et al., 2001), functional MRI (Song et al., 2006), susceptibility weighted imaging (SWI) (Haacke et al., 2004), myelin water imaging (MWI) (Alonso-Ortiz et al., 2015), etc., current MRI datasets often contain several modalities that provide complementary information, which should ideally be analyzed together, at the cost of even greater dimensionality. The capability of deep learning to automatically capture discriminative and abstract features through a hierarchical manner may prove particularly useful for reducing the high dimensionality of neuroimaging data to extract key patterns that are representative of important variations. Second, neuroimaging data often lacks labeled data, because labels typically require expert annotations, which can be time-consuming and expensive to obtain. Compared to the huge datasets of natural images often used in the machine learning community that often contain hundreds of thousands of labeled training samples, a neuroimaging dataset with several hundred labeled training samples is already considered large. The limited amount of labeled training data is a common cause of overfitting, making methodological development for neuroimaging application challenging. Many deep learning models can be trained in an unsupervised fashion to learn a representative feature set, which can then be used to initialize a supervised model that is trained with a smaller set of labeled images. By starting with a more generalized set of features, a supervised model can potentially be more robust to overfitting.

The current main applications of deep learning to neuroimaging are segmentation and classification, although there has also been some work on image registration. In neuroimaging, segmentation is typically used to extract desired structures or regions of the central nervous system (CNS) such as white matter (WM), gray matter (GM), cerebrospinal fluid (CSF), spinal cord, corpus callosum, etc., of which volume or shape can be subsequently used to perform diagnosis, monitor disease progress, or study its pathology. Neuroimage classification can be used to perform automatic computer-aided diagnosis, clinical prediction and early disease detection. Traditional machine learning techniques for neuroimage classification generally require hand-crafted features with domain knowledge or assumptions about disease pathology, which may be subject to bias. Several recent articles have shown that deep learning can automatically learn discriminative features for classification with little prior knowledge. The remainder of this section will describe the recent applications of deep learning for medical image registration, segmentation, and classification that are focused on neuroimaging.

3.2.1 DEFORMABLE IMAGE REGISTRATION USING DEEP-LEARNED FEATURES

Deformable image registration is an important process in many neurological studies for determining the anatomical correspondences which can be used, for example, for atlas-based segmentation of brain structures. The principle behind deformable image registration is to determine the optimal transformation that maximizes the feature similarities between two images, which often relies on user-selected features such as Gabor filters. Wu et al. (2013) proposed using an unsupervised two-layer stacked convolutional independent subspace analysis (ISA), which is an extension of independent component analysis (ICA), to directly learn the basis image filters that represent the training dataset. During image registration, the coefficients of these learned basis filters are used as morphological signatures to detect the spatial correspondences. When incorporated into existing registration methods, the data-adaptive features learned from the unsupervised deep learning framework gave rise to improved results over the hand-crafted features.

3.2.2 SEGMENTATION OF NEUROIMAGING DATA USING DEEP LEARNING

In machine learning, the most common approach to the image segmentation problem consists of two stages: hand-designed feature detectors are used to build feature vectors for each input in the first step, sometimes with prior knowledge such as spatial regularization or contour smoothness, and then the extracted features and target labels are used to train a supervised classifier to perform segmentation. This generally requires much labeled training data and good domain knowledge to design or select features, such as Gabor filters (Jain and Farrokhnia, 1990), Haar wavelet (Mallat, 1989), and SIFT (Lowe, 1999). However, such features are designed based on users' prior knowledge. With the hypothesis that the most effective features could be learned directly from training data, a number of researchers have recently adopted the deep learning framework as discussed below.

3.2.2.1 Hippocampus segmentation

Several methods have been developed for automatic hippocampus segmentation in MR images, as measurement of the hippocampus is useful for studying many neurological diseases including Alzheimer's disease (AD). However, segmentation accuracy is often limited due to the small size of the hippocampus and the complexity of surrounding structures. As similarly done in Wu et al. (2013), Kim et al. (2013) proposed integrating an unsupervised two-layer stacked convolutional ISA into a multi-atlas-based segmentation framework. The authors compared the traditional hand-crafted image features with the hierarchical feature representations learned from 7.0T MR images. They showed that the deep-learned feature representation improved the segmentation accuracy by about 3–4% on overlap metrics over the methods using hand-crafted features in the same segmentation framework. Guo et al. (2014) investigated using a two-layer SAE to learn the features for segmenting the hippocampus from infant T1- and T2-weighted (T1w and T2w) brain MR images. The deep-learned features were used to measure inter-patch similarity for sparse patch matching in a multi-atlas-based segmentation framework, and demonstrated an improvement of 4–8% in Dice similarity over features based on intensity, Haar wavelet, histogram of oriented gradients (HOG) (Dalal and Triggs, 2005) and local gradient patterns (Ojala et al., 2002).

3.2.2.2 Infant brain image segmentation

In studying early brain development in health and disease, segmenting infant brain images is a more challenging task than in the adult brain because infant WM and GM exhibit similar intensity levels in both T1w and T2w MR images. Zhang et al. (2015) proposed employing a deep CNN to perform segmentation of brain tissues on T1w, T2w and fractional anisotropy (FA) MR images. They trained a three-layer CNN using approximately 10,000 local patches extracted from all voxels in a training set of 10 brain images. The cross-entropy loss functional between the predicted and ground truth labels, and a three-way softmax layer were used to generate a posterior distribution over the three class labels.

3.2.2.3 Brain tumor segmentation

Automatic segmentation of brain tumors is a challenging problem because the tumors can appear randomly in the brain and have any kind of shape, size, and contrast. Havaei et al. (2015) proposed a fully automatic brain tumor segmentation method based on deep CNNs. The proposed method was designed to detect both low- and high-grade glioblastomas seen in MRI scans. The proposed deep CNNs learn contextual features at different scales and are fully convolutional, in contrast to traditional CNNs whose final layers are typically fully connected and therefore much more computationally demanding. In brain tumor segmentation, the training samples are highly unbalanced in that the healthy voxels comprise a large percentage of the total brain voxels. The authors tackled this problem by introducing a sequential training procedure in which the model was first trained with unbalanced training data using randomly sampled voxels, and then the top layers of the model were trained with balanced samples that contained the same number of tumor and healthy voxels.

3.2.3 CLASSIFICATION OF NEUROIMAGING DATA USING DEEP LEARNING

Classification of human neuroimaging data has been typically used to demonstrate the performance of a proposed hand-crafted feature (eg, a set of voxel intensities or size of particular regions of interest) or a feature selection method, both of which often require a thorough understanding of the disease by the user. Deep learning algorithms have recently attracted considerable attention from neuroimaging researchers due to the promise of automatic feature discovery for the tasks of computer-aided disease diagnosis and prognosis. Several recent studies have shown that deep learning methods can improve neuroimaging data classification by learning physiologically important image feature representations and discovering multimodal latent patterns in a data-driven way, as discussed below.

3.2.3.1 Schizophrenia diagnosis

Plis et al. (2014) adapted RBMs for performing schizophrenia diagnosis using the structural brain MR images from 198 schizophrenia patients and 191 matched controls. The MR images were aligned to brain templates and their gray matter was segmented, which resulted in 60,465 voxels per image. The gray matter voxels were vectorized and used to train a three-layer DBN. The first two layers had 50 hidden units in each layer and 100 hidden units were used for the third layer. They pretrained each layer via an unsupervised RBM and discriminatively fine-tuned the network by adding a softmax layer on top of the model using backpropagation. Using activations of the topmost hidden layers in the fine-tuned model with a 10-fold cross-validation, they trained and tested supervised classifiers, such as support vector machine (SVM) with the radial basis function kernel, logistic regression and k-nearest neighbors. The effect of the model depth on classification accuracy was investigated. The accuracy remained almost the same from depth 1 to depth 2 (66% and 62%, respectively, using the SVM classifier), but significantly improved for depth 3 (90% using the SVM classifier). Even though the model did not improve from depth 1 to depth 2, the model continued to learn useful transformations of the training data. This result strengthens the hypothesis that unsupervised pretraining can potentially lead to progressively more discriminative features at higher layers of data representation.

3.2.3.2 Huntington disease diagnosis

Plis et al. (2014) used the same three-layer DBN model described in Section 3.2.3.1 to investigate its potential for diagnosing Huntington disease. The dataset in the study was unbalanced, and consisted of 1.5 T and 3.0 T T1w MR images collected from 2641 patients and 859 healthy controls. Similarly to the previous study, the segmented gray matter voxels were utilized to train the deep learning network. The learned network performed binary classification as well as distinguished the patients by disease severity, which enabled spectral decomposition of the images using regression on the learned features.

3.2.3.3 Task identification using functional MRI dataset

ICA is the most widely used method for identifying the most salient signals in functional MRI data. Hjelm et al. (2014) proposed using a Gaussian-Bernoulli RBM model to isolate linear factors in functional brain imaging data by fitting a probability distribution model to the data, in order to identify functional networks. A voxel-by-time data matrix was utilized to train the Gaussian-Bernoulli RBM model. Various aspects of analyzing functional networks and temporal activations were considered for comparing between RBMs and ICA, which led the authors to conclude that an RBM can be used to perform functional network identification with accuracy that is equal to or greater than that of ICA.

3.2.3.4 Early diagnosis of Alzheimer's disease

Making an accurate early diagnosis of AD is particularly important because awareness of the severity and progression risks may enable early treatment. Suk et al. (2015) proposed a three-layer SAE model for classifying between AD and mild cognitive impairment (MCI), a prodromal stage of AD. Gray matter tissue volumes from MRI, mean signal intensities from PET, and biological measures from CSF samples were used as features for training modality-specific deep learning networks. The learned feature set was then reduced using group lasso (Yuan and Lin, 2006), which regularizes a linear regression model with L_1- and $L_{2,1}$-norm. Finally, a multikernel SVM (Gönen and Alpaydın, 2011) designed to learn the complementary information from multimodal data was trained with the learned MRI, PET, CSF features and labels to perform the following classification tasks: (1) AD vs. healthy normal control (NC); (2) MCI vs. NC; (3) AD vs. MCI;, and (4) MCI converter (MCI-C) vs. MCI nonconverter (MCI-NC). The proposed method demonstrated accuracy rates of 98.8%, 90.7%, 83.7%, and 83.3% for AD/NC, MCI/NC, AD/MCI, and MCI-C/MCI-NC classification, respectively, on the scans of 51 AD patients, 99 MCI patients (43 MCI-Cs and 56 MCI-NCs), and 52 NC subjects from the Alzheimer's Disease Neuroimaging Initiative (ADNI) dataset. A limitation of this study was that since structural and functional changes involved in AD can occur in multiple brain regions that do not necessarily correspond to user-defined regions of interest (ROIs), the features extracted from such user-defined ROIs may not be able to reflect small or subtle but potentially important pathological changes. Motivated by this, Suk et al. (2014) in a follow-up study proposed using a patch-based multimodal DBM framework (Salakhutdinov and Hinton, 2012) to learn a joint spatial feature representation from the paired 3D patches of MRI and PET images. In contrast to RBMs, the approximate inference procedure of DBMs is performed using two-way dependencies, that is, bottom-up and top-down, which allows DBMs to use higher-level information to learn intermediate-level features, thus creating potentially more accurate representations of the data (Salakhutdinov and Hinton, 2012). Unlike the previous work (Suk et al., 2015), CSF measures were not used in the study. The learned MRI-PET feature representation was then used to train an image-level supervised classifier based on weighted ensemble SVMs. The authors reported accuracy rates of 95.35%, 85.67%, and 74.58% for AD vs. NC, MCI vs. NC, and MCI-C vs. MCI-NC classification, respectively. Liu et al. (2014a) investigated a deep learning architecture consisting of an SAE with a softmax output layer for performing four-class classification simultaneously for the following labels: AD, NC, MCI-C, and MCI-NC. The mean MR and PET intensity values extracted from 83 brain ROIs were used as features. A multimodal layer was included in the deep learning framework in a denoising fashion, which hid one modality in some samples during training for regularization. The softmax layer had four units for representing the four AD categories. A mean accuracy of 53.79% for four-class classification was reported on 331 scans of the ADNI dataset.

3.2.3.5 High-level 3D PET image feature learning

Liu et al. (2014b) developed a framework based on SAEs to extract high-level ROI features from 3D PET images. The learned feature parameters were used as the encodings for content-based retrieval with the k-nearest-neighbor algorithm. The method was evaluated on mean average precision (MAP) using the leave-one-out paradigm with 331 3D PET images from the ADNI cohort. It was shown that the high-level PET ROI features extracted by deep learning can achieve an overall MAP of 56.13%, which outperformed the most widely used state-of-the-art data representation methods, such as Isomap (Tenenbaum et al., 2000) and elastic net (Shen et al., 2011).

3.3 FOCUS ON DEEP LEARNING IN MULTIPLE SCLEROSIS

3.3.1 MULTIPLE SCLEROSIS AND THE ROLE OF IMAGING

Multiple sclerosis (MS) is a chronic, degenerative disease of the brain and spinal cord. The clinical presentation of MS is very heterogeneous, and the range and severity of symptoms can vary greatly between patients. The clinical course of MS is highly unpredictable, but most patients are initially diagnosed as having relapsing remitting MS (RRMS), which is characterized by inflammatory attacks separated by variable periods of remission and recovery. The majority of RRMS patients will eventually transition into the secondary progressive MS (SPMS) phase, in which there is an unremitting and progressive accumulation of disability. There is currently no cure for MS. Existing therapies that focus on symptomatic management and prevention of further damage have variable degrees of effectiveness, although several recent breakthroughs are promising. MS pathology originates at the cellular level and many aspects are not well understood, but there are characteristic (but not specific) signs of tissue damage, the most recognizable of which are white matter lesions (WMLs) and brain atrophy, or shrinkage due to degeneration. These signs can be observed on MRI, which has become a vital tool to noninvasively monitor MS patients in the clinic and to advance the understanding of MS pathology. WML counts and volume and brain volume have become established imaging biomarkers for MS clinical trials, and there is promise for their use in routine clinical practice, but they generally only correlate modestly with clinical disability scores. The weak link between the image-based measures of MS pathology and disability scores is known as the "clinico-radiological paradox" of MS (Barkhof, 2002), and results in the low utility of current imaging biomarkers for the purposes of personalized medicine. There are a number of key reasons why the current imaging biomarkers do not have stronger quantitative relationships with clinical scores:

- Due to the wide range of symptoms, MS disability is difficult to score comprehensively in routine clinical practice. For example, the Kurtzke (1983) expanded disability status scale (EDSS) is the most commonly used clinical score, but does not account for cognitive impairment, which is a significant contributor to disability in the majority of MS patients (Chiaravalloti and DeLuca, 2008).
- Through neuroplasticity, the brain and spinal cord can adapt to damage in order to maintain functionality (Tomassini et al., 2012). As a result, clinically silent or subtle pathology is often present.
- Conventional MRI does not capture all aspects of MS pathology. For example, myelin is a nerve insulator that is critical for proper signal conduction and demyelination is a key pathological feature of MS in which white matter that appears normal on conventional MRI may actually have reduced myelin (Laule et al., 2004).
- The current established imaging biomarkers largely capture volumetric changes, which are important and relatively easy to compute, but do not reflect potentially important structural variations, such as shape changes in the brain and spatial dispersion of the lesions.
- Traditional statistical approaches like simple prediction models such as logistic regression are common tools for analysis. The general assumption behind these simple models is that the data is generated by a known stochastic data model, with the goodness-of-fit evaluated by residual analysis, but this works well only with a very low number of variables (Breiman, 2001b). Consequently, these traditional statistical approaches, which place a strong emphasis on interpretability, often come with the sacrifice of accuracy.

In this section, we summarize the recent work on deep learning methods for discovering image features relevant to MS, which hold the promise of overcoming the last two limitations listed above. In particular, we focus on two applications: segmentation of WMLs and modeling of disease variability. The overall goal is to investigate the potential of deep learning to automatically capture image features with as little user bias as possible (such as in the choice of model distribution or feature representation), in a way that is complementary to the traditional approach of proposing and validating imaging biomarkers based on biological hypotheses. By performing unsupervised learning on the very high-dimensional space of 3D brain MRIs, deep learning can potentially be a powerful method for generating hypotheses that can then be investigated with more traditional means to facilitate interpretation.

3.3.2 WHITE MATTER LESION SEGMENTATION

Focal lesions in the brain and spinal cord are one of the hallmarks of MS pathology, and are primarily visible in the white matter on structural MRIs. These lesions are observable as hyperintensities on T2w, proton density-weighted (PDw), or fluid-attenuated inversion recovery (FLAIR) scans, and as hypointensities, or "black holes" (García-Lorenzo et al., 2013) on T1w scans. Imaging biomarkers based on the identification of lesions, such as lesion count and lesion volume, have established their importance for assessing disease progression and treatment effect. However, lesions vary greatly in size, shape, intensity and location, which makes their automatic and accurate segmentation challenging. Many automatic methods have been proposed for the segmentation of MS lesions over the last two decades, most of which are described in recent surveys by García-Lorenzo et al. (2013) and Lladó et al. (2012). Lesion segmentation methods can be broadly classified into unsupervised and supervised methods. Unsupervised methods do not require a labeled dataset for training, but most will rather model healthy tissue (eg, via intensity clustering) and identify lesions as an outlier class. Supervised approaches typically start with a set of features, sometimes very large, which are usually defined or selected by the user. A training step with labeled data is used to determine which subset of features produce the most accurate segmentations, and those features are then used to identify lesion voxels in new images.

Given the ability to automatically learn useful features demonstrated by deep learning in various computer vision applications, it is not surprising that a number of MS lesion segmentation methods based on deep learning have been proposed. The potential advantage of deep learning is that the feature set would no longer need to be predetermined by the user, but rather learned directly from the training images. This is a useful property because it is difficult for a person to characterize the features that separate lesion voxels from those of healthy tissue. From the perspective of the deep learning researcher, the high-dimensionality of the input images, the difficulty of obtaining reliable ground truth (thereby making unsupervised learning even more important), and the high accuracy required for clinical practicality all make WML segmentation a worthy test application. In this section, we summarize several deep learning methods that have been recently proposed for MS lesion segmentation. We refrain from listing the values of the performance measures (eg, Dice coefficient) given in the cited papers, because these numbers cannot be compared across datasets, and a publicly available dataset with an adequate sample size for deep learning is sadly not yet available. Therefore we instead focus our summary on the overall observations made by the respective authors. Unless otherwise noted, the experiments described below used MRIs with a voxel size of approximately $1\,\text{mm} \times 1\,\text{mm} \times 3\,\text{mm}$.

3.3.2.1 Patch-based segmentation methods

Yoo et al. (2014) were the first to propose an automated feature learning approach for MS lesion segmentation. In this method, uniformly spaced and nonoverlapping 3D patches at two scales (9×9×3 and 15×15×5) were extracted from co-registered T2w and PDw images. Unsupervised feature learning was applied to the T2w and PDw images separately, using a basic architecture of a single RBM for the smaller patches and a two-layer DBN for the larger patches for each image type. Sigmoid hidden units and Gaussian visible units were used, with contrastive divergence for training. The learned features for each image pair were then concatenated into an approximately 5000-element vector that was then fed into a random forest, which was trained with voxel labels for supervised learning. The authors varied the number of image pairs used for unsupervised training from 100 to 1400, while keeping the number of labeled pairs constant at 100, with the hypothesis that feature learning from unlabeled images can improve performance. While the results were not definitive, there was a trend toward improved segmentation accuracy. Overall, this study established that automatic feature discovery with deep learning is a viable alternative to user preselection of features for WML segmentation.

In 2015, the International Symposium on Biomedical Imaging (ISBI) conference held a grand challenge on longitudinal MS lesion segmentation, and released a dataset composed of training data with longitudinal images (with manual delineations of WMLs produced by two raters) from five patients, and two test datasets consisting of longitudinal images from ten and five patients. Each longitudinal dataset included T1w, T2w, PDw, and FLAIR MRIs with three to five time points acquired on a 3T MR scanner. The T1w images had approximately a 1 mm cubic voxel resolution. The winner of the challenge was a team (Vaidya et al., 2015) that used 3D CNNs to automatically learn features from 3D four-channel patches of size 19 × 19 × 19. The basic network architecture was composed of two convolutional layers with 60 (4 × 4 × 4) and 60 (3 × 3 × 3) filters using the softplus activation function and average pooling (2×2×2), followed by a fully connected layer and finally a softmax layer for voxel-wise classification. Gradient descent using log-likelihood as the cost function and momentum was used for training. Since MS lesions typically comprise a very small percentage (<1%) of the voxels in an MRI volume, the patches were selected by dividing each image volume into subregions, and only using patches from those subregions that have greater than a given percentage of lesion voxels. This method obtained segmentation performance comparable to the variability between the two manual raters. It is notable that a deep learning method was able to win the challenge using such a small number of training images. Another method using CNNs (Ghafoorian and Platel, 2015) was also presented at the ISBI challenge. This method used 2D (32 × 32), four-channel patches, obtained in a sliding window manner and sampled to maintain the proportion between the positive and negative voxels. The network architecture was a CNN consisting of four convolutional layers with 15 (13 × 13), 25 (9 × 9), 60 (7 × 7), and 130 (3 × 3) filters, with no pooling, and a final logistic regression layer. SGD with a fixed learning rate was used for training. Although the highly competitive performance of the first method is a positive indication for the use of deep learning for WML segmentation, the difference in performance between the two similar CNN-based methods highlights the lack of standardization for the design and application of such models in neuroimaging.

3.3.2.2 Convolutional encoder network segmentation

The computational demands of deep learning methods have largely restricted the size of the input images, and subdivision into patches has been the most popular workaround for processing larger images such as MRI volumes. For WML segmentation, a patch-based strategy can even have some benefits such as the ability to selectively sample more representative regions. However, most patch-based methods are inefficient in that they perform many redundant computations in the overlapping regions of neighboring patches. While some methods have been proposed for speeding up patch-based networks (eg, Li et al., 2014, as used by Vaidya et al., 2015), some recent segmentation approaches have used fully convolutional networks (FCNs; Long et al., 2015), which only contain layers that can be framed as convolutions (eg, pooling and up sampling), to perform dense prediction by producing segmented output that is of the same dimensions as the original images. Brosch et al. (2015) proposed a 3D FCN to process entire MRI volumes for MS WML segmentation. The network used a convolutional layer with 32 ($9 \times 9 \times 5$) filters to extract features from the input layer at each voxel location, and a deconvolutional layer that used the extracted features to predict a lesion mask and thereby classify each voxel of the image in a single operation. The authors called this model a convolutional encoder network due to its similarity to a convolutional autoencoder, and applied an efficient Fourier-based training algorithm (Brosch and Tam, 2015) to perform end-to-end training, which enabled feature learning to be driven by segmentation performance. By processing entire MRI volumes instead of patches, the algorithm avoids redundant calculations, and therefore could scale up more efficiently with image resolution. To overcome the problem of unbalanced classes without selective voxel sampling, the authors proposed a new objective function based on a weighted combination of sensitivity and specificity ($\approx$1:10 ratio), reformulated to be error terms that allowed for stable gradient computations. Optimization was performed with SGD. The method was evaluated on a large dataset of PDw and T2w volumes from an MS clinical trial, acquired from 45 different scanning sites, of 500 subjects that the authors split equally into training and test sets. By varying the training sample size, the authors showed that approximately 100 scans were sufficient for this framework to learn to segment the test scans optimally. More recent work by the authors (Brosch et al., 2016) has shown that adding more layers can further improve segmentation performance. Overall, the FCN approach applied to full MRI volumes can be seen as a promising alternative to patch-based methods, especially where computational efficiency is a concern.

3.3.3 MODELING DISEASE VARIABILITY

Changes in brain morphology and white matter lesions are two hallmarks of MS pathology, but their variability beyond volumetrics is poorly characterized. To further the understanding of complex MS pathology, Brosch et al. (2014) proposed using DBNs to build a statistical model of brain images that can automatically discover spatial patterns of variability in brain morphology and lesion distribution. The test data was composed of MRIs of 474 MS patients, with each having a multimodal set of T1w, T2w, and PDw volumes with a resolution of $256\times256\times50$ voxels and a voxel size of 0.937 mm × 0.937 mm × 3.000 mm. In contrast to other methods for manifold learning, the DBN approach, with its capability for automatic feature learning, does not require a prebuilt proximity graph, which is particularly advantageous for modeling sparse and pseudorandom content such as lesions, because defining a suitable distance measure between lesion images would be very challenging. The proposed network consisted of a morphology DBN, a lesion DBN, and a joint DBN that modeled concurring morphological and lesion patterns. The input to the morphology DBN was a set of deformation fields computed by nonlinear registration of the T1w MRIs to a standard template. The input to the lesion DBN was a set of binary lesion masks (produced from the T2w and PDw scans) with the same transformations applied. Both the morphology and lesion DBNs were composed of three strided convolutional RBMs (sconvRBMs) and two dense RBMs with 16 and 2 hidden units. For the morphology DBN, the three sconvRBMs had stride sizes of $2\times2\times1$, $2\times2\times2$, $1\times1\times1$, filter sizes of $10\times10\times7$, $10\times10\times10$, $3\times5\times3$, and 32, 64, 32 filters, respectively. For the lesion DBN, the three sconvRBMs had stride sizes of $4\times4\times2$, $2\times2\times2$, $2\times2\times2$, filter sizes of $20\times20\times10$, $14\times14\times10$, $10\times14\times6$, and 32, 64, 64 filters, respectively. The joint DBN consisted of two pathways, each consisting of the first four layers of the morphology and lesion DBNs, respectively, and a fifth RBM layer with four hidden units, which replaced the fifth layer of the individual DBNs and combined the hidden unit activations of the fourth layer RBMs. Fig. 3.4 shows some images sampled from the learned manifolds. The results allowed the authors to make three main observations: (1) the model automatically discovered the classic patterns of MS pathology, such as enlarged ventricles and increased preventricular lesion load (Traboulsee and Li, 2008), as well as the more subtle ones, such as lesion load in specific structures (eg, brain stem); (2) the parameters of the joint model correlated stronger with MS clinical scores than the parameters of either individual model; and (3) the parameters of the individual models and the joint model correlated stronger with MS clinical scores than the traditional imaging biomarkers of brain volume and lesion volume. Overall, this study demonstrated that deep learning can be used to learn complex and clinically relevant features from brain images of MS patients, with very few assumptions by the user.

3.4 FUTURE RESEARCH NEEDS

We have reached an exciting time for deep learning, a field that is progressing very rapidly in academia, and also having well-publicized success in industry. For the neuroimaging researcher interested in applying deep learning, it seems almost impossible to keep up with the latest technical developments published in machine learning conferences and journals. However, it should be realized that despite research over the last several years that has produced very promising results and even

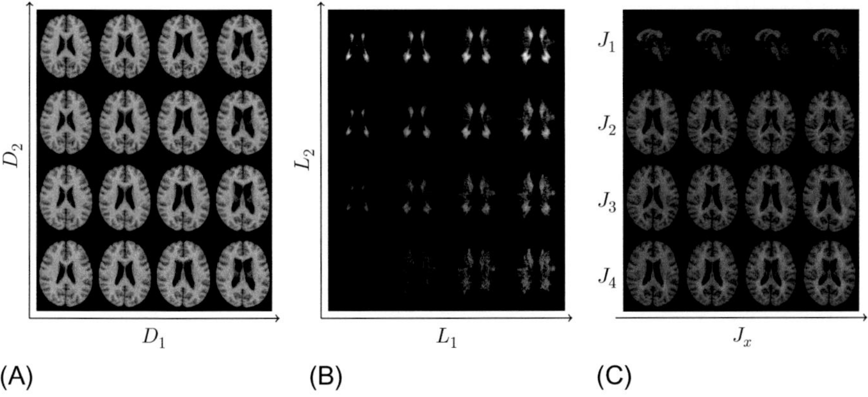

FIG. 3.4

Slices from generated volumes from the (A) morphology, (B) lesion, and (C) joint models. The morphology model captured ventricular enlargement (D_1) and decrease in brain size (D_2) as the main modes of variation. For the lesion model, L_1 captured an increase in lesion load throughout the WM, while L_2 captured primarily periventricular lesion load variations. The parameters of the joint model captured combinations of the variability found in the individual models.

some true breakthroughs, deep learning remains largely unvalidated for automated feature learning of brain images. This is because much of the work done by the machine learning community has been applied to much larger datasets of images with much lower dimensionality. This is evident, for example, from the fact that support for 3D convolutions has just recently been added to GPU libraries. Neuroimaging data has unique challenges, and much work still needs to be done to establish standards for designing and training deep networks, even just the basic ones described in this chapter. For example, determining the optimal number of layers, filter sizes, learning rates, and regularization strategy all need further investigation. Fortunately, there has been enough success so far to establish research momentum, and positive results will continue to come, and hopefully the advantages of deep learning seen in other fields will also be realized in neuroimaging. What we see as particularly promising are highly efficient models such as fully convolutional networks, because they allow full images to be processed, which facilitates the use of deeper models and larger datasets, as well as the experimentation with different network architectures.

ACKNOWLEDGMENTS

This work was supported by the Natural Sciences and Engineering Research Council of Canada, the Milan and Maureen Ilich Foundation, and the UBC Engineers-in-Scrubs Program. The authors gratefully acknowledge the valuable feedback from Drs. David Li and Anthony Traboulsee.

REFERENCES

Alonso-Ortiz, E., Levesque, I.R., Pike, G.B., 2015. MRI-based myelin water imaging: a technical review. Magn. Reson. Med. 73 (1), 70–81.

Barkhof, F., 2002. The clinico-radiological paradox in multiple sclerosis revisited. Curr. Opin. Neurol. 15 (3), 239–245.

Breiman, L., 2001a. Random forests. Mach. Learn. 45 (1), 5–32.

Breiman, L., 2001b. Statistical modeling: the two cultures. Stat. Sci. 16 (3), 199–231.

Brosch, T., Tam, R., 2015. Efficient training of convolutional deep belief networks in the frequency domain for application to high-resolution 2D and 3D images. Neural Comput. 27 (1), 211–227.

Brosch, T., Yoo, Y., Traboulsee, A., Li, D., Tam, R., 2014. Modeling the variability in brain morphology and lesion distribution in multiple sclerosis by deep learning. In: Proceedings of Medical Image Computing and Computer Assisted Intervention (MICCAI) Part II, pp. 463–470.

Brosch, T., Yoo, Y., Tang, L., Traboulsee, A., Li, D., Tam, R., 2015. Deep convolutional encoder networks for multiple sclerosis lesion segmentation. In: Proceedings of Medical Image Computing and Computer Assisted Intervention (MICCAI) Part III, pp. 3–11.

Brosch, T., Tang, L.Y.W., Yoo, Y., Li, D.K.B., Traboulsee, A., Tam, R., 2016. Deep 3D convolutional encoder networks with shortcuts for multiscale feature integration applied to multiple sclerosis lesion segmentation. In: IEEE Transactions on Medical Imaging, Special Issue on Deep Learning (in press).

Chiaravalloti, N.D., DeLuca, J., 2008. Cognitive impairment in multiple sclerosis. Lancet Neurol. 7 (12), 1139–1151.

Cho, K., Ilin, A., Raiko, T., 2011. Improved learning of Gaussian-Bernoulli restricted Boltzmann machines. In: Artificial Neural Networks and Machine Learning—ICANN 2011. Springer, New York, pp. 10–17.

Ciresan, D., Giusti, A., Schmidhuber, J., 2012. Deep neural networks segment neuronal membranes in electron microscopy images. In: Advances in Neural Information Processing Systems, pp. 1–9.

Cortes, C., Vapnik, V., 1995. Support-vector networks. Mach. Learn. 20 (3), 273–297.

Dalal, N., Triggs, B., 2005. Histograms of oriented gradients for human detection. In: Proceedings of the IEEE Computer Society Conference on Computer Vision and Pattern Recognition, pp. 886–893.

Dauphin, Y.N., de Vries, H., Chung, J., Bengio, Y., 2015. RMSProp and equilibrated adaptive learning rates for non-convex optimization. arXiv 1502.04390v1.

Deng, J., Dong, W., Socher, R., Li, L.J., Li, K., Fei-Fei, L., 2009. ImageNet: a large-scale hierarchical image database. In: Proceedings of the IEEE Conference on Computer Vision and Pattern Recognition. IEEE, Piscataway, NJ, pp. 248–255.

Duchi, J., Hazan, E., Singer, Y., 2011. Adaptive subgradient methods for online learning and stochastic optimization. J. Mach. Learn. Res. 12, 2121–2159.

Farley, B., Clark, W., 1954. Simulation of self-organizing systems by digital computer. Trans. IRE Prof. Group Inform. Theory 4 (4), 76–84.

Freund, Y., Haussler, D., 1992. Unsupervised learning of distributions on binary vectors using two layer networks. In: Proceedings of Advances in Neural Information Processing Systems, pp. 912–919.

Fukushima, K., 1980. Neocognitron: a self-organizing neural network model for a mechanism of pattern recognition unaffected by shift in position. Biol. Cybern. 36 (4), 193–202.

García-Lorenzo, D., Francis, S., Narayanan, S., Arnold, D.L., Collins, D.L., 2013. Review of automatic segmentation methods of multiple sclerosis white matter lesions on conventional magnetic resonance imaging. Med. Image Anal. 17 (1), 1–18.

Ghafoorian, M., Platel, B., 2015. Convolutional neural networks for MS lesion segmentation, method description of Diag team. In: Proceedings of IEEE International Symposium on Biomedical Imaging (ISBI): Grand Challenge in Longitudinal Multiple Sclerosis Lesion Segmentation.

Gönen, M., Alpaydın, E., 2011. Multiple kernel learning algorithms. J. Mach. Learn. Res. 12, 2211–2268.

Guo, Y., Wu, G., Commander, L.A., Szary, S., Jewells, V., Lin, W., Shen, D., 2014. Segmenting hippocampus from infant brains by sparse patch matching with deep-learned features. In: Medical Image Computing and Computer-Assisted Intervention—MICCAI 2014. Springer, New York, pp. 308–315.

Haacke, E.M., Xu, Y., Cheng, Y.C.N., Reichenbach, J.R., 2004. Susceptibility weighted imaging (SWI). Magn. Reson. Med. 52 (3), 612–618.

Havaei, M., Davy, A., Warde-Farley, D., Biard, A., Courville, A., Bengio, Y., Pal, C., Jodoin, P.M., Larochelle, H., 2015. Brain tumor segmentation with deep neural networks. arXiv 1505.03540.

Hinton, G.E., 2002. Training products of experts by minimizing contrastive divergence. Neural Comput. 14 (8), 1771–1800.

Hinton, G.E., 2010. A practical guide to training restricted Boltzmann machines. Momentum 9 (1), 926.

Hinton, G.E., Salakhutdinov, R., 2006. Reducing the dimensionality of data with neural networks. Science 313 (5786), 504–507.

Hinton, G.E., Osindero, S., Teh, Y.W., 2006. A fast learning algorithm for deep belief nets. Neural Comput. 18 (7), 1527–1554.

Hjelm, R.D., Calhoun, V.D., Salakhutdinov, R., Allen, E.A., Adali, T., Plis, S.M., 2014. Restricted Boltzmann machines for neuroimaging: an application in identifying intrinsic networks. NeuroImage 96, 245–260.

Hubel, D.H., Wiesel, T.N., 1962. Receptive fields, and binocular interaction and functional architecture in the cat's visual cortex. J. Physiol. 160 (1), 106.

Hubel, D.H., Wiesel, T.N., 1968. Receptive fields and functional architecture of monkey striate cortex. J. Physiol. 195 (1), 215–243.

Jain, A.K., Farrokhnia, F., 1990. Unsupervised texture segmentation using Gabor filters. In: Proceedings of the IEEE International Conference on Systems, Man and Cybernetics, pp. 14–19.

Kamnitsas, K., Chen, L., Ledig, C., Rueckert, D., Glocker, B., 2015. Multi-scale 3D convolutional neural networks for lesion segmentation in brain MRI. In: Proceedings of Ischemic Stroke Lesion Segmentation Challenge, pp. 13–16.

Kim, M., Wu, G., Shen, D., 2013. Unsupervised deep learning for hippocampus segmentation in 7.0 tesla MR images. In: Proceedings of Medical Image Computing and Computer Assisted Intervention (MICCAI) Machine Learning in Medical Imaging (MLMI) Workshop, pp. 1–8.

Kingma, D., Ba, J., 2014. Adam: a method for stochastic optimization. arXiv 1412.6980.

Krizhevsky, A., Sutskever, I., Hinton, G.E., 2012. Imagenet classification with deep convolutional neural networks. In: Proceedings of Advances in Neural Information Processing Systems, pp. 1097–1105.

Kurtzke, J.F., 1983. Rating neurologic impairment in multiple sclerosis: an expanded disability status scale (EDSS). Neurology 33 (11), 1444–1452.

Laule, C., Vavasour, I.M., Moore, G.R.W., Oger, J., Li, D.K.B., Paty, D.W., MacKay, A.L., 2004. Water content and myelin water fraction in multiple sclerosis. A T2 relaxation study. J. Neurol. 251 (3), 284–293.

Le Bihan, D., Mangin, J.F., Poupon, C., Clark, C.A., Pappata, S., Molko, N., Chabriat, H., 2001. Diffusion tensor imaging: concepts and applications. J. Mag. Reson. Imaging 13 (4), 534–546.

LeCun, Y., Boser, B., Denker, J.S., Henderson, D., Howard, R.E., Hubbard, W., Jackel, L.D., 1989. Backpropagation applied to handwritten zip code recognition. Neural Comput. 1 (4), 541–551.

LeCun, Y., Bottou, L., Bengio, Y., Haffner, P., 1998. Gradient-based learning applied to document recognition. Proc. IEEE 86 (11), 2278–2324.

LeCun, Y., Bengio, Y., Hinton, G., 2015. Deep Learning. Nature 521 (7553), 436–444.

Lee, H., Grosse, R., Ranganath, R., Ng, A.Y., 2009. Convolutional deep belief networks for scalable unsupervised learning of hierarchical representations. In: Proceedings of the 26th Annual International Conference on Machine Learning, pp. 609–616.

Lee, H., Grosse, R., Ranganath, R., Ng, A.Y., 2011. Unsupervised learning of hierarchical representations with convolutional deep belief networks. Commun. ACM 54 (10), 95–103.

Li, H., Zhao, R., Wang, X., 2014. Highly efficient forward and backward propagation of convolutional neural networks for pixelwise classification. arXiv 1412.4526 [cs].

Liu, S., Cai, W., Che, H., Pujol, S., Kikinis, R., Feng, D., Fulham, M., 2014a. Multi-modal neuroimaging feature learning for multi-class diagnosis of Alzheimer's disease. IEEE Trans. Biomed. Eng. 62 (4), 1132–1140.

Liu, S., Liu, S., Cai, W., Che, H., Pujol, S., Kikinis, R., Fulham, M., Feng, D., 2014b. High-level feature based PET image retrieval with deep learning architecture. J. Nucl. Med. 55 (Suppl. 1), 2028–2028.

Lladó, X., Oliver, A., Cabezas, M., Freixenet, J., Vilanova, J.C., Quiles, A., Valls, L., Ramió-Torrentà, L., Àlex Rovira, 2012. Segmentation of multiple sclerosis lesions in brain MRI: a review of automated approaches. Inform. Sci. 186 (1), 164–185.

Long, J., Shelhamer, E., Darrell, T., 2015. Fully convolutional networks for semantic segmentation. In: Proceedings of Computer Vision and Pattern Recognition (CVPR), pp. 3431–3440.

Lowe, D.G., 1999. Object recognition from local scale-invariant features. In: IEEE International Conference on Proceedings of Computer Vision, vol. 2, pp. 1150–1157.

Mallat, S.G., 1989. A theory for multiresolution signal decomposition: the wavelet representation. IEEE Trans. Pattern Anal. Mach. Intell. 11 (7), 674–693.

Nair, V., Hinton, G.E., 2010. Rectified linear units improve restricted Boltzmann machines. In: Proceedings of the 27th International Conference on Machine Learning, pp. 807–814.

Ojala, T., Pietikainen, M., Maenpaa, T., 2002. Multiresolution gray-scale and rotation invariant texture classification with local binary patterns. IEEE Trans. Pattern Anal. Mach. Intell. 24 (7), 971–987.

Plis, S.M., Hjelm, D.R., Salakhutdinov, R., Allen, E.A., Bockholt, H.J., Long, J.D., Johnson, H.J., Paulsen, J.S., Turner, J.A., Calhoun, V.D., 2014. Deep learning for neuroimaging: a validation study. Front. Neurosci. 8, Article 229.

Polyak, B.T., Juditsky, A.B., 1992. Acceleration of stochastic approximation by averaging. SIAM J. Control Optim. 30 (4), 838–855.

Raina, R., Madhavan, A., Ng, A.Y., 2009. Large-scale deep unsupervised learning using graphics processors. In: Proceedings of the 26th Annual International Conference on Machine Learning, pp. 873–880.

Rumelhart, D.E., Hinton, G.E., Williams, R.J., 1986. Learning representations by back-propagating errors. Nature 323, 533–536.

Sainath, T.N., Mohamed, A.R., Kingsbury, B., Ramabhadran, B., 2013. Deep convolutional neural networks for LVCSR. In: Proceedings of 2013 IEEE International Conference on Acoustics, Speech and Signal Processing (ICASSP). IEEE, Piscataway, NJ, pp. 8614–8618.

Salakhutdinov, R., Hinton, G., 2012. An efficient learning procedure for deep Boltzmann machines. Neural Comput. 24 (8), 1967–2006.

Scherer, D., Müller, A., Behnke, S., 2010. Evaluation of pooling operations in convolutional architectures for object recognition. In: Artificial Neural Networks-ICANN 2010. Springer, New York, pp. 92–101.

Shen, L., Kim, S., Qi, Y., Inlow, M., Swaminathan, S., Nho, K., Wan, J., Risacher, S.L., Shaw, L.M., Trojanowski, J.Q., et al., 2011. Identifying neuroimaging and proteomic biomarkers for MCI and AD via the elastic net. In: Multimodal Brain Image Analysis. Springer, New York, pp. 27–34.

Song, A.W., Huettel, S.A., McCarthy, G., 2006. Functional neuroimaging: basic principles of functional MRI. In: Cabeza, R., Kingstone, A. (Eds.), Handbook of Functional Neuroimaging of Cognition, Second ed. MIT Press, Cambridge, MA, pp. 21–52.

Suk, H.I., Lee, S.W., Shen, D., Alzheimer's Disease Neuroimaging Initiative, 2014. Hierarchical feature representation and multimodal fusion with deep learning for AD/MCI diagnosis. NeuroImage 101, 569–582.

Suk, H.I., Lee, S.W., Shen, D., Alzheimer's Disease Neuroimaging Initiative, 2015. Latent feature representation with stacked auto-encoder for AD/MCI diagnosis. Brain Struct. Funct. 220 (2), 841–859.

Tenenbaum, J.B., De Silva, V., Langford, J.C., 2000. A global geometric framework for nonlinear dimensionality reduction. Science 290 (5500), 2319–2323.

Tieleman, T., 2008. Training restricted Boltzmann machines using approximations to the likelihood gradient. In: Proceedings of the 25th International Conference on Machine Learning, pp. 1064–1071.

Tomassini, V., Matthews, P.M., Thompson, A.J., Fuglø, D., Geurts, J.J., Johansen-Berg, H., Jones, D.K., Rocca, M.A., Wise, R.G., Barkhof, F., Palace, J., 2012. Neuroplasticity and functional recovery in multiple sclerosis. Nat. Rev. Neurol. 8 (11), 635–646.

Traboulsee, A., Li, D.K., 2008. Conventional MR imaging. Neuroimaging Clin. N. Am. 18 (4), 651–673.

Vaidya, S., Chunduru, A., Muthuganapathy, R., Krishnamurthi, G., 2015. Longitudinal multiple sclerosis lesion segmentation using 3D convolutional neural networks. In: Proceedings of IEEE International Symposium on Biomedical Imaging (ISBI): Grand Challenge in Longitudinal Multiple Sclerosis Lesion Segmentation.

Vincent, P., Larochelle, H., Lajoie, I., Bengio, Y., Manzagol, P.A., 2010. Stacked denoising autoencoders: learning useful representations in a deep network with a local denoising criterion. J. Mach. Learn. Res. 11, 3371–3408.

Werbos, P., 1974. Beyond regression: new tools for prediction and analysis in the behavioral sciences. Ph.D. thesis, Harvard University, Cambridge, MA.

Wu, G., Kim, M., Wang, Q., Gao, Y., Liao, S., Shen, D., 2013. Unsupervised deep feature learning for deformable registration of MR brain images. In: Medical Image Computing and Computer-Assisted Intervention—MICCAI 2013. Springer, pp. 649–656.

Yoo, Y., Brosch, T., Traboulsee, A., Li, D., Tam, R., 2014. Deep learning of image features from unlabeled data for multiple sclerosis lesion segmentation. In: Proceedings of Medical Image Computing and Computer Assisted Intervention (MICCAI) Machine Learning in Medical Imaging (MLMI) Workshop, pp. 117–124.

Yuan, M., Lin, Y., 2006. Model selection and estimation in regression with grouped variables. J. R. Stat. Soc. B 68 (1), 49–67.

Zeiler, M.D., 2012. ADADELTA: an adaptive learning rate method. arXiv 1212.5701.

Zhang, W., Li, R., Deng, H., Wang, L., Lin, W., Ji, S., Shen, D., 2015. Deep convolutional neural networks for multi-modality isointense infant brain image segmentation. NeuroImage 108, 214–224.

CHAPTER

Machine learning and its application in microscopic image analysis

4

F. Xing, L. Yang
University of Florida, Gainesville, FL, United States

CHAPTER OUTLINE

4.1 INTRODUCTION

Microscopic image analysis in pathology is very important for image-based computer-aided diagnosis (CAD), which might provide potential support for early detection of diseases. Due to the advance in microscopic imaging techniques, a huge number of microscopic images are generated every day (Sommer and Gerlich, 2013), which prohibit data manual assessment. Therefore automated microscopic image analysis is in urgent need of digitized specimen evaluation. In addition, computerized approaches can eliminate the inter-observer variations and thus significantly improve the objectivity and reproducibility of the assessment (Foran et al., 2011). This will allow comparative study of diseases, and can potentially support decision making in diagnosis.

Machine learning is a subfield of computer science that aims to develop a set of algorithms to detect patterns within existing data and then uses these uncovered patterns to make predictions on new data (Murphy, 2012; Hastie et al., 2009; Bishop, 2007). Compared with nonlearning-based methods that are usually specifically designed for certain applications and not adaptive to other scenarios (eg, different biology assays), machine learning does not require manual software adaption, and more importantly, it can effectively handle high-dimensional data that are not easily

Machine Learning and Medical Imaging. http://dx.doi.org/10.1016/B978-0-12-804076-8.00004-9

modeled with a few parameters (Sommer and Gerlich, 2013). Machine learning has attracted a great deal of interest in the fields of computer vision, image processing, data mining, medical imaging, computational biology, etc. Specifically, there exist a large number of reports applying various machine learning techniques to automated data analysis in medicine, biomedical images, and digitized pathology specimens (Waljee and Higgins, 2010; Wernick et al., 2010; Tarca et al., 2007; Kourou et al., 2015), which can provide strong support for CAD. In this chapter, we will not attempt to comprehensively review the literature of machine learning-based CAD, but focus on machine learning applications in microscopic image analysis by using recent publications in our own research instead. We will introduce several popular machine learning techniques in digital pathology image analysis and their applications in nucleus/cell detection and segmentation on microscopic images.

4.2 DETECTION

Accurate nucleus/cell detection in microscopic images is a fundamental step for many subsequent computer-aided biomedical image analyses, such as nucleus/cell segmentation, counting, tracking, classification, etc. However, robust object localization, especially in those images exhibiting dense clusters and large variations in object scales, remains a challenging task (Quelhas et al., 2010). In the past few years, a number of methods including spatial filters (Al-Kofahi et al., 2010), kernel-based voting (Parvin et al., 2007; Qi et al., 2012; Xing et al., 2014), and graph partition (Bernardis and Yu, 2010; Zhang et al., 2014) have been proposed to automatically detect nuclei/cells in microscopic images. However, these approaches usually require careful parameter tuning to achieve desired performance on different types of images. This procedure of manual adjustment is tedious and limits the usage of those algorithms. By contrast, machine learning, especially supervised learning, learns processing rules from given image data instead of relying on manual parameter selection (Sommer and Gerlich, 2013). These learning methods have received a large amount of attention in microscopic image analysis (Arteta et al., 2012; Mualla et al., 2013; Cireşan et al., 2013). Currently the nucleus/cell detection is usually formulated as a pixel or superpixel-wise (region-wise) classification problem, and a specific model is learned to map data examples into discrete labels. There exist many classifiers with various feature representations reported in the literature, and here we mainly focus on two recently popular classifiers, support vector machine (SVM) and deep convolutional neural network (CNN), by explaining their applications in fluorescence, bright-field, and phase-contrast microscopy images.

4.2.1 SUPPORT VECTOR MACHINE

The SVM is a nonprobabilistic binary classifier, aiming to find a hyperplane with a maximal margin to separate high-dimensional data points (Cortes and Vapnik, 1995).

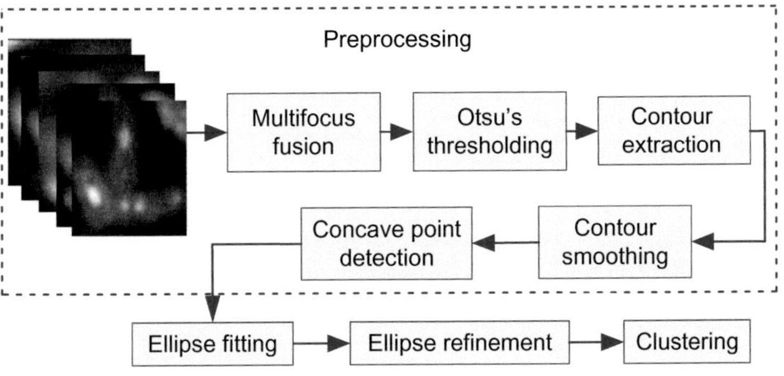

FIG. 4.1

An overview of the proposed myonucleus detection in isolated single muscle fiber fluorescence images.

Given a set of training data $\{(\boldsymbol{x}_i, y_i)\}_{i=1}^N$, where $\boldsymbol{x}_i \in R^p$ and $y_i \in \{1, -1\}$, SVM solves the following problem:

$$\min_{\boldsymbol{w},b,\boldsymbol{\xi}} \frac{1}{2}\boldsymbol{w}^{\mathrm{T}}\boldsymbol{w} + C\sum_{i=1}^{N} \xi_i, \text{ s.t. } y_i(\boldsymbol{w}^{\mathrm{T}}\phi(\boldsymbol{x}_i) + b) \geq 1 - \xi_i, \quad \xi_i \geq 0, \forall i, \tag{4.1}$$

where the $\phi(\boldsymbol{x})$ maps the data points into a high-dimensional space. $C > 0$ is a penalty parameter to control the violation, which is represented by the slack variables $\boldsymbol{\xi} = [\xi_1, \cdots, \xi_N]^{\mathrm{T}}$. By using the kernel trick (Scholkopf and Smola, 2001), SVM can produce nonlinear decision boundaries for robust classification.

We have applied a binary SVM classifier to automated myonucleus detection in isolated single muscle fiber fluorescence images (Su et al., 2014, 2013). The framework is shown in Fig. 4.1, which mainly consists of four steps: image preprocessing, robust ellipse fitting, SVM-based ellipse refinement, and mean-shift clustering based on geodesic inner distance. It first fuses a set of z-stack images with a normalized linear combination and extracts foreground edges by applying Otsu's method (Otsu, 1979) to the fused image, then fits a sufficient number of ellipse hypotheses using heteroscedastic errors-in-variable (HEIV) regression (Matei and Meer, 2000), next uses an SVM classifier with a set of sophisticated design features to select the good candidate fittings, and finally applies inner geodesic distance-based clustering to ellipse localization, which determines the locations of myonuclear centers.

4.2.1.1 Image preprocessing

In z-stack imaging, the microscope collects lights from both in-focus and out-of-focus planes. Therefore each image consists of in-focus objects as well as out-of-focus objects. Motivated by Hariharan et al. (2007), we have proposed a modified image fusion algorithm to compose a smooth multifocus image from the z-stack images. Let $i = 1, 2, \ldots, M$ denote the z-stack images. The gradient magnitude of each image is calculated as

$$Z_i(x, y) = \sqrt{I^2_{X(i)}(x, y) + I^2_{Y(i)}(x, y)}, \tag{4.2}$$

where $I^2_{X(i)}(x, y)$ and $I^2_{Y(i)}(x, y)$ are the horizontal and vertical gradients of the ith image, respectively. Different from Hariharan et al. (2007), which picks the pixel $I_i(x, y) = \text{argmax}_i Z_i(x, y)$ as the synthesized image intensity $I_s(x, y)$, we linearly combine all the pixel values in the z-stack to calculate $I_i(x, y)$:

$$g_i(x, y) = \frac{Z_i(x, y)^k}{\sum_{i=1}^{N} Z_i(x, y)^k}, \quad k = 3, \tag{4.3}$$

$$I_s(x, y) = \sum_{i=1}^{M} g_i(x, y) I_i(x, y). \tag{4.4}$$

Our image fusion algorithm can provide the final composed image with smooth boundaries, which greatly increase the robustness of the subsequence automatic detection procedure. We apply Otsu's method (Otsu, 1979) to the fused image for binarization and then contour extraction of myonuclei or myonucleus clumps. Thereafter we use an elliptical Fourier descriptor (Kuhl and Giardina, 1982) to smooth the contours by keeping the first 20 Fourier coefficients, and then exploit a concavity detection algorithm (Yang et al., 2008) to detect concave points, which will be used for subsequent ellipse fitting.

4.2.1.2 Robust ellipse fitting

Based on the detected concave points, we adopt the robust HEIV regression to generate fitting ellipses. Compared with some other traditional methods, HEIV has a weaker dependence on initialization and a faster convergence. The direct least-squares (DLS) method (Fitzgibbon et al., 1999) is biased when the input data points are a short low-curvature segment of a whole ellipse, the geometric distance minimization algorithm (Zhang, 1997) is sensitive to the initialization, and Taubin's method does not provide robust fitting either when only one short curve segment is available (Taubin, 1991). In our approach, an ellipse is modeled using an errors-in-variables model with Gaussian-distributed noise, which can be iteratively solved by reformulating it as a generalized eigenvalue problem.

In order to segment the touching myonuclei, robust ellipse fitting based on the HEIV regression model is performed on the contour pixels. Suppose that there are n concave points for a contour; the original object contour is intersected by these concave points into n segments. We fit candidate ellipses using the different combinations of the contour segments. Thus there are $\binom{n}{k}$ segment combinations if we choose k segments to fit one ellipse. We empirically set $k = \{1, 2, 3\}$ to guarantee that a sufficient number of ellipses using the HEIV regression are generated as input during the subsequent refinement procedure.

4.2.1.3 SVM-based ellipse refinement

Given the candidate ellipses calculated from Section 4.2.1.2, we propose to refine these ellipses and remove the false fittings using a supervised learning technique, an SVM classifier. In order to conduct SVM training and testing, we specifically design several groups of features for ellipse representation. In total we have extracted 42 features for each ellipse hypothesis, and they are summarized as follows.

Group 1

The most intuitive geometric feature to evaluate the ellipse fitting is to measure whether or not the fitting ellipses match with the boundaries of myonuclei. A set of morphological ratios is defined for this type of measurements. Given the fitting ellipse e and the myonucleus contour c, two overlapping ratios are calculated: $r_{aoc} = \frac{A_o}{A_c}$ and $r_{aoe} = \frac{A_o}{A_e}$, where A_c and A_e represent the region areas inside e and c, respectively, and A_o denotes the area of the overlapping region inside both e and c. In addition, the ratio $r_{oace} = \frac{A_o}{A_c+A_e}$ is also calculated.

Considering robustness, the pixel-wise overlapping ratios are measured as well. Let p_o denote the number of overlapping pixels between the fitting ellipse and the contour, p_c represent the number of contour pixels, and $r_{oc} = \frac{p_o}{p_c}$ be the contour pixel-level overlapping ratio. Based on r_{oc}, we design an iterative procedure to assign a match-quality score to each fitting ellipse. In each iteration, we count the contour pixels overlapping with an ellipse, and then sort the ellipses with respect to the number of their overlapping contour pixels. Within each iteration, the contour pixels that overlap with the highest ranked ellipse are removed, and r_{oc} is updated for each remaining ellipse. This iteration terminates when there is a small fraction (decided by a threshold) of contour pixels left. Note that, in each iteration, the ellipse that has the most overlapping pixels with the object contour will be assigned the highest rank. The match-quality of an ellipse is defined as the rank of this fitting ellipse calculated during each iteration. To improve efficiency, we only record the ranks of an ellipse in the first three iterations.

Group 2

The myonuclei are objects with certain biologically meaningful areas. Therefore the ellipse area A_e, the axis ratio (long-to-short) r_{axis}, and the perimeter p_c and the area A_c of the myonucleus contour are also considered as potential geometric features for classification.

Group 3

Concave point depth is a feature designed to measure the distance between concave points and fitting ellipses. This feature design is based on the observation that an accurate ellipse fitting should not have a concave point deeply inside the ellipse. For an ellipse e and a set of concave points q_j ($j = 1, \ldots, M$), the concave point depth d is defined as the sum of squares of the Euclidean distances from the concave points to ellipse e:

$$d = \sum_j dist^2(q_j, e),\ j \in \{j:\ q_j \text{ is in ellipse } e\}, \tag{4.5}$$

where $dist(\cdot)$ denotes the Euclidean distance from q_j to e.

Because the center of an accurate fitting ellipse should not be located near the myonucleus boundary, the distance between the ellipse center and the myonucleus boundary d_{ecc} is also calculated. This feature can help to remove suboptimal fitting ellipses whose centers are close to the object boundaries.

Group 4

The irregularity of the boundaries is defined as $r_{irg} = \frac{n_c}{p_c}$, where p_c denotes the perimeter of a myonucleus contour and n_c represents the number of concave points detected.

Group 5

A set of statistical features is computed to capture the relationships among the ellipses generated from the same contour. Assuming N ellipse fitting candidates are generated from the segments of an object contour, and $f_i, i = (1, 2, \ldots, N)$ represents one specific feature calculated for the ith ellipse e_i, we can generate the following statistics:

$$f_{1i} = f_i - \frac{1}{N}\sum_{i=1}^{N} f_i, \tag{4.6}$$

$$f_{2i} = f_i - \mathbf{median}(f_1, f_2, \ldots, f_N), \tag{4.7}$$

$$f_{3i} = f_i - \mathbf{max}(f_1, f_2, \ldots, f_N), \tag{4.8}$$

$$f_{4i} = f_i - \mathbf{min}(f_1, f_2, \ldots, f_N), \tag{4.9}$$

where the functions **median**$(\cdot)$, **max**$(\cdot)$, and **min**$(\cdot)$ compute the median, maximum, and minimum value of the input feature vector, respectively. These statistical features capture the inter-group variance between one fitting ellipse and the whole group of fitting ellipses for the same object contour. These statistical features are calculated for each feature only in Groups 1 and 3.

Group 6

The following two features are also considered in the classification algorithm: (1) distance from the fitting ellipse centroid to the object centroid and (2) average distance from one ellipse centroid to the centroids of all the other fitting ellipses for the same object contour.

With these specifically designed features, we train a binary SVM classifier to refine the generated candidate ellipses. The training data are manually annotated, with correct fitting ellipses as positive and false fitting as negative. For classification improvement, we can select only the most representative ellipses and the discriminative features (Su et al., 2014). During the testing stage, many false fitting ellipses

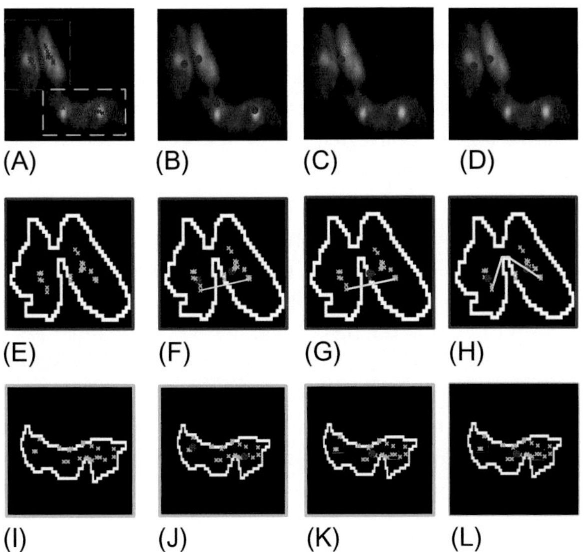

FIG. 4.2

The demonstration of the effectiveness of mean-shift clustering based on inner distance. Gold-standard myonuclei detection should identify two touching nuclei on the top and one single nuclei at the bottom (in total three myonuclei). This is a challenging problem because of the bright dots irregularly distributed inside the myonuclei, which is caused by the DAPI intercalating preferentially into heterochromatin. (A) The initial fitting ellipses' centers generated by the SVM classifier-based refinement. (B) The clustering results (red dots) based on Euclidean distance with bandwidth $bwd_C = 10$. (C) The clustering based on Euclidean distance with bandwidth $bwd_C = 14$. As presented in (B) and (C), it is difficult to find a unified bandwidth that can produce accurate detection results for both cases. However, using inner distance, a unified bandwidth ($bwd_C = 14$) can be used and the accurate detection results for both cases are shown in (D). (E)–(H) Edge images overlaid with SVM refinement results (green crosses) and clustering results (red dots) for the top-left patch in (A). A similar illustration is provided in (I)–(L) for the bottom-right patch in (A). The lines in (F) and (G) denote Euclidean distances between two sample seeds, and the line in (H) denotes an inner distance between two sample seeds.

can be removed, but multiple candidates still remain for a gold-standard myonucleus. Therefore postprocessing is required to achieve final myonucleus detection.

4.2.1.4 Inner geodesic distance-based clustering

In order to remove redundant ellipses, which crowd around a single myonucleus, we apply mean-shift clustering to the candidates after SVM refinement, as shown in Fig. 4.2. Instead of relying on the Euclidean distance, we use the inner geodesic distance for clustering to find the final myonucleus centers. Unlike the Euclidean distance, the inner distance is calculated as the length of the shortest connecting paths that only lay inside the contour. It builds a graph with the centers of candidate ellipses and the concave points as vertices and the links connecting these vertices inside the contour as edges, and then runs a shortest distance algorithm in the graph. Since the inputs of the mean-shift clustering are the coordinates of the points, the inner-distance matrix will be converted into a new coordinate system by harboring the origin at one of the two points and calculating the other relevant distances.

Inner distance is intuitively correct because the real distance between two seeds should be the paths within the object, instead of a direct line that might cross the cell boundaries (Yang et al., 2008). As shown in Fig. 4.2A, the Euclidean-distance based clustering will encounter some significant challenges in selecting proper clustering bandwidth (Fig. 4.2B or C). On the other hand, based on the inner geodesic distance, we can obtain correct clustering results using one unified bandwidth, as shown in Fig. 4.2D. Inner distance is proven to be quite effective in natural shape classification (Ling and Jacobs, 2007). The myonucleus detection using the proposed method on nine sample fused images is shown in Fig. 4.3, which demonstrates that hundreds of myonuclei are correctly located. We compare the proposed framework with three recent state-of-the-art cell detection methods: Laplacian-of-Gaussian filters (LoG)

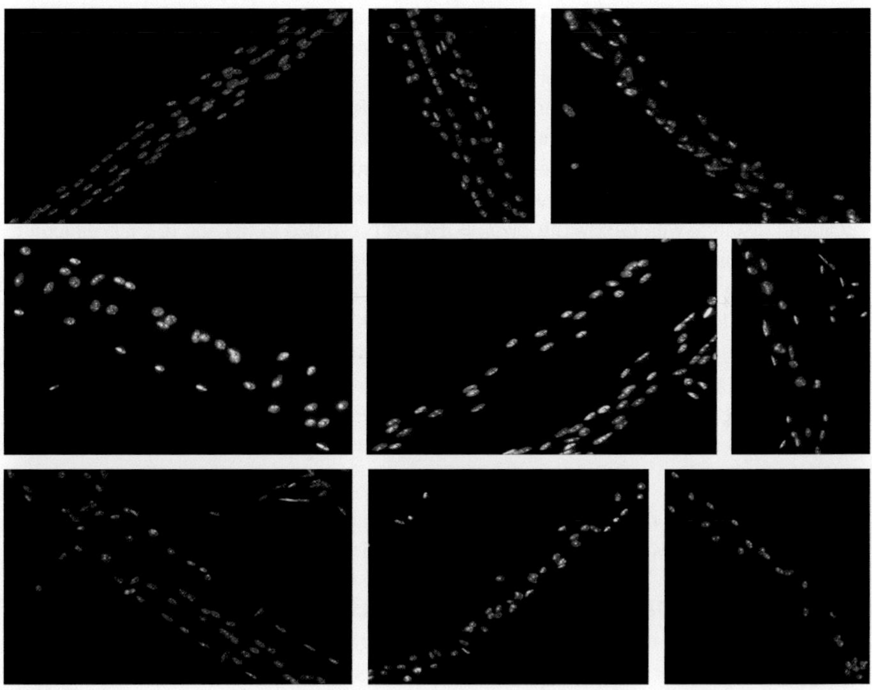

FIG. 4.3

The automated myonucleus detection results on nine randomly selected image patches.

Table 4.1 The Pixel-Wise Seed Detection Accuracy Compared With the Gold Standard

	Mean	Standard Deviation	Min	Max
LoG (Al-Kofahi et al., 2010)	3.72	4.22	0.55	7.99
IRV (Parvin et al., 2007)	4.28	3.86	0.24	**7.93**
SPV (Qi et al., 2012)	2.99	3.54	**0**	7.96
Proposed	**2.6**	**2.90**	0.17	**7.93**

(Al-Kofahi et al., 2010), iterative radial voting (IRV) (Parvin et al., 2007), and single-pass voting (SPV) (Qi et al., 2012). The pixel-wise detection accuracy using over 500 multifocus *z*-stack images (over 1500 myonuclei) is listed in Table 4.1. The best performance in terms of each metric is highlighted in bold. It is clear that the proposed method produces the best accuracy with respect to both mean errors and standard deviations, representing high detection accuracy and reliability.

4.2.2 DEEP CONVOLUTIONAL NEURAL NETWORK

Recently, deep learning-based models, especially CNN, have achieved outstanding performance in both natural (LeCun et al., 2015; Krizhevsky et al., 2012; Farabet et al., 2013) and medical (Liao et al., 2013; Wu et al., 2013; Cireşan et al., 2013) image analyses. In contrast to the conventional supervised learning techniques (eg, SVM) relying on hand-crafted features that need sophisticated design, CNN can automatically learn multilevel hierarchies of features that are invariant to irrelevant variations of samples while preserving relevant information (LeCun et al., 2010). Usually a CNN consists of successive pairs of convolutional and pooling layers, followed by several fully connected layers (LeCun et al., 1998). A convolutional layer learns a set of convolutional filters that will be used to calculate output feature maps, with all units in a feature map sharing the same weights. A max-pooling layer summarizes the activities and picks up the max values over a neighborhood region in each feature map (Hinton et al., 2012), which not only reduces feature dimensionality but also introduces local shift and translation invariance to the neural network. The convolutional-pooling layers are stacked to learn local hierarchical features, based on which the fully connected layers learn more higher level feature representation for classification. The last layer is often a sigmoid layer producing the probability distributions over categories (Krizhevsky et al., 2012; Farabet et al., 2013).

Instead of solving a pixel-wise classification problem, we have proposed a novel CNN-based structured regression model for robust nucleus and cell detection on breast cancer, pancreatic neuroendocrine tumor (NET), and HeLa cervical cancer microscopic images (Xie et al., 2015). We modify the conventional CNN by replacing the last layer (classifier) with a structured regression layer to encode topological information. Meanwhile, instead of working on the label space, regression on the proposed structured proximity space for patches is performed so that the centers of

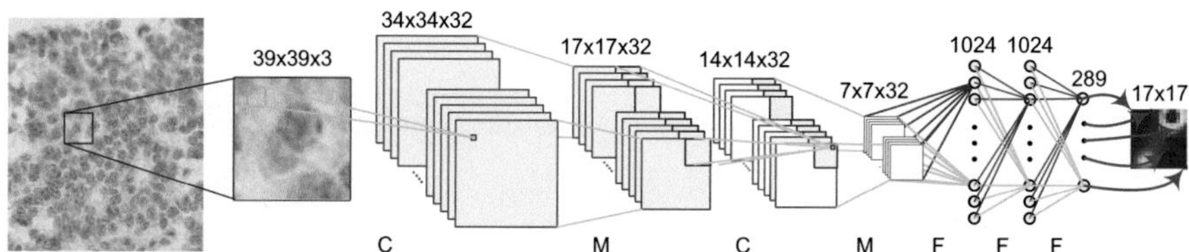

FIG. 4.4

The CNN architecture used in the proposed structured regression model on the NET dataset. C, M, and F represent the convolutional layer, max-pooling layer, and fully connected layer, respectively. The arrows from the last layer illustrate the mapping from the final layer's outputs to the proximity patch.

image patches are explicitly forced to get higher values than their neighbors. These proximity patches are then gathered from all testing image patches and fused to obtain the final proximity map, where the maximum positions indicate the object centroids. The proposed model, whose architecture is listed in Fig. 4.4, is able to handle touching cells, inhomogeneous background noises, and large variations in sizes and shapes, and exhibits superior performance over existing state-of-the-art methods.

4.2.2.1 CNN-based structured regression

Let $\mathcal{X}$ denote the patch space, which consists of a set of $d \times d \times c$ local image patches extracted from c-channel images I. It defines $\mathcal{M}$ as the proximity mask corresponding to image I, and computes the ijth entry in $\mathcal{M}$ as

$$\mathcal{M}_{ij} = \begin{cases} \frac{1}{1+\alpha D(i,j)} & \text{if } D(i,j) \leq r, \\ 0 & \text{otherwise,} \end{cases} \tag{4.10}$$

where $D(i,j)$ represents the Euclidean distance from pixel (i,j) to the manually annotated nucleus or cell center, r is a distance threshold (selected as 5 pixels), and α is the decay ratio (set as 0.8). An image patch $x \in \mathcal{X}$ centered at (u, v) is represented by a quintuple $\{u, v, d, c, I\}$, with the corresponding proximity patch $s \in \mathcal{V}^{d' \times d'}$ that can be represented as $\{u, v, d', \mathcal{M}\}$ (d' is not necessarily equal to d). We are given training data $\{(\boldsymbol{x}^i, \boldsymbol{y}^i) \in (\mathcal{X}, \mathcal{Y})\}_{i=1}^{\mathcal{N}}$, where $\mathcal{Y} \subset \mathcal{V}^{p \times 1}$ represents the output space of the structured regression model, and $p = d' \times d'$ denotes the number of units in the last layer. The outputs are $\boldsymbol{y}^i = \Gamma(\boldsymbol{s}^i)$, where $\Gamma : \mathcal{V}^{d' \times d'} \to \mathcal{Y}$ is a mapping function to represent the vectorization operation in column-wise order for a proximity patch. Define $\{\boldsymbol{\theta}_l\}_{l=1}^{L}$ as the parameters corresponding to each of the L layers, the training process of the structured regression model can be formulated as learning a mapping function F represented by $\{\boldsymbol{\theta}_l\}_{l=1}^{L}$, which will map the image space $\mathcal{X}$ to the output space $\mathcal{Y}$. Therefore the optimization problem can be formulated as

$$\arg\min_{\boldsymbol{\theta}_1,\ldots,\boldsymbol{\theta}_L} \frac{1}{\mathcal{N}} \sum_{i=1}^{\mathcal{N}} \mathcal{L}(\psi(\boldsymbol{x}^i;\boldsymbol{\theta}_1,\ldots,\boldsymbol{\theta}_L),\boldsymbol{y}^i), \tag{4.11}$$

where $\mathcal{L}$ is the user-defined loss function. For training sample $(\boldsymbol{x}^i,\boldsymbol{y}^i)$, it is defined as

$$\begin{aligned}\mathcal{L}(\psi(\boldsymbol{x}^i;\boldsymbol{\theta}_1,\ldots,\boldsymbol{\theta}_L),\boldsymbol{y}^i) = \mathcal{L}(\boldsymbol{o}^i,\boldsymbol{y}^i) &= \frac{1}{2}\sum_{j=1}^{p}(y^i_j+\lambda)(y^i_j-o^i_j)^2 \\ &= \frac{1}{2}\left\| (Diag(\boldsymbol{y}^i)+\lambda\mathbf{I})^{1/2}(\boldsymbol{y}^i-\boldsymbol{o}^i)\right\|_2^2,\end{aligned} \tag{4.12}$$

where $\mathbf{I}$ is an identity matrix of size $p \times p$, $\boldsymbol{o}^i = \psi(\boldsymbol{x}^i;\boldsymbol{\theta}_1,\ldots,\boldsymbol{\theta}_L)$ represents the output of the last layer, and $Diag(\boldsymbol{y}^i)$ denotes a diagonal matrix with the jth diagonal element equal to y^i_j.

Eq. (4.11) can be solved using the classical back-propagation algorithm. Since the nonzero region in the proximity patch is relatively small, our model might return a trivial solution. To alleviate this problem, we adopt a weighting strategy (Szegedy et al., 2013) to give more weights to the loss coming from the network's outputs corresponding to the nonzero area in the proximity patch. A small λ indicates strong penalization that is applied to errors coming from the outputs with low proximity values in the training data. Our model is different from Szegedy et al. (2013) which applies a bounding box mask regression approach on the entire image.

We choose the sigmoid activation function in the last layer, that is, $o^i_j = sigm(a^i_j)$ (a^i_j is the jth element of $\boldsymbol{a}^i$ representing the input of the last layer). The partial derivative of (4.12) with respect to the input of the jth unit in the last layer is given by

$$\frac{\partial\mathcal{L}(\boldsymbol{o}^i,\boldsymbol{y}^i)}{\partial a^i_j} = \frac{\partial\mathcal{L}(\boldsymbol{o}^i,\boldsymbol{y}^i)}{\partial o^i_j}\frac{\partial o^i_j}{\partial a^i_j} = (y^i_j+\lambda)(o^i_j-y^i_j)a^i_j(1-a^i_j). \tag{4.13}$$

Based on (4.13), we can evaluate the gradients of (4.11) with respect to the model's parameters in the same way as in LeCun et al. (1998). The optimization is conducted with the mini-batch stochastic gradient descent.

4.2.2.2 CNN architecture

The proposed structured regression model contains several convolutional layers (C), max-pooling layers (M), and fully connected layers (F). Fig. 4.4 illustrates the architectures and mapped proximity patches in the proposed model on the NET dataset. The detailed model configuration is: Input(39 × 39 × 3) − C(34 × 34 × 32) − M(17 × 17 × 32) − C(14 × 14 × 32) − M(7 × 7 × 32) − F(1024) − F(1024) − F(289). The input image size depends on cell scales, and a 39 × 39 patch is large enough to cover a single cell in NET images. Due to the small size of the input image patch, it is sufficient to stack two pairs of C-M layers for feature computation. Meanwhile, multiple F layers are designed to learn more higher level feature representation, which can benefit the final regression. The activation function of the last F (regression) layer is chosen as the sigmoid function, and an ReLu function is used for all the other F and C layers. The sizes of C and M layers are defined as *width* × *height* × *depth*, where *width* × *height* determines the dimensionality of each feature map and *depth* represents the number of feature maps. Since the input image size is relatively small, the filter size is chosen as 6 × 6 for the first convolutional layer and 3 × 3 for the other. The max-pooling layer uses a window of size 2 × 2 with a stride of 2, which has been widely adopted in current object detection algorithms and gives an encouraging performance. Similar CNN architectures are used for breast cancer and HeLa cervical cancer datasets, but with input patch sizes of 49 × 49 × 3 and 31 × 31 × 3, respectively.

4.2.2.3 Structured prediction fusion and cell localization

Given a testing image patch $x = (u, v, d, c, I)$, it is easy to get the corresponding proximity mask as $s = \Gamma^{-1}(\mathbf{y})$, where $\mathbf{y} \in \mathcal{Y}$ represents the model's output corresponding to $\boldsymbol{x}$. In the fusion process, s will cast a proximity value for every pixel that lies in the $d' \times d'$ neighborhood area of (u, v). For example, pixel $(u + i, v + j)$ in image I will get a prediction s_{ij} from pixel (u, v). In other words, each pixel actually receives $d' \times d'$ predictions from its neighboring pixels. To get the fused proximity map, we average all the predictions from its neighbors for each pixel to calculate its final proximity prediction. After this step, the cell localization can be easily obtained by finding the local maximum positions in the average proximity map. In order to reduce the running time of testing, we present a striding strategy for speed improvement. This is based on the observation that our model generates a $d' \times d'$ proximity patch for each testing patch, and it is feasible to skip a lot of pixels and only test the image patches at a certain stride ss $(1 \leq ss \leq d')$ without significantly sacrificing the accuracy.

4.2.2.4 Experimental results

The proposed model is evaluated on three datasets: 32 breast cancer, 60 pancreatic NET, and 22 phase-contrast HeLa cervical cancer images. All of the data are randomly split into halves for training and testing. For quantitative analysis, we define the gold-standard areas as circular regions within 5 pixels of each annotated cell center. A detected cell centroid is considered to be a true positive (*TP*) only if it lies within the gold-standard areas; otherwise, it is considered as a false positive (*FP*). Each *TP* is matched with the nearest ground-truth annotated cell center. The gold-standard cell centers that are not matched by any detected results are considered to be false negatives (*FN*). Based on these definitions, we can compute the precision (*P*), recall (*R*), and F_1-score as $P = \frac{TP}{TP+FP}$, $R = \frac{TP}{TP+FN}$, and $F_1 = \frac{2PR}{P+R}$, respectively.

We compare the proposed structured regression model (SR-1 means testing with a stride of 1 pixel) with four state-of-the-art methods, including nonoverlapping extremal regions selection (NERS) (Arteta et al., 2012), iterative radial voting (IRV) (Parvin et al., 2007), Laplacian-of-Gaussian filtering (LoG) (Al-Kofahi et al., 2010), and image-based tool for counting nuclei (ITCN) (Byun et al., 2006). In addition to precision, recall, and F_1-score, we also compute the mean and standard deviation of two terms: (1) the absolute difference $\mathbf{E_n}$ between the number of true positive and ground-truth annotations, and (2) the Euclidean distance $\mathbf{E_d}$ between the true positive and the corresponding annotations. The quantitative experimental results are reported in Table 4.2. It is obvious that our method provides better performance than the other methods in all three data sets, especially in terms of F_1-score. Our method also shows strong reliability with the lowest mean and standard deviations in $\mathbf{E_n}$ and $\mathbf{E_d}$ on NET and phase contrast data sets.

Table 4.2 The Comparative Cell Detection Results on Three Data Sets

Data	Methods	P	R	F_1	$\mu_d \pm \sigma_d$	$\mu_n \pm \sigma_n$
Breast cancer	SR-1	**0.919**	0.909	**0.913**	$\mathbf{3.151 \pm 2.049}$	4.8750 ± 2.553
	NERS (Arteta et al., 2012)	–	–	–	–	–
	IRV (Parvin et al., 2007)	0.488	0.827	0.591	5.817 ± 3.509	9.625 ± 4.47
	LoG (Al-Kofahi et al., 2010)	0.264	**0.95**	0.398	7.288 ± 3.428	$\mathbf{2.75 \pm 2.236}$
	ITCN (Byun et al., 2006)	0.519	0.528	0.505	7.569 ± 4.277	26.188 ± 8.256
NET	SR-1	0.864	**0.958**	**0.906**	$\mathbf{1.885 \pm 1.275}$	$\mathbf{8.033 \pm 10.956}$
	NERS (Arteta et al., 2012)	**0.927**	0.648	0.748	2.689 ± 2.329	32.367 ± 49.697
	IRV (Parvin et al., 2007)	0.872	0.704	0.759	2.108 ± 3.071	15.4 ± 14.483
	LoG (Al-Kofahi et al., 2010)	0.83	0.866	0.842	3.165 ± 2.029	11.533 ± 21.782
	ITCN (Byun et al., 2006)	0.797	0.649	0.701	3.643 ± 2.084	24.433 ± 40.82
Cervical cancer	SR-1	**0.942**	**0.972**	**0.957**	$\mathbf{2.069 \pm 1.222}$	$\mathbf{3.455 \pm 4.547}$
	NERS (Arteta et al., 2012)	0.934	0.901	0.916	2.174 ± 1.299	11.273 ± 11.706
	IRV (Parvin et al., 2007)	0.753	0.438	0.541	2.705 ± 1.416	58.818 ± 40.865
	LoG (Al-Kofahi et al., 2010)	0.615	0.689	0.649	3.257 ± 1.436	29.818 ± 16.497
	ITCN (Byun et al., 2006)	0.625	0.277	0.371	2.565 ± 1.428	73.727 ± 41.867

μ_d, σ_d represent the mean and standard deviation of $\mathbf{E_d}$ and μ_n, σ_n represent the mean and standard deviation of $\mathbf{E_n}$.

4.3 SEGMENTATION

Unlike nucleus or cell detection that only needs to locate the object centroids, segmentation aims to separate individual nuclei or cells by delineating their boundaries. It is a prerequisite for many quantitative image analyses including cellular characteristic description (eg, size, shape, texture, and other imagenomics), and thus might provide diagnosis and prognosis support for improved characterization and personalized treatment. However, it is not easy to achieve robust automated nucleus or cell segmentation. First, there usually exist a lot of noise in microscopic images, especially histopathological images; second, nuclei or cells exhibit significant scale and intracellular intensity variations; finally, many nuclei or cells are clustered or even partially overlap with each other such that no clear boundaries exist.

Many state-of-the-art approaches have been applied to nucleus or cell segmentation on specific medical images. The watershed transform was perhaps the most popular in the early days, but it is prone to oversegmentation and requires region merging (Lin et al., 2003; Wählby et al., 2002) or marker controlling (Zhou et al., 2005; Adiga et al., 2006). The graph cut framework (Chang et al., 2012; Lou et al., 2012) or Voronoi diagrams (Jones et al., 2005) are also used for cell segmentation, but they usually need high time cost for large-scale images or might fail to handle weak cell boundaries. Another type of widely used methods are deformable models (Xing and Yang, 2013; Qi et al., 2012), but the level set-based implementation (Ali and Madabhushi, 2012; Yan et al., 2008) is usually computationally expensive and probably produces undesired object topology changes. In addition, it often needs experts to carefully tune the parameters. On the contrary, the learning techniques (Kong et al., 2011; Janssens et al., 2013) do not require sophisticated parameter selection or predefined processing. In this section, we will introduce two machine learning-based techniques, random forests and sparsity-based dictionary learning, for nucleus or cell segmentation in microscopic images.

4.3.1 RANDOM FORESTS

A random forest (Breiman, 2001) $\mathcal{F} = \{T_t\}$ is an ensemble of decision trees T_t, which are trained independently on randomly selected samples $S = \{s_i = (\boldsymbol{x}_i \in \mathcal{X}, y_i \in \mathcal{Y})\}$, where $\mathcal{X}$ and $\mathcal{Y}$ denote the input features and output labels, respectively. A decision tree $T_t(\boldsymbol{x})$ produces the prediction results by recursively branching a feature sample $\boldsymbol{x} \in \mathcal{X}$ left or right down the tree until a leaf node is reached. For a decision forest, the predictions $T_t(\boldsymbol{x})$ from individual trees are combined together using an ensemble model. Majority voting and averaging are typical ensemble choices for classification and regression problems, respectively.

During the training of a decision tree, at each node n, a split function $h(\boldsymbol{x}, \boldsymbol{\theta}_n)$ is chosen to split the samples S_n into left S_n^L or right S_n^R. The split function $h(\boldsymbol{x}, \boldsymbol{\theta}_n)$ is optimized by maximizing the information gain:

$$\mathcal{I}(S_n) = H(S_n) - \left(\frac{|S_n^L|}{|S_n|} H(S_n^L) + \frac{|S_n^R|}{|S_n|} H(S_n^R) \right), \tag{4.14}$$

where $H(\cdot)$ is the class entropy function. The split function $h(\boldsymbol{x}, \boldsymbol{\theta}_n)$ can be an arbitrary classifier. A common choice is a stump function that is found to be computationally efficient and effective in practice (Dollar and Zitnick, 2014). The training procedure continues to split the samples until either a maximum depth is reached, or too few samples are left, or information gain falls below a certain threshold.

We have proposed a structured edge detection algorithm based on random forests for muscle cell segmentation on hematoxylin and eosin (H&E)-stained microscopic images (Liu et al., 2015). The structured edge detection, which can better capture inherent muscle image edge structures, is achieved by extending a random decision forest framework. It is noted in Kontschieder et al. (2011) and Dollar and Zitnick (2014) that, by storing structure information instead of class probabilities at the leaf nodes of the random decision trees, random decision forests can be conveniently used for structured learning. Therefore edge masks rather than edge probability values will be stored at the leaf nodes in our proposed structured edge detection algorithm. In order to accurately segment each muscle cell exhibiting both strong and weak boundaries, a hierarchical segmentation method is proposed, which takes a set of partitions produced by using a segmentation algorithm with varying parameters as inputs, and selects a best subset of nonoverlapping partition regions as the final results.

An overview of our proposed muscle image segmentation algorithm is shown in Fig. 4.5. Given an image patch, (1) an edge map is generated by the proposed structured edge detection algorithm; (2) an Ultrametric Contour Map (UCM) (Arbelaez et al., 2011) is constructed and a set of segmentation candidates is generated by adjusting the thresholds of UCM; and (3) an efficient dynamic programming-based subset selection algorithm is then used to choose the best regions for muscle image segmentation based on a constructed tree graph.

4.3.1.1 Structured edge detection

Since a decision tree classifier generates the actual prediction at the leaf nodes, more information (instead of only class likelihoods) can be stored at the leaf nodes. For example, in Kontschieder et al. (2011), structured class label information is stored at leaf nodes for semantic image segmentation. Similar to Dollar and Zitnick (2014) and Chen et al. (2015), we have stored edge structure information at the leaf nodes for structured muscle image edge detection. Different from traditional edge detection algorithms (Arbelaez et al., 2011), which take an image patch x as an input and compute the probability of the edge existence at the center pixel p, the output of our proposed structured edge detection algorithm is an edge mask around the central pixel p instead of the likelihood value. After the decision tree is learned, the median or mean of the edge masks sent to the leaf node will be stored as the leaf node output, as shown in Fig. 4.5.

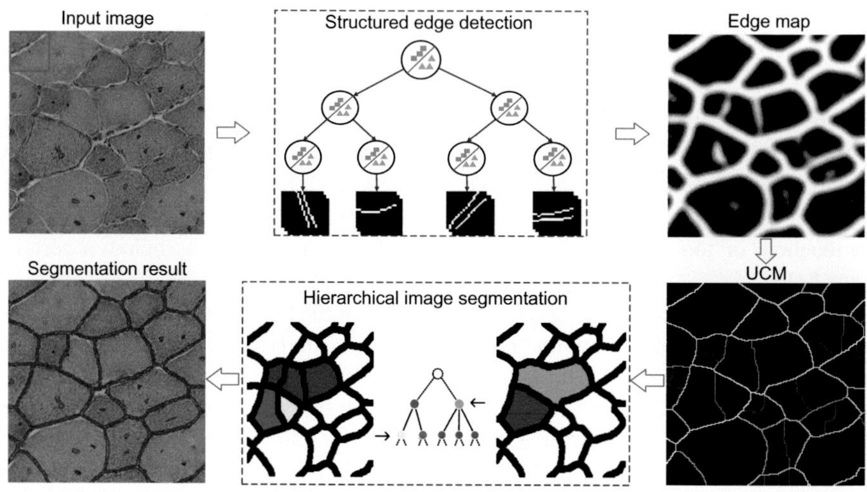

FIG. 4.5

An overview of the proposed muscle image segmentation algorithm.

The information gain criterion in Eq. (4.14) is effective in practice for decision tree training. In order to follow this criterion, the edge masks must be explicitly assigned proper class labels at each internal node of the tree during the training stage. One straightforward idea is to group the edge masks at a node into several clusters by an unsupervised clustering algorithm such as k-means or mean-shift (Comaniciu and Meer, 2002), and then treat each cluster *id* as the class label for the sample belonging to that cluster. However, the edge masks $\mathcal{Y}$ do not reside in the Euclidean space so that direct grouping may not generate desired results. In addition, clustering in a high-dimension space ($y \in \mathbb{R}^{256\times 1}$ for a 16×16 edge mask) is computationally expensive. To address this problem, we propose to reduce the high-dimension edge masks $\mathcal{Y} \in \mathbb{R}^n$ to a lower dimensional subspace $\mathcal{Z} \in \mathbb{R}^m$ ($m << n$) using an autoencoder (Hinton and Salakhutdinov, 2006) before clustering the edge masks. For notation convenience, we use the matrix form and vector form of edge mask space $\mathcal{Y}$ interchangeably in this section.

Although the transformed data $z \in \mathcal{Z}$ is used to choose a split function $h(\boldsymbol{x}, \boldsymbol{\theta}_n)$ during the training of the decision tree, only the original edge masks are stored at leaf nodes for the prediction. Several sample edge masks learned and stored at the leaf nodes are shown in Fig. 4.6. As one can tell, many edge structures are unique for muscle cell boundaries, which demonstrates the effectiveness of the structured edge detection procedure. The proposed structured edge detection algorithm takes a 32×32 image patch as input and generates a 16×16 edge mask around the input's center pixel. The image patch is represented with the same high-dimensional feature used in Dollar and Zitnick (2014) and Arbelaez et al. (2014), and it is effective and computationally efficient. In total, two million samples are randomly generated to

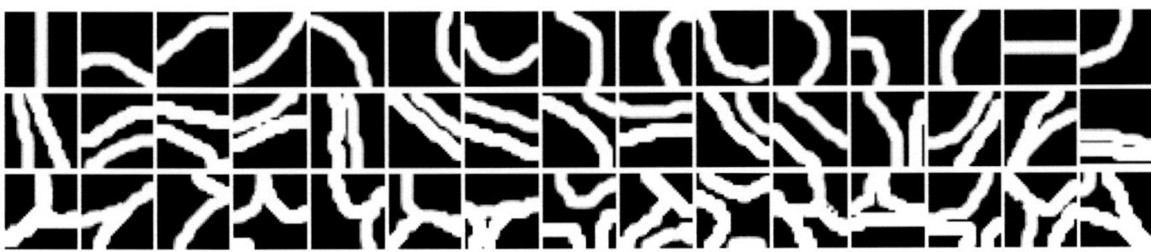

FIG. 4.6

Several sample edge masks learned and stored at the leaf nodes of the random decision trees.

train the structured decision random forest, which consists of eight decision trees. The autoencoder model used in our work consists of an encoder with layers of sizes $(16 \times 16) - 512 - 256 - 30$ and a symmetric decoder. The autoencoder model is trained once offline and applied to all decision trees, and the data compression is only performed at the root node.

4.3.1.2 Hierarchical image segmentation

Recently, the hierarchical strategy has been successfully applied to image segmentation (Farabet et al., 2013; Uzunbaş et al., 2014). In general, the hierarchical image segmentation consists of two steps: candidate region generation and selection. Specifically, a collection of segmentation candidates is first generated by running some existing segmentation algorithms with different parameters. Usually, an undirected graph is constructed from these partition candidates, in which an edge exists between two touching or overlapping regions. Next, based on some domain-specific criteria, a subset of nonoverlapping regions is selected as the final segmentation results. For example, Felzenszwalb's method (Felzenszwalb and Huttenlocher, 2004) with multiple levels is used to generate the segmentation candidate pool, and an optimal purity cover algorithm (Farabet et al., 2013) can be adopted to select the most representative regions. In Uzunbaş et al. (2014), the watershed segmentation method with different thresholds gives a collection of partitions, and then a conditional random field (CRF)-based learning algorithm is utilized to find the best ensembles as final segmentation.

In our implementation, an UCM (Arbelaez et al., 2011; Arbelaez, 2006), which defines a duality between closed, nonself-intersecting weighted contours and a hierarchy of regions, is used to generate a pool of segmentation candidates. Because of the nice property of UCM where the segmentation results using different thresholds are nested into one another, we can construct a tree graph for this pool of segmentation candidates. The final step is to solve this tree graph-based problem using dynamic programming.

Given a set of segmentation candidates generated with different thresholds using UCM, an undirected and weighted tree graph, $G = (V, E, \mathbf{w})$, is constructed, where $V = \{v_i, i = 1, 2, \ldots, n\}$ represents the nodes with each v_i corresponding to a segmented region S_i. E denotes the edges of the graph. The $w(v_i)$ is learned via a general random decision forest classifier to represent the likelihood of S_i as a real muscle cell. An adjacent matrix $A = \{a_{ij} | i,j = 1, \ldots, n\}$ is then built with $a_{ij} = 1$ if $S_i \subset S_j$ or $S_j \subset S_i$, and otherwise 0. Denote $\mathbf{x} \in \{0,1\}^n$ the indicator vector, where its element is equal to 1 if the corresponding node is selected, otherwise 0. Finally, the constrained subset selection problem is formulated as

$$\mathbf{x}^* = \arg\max_{\mathbf{x} \in \mathcal{X}} \mathbf{w}^{\mathrm{T}} \mathbf{x}, \quad (4.15)$$

where $\mathcal{X}$ denotes all possible valid configurations of $\mathbf{x}$. Considering the special tree graph structure, we can efficiently solve (4.15) via the dynamic programming approach with a bottom-up strategy.

In order to ensure that solving (4.15) selects the desired regions, each candidate region (node) must be assigned an appropriate muscle cell likelihood score w. In our algorithm, each candidate region is discriminatively represented with a feature vector that consists of a descriptor to model the convexity of the shape, and two histograms to describe the gradient magnitudes of the pixels on the cell boundary and inside the cell region. These morphological features are proposed based on the following observations: (1) the shape of a muscle cell is nearly convex; (2) the cell boundaries often exhibit higher gradient magnitudes; (3) the intensities within the cell regions should be relatively homogeneous.

4.3.1.3 Experimental results

We have tested the proposed approach on 120 H&E-stained muscle cell images captured at 10× magnification. Each image contains around 200 muscle cells. The images are randomly split into two sets of equal size, one for training and the other for testing. The gold standard of each individual muscle cell is manually annotated. To quantitatively analyze the pixel-wise segmentation accuracy, we calculate the precision $P = \frac{|S \cap G|}{|S|}$, recall $R = \frac{|S \cap G|}{|R|}$, and F_1-score $= \frac{|2PR|}{|P+R|}$, where S denotes the segmentation result and G is the gold standard.

We compare the proposed method with three state-of-the-art methods: isoperimetric graph partition (ISO) (Grady and Schwartz, 2006), global probability of boundary detector (gPb) (Arbelaez et al., 2011), and a six-layer deep CNN. The comparative boxplots of F_1-scores are shown in Fig. 4.7. We can see that our structured edge detection-based segmentation algorithm outperforms the others on digitized muscle specimens. The quantitative comparative results are shown in Table 4.3, where we report the average and standard variance of F_1-score, precision, and recall. In the table, we provide the proposed structured detection-based segmentation algorithm with and without using the presented hierarchical segmentation (denoted as Prop. w. H. and Prop. w.o. H., respectively). It is worth noting that the proposed structured edge detection algorithm indeed performs better than DCNN for the muscle image dataset. One potential reason is that we do not have sufficient training data to learn a DCNN model that is sufficient to capture all the edge variations. A larger DCNN model with more training data might be able to achieve better performance. With

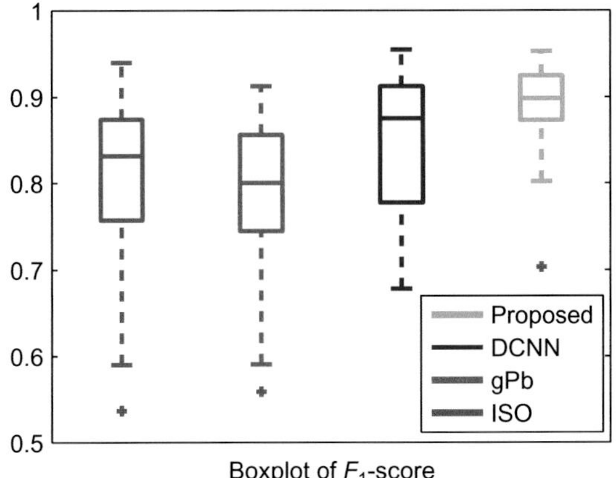

FIG. 4.7

The comparative segmentation results of the proposed muscle image segmentation algorithm with three state-of-the-art methods.

Table 4.3 The Pixel-Wise Segmentation Accuracy

	F_1-Score		Prec.		Rec.	
Method	**Mean**	**Std**	**Mean**	**Std**	**Mean**	**Std**
ISO (Grady and Schwartz, 2006)	0.8050	0.0993	0.8988	0.0589	0.7429	0.1369
gPb (Arbelaez et al., 2011)	0.7904	0.0780	0.9123	0.0515	0.7011	0.0962
DCNN	0.8388	0.1073	**0.9464**	**0.0411**	0.7666	0.1441
Prop. w.o. H.	0.8815	0.0523	0.8861	0.0551	0.8783	0.0587
Prop. w. H.	**0.8974**	**0.0422**	0.9078	0.0433	**0.8888**	**0.0543**

respect to speed, the current DCNN model takes 28 seconds to generate an edge map for segmentation on an image of size 1024 × 768, which is much slower than our structured edge detection, which takes less than 1 s.

4.3.2 SPARSITY-BASED DICTIONARY LEARNING

Sparse representation has received a great deal of attention in the fields of machine learning and computer vision due to its state-of-the-art performance in various applications including object segmentation. The K-SVD algorithm (Aharon et al., 2006), which takes its name from K computations of singular value decomposition (SVD) for dictionary basis update (Rubinstein et al., 2010), and its variants are widely used for dictionary learning. K-SVD learns a dictionary from a set of data points and emphasizes its generative power, and the purpose is to minimize the overall reconstruction errors. On the other hand, supervised dictionary learning techniques (Mairal et al., 2008; Zhang and Li, 2010; Jiang et al., 2011) have also been proposed to enforce its discriminative power, which can benefit object classification. Recently a selection-based dictionary learning algorithm was reported in Liu et al. (2013), which selects most representative data points as dictionary bases.

Defining $X = \{\boldsymbol{x}_i | i = 1 \cdots N\}$ as a set of N input data points with p-dimensions, the dictionary learning can be formulated as finding a repository $\boldsymbol{\Phi} \in R^{p \times K}$ for a sparse representation of X:

$$\min_{\boldsymbol{\Phi}, \{\boldsymbol{\alpha}_i\}_{i=1,2,\ldots,N}} \sum_{i=1}^{N} ||\boldsymbol{x}_i - \boldsymbol{\Phi}\boldsymbol{\alpha}_i||_2^2 + \lambda||\boldsymbol{\alpha}_i||_m, \tag{4.16}$$

where $\boldsymbol{\alpha}_i$ represents the ith sparse coefficient, $m = 0$ or 1 indicates using ℓ_0 or ℓ_1 penalty, and λ is the regularization parameter controlling the sparsity of α_i. For sparse coefficient computation, orthogonal matching pursuit (OMP) (Tropp, 2004) can be applied to solve the sparse representation problem for ℓ_0 penalty. When using ℓ_1 penalty, it is a Lasso problem (Tibshirani, 1994), which can be efficiently solved by the LARS algorithm (Efron et al., 2004).

Segmentation models with shape priors can handle weak or misleading object boundaries so that they can significantly improve the accuracy (Cootes et al., 1995; Zhang et al., 2012). Recently the sparse shape model has been shown to be more effective than the principal component analysis-based shape prior due to its insensitiveness to object occlusion (Zhang et al., 2011, 2012). However, using all training shapes is inefficient during sparse reconstruction on a large dataset at run-time, and thus data summarization or dictionary learning is usually required for efficient runtime optimization. We have presented a robust sparsity-based dictionary learning algorithm for nucleus shape modeling, which is incorporated into an integrated framework for automated nucleus segmentation in microscopic images (Xing and Yang, 2013; Xing et al., 2016). In contrast to K-SVD (Aharon et al., 2006), this method directly selects the most representative nucleus shapes from the training dataset as dictionary bases. The robustness of the dictionary learning method is achieved by minimizing an integrated square error with a sparse constraint. In order to simultaneously and efficiently segment multiple nuclei, we combine the top-down shape prior model and a bottom-up deformable model with locality and repulsion constraints. Given a microscopic image, it begins with a deep CNN model to generate a probability map, on which an iterative region merging approach is performed for shape initializations. Next, it alternately performs shape deformation using an efficient local repulsive deformable model, and shape inference using the shape prior derived from the sparse shape model. The flowchart of nucleus segmentation is shown in Fig. 4.8.

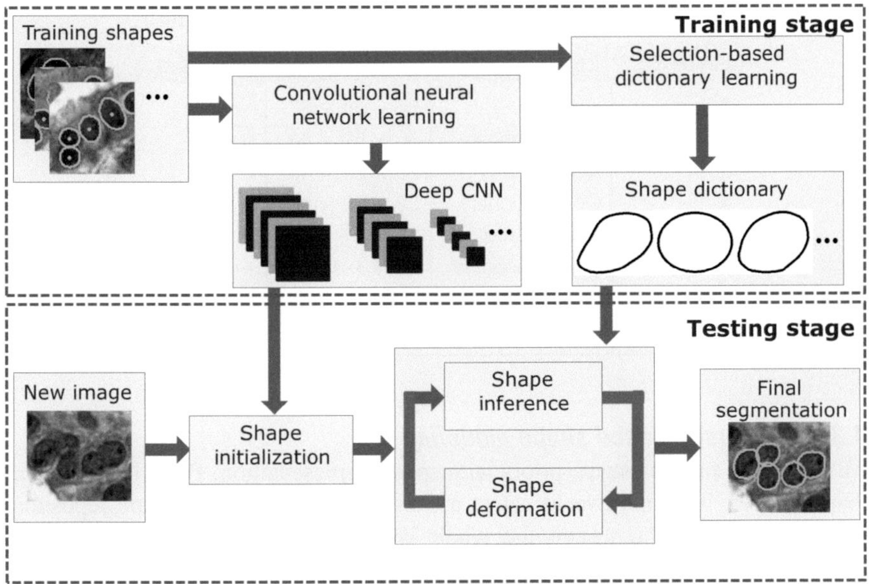

FIG. 4.8

The flowchart of the proposed framework for automated nucleus segmentation in microscopic images.

4.3.2.1 CNN-based shape initialization

In order to facilitate the subsequent contour evolution step, we need to obtain robust initializations. For this purpose, we first learn a deep CNN to generate a probability map which presents nucleus regions, and then apply an iterative region merging algorithm to the map to create the initial shape markers. One of the CNN structures (for breast cancer microscopic images) used in our algorithm is listed in Table 4.4. The last layer is a softmax layer producing probability distributions over classes (Farabet et al., 2013). Our CNN is trained using raw pixel values (the YUV color space) of small image patches with a certain size, centered on the pixel itself. In the testing stage, we apply the learned CNN model to small patches cropped from a new testing image using a pixel-wise sliding window of the same size as training patches. For each testing image, the CNN model with a sliding-window technique creates a probability map where each pixel is assigned a probability of being close to nucleus centers, and those lying in the background would have lower probabilities. Furthermore, an iterative region merging algorithm (Xing et al., 2016) is applied to the probability map to generate initial shapes for the subsequent shape deformation.

Table 4.4 One of the CNN Structure in Our Algorithm

Layer No.	Layer Type	Output Size	Filter Size
1	Input	55 × 55 × 3	—
2	Convolutional	50 × 50 × 48	6 × 6
3	Max-pooling	25 × 25 × 48	2 × 2
4	Convolutional	22 × 22 × 48	4 × 4
5	Max-pooling	11 × 11 × 48	2 × 2
6	Fully connected	1024 × 1	—
7	Fully connected	1024 × 1	—
8	Output	2 × 1	—

4.3.2.2 Sparsity-based shape modeling

We model the nucleus shape priors with sparse representation. For a large nucleus shape dataset, it is intuitive to select a subset of the data as a shape repository that can sufficiently represent the whole dataset. This summarization can help remove outliers that are not true representatives of the dataset and might reduce the computational time for runtime optimization due to the decreased object-space dimension (Elhamifar et al., 2012). Based on these considerations, we propose a novel selection-based dictionary learning method for sparse representation by minimizing a locality-constrained integrated squared error (ISE). Scott (2001) has shown that minimizing the ISE is analogous to minimizing the objective function: $\int g(\boldsymbol{x}|\boldsymbol{\theta})^2 d\boldsymbol{x} - \frac{2}{N}\sum_{i=1}^{N} g(\boldsymbol{x}_i|\boldsymbol{\theta})$, where $g(\boldsymbol{x}|\boldsymbol{\theta})$ is a parametric model with parameter $\boldsymbol{\theta}$ and N is the number of data points $\{\boldsymbol{x}_i\}_{i=1}^{N}$. In a sparse shape model with N nucleus shapes $\{v_i\}_{i=1}^{N}$ aligned by Procrustes analysis, we have $v_i = \boldsymbol{B}\boldsymbol{\alpha}_i + \boldsymbol{\epsilon}_i$, where $B = [\boldsymbol{b}_1\, \boldsymbol{b}_2, \ldots, \boldsymbol{b}_K]$ ($\{\boldsymbol{b}_k \in R^{2m}\}_{k=1}^{K}$ are bases) is the shape dictionary, $\boldsymbol{\alpha}_i$ denotes the sparse coefficient, and $\boldsymbol{\epsilon}_i$ is the residual for the ith shape. Therefore we can model the residual density with function $g(\boldsymbol{\epsilon}|\boldsymbol{\theta})$ and minimize the objective function as follows:

$$\min_{\boldsymbol{\theta}} J(\boldsymbol{\theta}) = \min_{\boldsymbol{\theta}} \left[\left(\int g(\boldsymbol{\epsilon}|\boldsymbol{\theta})^2\, d\boldsymbol{\epsilon} - \frac{2}{N}\sum_{i=1}^{N} g(\boldsymbol{\epsilon}_i|\boldsymbol{\theta}) \right) + \lambda \sum_{i=1}^{N}\sum_{k=1}^{K} |\boldsymbol{\alpha}_{ik}|\,||v_i - \boldsymbol{b}_k||^2 \right], \quad \text{s.t. } 1^{\mathrm{T}}\boldsymbol{\alpha}_i = 1, \quad \forall i, \tag{4.17}$$

where $\boldsymbol{\epsilon}_i = v_i - \boldsymbol{B}\boldsymbol{\alpha}_i$ and $\boldsymbol{\alpha}_i = [\alpha_{i1}\ \alpha_{i2}\ \ldots\ \alpha_{iK}]^{\mathrm{T}}$. The first two terms form the L_2E criteria, which is robust to outliers (Scott, 2001). The last term constrains local representation of bases with weighted sparse codes, and is used to encourage each nucleus to be sufficiently represented by its neighboring dictionary bases to preserve similarity, which is essential in the sparse reconstruction. The constraint $1^{\mathrm{T}}\boldsymbol{\alpha}_i = 1, \forall i$, ensures the shift invariance. The residual is modeled with multivariate normal distribution: $\boldsymbol{\epsilon}_i \sim N(0, \sigma^2 I_{2m})$. In this way $g(\boldsymbol{\epsilon}_i|\boldsymbol{\theta}) = \xi\phi(\boldsymbol{\epsilon}_i|0, \sigma^2 I_{2m})$, where ξ denotes the percentage of the inlier shapes that need to be estimated and ϕ is the probability density function of a multivariate normal distribution. Based on (4.17), the dictionary B and sparse coefficients $\{\boldsymbol{\alpha}_i\}_{i=1}^N$ can be calculated by estimating $\boldsymbol{\theta} = \{\xi, B, \boldsymbol{\alpha}_1, \boldsymbol{\alpha}_2, \ldots, \boldsymbol{\alpha}_N, \sigma^2\}$.

Eq. (4.17) can be solved by performing dictionary basis selection and coefficient computation alternately. As $J(\boldsymbol{\theta})$ in Eq. (4.17) is differentiable with respect to $\{\boldsymbol{b}_k\}_{k=1}^K$, projection-based gradient descent is utilized for minimization to update the bases, which are directly selected shapes within each iteration. For coefficient calculation, we keep the dictionary fixed. Based on the sparse reconstruction criterion, the sparse coding objective function can be rewritten as

$$\min_{\{\boldsymbol{\alpha}_i\}_{i\in A}} \left[\sum_{i\in A} ||v_i - B\boldsymbol{\alpha}_i||^2 + \lambda \sum_{k=1}^K |\alpha_{ik}| ||v_i - \boldsymbol{b}_k||^2 \right], \ s.t.\ 1^{\mathrm{T}}\boldsymbol{\alpha}_i = 1, \quad i \in A, \tag{4.18}$$

where A is the set of indices corresponding to estimated inlier shapes. Locality-constrained linear coding (LLC) (Wang et al., 2010) is applied to Eq. (4.18) for coefficient computation, where the neighboring bases are defined in terms of the Euclidean distances between the shape and dictionary bases. Due to large shape variations of nuclei, it would be more effective to build multiple subpopulation shape prior models based on clustered shapes (Xing et al., 2016).

4.3.2.3 Local repulsive deformable model

Combined with shape prior modeling, we propose an efficient shape deformation method based on the Chan-Vese model (Chan and Vese, 2001) for nucleus segmentation. In order to enhance the robustness, we add an edge detector into the original Chan-Vese model combined with the region-based data fitting term to better move contours towards nucleus boundaries. In addition, we introduce a repulsive term (Zimmer and Olivo-Marin, 2005) to handle touching or overlapping nuclei. More importantly, we observe that each nucleus is often surrounded by a limited number of adjacent nuclei, and only its neighboring nuclei make dominant repulsive contributions to its shape deformation during contour evolution. Therefore for each nucleus we do not need to calculate the repulsion from all the other nuclei on the image, but only from its nearest neighbors in a local coordinate system. In this way, the computational cost will be significantly reduced. Consider an image I containing N nuclei, denoted by $v_i (i = 1, \ldots, N)$ with v_i representing the ith contour. The energy function for nucleus segmentation combining the driving and repulsive mechanisms thus can be expressed as follows:

$$\lambda_1 \sum_{i=1}^N \int_{\Omega_i} (I(\mathbf{x}) - u_i)^2 \, d\mathbf{x} + \lambda_2 \int_{\Omega_0} (I(\mathbf{x}) - u_0)^2 \, d\mathbf{x}$$
$$+ \lambda_3 \sum_{i=1}^N \int_0^1 \mathrm{e}(v_i(s)) \, ds + \omega \sum_{i=1}^N \sum_{j\in V_i} \int_{\Omega_i \cap \Omega_j} 1 \, d\mathbf{x} + \sum_{i=1}^N \gamma |v_i|, \tag{4.19}$$

where Ω_i and Ω_0 represent the region inside v_i and outside all the contours (background), and u_i and u_0 represent the average intensity of Ω_i and Ω_0, respectively. The third term with $e(v_i(s))$ is the edge detector and is chosen as $-||\nabla I(v_i(s))||^2$ ($s \in [0, 1]$ is the parameter for contour representation), and the fourth term denotes the repulsion preventing contours from crossing with each other. The last term represents the length of v_i.

Using the Euler-Lagrange equations associated with the minimization of (4.19), we can get the following evolution equation:

$$\begin{aligned}\frac{\partial v_i}{\partial t} =& \left|\frac{\partial v_i}{\partial s}\right| \mathbf{n}_i(-\lambda_1(I-u_i)^2 + \lambda_2(I-u_0)^2 - \lambda_3 \nabla e(v_i) \\ &- \omega \sum_{j \in V_i} z_j(v_i) + \gamma \rho(v_i)),\end{aligned} \tag{4.20}$$

where $\mathbf{n}_i$ is the normal unit vector of v_i, and $z_j(\mathbf{x})$ represents the indicator function: $z_j(\mathbf{x}) = 1$ if $\mathbf{x} \in \Omega_j$, otherwise 0. $\rho(\cdot)$ denotes the curvature. Given the initial shapes, we can iteratively evolve the contours toward desired nucleus boundaries.

Given initial contours, the proposed segmentation framework alternately performs shape deformation with the repulsive active contour model and shape inference with the sparse shape prior model. The shapes always expand from inside nuclei, one per nucleus, and evolve towards nucleus boundaries. In the active contour model, contours move based on image appearance information until Eq. (4.20) reaches a stable state, where the associated energy function achieves a minimum value; in the shape inference stage, contours evolve based on the high-level shape prior to constrain the shapes. This alternative operation scheme of combining bottom-up and top-down information has been successfully applied to biomedical image segmentation (Zhang et al., 2011, 2012).

4.3.2.4 Experimental results

The proposed framework has been extensively tested on three types of pathology specimens: brain tumor, pancreatic neuroendocrine tumor (NET), and breast cancer. The segmentation results using our method on three sample slide digitized images are shown in Fig. 4.9. As we can see, thousands of nuclei are correctly segmented. For quantitative analysis, we compare the proposed method with six state-of-the-art methods: mean shift (MS) (Comaniciu and Meer, 2002), isoperimetric graph partition (ISO) (Grady and Schwartz, 2006), superpixel (SUP) (Mori, 2005), marker-based watershed (MWS), graph-cut and coloring (GCC) (Al-Kofahi et al., 2010), and repulsive level set (RLS) (Qi et al., 2012), on 30 brain tumor, 51 NET, and 35 breast cancer images. In order to quantitatively analyze the pixel-wise segmentation accuracy, we apply multiple metrics, including Dice similarity coefficient (*DSC*), Hausdorff distance (*HD*), and mean absolute distance (*MAD*), to the evaluation of the algorithms. Letting Ω_{sr} and Ω_{gs} represent the regions inside the automatic segmentation contour v_{sr} and the gold standard contour v_{gs}, respectively, the metrics are defined as (Zhou et al., 2013).

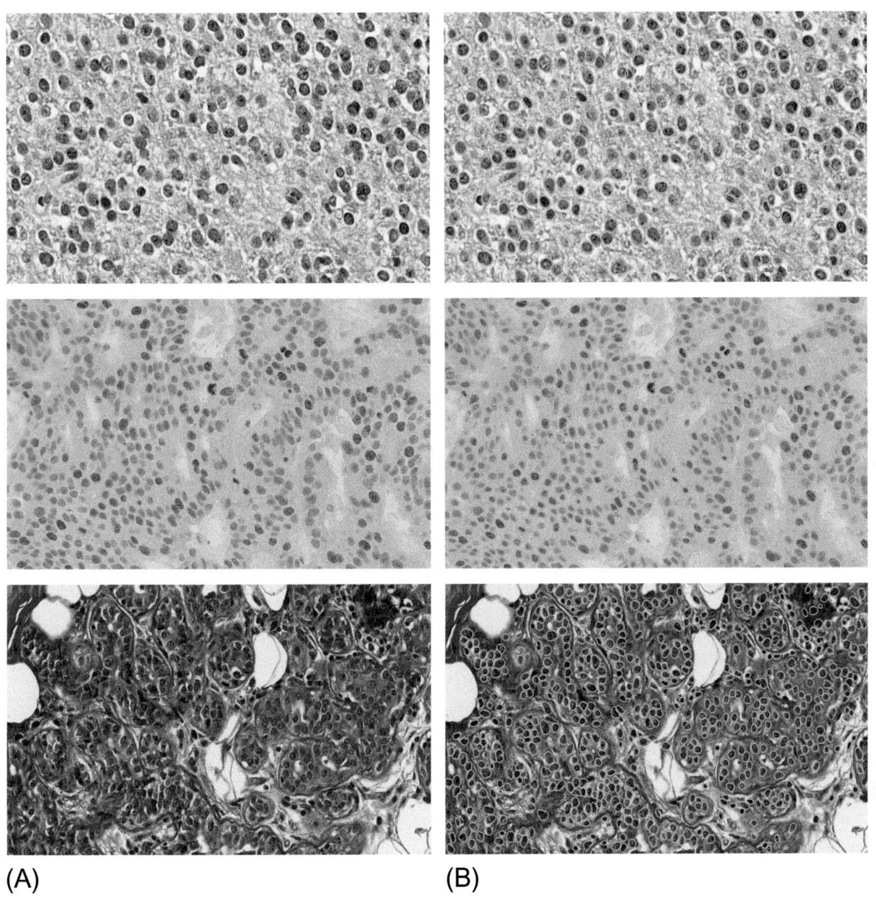

FIG. 4.9

Segmentation results using the proposed method on three sample images of the brain tumor (top), NET (middle), and breast cancer (bottom) datasets. Note that all the nuclei in the connected regions touching image boundaries are ignored. (A) Original images. (B) Segmentation results.

$$\begin{aligned}
DSC &= \frac{2|\Omega_{sr} \cap \Omega_{gs}|}{|\Omega_{sr}| + |\Omega_{gs}|}, \\
HD &= \max\{\sup_{s} d(v_{sr}(s), v_{gs}), \sup_{s} d(v_{gs}(s), v_{sr})\}, \\
MAD &= \frac{\int_0^1 d(v_{sr}(s), v_{gs})|v'_{sr}(s)|ds}{2|v_{sr}|} + \frac{\int_0^1 d(v_{gs}(s), v_{sr})|v'_{gs}(s)|ds}{2|v_{gs}|},
\end{aligned} \tag{4.21}$$

Table 4.5 Comparative Pixel-Wise Segmentation Accuracy on Brain Tumor, NET, and Breast Cancer Datasets

	Brain Tumor			NET			Breast Cancer		
Brain	***DSC***	***HD***	***MAD***	***DSC***	***HD***	***MAD***	***DSC***	***HD***	***MAD***
MS	0.74	9.98	5.23	0.66	7.01	4.35	0.49	20.83	13.17
ISO	0.70	10.15	6.63	0.48	10.02	8.14	0.56	17.19	11.59
SUP	0.75	10.89	5.32	0.75	6.62	3.77	0.68	17.14	9.32
GCC	0.81	7.07	3.85	0.61	6.37	5.00	0.59	16.84	10.75
MWS	0.81	7.21	3.57	0.82	4.12	2.33	0.73	11.12	6.66
RLS	0.80	8.51	4.59	0.84	2.71	2.26	0.77	10.50	6.30
Proposed	0.85	5.06	3.26	0.92	2.41	1.58	0.80	8.60	6.24

where $d(v_{sr}(s), v_{gs})$ denotes the minimum distance from point s to the contour v_{gs}, sup means the supremum, and $|v_{sr}|$ represents the length of v_{sr}. A large DSC, or a small HD/MAD, indicates high segmentation accuracy. Table 4.5 displays the DSC, HD, and MAD values using MS (Comaniciu and Meer, 2002), ISO (Grady and Schwartz, 2006), SUP (Mori, 2005), MWS, GCC (Al-Kofahi et al., 2010), RLS (Qi et al., 2012), and the proposed method. It is clear that the proposed method provides the best performance, especially in terms of HD that calculates the largest error for each segmentation.

4.4 SUMMARY

This chapter presents several popular machine learning techniques and their applications in nucleus/cell detection and segmentation in microscopic images. Specifically, it discusses a SVM-based myonucleus detection in z-stack fluorescence images, and a deep CNN-based regression model for nucleus detection in bright-field and phase-contrast microscopy images. Due to the encouraging performance of the deep learning on object detection, recently deep neural network has attracted much interest. In addition, it introduces a random forest-based structured edge detection algorithm for muscle cell segmentation in H&E-stained images, and a sparsity-based dictionary learning approach for object shape prior modeling, which is used for nucleus segmentation in both H&E- and Ki-67-stained histopathology images. There exist many other machine learning-based software tools for microscopic image analysis, and more details can be found in Sommer and Gerlich (2013) and Eliceiri et al. (2012).

Currently one challenging problem is to design an efficient and effective algorithm for nucleus/cell detection and segmentation in large-scale microscopic images, such as whole slide imaging (WSI) specimens. WSI provides richer information that might better support CAD, and thus it is necessary to conduct image analysis on WSI specimens. However, these WSI images usually have billions of pixels, which might provide an obstacle for many algorithms. Therefore designing a scalable algorithm that can adapt to large-scale images could attract increasing attention in microscopic image analysis.

REFERENCES

Adiga, U., Malladi, R., Fernandez-Gonzalez, R., de Solorzano, C.O., 2006. High-throughput analysis of multispectral images of breast cancer tissue. IEEE Trans. Image Process. 15 (8), 2259–2268.

Aharon, M., Elad, M., Bruckstein, A., 2006. K-SVD: an algorithm for designing overcomplete dictionaries for sparse representation. IEEE Trans. Signal Process. 54 (11), 4311–4322.

Al-Kofahi, Y., Lassoued, W., Lee, W., Roysam, B., 2010. Improved automatic detection and segmentation of cell nuclei in histopathology images. IEEE Trans. Biomed. Eng. 57 (4), 841–852.

Ali, S., Madabhushi, A., 2012. An integrated region-, boundary-, shape-based active contour for multiple object overlap resolution in histological imagery. IEEE Trans. Med. Imaging 31 (7), 1448–1460.

Arbelaez, P., 2006. Boundary extraction in natural images using ultrametric contour maps. In: Conference on Computer Vision and Pattern Recognition Workshop—CVPRW'06, pp. 182–182.

Arbelaez, P., Maire, M., Fowlkes, C., Malik, J., 2011. Contour detection and hierarchical image segmentation. IEEE Trans. Pattern Anal. Mach. Intell. 33 (5), 898–916.

Arbelaez, P., Pont-Tuset, J., Barron, J., Marques, F., Malik, J., 2014. Multiscale combinatorial grouping. In: 2014 IEEE Conference on Computer Vision and Pattern Recognition (CVPR), pp. 328–335.

Arteta, C., Lempitsky, V., Noble, J.A., Zisserman, A., 2012. Learning to detect cells using non-overlapping extremal regions. In: Medical Image Computing and Computer-Assisted Intervention—MICCAI 2012, vol. 7510, pp. 348–356.

Bernardis, E., Yu, S., 2010. Finding dots: segmentation as popping out regions from boundaries. In: 2010 IEEE Conference on Computer Vision and Pattern Recognition (CVPR), pp. 199–206.

Bishop, C., 2007. Pattern Recognition and Machine Learning. Springer, New York.

Breiman, L., 2001. Random forests. Mach. Learn. 45 (1), 5–32.

Byun, J.Y., Verardo, M.R., Sumengen, B., Lewis, G.P., Manjunath, B.S., Fisher, S.K., 2006. Automated tool for the detection of cell nuclei in digital microscopic images: application to retinal images. Mol. Vis. 12, 949–960.

Chan, T.F., Vese, L.A., 2001. Active contours without edges. IEEE Trans. Image Process. 10 (2), 266–277.

Chang, H., Han, J., Spellman, P.T., Parvin, B., 2012. Multireference level set for the characterization of nuclear morphology in glioblastoma multiforme. IEEE Trans. Biomed. Eng. 59 (12), 3460–3467.

Chen, Y., Yang, J., Yang, M., 2015. Extracting image regions by structured edge prediction. In: 2015 IEEE Winter Conference on Applications of Computer Vision (WACV), pp. 1060–1067.

Cireşan, D.C., Giusti, A., Gambardella, L.M., Schmidhuber, J., 2013. Mitosis detection in breast cancer histology images with deep neural networks. In: Medical Image Computing and Computer-Assisted Intervention—MICCAI 2013, vol. 8150, pp. 411–418.

Comaniciu, D., Meer, P., 2002. Mean shift: a robust approach toward feature space analysis. IEEE Trans. Pattern Anal. Mach. Intell. 24 (5), 603–619.

Cootes, T.F., Taylor, C.J., Cooper, D.H., Graham, J., 1995. Active shape models—their training and application. Comput. Vis. Image Und. 61 (1), 38–59.

Cortes, C., Vapnik, V., 1995. Support-vector networks. Mach. Learn. 20 (3), 273–297.

Dollar, P., Zitnick, C.L., 2014. Fast edge detection using structured forests, pp. 1–12. arXiv 1406.5549 [cs.CV].

Efron, B., Hastie, T., Johnstone, I., Tibshirani, R., 2004. Least angle regression. Ann. Stat. 32 (2), 407–499.

Elhamifar, E., Sapiro, G., Vidal, R., 2012. See all by looking at a few: sparse modeling for finding representative objects. In: 2012 IEEE Conference on Computer Vision and Pattern Recognition (CVPR), pp. 1600–1607.

Eliceiri, K.W., Berthold, M.R., Goldberg, I.G., Ibanez, L., Manjunath, B.S., Martone, M.E., Murphy, R.F., Peng, H., Plant, A.L., Roysam, B., Stuurman, N., Swedlow, J.R., Tomancak, P., Carpenter, A.E., 2012. Biological imaging software tools. Nat. Methods 9 (7), 697–710.

Farabet, C., Couprie, C., Najman, L., LeCun, Y., 2013. Learning hierarchical features for scene labeling. IEEE Trans. Pattern Anal. Mach. Intell. 35 (8), 1915–1929.

Felzenszwalb, P.F., Huttenlocher, D.P., 2004. Efficient graph-based image segmentation. Int. J. Comput. Vis. 59 (2), 167–181.

Fitzgibbon, A.W., Pilu, M., Fisher, R.B., 1999. Direct least-squares fitting of ellipses. IEEE Trans. Pattern Anal. Mach. Intell. 21 (5), 476–480.

Foran, D.J., Yang, L., Chen, W., Hu, J., Goodell, L.A., Reiss, M., Wang, F., Kurc, T., Pan, T., Sharma, A., Saltz, J.H., 2011. Imageminer: a software system for comparative analysis of tissue microarrays using content-based image retrieval, high-performance computing, and grid technology. J. Am. Med. Inform. Assoc. 18 (4), 403–415.

Grady, L., Schwartz, E., 2006. Isoperimetric graph partitioning for image segmentation. IEEE Trans. Pattern Anal. Mach. Intell. 28 (3), 469–475.

Hariharan, H., Koschan, A., Abidi, M., 2007. Multifocus image fusion by establishing focal connectivity. In: IEEE International Conference on Image Processing—ICIP 2007, vol. 3, pp. 321–324.

Hastie, T., Tibshirani, R., Friedman, J., 2009. The Elements of Statistical Learning: Data Mining, Inference, and Prediction, second ed. Springer, New York.

Hinton, G.E., Salakhutdinov, R.R., 2006. Reducing the dimensionality of data with neural networks. Science 313 (5786), 504–507.

Hinton, G.E., Srivastava, N., Krizhevsky, A., Sutskever, I., Salakhutdinov, R., 2012. Improving neural networks by preventing coadaptation of feature detectors, pp. 1–18. arXiv 1207.0580 [cs.NE].

Janssens, T., Antanas, L., Derde, S., Vanhorebeek, I., den Berghe, G.V., Grandas, F.G., 2013. Charisma: an integrated approach to automatic H&E-stained skeletal muscle cell segmentation using supervised learning and novel robust clump splitting. Med. Image Anal. 17 (8), 1206–1219.

Jiang, Z., Lin, Z., Davis, L.S., 2011. Learning a discriminative dictionary for sparse coding via label consistent k-SVD. In: 2011 IEEE Conference on Computer Vision and Pattern Recognition (CVPR), pp. 1697–1704.

Jones, T.R., Carpenter, A.E., Golland, P., 2005. Voronoi-based segmentation of cells on image manifolds. In: Computer Vision for Biomedical Image Applications (CVBIA), vol. 3765, pp. 535–543.

Kong, H., Gurcan, M., Belkacem-Boussaid, K., 2011. Partitioning histopathological images: an integrated framework for supervised color-texture segmentation and cell splitting. IEEE Trans. Med. Imaging 30 (9), 1661–1677.

Kontschieder, P., Rota Bulo, S., Bischof, H., Pelillo, M., 2011. Structured class-labels in random forests for semantic image labelling. In: 2011 IEEE International Conference on Computer Vision (ICCV), pp. 2190–2197.

Kourou, K., Exarchos, T.P., Exarchos, K.P., Karamouzis, M.V., Fotiadis, D.I., 2015. Machine learning applications in cancer prognosis and prediction. Comput. Struct. Biotech. J. 13, 8–17.

Krizhevsky, A., Sutskever, I., Hinton, G.E., 2012. Imagenet classification with deep convolutional neural networks. Adv. Neural Inform. Process. Syst. 25, 1097–1105.

Kuhl, F.P., Giardina, C.R., 1982. Elliptic Fourier features of a closed contour. Comput. Graph. Image Process. 18 (3), 236–258.

LeCun, Y., Bottou, L., Bengio, Y., Haffner, P., 1998. Gradient-based learning applied to document recognition. Proc. IEEE 86 (11), 2278–2324.

LeCun, Y., Kavukcuoglu, K., Farabet, C., 2010. Convolutional networks and applications in vision. In: Proceedings of 2010 IEEE International Symposium on Circuits and Systems (ISCAS), pp. 253–256.

LeCun, Y., Bengio, Y., Hinton, G., 2015. Deep learning. Nature 521 (28), 436–444.

Liao, S., Gao, Y., Oto, A., Shen, D., 2013. Representation learning: a unified deep learning framework for automatic prostate MR segmentation. In: Medical Image Computing and Computer-Assisted Intervention—MICCAI 2013, vol. 8150, pp. 254–261.

Lin, G., Adiga, U., Olson, K., Guzowski, J.F., Barnes, C.A., Roysam, B., 2003. A hybrid 3d watershed algorithm incorporating gradient cues and object models for automatic segmentation of nuclei in confocal image stacks. Cytometry A 56A (1), 23–36.

Ling, H., Jacobs, D.W., 2007. Shape classification using the inner-distance. IEEE Trans. Pattern Anal. Mach. Intell. 29 (2), 286–299.

Liu, B., Huang, J., Kulikowski, C., Yang, L., 2013. Robust visual tracking using local sparse appearance model and k-selection. IEEE Trans. Pattern Anal. Mach. Intell. 35 (12), 2968–2981.

Liu, F., Xing, F., Zhang, Z., Mcgough, M., Yang, L., 2015. Robust muscle cell quantification using structured edge detection and hierarchical segmentation. In: Medical Image Computing and Computer-Assisted Intervention—MICCAI 2015, vol. 9351, pp. 324–331.

Lou, X., Koethe, U., Wittbrodt, J., Hamprecht, F., 2012. Learning to segment dense cell nuclei with shape prior. In: 2012 IEEE Conference on Computer Vision and Pattern Recognition (CVPR), pp. 1012–1018.

Mairal, J., Bach, F., Ponce, J., Sapiro, G., Zisserman, A., 2008. Discriminative learned dictionaries for local image analysis. In: 2008 IEEE Conference on Computer Vision and Pattern Recognition (CVPR), pp. 1–8.

Matei, B., Meer, P., 2000. A general method for errors-in-variables problems in computer vision. In: 2000 IEEE Conference on Computer Vision and Pattern Recognition (CVPR), vol. 2, pp. 18–25.

Mori, G., 2005. Guiding model search using segmentation. In: 2005 IEEE International Conference on Computer Vision (ICCV), vol. 2, pp. 1417–1423.

Mualla, F., Scholl, S., Sommerfeldt, B., Maier, A., Hornegger, J., 2013. Automatic cell detection in bright-field microscope images using sift, random forests, and hierarchical clustering. IEEE Trans. Med. Imaging 32 (12), 2274–2286.

Murphy, K.P., 2012. Machine Learning: A Probabilistic Perspective, first ed. MIT Press, Cambridge, MA.

Otsu, N., 1979. A threshold selection method from gray-level histograms. IEEE Trans. Syst. Man Cybern. 9 (1), 62–66.

Parvin, B., Yang, Q., Han, J., Chang, H., Rydberg, B., Barcellos-Hoff, M.H., 2007. Iterative voting for inference of structural saliency and characterization of subcellular events. IEEE Trans. Image Process. 16 (3), 615–623.

Qi, X., Xing, F., Foran, D.J., Yang, L., 2012. Robust segmentation of overlapping cells in histopathology specimens using parallel seed detection and repulsive level set. IEEE Trans. Biomed. Eng. 59 (3), 754–765.

Quelhas, P., Marcuzzo, M., MendoncÌğa, A., Campilho, A., 2010. Cell nuclei and cytoplasm joint segmentation using the sliding band filter. IEEE Trans. Med. Imaging 29 (8), 1463–1473.

Rubinstein, R., Bruckstein, A.M., Elad, M., 2010. Dictionaries for sparse representation modeling. Proc. IEEE 98 (6), 1045–1057.

Scholkopf, B., Smola, A.J., 2001. Learning With Kernels: Support Vector Machines, Regularization, Optimization, and Beyond. MIT Press, Cambridge, MA.

Scott, D.W., 2001. Parametric statistical modeling by minimum integrated squared error. Technometrics 43, 274–285.

Sommer, C., Gerlich, D.W., 2013. Machine learning in cell biology—teaching computers to recognize phenotypes. J. Cell Sci. 126 (24), 5529–5539.

Su, H., Xing, F., Lee, J.D., Peterson, C.A., Yang, L., 2013. Learning based automatic detection of myonuclei in isolated single skeletal muscle fibers using multi-focus image fusion. In: 2013 IEEE 10th International Symposium on Biomedical Imaging (ISBI), pp. 432–435.

Su, H., Xing, F., Lee, J.D., Peterson, C.A., Yang, L., 2014. Automatic myonuclear detection in isolated single muscle fibers using robust ellipse fitting and sparse representation. IEEE/ACM Trans. Comput. Biol. Bioinform. 11 (4), 714–726.

Szegedy, C., Toshev, A., Erhan, D., 2013. Deep neural networks for object detection. Adv. Neural Inform. Process. Syst. 26, 2553–2561.

Tarca, A.L., Carey, V.J., Chen, X.W., R., R., Drăghici, S., 2007. Machine learning and its applications to biology. PLoS Comput. Biol. 3 (6), e116.

Taubin, G., 1991. Estimation of planar curves, surfaces, and nonplanar space curves defined by implicit equations with applications to edge and range image segmentation. IEEE Trans. Pattern Anal. Mach. Intell. 13 (11), 1115–1138.

Tibshirani, R., 1994. Regression shrinkage and selection via the lasso. J. R. Stat. Soc. B 58, 267–288.

Tropp, J.A., 2004. Greed is good: algorithmic results for sparse approximation. IEEE Trans. Inform. Theory 50 (10), 2231–2242.

Uzunbaş, M., Chen, C., Metaxsas, D., 2014. Optree: a learning-based adaptive watershed algorithm for neuron segmentation. In: Medical Image Computing and Computer-Assisted Intervention—MICCAI 2014, vol. 8673, pp. 97–105.

Wählby, C., Lindblad, J., Vondrus, M., Bengtsson, E., Björkesten, L., 2002. Algorithms for cytoplasm segmentation of fluorescence labelled cells. Anal. Cell. Pathol. 24 (2), 101–111.

Waljee, A.K., Higgins, P.D., 2010. Machine learning in medicine: a primer for physicians. Am. J. Gastroenterol. 105 (6), 1224–1226.

Wang, J., Yang, J., Yu, K., Lv, F., Huang, T., Gong, Y., 2010. Locality-constrained linear coding for image classification. In: 2010 IEEE Conference on Computer Vision and Pattern Recognition (CVPR), pp. 3360–3367.

Wernick, M.N., Yang, Y., Brankov, J.G., Yourganov, G., Strother, S.C., 2010. Machine learning in medical imaging. IEEE Signal Process. Mag. 27 (4), 25–38.

Wu, G., Kim, M., Wang, Q., Gao, Y., Liao, S., Shen, D., 2013. Unsupervised deep feature learning for deformable registration of MR brain images. In: Medical Image Computing and Computer-Assisted Intervention—MICCAI 2013, vol. 8150, pp. 649–656.

Xie, Y., Xing, F., Kong, X., Su, H., Yang, L., 2015. Beyond classification: structured regression for robust cell detection using convolutional neural network. In: Medical Image Computing and Computer-Assisted Intervention—MICCAI 2015, vol. 9351, pp. 358–365.

Xing, F., Yang, L., 2013. Robust selection-based sparse shape model for lung cancer image segmentation. In: Medical Image Computing and Computer-Assisted Intervention—MICCAI 2013, vol. 8151, pp. 404–412.

Xing, F., Su, H., Neltner, J., Yang, L., 2014. Automatic ki-67 counting using robust cell detection and online dictionary learning. IEEE Trans. Biomed. Eng. 61 (3), 859–870.

Xing, F., Xie, Y., Yang, L., 2016. An automatic learning-based framework for robust nucleus segmentation. IEEE Trans. Med. Imaging 35 (2), 550–566.

Yan, P., Zhou, X., Shah, M., Wong, S.T.C., 2008. Automatic segmentation of high-throughput RNAi fluorescent cellular images. IEEE Trans. Inform. Tech. Biomed. 12 (1), 109–117.

Yang, L., Tuzel, O., Meer, P., Foran, D.J., 2008. Automatic image analysis of histopathology specimens using concave vertex graph. In: Medical Image Computing and Computer-Assisted Intervention—MICCAI 2008, vol. 5241, pp. 833–841.

Zhang, Z., 1997. Parameter estimation techniques: a tutorial with application to conic fitting. Image Vis. Comput. 15 (1), 59–76.

Zhang, Q., Li, B., 2010. Discriminative k-SVD for dictionary learning in face recognition. In: 2010 IEEE Conference on Computer Vision and Pattern Recognition (CVPR), pp. 2691–2698.

Zhang, S., Zhan, Y., Dewan, M., Huang, J., Metaxas, D.N., Zhou, X.S., 2011. Deformable segmentation via sparse shape representation. In: Medical Image Computing and Computer-Assisted Intervention—MICCAI 2011, pp. 451–458.

Zhang, S., Zhan, Y., Metaxas, D.N., 2012. Deformable segmentation via sparse shape representation and dictionary learning. Med. Image Anal. 16 (7), 1385–1396.

Zhang, C., Yarkony, J., Hamprecht, F.A., 2014. Cell detection and segmentation using correlation clustering. In: Medical Image Computing and Computer-Assisted Intervention—MICCAI 2014, vol. 8673, pp. 9–16.

Zhou, X., Liu, K.Y., Bradley, P., Perrimon, N., Wong, S.T.C., 2005. Towards automated cellular image segmentation for RNAI genome-wide screening. In: Medical Image Computing and Computer-Assisted Intervention—MICCAI 2005, vol. 3749, pp. 885–892.

Zhou, X., Huang, X., Duncan, J.S., Yu, W., 2013. Active contour with group similarity. In: 2013 IEEE Conference on Computer Vision and Pattern Recognition (CVPR), pp. 2969–2976.

Zimmer, C., Olivo-Marin, J.C., 2005. Coupled parametric active contours. IEEE Trans. Pattern Anal. Mach. Intell. 27 (11), 1838–1842.

CHAPTER

Sparse models for imaging genetics

5

J. Wang[1], T. Yang[2], P. Thompson[3], J. Ye[1]

University of Michigan, Ann Arbor, MI, United States[1] Arizona State University, Tempe, AZ, United States[2] University of Southern California, Los Angeles, CA, United States[3]

CHAPTER OUTLINE

5.1 INTRODUCTION

Imaging genetics studies neuroimaging-related genetic variation. In the past decade, neuroimaging techniques—for example, computed tomography (CT), magnetic resonance imaging (MRI), functional MRI (fMRI), and positron emission tomography (PET)—provide both anatomical and functional visualizations of the nervous system, which greatly advance modern medicine, neuroscience, and psychology. As an emerging promising technique, imaging genetics research has attracted extensive attention. With the integration of molecular genetics and disorder-related neuroimaging phenotypes, imaging genetics provides a unique opportunity to reveal the impact

Machine Learning and Medical Imaging. http://dx.doi.org/10.1016/B978-0-12-804076-8.00005-0

of genetic variation in neuroimaging, that is, how individual differences in single nucleotide polymorphisms (SNPs) affect brain development, structure, and function (Hariri et al., 2006; Thompson et al., 2013). Molecular geneticists believe that some common genetic variants in SNPs may lead to common disorders (Cirulli and Goldstein, 2010). Moreover, as another benefit of exploiting neuroimaging in genetics, imaging phenotypes are closer to the biology of genetic function (Meyer-Lindenberg, 2012) than disease or cognitive phenotypes.

Previous studies show the great promise of imaging genetics. For example, the $\epsilon 4$ allele of apolipoprotein E (ApoE4) is one of the well-known genetic risk factors for Alzheimer's disease (AD). From a neuroimaging perspective, the degeneration of brain tissue of ApoE4 carriers is faster as they age; young adult ApoE4 carriers often exhibit thinner cortical gray matter than noncarriers (Shaw et al., 2007). It has been verified in a series of genome-wide association (GWA) studies of AD that ApoE4 is strongly associated with the volumes of key brain regions, such as the hippocampus and entorhinal cortex (Potkin et al., 2009; Stein et al., 2012; Yang et al., 2015). Recent worldwide consortium efforts, such as ENIGMA (Enhancing Neuroimaging Genetics through Meta-Analysis (Stein et al., 2012)) and CHARGE (Cohorts for Heart and Aging Research in Genomic Epidemiology (Bis et al., 2012; Psaty et al., 2009)), enable us to detect robust common neuroimaging-genetic associations (Medland et al., 2014).

Imaging genetic studies are challenging in practice due to the relatively small number of subjects and extremely high dimensionality of imaging as well as genetic data. Neuroimaging data, for example, contains hundreds of thousands of voxels. Advances in modern sequencing techniques lead to huge scale (whole) genome sequencing data with tens of millions of SNPs. However, most traditional statistical methods are intended for low-dimensional data sets (James et al., 2013), in which the number of subjects is much larger than the number of features. This significantly limits the practical usage of traditional methods to the high-dimensional imaging data sets, as they are prone to overfitting.

The high-dimensional data sets involved in many imaging genetics studies confront researchers and scientists with an urgent need for novel methods that can effectively uncover the predictive patterns from these types of data. A useful observation from many real-world applications is that data with complex structures often has sparse underlying representations. More specifically, although the data may have millions of features, it may be well interpreted by a few of the most relevant explanatory features. For example, the neural representation of natural scenes in the visual cortex is sparse, as only a small number of neurons are active at a given instant (Vinje and Gallant, 2000); images have very sparse representations with respect to an overcomplete dictionary because they lie on or close to low-dimensional subspaces or submanifolds (Wright et al., 2010); although humans have millions of SNPs, only a small number of them are relevant to certain diseases such as leukemia and Alzheimer's disease (Golub et al., 1999; Guyon et al., 2002; Mu and Gage, 2011). Moreover, sparsity has been shown to be an effective approach to alleviate overfitting, from which most traditional statistical methods suffer. Therefore finding sparse representations is particularly important in discovering the underlying mechanisms of many complex systems.

As an emerging and powerful technique, sparse models have attracted increasing research interest in image genetics in the past decade. As well as their robustness to overfitting, sparse models are also promising in enhancing the interpretability of the model by automatically identifying a small subset of features that can best explain the outcome. Indeed, we can categorize existing methodological approaches for imaging genetics into three classes (Thompson et al., 2013).

The first one is the so-called *univariate-imaging univariate-genetic association* analysis that performs a univariate statistical test on each SNP-voxel pair individually. This type of approach has been widely used in previous GWA studies. However, these approaches fail to reveal scenarios such as SNP-SNP interactions and the joint effects of multiple SNPs, which occur commonly in gene expression (Dinu et al., 2012; Cornelis et al., 2009; Singh et al., 2011; Yang et al., 2012). In addition, it is worth mentioning that this kind of analysis is computationally inefficient.

The second class is the *univariate-imaging multivariate-genetic association* method. Based on a candidate imaging phenotype, a common multivariate approach utilizes sparse models, for example, Lasso (least absolute shrinkage and selection operator (Tibshirani, 1996; Yang et al., 2015)), to perform simultaneous model fitting and variable (causal SNPs) selection. Moreover, by incorporating biological prior knowledge such as linkage disequilibrium (LD) information, we can employ group Lasso to locate groups of candidate SNPs (Wang et al., 2012; Yuan and Lin, 2006). In the sequel, tree-structured group Lasso can also be applied if the hierarchical structure of SNPs is further available (Liu and Ye, 2010).

The third class of methodology in imaging genetics is *joint multivariate association* analysis, for example, canonical correlation analysis (CCA) and partial least squares (PLS) regression. However, a clear drawback of this kind of approach is that the detected genetic variants and imaging features may not be immediately related to a disorder (Batmanghelich et al., 2013).

In this chapter, we focus on univariate-imaging multivariate-genetic association studies in imaging genetics. We first introduce two simple sparse models, that is, Lasso and sparse logistic regression, in Section 5.2. Then, in Section 5.3, we introduce a series of popular structured sparse methods, which incorporate some prior knowledge. We will also review some popular optimization algorithms in Section 5.4. In Section 5.5, we pay particular attention to a suite of novel techniques, that is, screening rules, for sparse models (Hastie et al., 2015; Wang et al., 2015b), which can improve the computational efficiency by several orders of magnitude.

5.2 BASIC SPARSE MODELS

To illustrate the basic idea of sparse models, in this section we introduce two simple but widely used sparse models: Lasso (Tibshirani, 1996) that is for regression and sparse logistic regression (Sun et al., 2009; Wu et al., 2009; Zhu and Hastie, 2004) that is for classification.

Suppose that the training samples contain N observations with p features. We denote the outcome by a vector $\mathbf{y} \in \mathbb{R}^p$ and the feature matrix by $\mathbf{X} \in \mathbb{R}^{N \times p}$. By convention, each row $\mathbf{x}^i \in \mathbb{R}^p$, $i = 1, \ldots, N$, of $\mathbf{X}$ represents a data sample and each column $\mathbf{x}_j \in \mathbb{R}^N$, $j = 1, \ldots, p$, of $\mathbf{X}$ represents a feature. In this chapter, we mainly focus on linear models $h : \mathbb{R}^p \to \mathbb{R}$ with

$$h(\mathbf{x}) = \beta^{\mathrm{T}}\mathbf{x}, \tag{5.1}$$

where $\beta \in \mathbb{R}^p$ is the coefficient vector that needs to be estimated.

Many traditional regression and classification methods like least squares and logistic regression are developed for low-dimensional data sets (James et al., 2013), in which the number of observations N is much larger than the number of features. However, as new technologies have advanced in the past two decades, we are frequently confronted with extremely high-dimensional data sets (like fMRI, PET, and GWAS), in which the number of features p is much greater than the number of samples N. Directly applying traditional regression or classification methods to the high-dimensional data sets may be inappropriate. Take the least squares regression as an example. When $p \gg N$, we can find a regression hyperplane that fits the data exactly (the training error is zero). In many applications, a perfect fit on the training data usually implies overfitting, which may lead to poor performance on the testing data.

Regularization has been shown to be a promising approach to alleviate overfitting. Many sparse models estimate the coefficient vector β by incorporating various sparse-inducing regularizers:

$$\min_{\beta} f(\beta) = \ell(\beta) + \lambda\Omega(\beta), \tag{5.2}$$

where $\ell(\beta)$ is a loss function measuring the fitness of the model on the training data, $\Omega(\beta)$ is the regularizer penalizing the complexity of the model, and $\lambda > 0$ is a regularization parameter controlling the trade-off between the loss $\ell(\cdot)$ and the penalty $\Omega(\cdot)$. We note that the sparse-inducing penalty $\Omega(\cdot)$ is a typically nonsmooth function of the coefficient vector.

Lasso is a widely used regression technique to find sparse representations of a given signal with respect to a set of basis vectors. Standard Lasso employs least squares loss and $\Omega(\beta) = \|\beta\|_1$ as its regularizer, that is,

$$\min_{\beta} \frac{1}{2}\|\mathbf{y} - \mathbf{X}\beta\|^2 + \lambda\|\beta\|_1. \tag{5.3}$$

Due to the ℓ_1-norm penalty, many components of the solution vector of Lasso are zero when the value of λ is large. The features corresponding to these nonzero components are considered to be important to explain the outcome. Therefore, in a wide range of real applications, Lasso serves as an effective feature selection method and has achieved great success (Bruckstein et al., 2009; Chen et al., 2001; Candès, 2006; Wright et al., 2010; Zhao and Yu, 2006).

Similar to Lasso, sparse logistic regression also employ the ℓ_1-norm regularization, while utilizing the logistic loss. Specifically, sparse logistic regression takes the form:

$$\min_{\beta} \sum_{i=1}^{N} \log\left(\frac{1}{1+\mathrm{e}^{-y_i(\beta^{\mathrm{T}}\mathbf{x}^i)}}\right) + \lambda\|\beta\|_1. \tag{5.4}$$

Sparse logistic regression has received much attention in the last few years and the interest in it is growing (Sun et al., 2009; Wu et al., 2009; Zhu and Hastie, 2004) due to the increasing prevalence of high-dimensional data. The popularity of sparse logistic regression is also due to the fact that it can simultaneously achieve the goals of classification and feature selection.

5.3 STRUCTURED SPARSE MODELS

A major drawback of Lasso and sparse logistic regression is that they do not take the feature structure into account. In other words, the sparse representation obtained by Lasso or sparse logistic regression remains the same if we shuffle the features. However, in many real applications, this is undesirable, as the features frequently exhibit certain intrinsic structures, for example, trees, graphs, spatial or temporal smoothness, and disjoint/overlapping groups.

In this section, we introduce several structured sparse models, which incorporate different prior knowledge of feature structures by carefully designed sparse-inducing regularizers.

5.3.1 GROUP LASSO AND SPARSE GROUP LASSO

In many applications, the features form groups or clusters. For example, features with discrete values are usually transformed into groups of dummy variables; in the study of Alzheimer's disease, we divide the voxels of the PET images into a set of nonoverlapping groups according to the brain regions. To select groups of features, Yuan and Lin (2006) proposed the nonoverlapping group Lasso, in which the groups do not share features. Assume that the features are partitioned into k disjoint groups $\{G_1, \ldots, G_k\}$, where G_i contains the indices of features belonging to the ith group. The regularizer of group Lasso takes the form:

$$\Omega_{\mathrm{gL}}(\beta) = \sum_{i=1}^{k} w_i \|\beta_{G_i}\|_q, \tag{5.5}$$

where w_i is the weight for the ith group, $\mathbf{X}_{G_i}$ is the submatrix whose columns consist of the features belonging to the ith group, and $\|\cdot\|_q$ with $q > 1$ is the

ℓ_q-norm (the value of q is usually set to be 2 or ∞) (Wang et al., 2013). Group Lasso has been widely used in applications with group structure available, for example, regression (Kowalski, 2009; Negahban and Wainwright, 2008; Yuan and Lin, 2006), classification (Meier et al., 2008), joint covariate selection for group selection (Obozinski et al., 2007), and multitask learning (Argyriou et al., 2008; Liu et al., 2009; Quattoni et al., 2009).

Group Lasso performs group selection. However, for some applications, it is desirable to identify features within each group that exhibit the strongest effects. To achieve this goal, sparse-group Lasso (SGL) (Friedman et al., 2010; Simon et al., 2013) combines the Lasso (Tibshirani, 1996) and group Lasso (Yuan and Lin, 2006) penalties to identify important groups and features simultaneously. Specifically, the sparse group Lasso penalty can be written as follows:

$$\Omega_{\mathrm{SGL}}(\beta) = \alpha\|\beta\|_1 + (1-\alpha)\sum_{i=1}^{k} w_i\|\beta_{G_i}\|_q, \tag{5.6}$$

where $\alpha \in [0, 1]$ balances the sparsity in the feature level and the sparsity in the group level. In recent years, SGL has found great success in many real-world applications, including but not limited to machine learning (Vidyasagar; Yogatama and Smith, 2014), signal processing (Sprechmann et al., 2011), bioinformatics (Peng et al., 2010), etc.

5.3.2 OVERLAPPING GROUP LASSO AND TREE LASSO

Group Lasso assumes that the feature groups are disjoint. However, in certain applications, some features may be shared across different groups. For example, in the study of biologically meaningful gene/proteins, we say that the proteins/genes in the same groups are related in the sense that: (1) the proteins/genes appear in the same pathway; (2) the proteins/genes belong to the same Gene Ontology (GO) term (Ashburner et al., 2000; Harris et al., 2004); (3) the proteins/genes are related from gene set enrichment analysis (GSEA) (Subramanian et al., 2005). As the same gene may be involved in different pathways, it may be shared across different groups. The overlapping group Lasso penalty (Zhao et al., 2009) takes the form:

$$\Omega_{\mathrm{ogL}}(\beta) = \alpha\|\beta\|_1 + (1-\alpha)\sum_{i=1}^{k} w_i\|\beta_{G_i}\|_q, \tag{5.7}$$

where $\alpha \in [0, 1]$, w_i is the nonnegative weight for the ith group, and G_i consists of the feature indices from the ith group. The difference that distinguishes (5.7) and (5.6) is that G_i may overlap with G_j for $i \neq j$.

A particularly interesting special case of overlapping group Lasso is the so-called tree structured group Lasso (tgLasso) (Kim and Xing, 2010; Zhao et al., 2009). In some applications, the data may exhibit hierarchical tree-structured sparse patterns among features. For example, based on the spatial locality (Liu and Ye, 2010), we can represent an image by a tree whose leaf node corresponds to a single feature (pixel) and whose internal node corresponds to a group of features (pixels). Another interesting application of tgLasso is to identify risk SNPs regarding AD from the GWAS data (Li et al., 2016). It is known that we can measure the association of alleles at different loci by linkage disequilibrium (LD). Thus by taking the LD information and the chromosomal loci of SNPs into account, we can build the tree structure for SNPs. If the tree structure is available, the tree-structured group Lasso penalty takes the form:

$$\Omega_{\mathrm{tgL}}(\beta) = \sum_{i,j} w_j^i \|\beta_{G_j^i}\|_q, \tag{5.8}$$

where G_j^i is the group of features corresponding to the jth node of depth i and w_j^i is the positive weight for G_j^i. We note that every node in the tree is a superset of its descendant nodes. Therefore if the features in a node are absent from the sparse representation, so are the features in all its descendant nodes.

5.3.3 FUSED LASSO AND GRAPH LASSO

In many applications, the data come with spatial or temporal smoothness. For example, in the study of arrayCGH (Tibshirani et al., 2005; Tibshirani and Wang, 2008), the features—the DNA copy numbers along the genome—have the natural spatial order. The fused Lasso penalty encodes the structure of smoothness by penalizing the differences between the adjacent coefficients, that is,

$$\Omega_{\mathrm{fL}}(\beta) = \alpha\|\beta\|_1 + (1-\alpha)\sum_{i=1}^{p-1} |\beta_i - \beta_{i-1}|, \tag{5.9}$$

where $\alpha \in [0, 1]$. We can see that the fused Lasso penalty would lead to solutions in which adjacent components are close or identical to each other.

In certain applications, the data may exhibit a more complex smoothness structure. Specifically, the features may form an undirected graph structure, in which connected features may share some common properties. For example, much biological evidence suggested that genes tend to work in groups if they have similar biological functions (Li and Li, 2008). This prior knowledge can be encoded by a graph, in which each node represents a gene and the edges denote the regulatory relationships between genes. Many recent works have shown that the structure information encoded as a graph can significantly improve the predictive power of the model. Let (V, E) be a given graph, where V denotes the set of nodes and E denotes the edges. By noting that an open chain is a special example of a graph, we can generalize the fused Lasso penalty to a graph Lasso penalty—known as the ℓ_1 graph Lasso—as follows:

$$\Omega_{\mathrm{grL}}^{\ell_1}(\beta) = \alpha\|\beta\|_1 + (1-\alpha)\sum_{(i,j)\in E} |\beta_i - \beta_j|, \tag{5.10}$$

where the second term penalizes the difference between the coefficients of connected features. Thus the coefficients of connected features tend to be close or identical to each other. The graph Lasso penalty in (5.10) entails a significant computational challenge as both terms are nonsmooth and the graph structure may be complicated. An efficient alternative, called Laplacian Lasso (or ℓ_2 graph Lasso), employs the following penalty:

$$\Omega_{\mathrm{grL}}^{\ell_2}(\beta) = \alpha\|\beta\|_1 + (1-\alpha)\beta^{\mathrm{T}} L\beta, \tag{5.11}$$

where L is the graph Laplacian matrix (Belkin and Niyogi, 2003; Chung, 1997). The graph Laplacian matrix is positive semidefinite and well captures the local geometric structure of the data. We note that the Laplacian Lasso penalty reduces to the elastic net penalty (Zou and Hastie, 2005) if the Laplacian matrix is an identity matrix. As the second term of the Laplacian penalty is quadratic, we can incorporate it with the least square loss, and thus many solvers for Lasso are applicable to the Laplacian Lasso.

In view of (5.10) and (5.11), both ℓ_1 graph Lasso and Laplacian Lasso encourage positive correlation of the coefficients of connected features, that is, they tend to have the same sign. However, in certain applications, connected features may be negatively correlated. To deal with this challenge, GFLasso incorporates the sample correlation into the penalty:

$$\Omega_{\mathrm{GFL}}(\beta) = \alpha\|\beta\|_1 + (1-\alpha)\sum_{(i,j)\in E} |\beta_i - \mathrm{sign}(r_{i,j})\beta_j|, \tag{5.12}$$

where $r_{i,j}$ is the sample correlation between the ith and jth features. We can see that GLLasso encourages positive correlation between connected features if $r_{i,j} > 0$ and negative correlation if $r_{i,j} < 0$. However, if the sample correlation is inaccurate, GFLasso may introduce additional bias.

An alternative, which is called graph OSCAR (GOSCAR), avoids the usage of the sample correlation. The GOSCAR penalty can be written as

$$\Omega_{\mathrm{GFL}}(\beta) = \alpha\|\beta\|_1 + (1-\alpha)\sum_{(i,j)\in E} \max\{|\beta_i|, |\beta_j|\}. \tag{5.13}$$

The ℓ_∞ penalty encourages the magnitude of the coefficients of connected features to be close or identical to each other. However, the ℓ_∞ penalty may overpenalize the coefficients, leading to additional bias. This motivates a nonconvex version of graph Lasso penalty:

$$\Omega_{\mathrm{ncFGS}}(\beta) = \alpha\|\beta\|_1 + (1-\alpha)\sum_{(i,j)\in E} ||\beta_i| - |\beta_j||, \tag{5.14}$$

which can reduce the bias in many applications compared to the aforementioned convex penalties. Similar to GOSCAR, ncFGS penalty does not make use of the sample correlation either.

5.4 OPTIMIZATION METHODS

Many sparse models in the form of (5.2) are nonsmooth and nondifferentiable, which imposes a serious challenge to the corresponding optimization algorithms. In the past few years, as sparse models have become increasingly popular, extensive research efforts have been devoted to developing efficient solvers for the sparse models. In this section, we briefly review two particularly popular first-order methods: proximal gradient descent and accelerated gradient methods, which are especially useful for large-scale problems.

5.4.1 PROXIMAL GRADIENT DESCENT

In this section, we briefly review the well-known proximal gradient descent algorithm for (5.2). For many sparse models, the loss function $\ell(\cdot)$ is convex and differentiable, and the regularizer $\Omega(\cdot)$ is convex but nondifferentiable. The major challenge in developing optimization algorithms for (5.2) is due to the nondifferentiable regularizer $\Omega(\cdot)$.

The key idea (Beck and Teboulle, 2009; Hastie et al., 2015) of proximal gradient descent is that, in each iteration, we minimize a local approximation of $f(\cdot)$ consisting of the nondifferentiable component $\Omega(\cdot)$ and a linear approximation of the differentiable component $\ell(\cdot)$. Specifically, in the kth iteration, we update β^k by the following generalized gradient update:

$$\beta^{k+1} = \operatorname*{argmin}_{\beta} \left\{ \ell(\beta^k) + \langle \nabla \ell(\beta^k), \beta - \beta^k \rangle + \frac{1}{2t^k} \|\beta - \beta^k\|^2 + \Omega(\beta) \right\}. \tag{5.15}$$

For a convex function h, we can define the proximal map:

$$\mathbf{prox}_h(u) = \operatorname*{argmin}_{\mathbf{v}} \left\{ \frac{1}{2} \|\mathbf{v} - \mathbf{u}\|^2 + h(\mathbf{v}) \right\}. \tag{5.16}$$

Then, it follows that

$$\beta^{k+1} = \mathbf{prox}_{t^k \Omega} \left(\beta^k - t^k \nabla \ell(\beta^k) \right). \tag{5.17}$$

Sufficient conditions (Nesterov, 2007) for the convergence of the update in (5.17) are as follows:

1. The gradient of the differentiable component $\ell(\cdot)$ is Lipschitz continuous, that is, for any $\beta, \beta' \in \mathbb{R}^p$, the following inequality holds:

$$\|\nabla \ell(\beta) - \nabla \ell(\beta')\|_2 \leq L \|\beta - \beta'\|_2. \tag{5.18}$$

2. The step size t^k is a constant that satisfies $t^k \in (0, 1/L]$.

Then, it can be shown that

$$f(\beta^k) - f(\beta^*) \leq \frac{L\|\beta^0 - \beta^*\|^2}{2k}, \tag{5.19}$$

where β^* is an optimal solution. Thus (5.19) implies that the proximal gradient descent in (5.17) leads to a convergence rate of $O(1/k)$.

5.4.2 ACCELERATED GRADIENT METHOD

When the proximal mapping in (5.17) can be computed efficiently, the proximal gradient descent approach is a very popular tool in solving the corresponding sparse models, especially for large-scale problems. However, the convergence can be slow for certain objective functions, as the update by proximal gradient descent may lead to an undesirable type of zig-zagging behavior from step to step (Hastie et al., 2015). To improve the convergence property, Nesterov (Nesterov, 1983, 2007) proposed a class of accelerated gradient methods with a convergence rate $O(1/k^2)$. We summarize the accelerated gradient method in Algorithm 5.1.

ALGORITHM 5.1 ACCELERATED GRADIENT METHOD

Input: A constant $t \in (0, 1/L]$, where L is a Lipschitz constant of $\nabla\ell$.

Set $\beta^0 = \theta^1 \in \mathbb{R}^p$, $s^1 = 1$, and $k = 1$.

while termination condition is not satisfied **do**

$\quad \beta^k = \mathbf{prox}_{t\Omega}(\theta^k - t\nabla\ell(\theta^k))$,

$\quad s^{k+1} = \frac{1+\sqrt{1+4(s^k)^2}}{2}$,

$\quad \theta^{k+1} = \beta^k + \left(\frac{s^k - 1}{s^{k+1}}\right)(\beta^k - \beta^{k-1})$,

$\quad k = k + 1$.

end while

Let β^k be generated by Algorithm 5.1. Then, it is shown that

$$f(\beta^k) - f(\beta^*) \leq \frac{2L\|\beta^0 - \beta^*\|^2}{(k+1)^2}, \tag{5.20}$$

where β^* is an optimum.

We note that, besides the convergence rates, a key difference—that distinguishes the accelerated gradient method from proximal gradient descent—is that the function values computed by the former may be increasing, that is, $f(\beta^{k+1})$ may be larger than $f(\beta^k)$, while they keep decreasing for the latter.

5.5 SCREENING

In the past few years, many algorithms have been proposed to efficiently solve the sparse models. However, when the feature dimension is extremely large, the applications of sparse models to large-scale problems remain challenging due to their nondifferentiable and complicated regularizers.

In the past few years, the idea of screening (El Ghaoui et al., 2012; Tibshirani et al., 2012; Wang et al., 2015b) has been found to be a very promising approach to improve the efficiency of sparse models. Essentially, screening aims to quickly identify the zero coefficients in the sparse solutions by simple testing rules. Then, we can remove the corresponding features from the optimization without sacrificing accuracy. Thus the size of the data matrix can be significantly reduced, leading to substantial savings in computational cost and memory usage. In many applications, the speedup gained by screening methods can be several orders of magnitude.

In this section, we focus on the screening method for Lasso. We also briefly review some screening methods for other, more complicated sparse models.

5.5.1 SCREENING FOR LASSO

We can roughly divide existing screening methods of Lasso into two categories: the heuristic screening methods and the safe screening methods.

As implied by the name, the heuristic screening methods may mistakenly discard features that have nonzero coefficients in the sparse representations. This type of method includes SIS (Fan et al., 2008) and strong rule (Tibshirani et al., 2012). SIS removes features based on the correlation between features and the outcome, but not from the perspective of optimization. Strong rule assumes that the inner products between features and the residue are nonexpansive (Bauschke and Combettes, 2011) with respect to the parameter values. However, this assumption may not hold in real applications. Thus strong rule needs to postprocess the results by KKT conditions to check if it makes mistakes.

In contrast to the heuristic screening methods, the safe screening methods can guarantee that the coefficients of the discarded features are zero in the solution vector. Existing safe screening methods include SAFE (El Ghaoui et al., 2012), DOME (Xiang and Ramadge, 2012), and EDPP (Wang et al., 2015b), which are inspired by the KKT conditions.

We note that, although heuristic in theory, strong rule seldom makes mistakes in practice and it significantly outperforms many safe screening methods like SAFE and DOME. Therefore, in this section, we focus on the EDPP screening rule, whose performance is comparable to or even better than strong rule. For details of EDPP (see Wang et al., 2015b).

5.5.1.1 Background

Recall that the Lasso problem is given by (5.3). It is known that the dual problem of Lasso is equivalent to

$$\inf_{\theta} \left\{ \frac{1}{2} \left\| \theta - \frac{\mathbf{y}}{\lambda} \right\|_2^2 : |\mathbf{x}_i^T \theta| \leq 1, i = 1, 2, \ldots, p \right\}. \tag{5.21}$$

Let $\beta^*(\lambda)$ and $\theta^*(\lambda)$ be the optimal solutions of problems (5.3) and (5.21) respectively, and $\mathcal{F}$ be the feasible set of problems (5.21). For notational convenience, we define the projection operator by

$$\mathbf{P}_C(\mathbf{w}) = \underset{\mathbf{u} \in C}{\text{argmin}} \|\mathbf{u} - \mathbf{w}\|_2, \tag{5.22}$$

where C is a closed and convex set. We can see that $\theta^*(\lambda)$ is the projection of $\mathbf{y}/\lambda$ onto $\mathcal{F}$, that is,

$$\theta^*(\lambda) = \mathbf{P}_{\mathcal{F}} \left(\frac{\mathbf{y}}{\lambda} \right). \tag{5.23}$$

Moreover, the primal optimum and dual optimum are related by the KKT conditions:

$$\mathbf{y} = \mathbf{X}\beta^*(\lambda) + \lambda\theta^*(\lambda), \tag{5.24}$$

$$(\theta^*(\lambda))^{\mathrm{T}} \mathbf{x}_i \in \begin{cases} \text{sign}([\beta^*(\lambda)]_i), & \text{if } [\beta^*(\lambda)]_i \neq 0, \\ [-1, 1], & \text{if } [\beta^*(\lambda)]_i = 0, \end{cases} \tag{5.25}$$

where $[\cdot]_k$ denotes the kth component.

Inspired by the KKT condition in Eq. (5.25), we can see that

$$|(\theta^*(\lambda))^{\mathrm{T}} \mathbf{x}_i| < 1 \Rightarrow [\beta^*(\lambda)]_i = 0, \text{ie}, \mathbf{x}_i \text{ is an inactive feature.} \tag{R1}$$

Thus we can potentially utilize (R1) to identify the inactive features for the Lasso problem. However, by a closer look at (R1), we can see that (R1) is not applicable to identify the inactive features, as it involves $\theta^*(\lambda)$. Inspired by the SAFE rules (El Ghaoui et al., 2012), we can relax (R1) as follows:

$$\sup_{\theta \in \Theta} |\mathbf{x}_i^{\mathrm{T}} \theta| < 1 \Rightarrow [\beta^*(\lambda)]_i = 0, \text{ie}, \mathbf{x}_i \text{ is an inactive feature,} \tag{R1'}$$

where Θ is a set that contains $\theta^*(\lambda)$.

Thus, without the knowledge of $\theta^*(\lambda)$, (R1') implies that an estimation of the dual optimum is sufficient to develop an applicable screening rule for the Lasso problem. Nevertheless, in view of (R1) and (R1'), we can see that a small region Θ implies an accurate estimation of $\theta^*(\lambda)$, and thus a more aggressive screening rule for identifying the inactive features.

A useful consequence of (R1) is that we can find the smallest value of λ such that $\beta^*(\lambda) = 0$. Indeed, we have (Wang et al., 2015b)

$$\lambda \geq \lambda_{\max} = \|\mathbf{X}^{\mathrm{T}}\mathbf{y}\|_\infty \Leftrightarrow \beta^*(\lambda) = 0. \tag{5.26}$$

5.5.1.2 Enhanced DPP (EDPP) screening rules

Following (R1'), the framework of the EDPP screening rule (Wang et al., 2015b) for Lasso can be divided into the following three steps:

1. We first estimate a region Θ which contains the dual optimum $\theta^*(\lambda)$.
2. We solve the maximization problem in (R1'), that is, $\sup_{\theta\in\Theta} |\mathbf{x}_i^T\theta|$.
3. By plugging in the upper bound we find, in the last step, that it is straightforward to develop the screening rule based on (R1').

The key step of EDPP is the estimation of the dual optimum, which determines the performance of the screening rule. Based on the geometric properties of the dual problem, EDPP provides a very accurate estimation of the dual optimum.

The first geometric property that EDPP utilizes is the so-called firmly nonexpansiveness of the projection operators.

Theorem 5.1 (Bauschke and Combettes, 2011). *Let C be a nonempty closed convex subset of a Hilbert space $\mathcal{H}$. Then the projection operator defined in Eq. (5.22) is continuous and firmly nonexpansive. In other words, for any $\mathbf{w}_1, \mathbf{w}_2 \in \mathcal{H}$, we have*

$$\|\mathbf{P}_C(\mathbf{w}_1) - \mathbf{P}_C(\mathbf{w}_2)\|_2^2 + \|(Id - \mathbf{P}_C)(\mathbf{w}_1) - (Id - \mathbf{P}_C)(\mathbf{w}_2)\|_2^2 \le \|\mathbf{w}_1 - \mathbf{w}_2\|_2^2,$$

where Id is the identity operator.

Another useful geometric property of the projection operators is related to the projection of rays.

Lemma 5.1 (Bauschke and Combettes, 2011). *Let C be a nonempty closed convex subset of a Hilbert space $\mathcal{H}$. For a point $\mathbf{w} \in \mathcal{H}$, let $\mathbf{w}(t) = \mathbf{P}_C(\mathbf{w}) + t(\mathbf{w} - \mathbf{P}_C(\mathbf{w}))$. Then, the projection of the point $\mathbf{w}(t)$ is $\mathbf{P}_C(\mathbf{w})$ for all $t \ge 0$, that is,*

$$\mathbf{P}_C(\mathbf{w}(t)) = \mathbf{P}_C(\mathbf{w}), \quad \forall t \ge 0.$$

Based on Theorem 5.1 and Lemma 5.1, EDPP estimates the dual optimum as follows.

Theorem 5.2 (Wang et al., 2015b). *For the Lasso problem, suppose that the dual optimal solution $\theta^*(\cdot)$ at $\lambda_0 \in (0, \lambda_{\max}]$ is known. For any $\lambda \in (0, \lambda_0]$, let us define*

$$\mathbf{x}_* = \operatorname{argmax}_{\mathbf{x}_i} |\mathbf{x}_i^T\mathbf{y}|, \tag{5.27}$$

$$\mathbf{v}_1(\lambda_0) = \begin{cases} \dfrac{\mathbf{y}}{\lambda_0} - \theta^*(\lambda_0), & \text{if } \lambda_0 \in (0, \lambda_{\max}), \\ \operatorname{sign}(\mathbf{x}_*^T\mathbf{y})\mathbf{x}_*, & \text{if } \lambda_0 = \lambda_{\max}, \end{cases} \tag{5.28}$$

$$\mathbf{v}_2(\lambda, \lambda_0) = \frac{\mathbf{y}}{\lambda} - \theta^*(\lambda_0), \tag{5.29}$$

$$\mathbf{v}_2^{\perp}(\lambda, \lambda_0) = \mathbf{v}_2(\lambda, \lambda_0) - \frac{\langle \mathbf{v}_1(\lambda_0), \mathbf{v}_2(\lambda, \lambda_0)\rangle}{\|\mathbf{v}_1(\lambda_0)\|_2^2}\mathbf{v}_1(\lambda_0). \tag{5.30}$$

Then, the dual optimal solution $\theta^*(\lambda)$ can be estimated as follows:

$$\left\|\theta^*(\lambda) - \left(\theta^*(\lambda_0) + \frac{1}{2}\mathbf{v}_2^{\perp}(\lambda, \lambda_0)\right)\right\|_2 \le \frac{1}{2}\|\mathbf{v}_2^{\perp}(\lambda, \lambda_0)\|_2.$$

For notational convenience, let

$$\begin{aligned}
\mathbf{o}(\lambda, \lambda_0) &= \theta^*(\lambda_0) + \frac{1}{2}\mathbf{v}_2^{\perp}(\lambda, \lambda_0),\\
r(\lambda, \lambda_0) &= \frac{1}{2}\|\mathbf{v}_2^{\perp}(\lambda, \lambda_0)\|_2,\\
\Theta(\lambda, \lambda_0) &= \{\theta : \|\theta - \mathbf{o}(\lambda, \lambda_0)\| \le r(\lambda, \lambda_0)\}.
\end{aligned}$$

Theorem 5.2 implies that

$$\theta^*(\lambda) \in \Theta(\lambda, \lambda_0). \tag{5.31}$$

By Cauchy-Schwartz inequality, it is easy to solve the optimization problem $\max_{\theta\in\Theta(\lambda,\lambda_0)} |\mathbf{x}_i^{\mathrm{T}}\theta|$. Thus, by (R1') and Eq. (5.24), we immediately have the following EDPP screening rules.

Theorem 5.3. *For the Lasso problem, assume that the dual optimum $\theta^*(\cdot)$ at $\lambda_0 \in (0, \lambda_{\max}]$ is known, and $\lambda \in (0, \lambda_0]$. Then, we have $[\beta^*(\lambda)]_i = 0$ if the following holds:*

$$\left|\mathbf{x}_i^T\left(\theta^*(\lambda_0) + \frac{1}{2}\mathbf{v}_2^{\perp}(\lambda, \lambda_0)\right)\right| < 1 - \frac{1}{2}\|\mathbf{v}_2^{\perp}(\lambda, \lambda_0)\|_2\|\mathbf{x}_i\|_2.$$

In many real applications, the optimal values of the parameters are usually unknown. To determine an appropriate parameter value, commonly used approaches such as cross-validation and stability selection solve the optimization problem along a grid of parameter values, which can be very time consuming. Motivated by the ideas of (Tibshirani et al., 2012; El Ghaoui et al., 2012), we can develop a sequential version of EDPP rules. Specifically, if we need to solve the Lasso problem along a sequence of parameter values $\lambda_1 > \lambda_2 > \cdots > \lambda_m$, we can first apply EDPP to discard inactive features for the Lasso problem with parameter value being λ_1. After solving the reduced optimization problem at λ_1, we obtain the exact solution $\beta^*(\lambda_1)$. Then, by Eq. (5.24), we can find $\theta^*(\lambda_1)$. According to (R1'), once we know the optimal dual solution $\theta^*(\lambda_1)$, we can construct a new screening rule to identify inactive features for Lasso at λ_2 based on $\theta^*(\lambda_1)$. By repeating the above process, we obtain the sequential version of the EDPP rule.

We formulate the sequential version of EDPP as follows.

Corollary 5.1 (EDPP). *For the Lasso problem, suppose that we are given a sequence of parameter values $\lambda_{\max} = \lambda_0 > \lambda_1 > \cdots > \lambda_{\mathcal{K}}$. Then for any integer $0 \le k < \mathcal{K}$, we have $[\beta^*(\lambda_{k+1})]_i = 0$ if $\beta^*(\lambda_k)$ is known and the following holds:*

$$\left|\mathbf{x}_i^T\left(\frac{\mathbf{y} - \mathbf{X}\beta^*(\lambda_k)}{\lambda_k} + \frac{1}{2}\mathbf{v}_2^{\perp}(\lambda_{k+1}, \lambda_k)\right)\right| < 1 - \frac{1}{2}\|\mathbf{v}_2^{\perp}(\lambda_{k+1}, \lambda_k)\|_2\|\mathbf{x}_i\|_2. \tag{5.32}$$

5.5.1.3 Applications of EDPP to imaging genetics

In this section, we apply Lasso to identify potential risk SNPs—that are related to AD imaging phenotypes—from the ADNI WGS data by stability selection. As stability selection usually involves solving the Lasso problem many times, this process can be very time consuming. Thus we utilize EDPP to speedup the computations. We can see that the speed up gained by EDPP can be orders of magnitude. For more discussions, see Yang et al. (2015).

The ADNI WGS data contains 717 subjects. We choose the baseline hippocampal volume to be the response. To illustrate the performance of EDPP in terms of speedup, we vary the number of features p from 0.1 million to one million with a step size of 0.1 million. For each value of p, we solve the Lasso problems along a sequence of 100 parameter values equally spaced in the logarithmic scale of $\lambda/\lambda_{\max}$ from 1.0 to 0.05.

Fig. 5.1 reports the speedup gained by EDPP for data sets with different dimensions. We can observe that the speedup is up to 406 times. Moreover, Fig. 5.1 shows that the speedup gained by EDPP increases with the feature dimension growth. This implies that EDPP is a promising approach to improve the efficiency of Lasso, especially for large-scale data sets.

We next apply Lasso to explore the imaging genetics association between imaging phenotypes and SNPs from the ADNI WGS SNPs data with 329 subjects and 5,906,152 features. We utilize EDPP to facilitate the computation of Lasso problems. Specifically, for each of the enthorhinal cortex (EC) and hippocampus (HIPP) brain regions, we choose the volume at baseline and volume changes over a 24-month interval as the response vectors. We employ stability selection (Meinshausen and

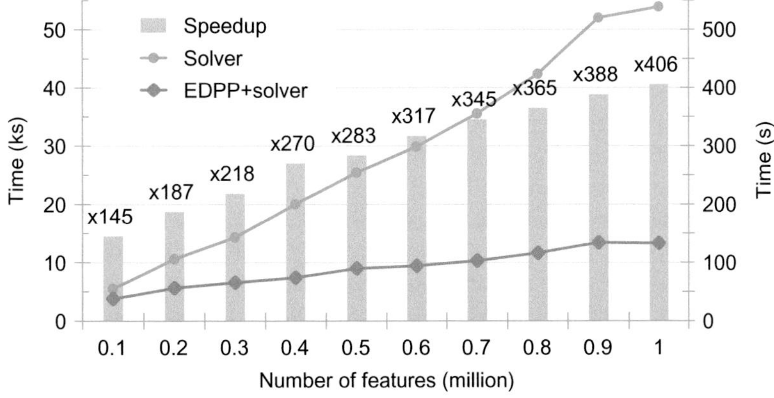

FIG. 5.1

Speedup gained by EDPP (Yang et al., 2015). Run time of solver and solver combined with EDPP are in units of kiloseconds and seconds, respectively.

Bühlmann, 2010) to identify the risk SNPs. For each response vector, we perform 100 simulations. In each simulation, we first randomly select half of the samples, and then we apply the solver combined with EDPP to solve the Lasso problem along a sequence of 100 parameter values equally spaced on the logarithmic scale of λ/λ_{max} from 1.0 to 0.05.

Tables 5.1 and 5.2 report the top 10 SNPs that are most frequently selected for each outcome. Table 5.1 shows that the genes corresponding to the selected top 10 SNPs contain those genes that are implicated in the risk of AD or other neuropsychiatric disorders. For example, APOE—that ranks among the top predictors of baseline enthorhinal and hippocampal volumes—is a top AD risk gene. This is consistent with the observations from much of the existing literature that APOE has an effect on temporal lobe structures not just in old age but also in children and adolescents (Shaw et al., 2007). We can also observe from Table 5.2 that the fourth ranked gene related to hippocampal volume change is CACNA1C. The gene CACNA1C is involved in calcium channel function that is associated with DTI measures in the hippocampus. It is known (Strohmaier et al., 2013) that CACNA1C is among the top genes associated with anxiety, depression, and obsessive compulsive disorder. Moreover, as shown in Table 5.2, Lasso identifies a gene named BACE2 that is a close homolog of BACE1. The existing literature shows that BACE1 encodes a key enzyme involved in the cellular pathways of AD.

5.5.2 SCREENING METHODS FOR OTHER SPARSE MODELS

As an emerging and promising technique in dealing with large-scale problems, *screening* has achieved great success in improving the efficiency of many popular sparse models, for example, Lasso (El Ghaoui et al., 2012; Tibshirani et al., 2012; Wang et al., 2015b; Xiang et al., 2011), nonnegative Lasso (Wang and Ye, 2014), group Lasso (Tibshirani et al., 2012; Wang et al., 2015b), mixed-norm regression (Wang et al., 2013), ℓ_1-regularized logistic regression (Wang et al., 2014b), sparse-group Lasso (Wang and Ye, 2014), tree-structured group Lasso (Wang and Ye, 2015), fused Lasso (Wang et al., 2015a), support vector machine (SVM) (Ogawa et al., 2013; Wang et al., 2014a), and least absolute deviation (LAD) (Wang et al., 2014a). The speedup gained by screening rules can be orders of magnitude.

We note that the framework of EDPP is very flexible. We have extended EDPP to many popular sparse models including all of the aforementioned ones. The package, named DPC (Dual Projection to Convex sets), is available at http://dpc-screening.github.io/. Interestingly, the recently proposed screening method for tree-structured group Lasso, MLFre (Wang and Ye, 2015), covers EDPP for Lasso as a special case, as Lasso, group Lasso, and sparse group Lasso are all special cases of tree-structured group Lasso.

Moreover, we can easily implement the screening rule EDPP and its variants in parallel, as the identification of inactive features/groups is independent from each other. This makes EDPP and its variants particularly suitable for distributed computation for data privacy and more efficiency (Li et al., 2016).

Table 5.1 Top 10 SNPs Associated With Baseline Volumes

	EC Baseline				HIPP Baseline			
	Chr	**Pos**	**RS_ID**	**Gene**	**Chr**	**Pos**	**RS_ID**	**Gene**
Rank 1	chr6	72869836	rs201890142	RIMS1	chr10	71969989	rs12412466	PPA1
Rank 2	chr19	15136345	Unknown	Unknown	chr19	45411941	rs429358	APOE
Rank 3	chr1	142555416	rs6672189	Unknown	chr11	11317240	rs10831576	GALNT18
Rank 4	chr19	45411941	rs429358	APOE	chr4	49147785	rs151073945	Unknown
Rank 5	chr2	96630311	rs369756382	ANKRD36C	chr8	145158607	rs34173062	MAF1
Rank 6	chr2	95552943	rs199536016	LOC442028	chr6	168107162	rs71573413	Unknown
Rank 7	chr2	95552986	rs200710055	LOC442028	chrX	143403234	rs4825209	Unknown
Rank 8	chr1	142545571	LOC442028	Unknown	chr2	231846840	rs4973360	Unknown
Rank 9	chr13	94122100	rs76403280	GPC6	chr14	20710095	rs35055545	OR11H4
Rank 10	chr5	97913219	rs202036446	Unknown	chr6	69775654	rs2343398	BAI3

Table 5.2 Top 10 SNPs Associated With Volume Changes

	EC Changes				HIPP Changes			
	Chr	**Pos**	**RS_ID**	**Gene**	**Chr**	**Pos**	**RS_ID**	**Gene**
Rank 1	chr7	7333294	rs1317198	Unknown	chr15	23060281	rs11636690	NIPA1
Rank 2	chr10	9216956	rs1149952	Unknown	chr21	42613255	rs74977559	BACE2
Rank 3	chr6	115278412	rs146156795	Unknown	chr9	91366940	rs79543088	Unknown
Rank 4	chr10	117854524	rs2530339	GFRA1	chr12	2342248	rs7303977	CACNA1C
Rank 5	chr8	6557130	rs2912047	LOC100507530	chr6	169666147	rs6605518	Unknown
Rank 6	chr12	23117263	rs12581794	Unknown	chr18	48740822	rs34794713	Unknown
Rank 7	chr16	77841030	rs16946521	VAT1L	chr13	102307271	rs9518474	ITGBL1
Rank 8	chr3	175784869	rs9845573	Unknown	chrX	2372296	rs7889210	DHRSX
Rank 9	chr4	7494498	rs4308363	SORCS2	chr4	174673036	rs12646029	LOC101928478
Rank 10	chr13	102616850	rs17502999	FGF14	chr4	170638416	rs149287207	CLCN3

5.6 CONCLUSIONS

In this chapter, we review many popular sparse models for imaging genetics. Due to their capability of incorporating various prior knowledge, sparse models are very effective in identifying the predictors that exhibit the strongest effects on the imaging phenotypes. However, as the regularizers of the sparse models are usually nonsmooth and complex, applications of sparse models to large-scale problems entail great challenges to existing optimization algorithms. To deal with this challenge, we introduce a suite of novel techniques, called sparse screening, to effectively scale existing algorithms to large-scale problems. This usually leads to substantial savings in memory usage and the resulting speedup can be several orders of magnitude. Thus we expect that sparse screening will be a powerful tool in facilitating the research of imaging genetics.

REFERENCES

Argyriou, A., Evgeniou, T., Pontil, M., 2008. Convex multi-task feature learning. Mach. Learn. 73 (3), 243–272.

Ashburner, M., Ball, C., Blake, J., Botstein, D., Butler, H., 2000. Gene ontology: tool for the unification of biology. the gene ontology consortium. Nat. Genet. 25, 25–29.

Batmanghelich, N.K., Dalca, A.V., Sabuncu, M.R., Golland, P., 2013. Joint modeling of imaging and genetics. In: Information Processing in Medical Imaging. Springer, Heidelberg, pp. 766–777.

Bauschke, H.H., Combettes, P.L., 2011. Convex Analysis and Monotone Operator Theory in Hilbert Spaces. Springer, New York.

Beck, A., Teboulle, M., 2009. A fast iterative shrinkage-thresholding algorithm for linear inverse problems. SIAM J. Imaging Sci. 2 (1), 183–202.

Belkin, M., Niyogi, P., 2003. Laplacian eigenmaps for dimensionality reduction and data representation. Neural Comput. 15, 1373–1396.

Bis, J., et al., 2012. Common variants at 12q14 and 12q24 are associated with hippocampal volume. Nat. Genet. 44 (5), 545–551.

Bruckstein, A., Donoho, D., Elad, M., 2009. From sparse solutions of systems of equations to sparse modeling of signals and images. SIAM Rev. 51, 34–81.

Candès, E.J., 2006. Compressive sampling. In: Proceedings of the International Congress of Mathematicians, Madrid, Spain, vol. 3, pp. 1433–1452.

Chen, S.S., Donoho, D.L., Saunders, M.A., 2001. Atomic decomposition by basis pursuit. SIAM Rev. 43, 129–159.

Chung, F., 1997. Spectral Graph Theory. American Mathematical Society Providence, RI.

Cirulli, E.T., Goldstein, D.B., 2010. Uncovering the roles of rare variants in common disease through whole-genome sequencing. Nat. Rev. Genet. 11 (6), 415–425.

Cornelis, M., Qi, L., Zhang, C., Kraft, P., Manson, J., Cai, T., Hunter, D., Hu, F., 2009. Joint effects of common genetic variants on the risk for type 2 diabetes in US men and women of European ancestry. Ann. Intern. Med. 150, 541–550.

Dinu, I., Mahasirimongkol, S., Liu, Q., Yanai, H., Eldin, N., Kreiter, E., Wu, X., Jabbari, S., Tokunaga, K., Yasui, Y., 2012. SNP-SNP interactions discovered by logic regression explain Crohns disease genetics. PLoS ONE 7, e43035.

El Ghaoui, L., Viallon, V., Rabbani, T., 2012. Safe feature elimination in sparse supervised learning. Pac. J. Optim. 8, 667–698.

Fan, R., Chang, K., Lv, J., 2008. Sure independence screening for ultrahigh dimensional feature spaces. J. R. Stat. Soc. B 70, 849–911.

Friedman, J., Hastie, T., Tibshirani, R., 2010. A note on the group lasso and a sparse group lasso. arXiv preprint, arXiv 1001.0736.

Golub, T.R., Slonim, D.K., Tamayo, P., Huard, C., Gaasenbeek, M., Mesirov, J.P., Coller, H., Loh, M.L., Downing, J.R., Caligiuri, M.A., Bloomfield, C.D., 1999. Molecular classification of cancer: class discovery and class prediction by gene expression monitoring. Science 286 (5439), 531–537.

Guyon, I., Weston, J., Barnhill, S., Vapnik, V., 2002. Gene selection for cancer classification using support vector machines. Mach. Learn. 46 (1–3), 389–422.

Hariri, A.R., Drabant, E.M., Weinberger, D.R., 2006. Imaging genetics: perspectives from studies of genetically driven variation in serotonin function and corticolimbic affective processing. Biol. Psychiat. 59 (10), 888–897.

Harris, M., et al., 2004. The gene ontology database and informatics resource. Nucleic Acids Res. 32, 258–261.

Hastie, T., Tibshirani, R., Wainwright, M., 2015. Statistical Learning With Sparsity: The Lasso and Generalizations. CRC Press, Boca Raton, FL.

James, G., Witten, D., Hastie, T., Tibshirani, R., 2013. An Introduction to Statistical Learning, vol. 112. Springer, New York.

Kim, S., Xing, E.P., 2010. Tree-guided group lasso for multi-task regression with structured sparsity. In: Proceedings of the 27th International Conference on Machine Learning (ICML-10), pp. 543–550.

Kowalski, M., 2009. Spare regression using mixed norms. Appl. Comput. Harmon. Anal. 27, 303–324.

Li, C., Li, H., 2008. Network-constrained regularization and variable selection for analysis of genomic data. Bioinformatics 24, 1175–1182.

Li, Q., Yang, T., Zhan, L., Hibar, D., Jananshad, N., Ye, J., Thompson, P., Wang, J., 2015. Large-scale collaborative genetic studies of risk SNPs for Alzheimer's disease across multiple institutions (under submission).

Li, Y., Wang, J., Yang, T., Chen, J., Liu, L., Thompson, P., Ye, J., 2016. Detection of Alzheimer's disease risk factors by tree-structured group lasso screening. In: 2016 IEEE 13th International Symposium on Biomedical Imaging (ISBI), in press.

Liu, J., Ye, J., 2010. Moreau-Yosida regularization for grouped tree structure learning. In: Lafferty, J.D., Williams, C.K.I., Shawe-Taylor, J., Zemel, R.S., Culotta, A. (Eds.), Advances in Neural Information Processing Systems 23. Curran Associates, Inc., Red Hook, NY, pp. 1459–1467.

Liu, H., Palatucci, M., Zhang, J., 2009. Blockwise coordinate descent procedures for the multi-task lasso, with applications to neural semantic basis discovery. In: Proceedings of the 26th Annual International Conference on Machine Learning, ICML '09, Montreal, Quebec, Canada. ACM, New York, NY, pp. 649–656. ISBN 978-1-60558-516-1

Medland, S.E., Jahanshad, N., Neale, B.M., Thompson, P.M., 2014. Whole-genome analyses of whole-brain data: working within an expanded search space. Nat. Neurosci. 17 (6), 791–800.

Meier, L., Geer, S., Bühlmann, P., 2008. The group lasso for logistic regression. J. R. Stat. Soc. B 70, 53–71.

Meinshausen, N., Bühlmann, P., 2010. Stability selection. J. R. Stat. Soc. B 72, 417–473.

Meyer-Lindenberg, A., 2012. The future of fMRI and genetics research. NeuroImage 62 (2), 1286–1292.

Mu, Y., Gage, F., 2011. Adult hippocampal neurogenesis and its role in Alzheimers disease. Mole. Neurodegen. 6, 85.

Negahban, S., Wainwright, M., 2008. Joint support recovery under high-dimensional scaling: benefits and perils of $\ell_{1,\infty}$-regularization. In: Advances in Neural Information Processing Systems, pp. 1161–1168.

Nesterov, Y., 1983. A method for solving a convex programming problem with convergence rate $1/k^2$. Sov. Math. Dokl. 27 (2), 372–376.

Nesterov, Y., 2007. Gradient methods for minimizing composite objective function. Center for Operations Research and Econometrics (CORE), Université Catholique de Louvain. CORE Discussion Papers No. 2007076, http://EconPapers.repec.org/RePEc:cor:louvco:2007076.

Obozinski, G., Taskar, B., Jordan, M.I., 2007. Joint covariate selection for grouped classification. Statistics Department, UC Berkeley.

Ogawa, K., Suzuki, Y., Takeuchi, I., 2013. Safe screening of non-support vectors in pathwise SVM computation. In: Proceedings of the 30th International Conference on Machine Learning, pp. 1382–1390.

Peng, J., Zhu, J., Bergamaschi, A., Han, W., Noh, D., Pollack, J., Wang, P., 2010. Regularized multivariate regression for identifying master predictors with application to integrative genomics study of breast cancer. Ann. Appl. Stat. 4, 53–77.

Potkin, S.G., Guffanti, G., Lakatos, A., Turner, J.A., Kruggel, F., Fallon, J.H., Saykin, A.J., Orro, A., Lupoli, S., Salvi, E., et al., 2009. Hippocampal atrophy as a quantitative trait in a genome-wide association study identifying novel susceptibility genes for Alzheimer's disease. PLoS ONE 4 (8), e6501.

Psaty, B.M., O'Donnell, C.J., Gudnason, V., Lunetta, K.L., Folsom, A.R., Rotter, J.I., Uitterlinden, A.G., Harris, T.B., Witteman, J.C., Boerwinkle, E., et al., 2009. Cohorts for heart and aging research in genomic epidemiology (charge) consortium design of prospective meta-analyses of genome-wide association studies from 5 cohorts. Circ. Cardiovasc. Genet. 2 (1), 73–80.

Quattoni, A., Carreras, X., Collins, M., Darrell, T., 2009. An efficient projection for $\ell_{1,\infty}$, infinity regularization. In: Proceedings of the 26th Annual International Conference on Machine Learning, ICML '09, Montreal, Quebec, Canada. ACM, New York, NY, pp. 857–864. ISBN 978-1-60558-516-1.

Shaw, P., Lerch, J.P., Pruessner, J.C., Taylor, K.N., Rose, A.B., Greenstein, D., Clasen, L., Evans, A., Rapoport, J.L., Giedd, J.N., 2007. Cortical morphology in children and adolescents with different apolipoprotein e gene polymorphisms: an observational study. Lancet Neurol. 6 (6), 494–500.

Simon, N., Friedman., J., Hastie., T., Tibshirani, R., 2013. A sparse-group lasso. J. Comput. Graph. Stat. 22, 231–245.

Singh, M., Singh, P., Juneja, P., Singh, S., Kaur, T., 2011. SNP–SNP interactions within APOE gene influence plasma lipids in postmenopausal osteoporosis. Rheumat. Int. 31, 421–423.

Sprechmann, P., Ramírez, I., Sapiro., G., Eldar, Y., 2011. C-HiLasso: a collaborative hierarchical sparse modeling framework. IEEE Trans. Signal Process. 59, 4183–4198.

Stein, J.L., Medland, S.E., Vasquez, A.A., Hibar, D.P., Senstad, R.E., Winkler, A.M., Toro, R., Appel, K., Bartecek, R., Bergmann, Ø., et al., 2012. Identification of common variants associated with human hippocampal and intracranial volumes. Nat. Genet. 44 (5), 552–561.

Strohmaier, J., Amelang, M., Hothorn, L.A., Witt, S.H., Nieratschker, V., Gerhard, D., Meier, S., Wust, S., Frank, J., Loerbroks, A., Rietschel, M., Sturmer, T., Schulze, T.G., 2012. The psychiatric vulnerability gene CACNA1C and its sex-specific relationship with personality traits, resilience factors and depressive symptoms in the general population. Mol. Psychiatry 18 (5), 607–613.

Subramanian, A., Tamayo, P., Mootha, V.K., Mukherjee, S., Ebert, B.L., Gillette, M.A., Paulovich, A., Pomeroy, S.L., Golub, T.R., Lander, E.S., Mesirov, J.P., 2005. Gene set enrichment analysis: A knowledge-based approach for interpreting genomewide expression profiles. Proc. Natl. Acad. Sci. USA 102 (43), 15545–15550.

Sun, L., Liu, J., Chen, J., Ye, J., 2009. Efficient recovery of jointly sparse vectors. In: Bengio, Y., Schuurmans, D., Lafferty, J.D., Williams, C.K.I., Culotta, A. (Eds.), Advances in Neural Information Processing Systems 22. Curran Associates, Inc., Red Hook, NY, pp. 1812–1820.

Thompson, P.M., Ge, T., Glahn, D.C., Jahanshad, N., Nichols, T.E., 2013. Genetics of the connectome. NeuroImage 80, 475–488.

Tibshirani, R., 1996. Regression shrinkage and selection via the lasso. J. R. Stat. Soc. B 58, 267–288.

Tibshirani, R., Wang, P., 2008. Spatial smoothing and hot spot detection for CGH data using the fused lasso. Biostatistics 9 (1), 18–29.

Tibshirani, R., Saunders, M., Rosset, S., Zhu, J., Knight, K., 2005. Sparsity and smoothness via the fused lasso. J. R. Stat. Soc. B 67 (1), 91–108.

Tibshirani, R., Bien, J., Friedman, J., Hastie, T., Simon, N., Taylor, J., Tibshirani, R., 2012. Strong rules for discarding predictors in lasso-type problems. J. R. Stat. Soc. B 74, 245–266.

Vidyasagar, M., 2014. Machine learning methods in the computational biology of cancer. Proc. R. Soc. Lond. A 471 (2173), 20140805. http://dx.doi.org/10.1098/rspa.2014.0805.

Vinje, W., Gallant, J., 2000. Sparse coding and decorrelation in primary visual cortex during natural vision. Science 287, 1273–1276.

Wang, J., Ye, J., 2014. Two-layer feature reduction for sparse-group lasso via decomposition of convex sets. In: Ghahramani, Z., Welling, M., Cortes, C., Lawrence, N.D., Weinberger, K.Q. (Eds.), Advances in Neural Information Processing Systems 27. Curran Associates, Inc., Red Hook, NY, pp. 2132–2140.

Wang, J., Ye, J., 2015. Multi-layer feature reduction for tree structured group lasso via hierarchical projection. In: Cortes, C., Lawrence, N.D., Lee, D.D., Sugiyama, M., Garnett, R. (Eds.), Advances in Neural Information Processing Systems 28. Curran Associates, Inc., Red Hook, NY, pp. 1279–1287.

Wang, H., Nie, F., Huang, H., Risacher, S.L., Saykin, A.J., Shen, L., et al., 2012. Identifying disease sensitive and quantitative trait-relevant biomarkers from multidimensional heterogeneous imaging genetics data via sparse multimodal multitask learning. Bioinformatics 28 (12), i127–i136.

Wang, J., Jun, J., Ye, J., 2013. Efficient mixed-norm regularization: algorithms and safe screening methods. CoRR, abs/1307.4156. http://arxiv.org/abs/1307.4156.

Wang, J., Wonka, P., Ye, J., 2014a, Scaling SVM and least absolute deviations via exact data reduction. In: Proceedings of the 31st International Conference on Machine Learning, ICML 2014, Beijing, China, June 21–26, 2014, pp. 523–531.

Wang, J., Zhou, J., Liu, J., Wonka, P., Ye, J., 2014b. A safe screening rule for sparse logistic regression. In: Ghahramani, Z., Welling, M., Cortes, C., Lawrence, N.D., Weinberger, K.Q. (Eds.), Advances in Neural Information Processing Systems 27. Curran Associates, Inc., Red Hook, NY, pp. 1053–1061.

Wang, J., Fan, W., Ye, J., 2015a. Fused lasso screening rules via the monotonicity of subdifferentials. IEEE Trans. Pattern Anal. Mach. Intell. 37 (9), 1806–1820.

Wang, J., Wonka, P., Ye, J., 2015b. Lasso screening rules via dual polytope projection. J. Mach. Learn. Res. 16, 1063–1101. http://jmlr.org/papers/v16/wang15a.html.

Wright, J., Ma, Y., Mairal, J., Sapiro, G., Huang, T.S., Yan, S., 2010. Sparse representation for computer vision and pattern recognition. Proc. IEEE 98, 1031–1044.

Wu, T.T., Chen, Y.F., Hastie, T., Sobel, E., Lange, K., 2009. Genomewide association analysis by lasso penalized logistic regression. Bioinformatics 25, 714–721.

Xiang, Z.J., Ramadge, P.J., 2012. Fast lasso screening tests based on correlations. In: 2012 IEEE International Conference on Acoustics, Speech and Signal Processing (ICASSP), pp. 2137–2140. http://dx.doi.org/10.1109/ICASSP.2012.6288334.

Xiang, Z.J., Xu, H., Ramadge, P.J., 2011. Learning sparse representations of high dimensional data on large scale dictionaries. In: Shawe-Taylor, J., Zemel, R.S., Bartlett, P.L., Pereira, F., Weinberger, K.Q. (Eds.), Advances in Neural Information Processing Systems 24. Curran Associates, Inc., Red Hook, NY, pp. 900–908.

Yang, J., Ferreira, T., Morris, A., Medland, S., Madden, P., Heath, A., Martin, N., Montgomery, G., Weedon, M., Loos, R., Frayling, T., McCarthy, M., Hirschhorn, J., Goddard, M., Visscher, P., 2012. Conditional and joint multiple-SNP analysis of GWAS summary statistics identifies additional variants influencing complex traits. Nat. Genet. 44, 369–375.

Yang, T., Wang, J., Sun, Q., Hibar, D., Jahanshad, N., Liu, L., Wang, Y., Zhan, L., Thompson, P., Ye, J., 2015, April. Detecting genetic risk factors for Alzheimer's disease in whole genome sequence data via lasso screening. In: IEEE 12th International Symposium on Biomedical Imaging (ISBI), pp. 985–989.

Yogatama, D., Smith, N.A., 2014. Linguistic structured sparsity in text categorization. In: Proceedings of the 52nd Annual Meeting of the Association for Computational Linguistics (Vol. 1: Long Papers), Baltimore, MD. Association for Computational Linguistics, Berlin, Germany, pp. 786–796.

Yuan, M., Lin, Y., 2006. Model selection and estimation in regression with grouped variables. J. R. Stat. Soc. B 68, 49–67.

Zhao, P., Yu, B., 2006. On model selection consistency of lasso. J. Mach. Learn. Res. 7, 2541–2563.

Zhao, P., Rocha, G., Yu, B., 2009. The composite absolute penalties family for grouped and hierarchical variable selection. Ann. Stat. 37 (6A), 3468–3497.

Zhu, J., Hastie, T., 2004. Classification of gene microarrays by penalized logistic regression. Biostatistics 5, 427–443.

Zou, H., Hastie, T., 2005. Regularization and variable selection via the elastic net. J. R. Stat. Soc. B 67, 301–320.

CHAPTER

Dictionary learning for medical image denoising, reconstruction, and segmentation

T. Tong, J. Caballero, K. Bhatia, D. Rueckert
Imperial College London, London, United Kingdom

CHAPTER OUTLINE

6.1 INTRODUCTION

Throughout the past decades, different forms of representation have emerged, leading in recent years to dictionary learning (DL). DL is the term given to the search for optimal sparse signal transforms which are obtained through a training stage, which is a radically different approach to signal modeling compared to hand-crafted

Machine Learning and Medical Imaging. http://dx.doi.org/10.1016/B978-0-12-804076-8.00006-2

signal models such as wavelets. It has been successfully applied to numerous image processing tasks such as denoising, super-resolution, and segmentation, and led to state-of-the-art results in recent years. In what follows we will present a brief history of signal transforms leading to DL and a summary of some of the DL algorithms available. Further details can be found in two excellent reviews (Rubinstein et al., 2008; Tosic and Frossard, 2011).

The choice of a transform for signal representation is crucial and involves a number of compromises. The use of orthogonal or bi-orthogonal transforms has long been favored because transform coefficients are given by a simple inner product between the signal and the transform or the transform inverse, respectively. However, the use of complete bases has limitations in representation flexibility as some signals may not be well encompassed by their modeling. The desire for greater flexibility at the expense of mathematical complexity drove the switch from complete transform bases to overcomplete dictionaries, and from transform functions to dictionary atoms.

6.1.1 THE CONVENIENCE OF ORTHOGONAL TRANSFORMS

One of the most recurrent signal analysis tools is the Fourier transform, which was greatly popularized in the 1960s with the emergence of the fast Fourier transform proposed in Cooley and Tukey (1965). The decomposition of a signal into its global frequency content can sparsely represent uniformly smooth signals, but is very inefficient for capturing discontinuities given that their energy is spread among several frequency coefficients. Sharp discontinuities are rare in natural signals, but the periodic assumption of finite signals for the computation of its transform artificially creates them at the signal boundary. This naturally led to the use of the discrete cosine transform (DCT), which avoids this phenomenon by assuming odd periodicity and is the core ingredient of the JPEG image compression standard discussed in Wallace (1992).

The following decades of the 1970s and 1980s centered the search of data simplicity on the data itself. Statistical tools such as principal component analysis as shown in Jolliffe (2005) and most notably the Karhunen-Loève transform as revisited by Mallat (1999) gained interest as they reduced the complexity of the signal on a low-dimensional subspace with minimum l_2-norm error. Using the first few eigenvectors of the eigenvalue decomposition of a signal's covariance matrix, it can be seen as a low-dimensional Gaussian data fit. Although it is more powerful as a data sparsifier than the Fourier transform, it is considerably more complex given its data-driven nature.

During the 1980s, it became clear that the search for simpler, sparser representations required the departure from restrictive linear transforms leading to the design of nonlinear transforms, where the support of nonzero coefficients is signal specific. Two major concepts are at the origin of wavelet design, emerging with the specific purpose of nonlinear sparse coding for natural signals: localization and multiresolution.

The Fourier transform allows the identification of the different frequency content of a signal, but it does not reveal where in time or space this content can be found. This lack of localization hinders compact signal representation, and results of this realization were the short-time Fourier transform and the use of Gabor filters as originally proposed by Gabor (1946). Multiresolution analysis was the consequence of noticing how natural signals exhibit fractal-like patterns, which repeat at different scales. Multiscale wavelet analysis was introduced in Grossmann and Morlet (1984) as the scaling and translation of a single function of finite support which could be designed to form an orthogonal basis. In Mallat (1999) this concept was later extended for optimal 1D multiresolution signal analysis, and most importantly, fast algorithms for wavelet decomposition were proposed enabling their practical use. Even though at higher dimensions wavelet analysis loses its optimality, these advances were adopted in the newer JPEG2000 image compression mechanism as described in Skodras et al. (2001).

Wavelet analysis is also limited in the lack of adaptability and geometric invariance. The orthogonality condition limits the range of temporal or spatial support of the functions which, if violated, allows for greater flexibility in representation. Wavelet packets were suggested by Coifman et al. (1992) as an extension to wavelets which, given a signal, could be reduced to the optimal orthogonal subset, gaining adaptability but keeping the attractive properties of orthogonal wavelets. Translation and rotation sensitivity are further drawbacks of standard wavelet transforms which were deemed unavoidable by orthogonal transforms in Simoncelli et al. (1992). Thus began work on overcomplete transforms with early examples such as the stationary wavelet transform in Beylkin (1992) seeking geometric invariance.

6.1.2 THE FLEXIBILITY OF OVERCOMPLETE DICTIONARIES

With the development in the 1990s of greedy algorithms for sparse solutions and the influential discovery that this problem could be approximated by the tractable l_1 relaxation, the use of overcomplete frames adopting the name of dictionaries was popularized. Allowing multiple representations of the same signal in a dictionary of atoms opened new perspectives in coding design, which could now be driven by a cost function, and markedly separated the task of dictionary design from signal coding. Simple concatenations of bases could overcome what used to be fundamental limitations. For instance, a Fourier transform was unable to compactly represent discontinuities, but concatenating it with a Dirac basis could immediately solve this problem.

Abandoning orthogonality paved the way for creative dictionary design. Two trends can currently be identified: the design of analytic dictionaries and data-driven adaptive dictionaries (Aharon et al., 2006; Mairal et al., 2009). The former approach relies on a mathematical model of the data to generate the dictionary and is usually characterized by efficient mechanisms to implicitly compute transform coefficients, as well as robust theoretical guarantees for signal approximation. Some examples of this category are curvelets as proposed in Candès and Donoho (1999), contourlets from Do and Vetterli (2002) and bandelets discussed in Le Pennec and Mallat (2005). Data-driven dictionary design is more recent, and draws from example observations of a signal to obtain an optimal representation. Adaptive dictionaries are powerful as there is no reason to believe a single dictionary should be optimal for all kinds of signals, but come at the price of increased processing complexity and weaker theoretical guarantees. The search for optimal dictionaries for a specific set of training signals is known as the DL problem.

6.2 SPARSE CODING AND DICTIONARY LEARNING

This section introduces some of the most important DL algorithms. In addition, sparse coding techniques are described before the introduction of DL techniques as they form an essential part of DL.

6.2.1 SPARSE CODING

The technique of finding a representation for a given signal with a small number of significant coefficients is referred to as sparse coding. To describe signal sparse coding we can rely on the sparse synthesis model, which is characterized by two properties: (1) synthesis: signals are assumed to be a linear combination of basis functions and (2) sparsity: the coding vector defining the synthesis of a signal from basis functions is sparse. Assume that the set of basis functions $\mathbf{D} \in \mathbb{R}^{M\times K}$ is given, which is called a dictionary. $\mathbf{D}$ can be predefined (eg, overcomplete wavelets) or learnt. Each column in $\mathbf{D}$ is called an atom. Here, the dictionary $\mathbf{D}$ has K atoms and each atom has M elements. A signal $\mathbf{y} \in \mathbb{R}^{M\times 1}$ can then be represented as a linear combination of atoms in $\mathbf{D}$, which is formulated as

$$\mathbf{y} = \mathbf{D}\boldsymbol{\gamma}. \tag{6.1}$$

To seek a sparse solution, different regularization schemes have been proposed to impose prior information over the coding coefficients $\boldsymbol{\gamma}$. One commonly used regularization scheme adds the minimum l_0-norm constraint on the coefficients, minimizing the number of nonzero entries in $\boldsymbol{\gamma}$. The linear model is then formulated as

$$\min_{\boldsymbol{\gamma}} \|\mathbf{y} - \mathbf{D}\boldsymbol{\gamma}\|_2 \quad \text{s.t.} \quad \|\boldsymbol{\gamma}\|_0 \leq S \tag{6.2}$$

or

$$\min_{\boldsymbol{\gamma}} \|\boldsymbol{\gamma}\|_0 \quad \text{s.t.} \quad \|\mathbf{y} - \mathbf{D}\boldsymbol{\gamma}\|_2 \leq \varepsilon, \tag{6.3}$$

which represent the sparsity constrained and the error constrained l_0 norm problems, respectively. Both problems are equivalent in that it is possible to select an error ε that will provide the same solution as a maximum sparsity S. The use of the l_0-norm as a measure of sparsity makes the problem nonconvex and finding the exact solution for γ in the above equations is NP-hard (Elad, 2010).

Algorithms for finding approximate solutions have been extensively investigated. A very well-known algorithm is orthogonal matching pursuit (OMP) (Mallat and Zhang, 1993). OMP is a greedy approach that proposes to iteratively seek the locally optimal choice in the hope of approximating the global minimum. This is a reasonable compromise for the sparse approximation problem given that the global solution to the NP-hard problem is practically unreachable, but sequentially deciding the entries in γ that will minimize the approximation error is computationally very cheap. Various extensions such as compressive sampling OMP (CoSaMP) (Needell and Tropp, 2009), regularized OMP (ROMP) (Needell and Vershynin, 2010), and stagewise OMP (StOMP) (Donoho et al., 2012) have been proposed to accelerate the convergence of OMP.

A major inconvenience for solving Eqs. (6.2) and (6.3) is that they are nonconvex problems. An alternative is to look for the solution to the closest problem that is convex, for which the vast literature on convex optimization would immediately apply. This was proposed in Chen et al. (1998) with the relaxation of the l_0-norm by the l_1-norm, targeting the solutions to

$$\min_{\gamma} \|\mathbf{y} - \mathbf{D}\gamma\|_2 \quad \text{s.t.} \quad \|\gamma\|_1 \leq S \tag{6.4}$$

or

$$\min_{\gamma} \|\gamma\|_1 \quad \text{s.t.} \quad \|\mathbf{y} - \mathbf{D}\gamma\|_2 \leq \varepsilon, \tag{6.5}$$

which is known as the basis pursuit (BP) problem. This problem, in constrained form or as an l_1 regularized least-squared problem, can be efficiently solved with different approaches, such as homotopy methods (Tibshirani, 1996), coordinate-wise descent methods (Friedman et al., 2007), Bregman iterative methods (Osher et al., 2005), and iterative shrinkage methods (Bioucas-Dias and Figueiredo, 2008).

6.2.2 DICTIONARY LEARNING PROBLEM

In sparse coding, it is assumed that the overcomplete dictionary $\mathbf{D}$ is given or known a priori. The dictionary can be directly chosen as a set of training signals or a prespecified basis such as overcomplete wavelets, curvelets, contourlets, and short-time Fourier transforms. Recent research has focused on learning an overcomplete dictionary based on a set of training signals rather than choosing a prespecified dictionary. Given a set of training signals $\mathbf{Y} = [\mathbf{y}_1, \mathbf{y}_2, \ldots, \mathbf{y}_N] \in \mathbb{R}^{M \times N}$, it is assumed that there exists a dictionary $\mathbf{D}$ that can sparsely represent each signal in $\mathbf{Y}$. The process of DL is then formulated as

$$\min_{\mathbf{D},\mathbf{\Gamma}} \|\mathbf{Y} - \mathbf{D}\mathbf{\Gamma}\|_F^2 \quad \text{s.t.} \quad \|\gamma_{\mathbf{i}}\|_0 \leq S \;\; \forall i. \tag{6.6}$$

Here each column in $\mathbf{\Gamma} \in \mathbb{R}^{K \times N}$ contains the coding coefficients corresponding to each training signal $\mathbf{y}_i$. It is a nonconvex problem to optimize Eq. (6.6) over the dictionary $\mathbf{D}$ and sparse coefficients $\mathbf{\Gamma}$ jointly (Aharon et al., 2006). This problem is intricately related to sparse coding, but has the additional difficulty that, on top of finding a sparse code, the dictionary for sparse representation has to be simultaneously estimated. Commonly, the problem is simplified by solving for the sparse code and the dictionary separately, and iteratively alternating their solutions until convergence. The method of optimal directions (MOD) (Engan et al., 1999) and the K-SVD (Aharon et al., 2006) are two efficient algorithms to learn dictionaries which utilize variants of this iterative optimization strategy. In practice, it has been observed that K-SVD converges with fewer iterations than MOD (Aharon et al., 2006). In the next section, we will give a detailed introduction to the K-SVD algorithm.

6.2.3 K-SVD DICTIONARY LEARNING

The K-SVD algorithm is inspired from the k-means clustering algorithm, which is also an NP-hard problem. The aim of k-means clustering is to partition all the signals into K clusters, in which each training signal belongs to the cluster with the nearest mean. It employs an iterative approach to find the solution of K clusters and there are two steps at each iteration: In the first step, each training signal is assigned to its nearest cluster; in the second step, the K clusters are updated as the centroids of their assigned training signals. The K-SVD follows a similar iterative two-step process to learn dictionary atoms and find sparse solutions of the training signals using those atoms. The dictionary is first initialized with a traditional data sparsifier, such as an overcomplete DCT dictionary. We can then look for the sparse coding matrix $\mathbf{\Gamma}$ by keeping the dictionary fixed and sparsely coding each training signal independently with an OMP coding stage. Then, the K atoms in the dictionary are updated separately in a dictionary update stage with $\mathbf{\Gamma}$ fixed. The K-SVD algorithm decomposes the penalty term by questioning one atom $\mathbf{d}_k$ and its associated sparse codes given in row k of $\mathbf{\Gamma}$. Denoting this row vector as $\boldsymbol{\gamma}_t^k$, the penalty can be rewritten as

$$\begin{aligned}
\|\mathbf{Y} - \mathbf{D}\mathbf{\Gamma}\|_F^2 &= \left\| \mathbf{Y} - \sum_{i=1}^{K} \mathbf{d}_k \boldsymbol{\gamma}_t^k \right\|_F^2 \\
&= \left\| \left(\mathbf{Y} - \sum_{j \neq k} \mathbf{d}_j \boldsymbol{\gamma}_t^j \right) - \mathbf{d}_k \boldsymbol{\gamma}_t^k \right\|_F^2 \\
&= \|\mathbf{E}^k - \mathbf{d}_k \boldsymbol{\gamma}_t^k\|_F^2,
\end{aligned} \tag{6.7}$$

where $\mathbf{E}^k$ would be the approximation error if atom $\mathbf{d}_k$ were to be removed from the dictionary.

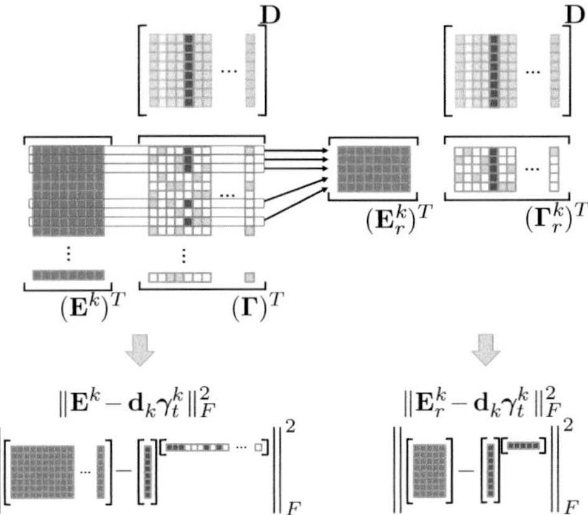

FIG. 6.1

K-SVD dictionary update step. The left-hand side of the figure shows the decomposition of the approximation error in $\mathbf{E}^k$ and the contribution from $\mathbf{d}_k$ and $\boldsymbol{\gamma}_t^k$ through a rank-1 matrix. Updating $\mathbf{d}_k$ and $\boldsymbol{\gamma}_t^k$ directly with an SVD decomposition of $\mathbf{E}^k$ does not guarantee the maintenance of sparsity in $\boldsymbol{\gamma}_t^k$. Instead, shrinking the matrices as shown on the right-hand side of the figure solves this problem as only the active support of $\boldsymbol{\gamma}_r^k$ is updated.

With this separation, the approximation $\mathbf{D\Gamma}$ has been divided into K rank-1 matrices, only one of which is being questioned for update. Finding the rank-1 approximation of $\mathbf{E}_k$ through singular value decomposition (SVD) and using it for the update of $\mathbf{d}_k$ and $\mathbf{y}_t^k$ would be the optimal update step, but this is likely to fill row $\mathbf{y}_t^k$, which we would like to keep as a sparse vector. A simple solution is to only consider the indices $\mathbf{w}_k$ of nonzero entries in $\mathbf{y}_t^k$, and define the shrunken vector $\mathbf{y}_r^k$ and matrix $\mathbf{E}_r^k$. The rank-1 approximation of this new error matrix $\mathbf{E}_r^k$ provides then the optimal update of $\mathbf{d}_k$ and $\mathbf{y}_r^k$ while the sparse coding support is either unchanged or reduced. This shrinkage operation is illustrated in Fig. 6.1, and the full K-SVD algorithm is summarized in Algorithm 6.1. Although OMP is used for the sparse coding stage, the K-SVD algorithm is flexible and can work with other sparse coding methods (Aharon et al., 2006).

ALGORITHM 6.1 THE K-SVD DICTIONARY LEARNING ALGORITHM

Require: A set of training signals $\mathbf{Y} = [\mathbf{y}_1, \mathbf{y}_2, \ldots, \mathbf{y}_n] \in \mathbb{R}^{M\times N}$.
Ensure: An overcomplete dictionary $\mathbf{D} \in \mathbb{R}^{M\times K}$ and sparse coding coefficients $\boldsymbol{\Gamma} \in \mathbb{R}^{K\times N}$.

```
Initialize the dictionary D with K randomly selected training signals
while converged do
    Sparse Coding:
    for each training signal y_i ∈ Y, use OMP to compute the corresponding coding
    coefficients γ_i: do
```

$$\min_{\boldsymbol{\gamma}_i} \|\boldsymbol{\gamma}_i\|_0 \quad \text{s.t. } \mathbf{y}_i = \mathbf{D}\boldsymbol{\gamma}_i, i = 1, \ldots, N$$

end for

Dictionary Update:

for $k = 1, \ldots, K$, update the kth atom $\mathbf{d}_k$ of $\mathbf{D}$ and the kth row $\boldsymbol{\gamma}_t^k$ of the coding coefficients $\boldsymbol{\Gamma}$: **do**

Find the groups that use $\mathbf{d}_k$: $\mathbf{w}_k = \{i \in \{1, \ldots, N\} : \boldsymbol{\gamma}_t^k(i) \neq 0\}$, and $\boldsymbol{\gamma}_r^k$ is obtained by discarding zero entries in $\boldsymbol{\gamma}_t^k$.

Compute representation error matrix: $\mathbf{E}^k = \mathbf{Y} - \sum_{i \neq k} \mathbf{d}_i \boldsymbol{\gamma}_t^k$.

Obtain $\mathbf{E}_r^k$ by selecting the columns of $\mathbf{E}^k$ corresponding to $\mathbf{w}_k$.

Apply SVD decomposition $\mathbf{E}_r^k = \mathbf{U}\boldsymbol{\Sigma}\mathbf{V}^t$, update the atom $\mathbf{d}_k$ with the first column of $\mathbf{U}$, and update $\boldsymbol{\gamma}_r^k$ with the first column of $\mathbf{V}$ multiplied by $\Sigma(1, 1)$.

end for

end while

Despite the fact that the K-SVD algorithm converges quickly, it is still computationally expensive at each iteration as an SVD decomposition must be calculated K times and all the N training signals are used for sparse coding at each iteration. This task is computationally expensive and relies on high memory use, especially when the set of training signals is large. Some modifications to the original algorithm have been presented that can alleviate computational cost, such as the use of Batch-OMP for K-SVD as presented in Rubinstein et al. (2008), and other strategies differing from K-SVD have also emerged. In the next section, we will introduce an online DL approach that can effectively solve this large-scale learning problem.

ALGORITHM 6.2 ONLINE DICTIONARY LEARNING ALGORITHM

Require: A set of training signals $\mathbf{Y} = [\mathbf{y}_1, \mathbf{y}_2, \ldots, \mathbf{y}_N] \in \mathbb{R}^{M \times N} \sim p(\mathbf{y})$, sparsity weight λ, T (number of iterations).

Ensure: An overcomplete dictionary $\mathbf{D} \in \mathbb{R}^{M \times K}$.

Initialize the dictionary $\mathbf{D}_0$ with K randomly selected training signals, $\mathbf{A}_0 \leftarrow 0$, $\mathbf{B}_0 \leftarrow 0$.

for $t = 1$ to T **do**

Draw $\mathbf{y}_t$ from $p(\mathbf{y})$.

Sparse Coding (Lasso):

$$\hat{\boldsymbol{\gamma}}_t = \arg\min_{\boldsymbol{\gamma}_t} \|\mathbf{y}_t - \mathbf{D}_{t-1}\boldsymbol{\gamma}_t\|_2^2 + \lambda\|\boldsymbol{\gamma}_t\|_1.$$

$\mathbf{A}_t \leftarrow \mathbf{A}_{t-1} + \hat{\boldsymbol{\gamma}}_t \hat{\boldsymbol{\gamma}}_t^T$, $\mathbf{B}_T \leftarrow \mathbf{B}_{t-1} + \mathbf{y}_t \hat{\boldsymbol{\gamma}}_t^T$.

Dictionary Update: compute $\mathbf{D}_t$ with $\mathbf{D}_{t-1}$ as initialization

$$\mathbf{D}_t = \arg\min_{\mathbf{D}} \frac{1}{t} \sum_{i=1}^{t} \left(\|\mathbf{y}_i - \mathbf{D}\boldsymbol{\gamma}_i\|_2^2 + \lambda \|\boldsymbol{\gamma}_i\|_1 \right)$$

$$= \arg\min_{\mathbf{D}} \frac{1}{t} \left(\frac{1}{2} Tr\left(\mathbf{D}^T \mathbf{D} \mathbf{A}_t\right) - Tr\left(\mathbf{D}^T \mathbf{B}_t\right) \right)$$

end for

6.2.4 ONLINE DICTIONARY LEARNING

A stochastic online learning algorithm was proposed in Mairal et al. (2009) in order to learn dictionaries for a large set of training signals. A relaxed version of the objective function for DL using the l_1-norm is formulated as

$$\begin{aligned}\left\langle \hat{\mathbf{D}}, \hat{\mathbf{\Gamma}} \right\rangle &= \underset{\mathbf{D},\mathbf{\Gamma}}{\arg\min} \|\mathbf{Y} - \mathbf{D}\mathbf{\Gamma}\|_2^2 + \lambda \|\mathbf{\Gamma}\|_1 \\ &= \underset{\mathbf{D},\mathbf{\Gamma}}{\arg\min} \frac{1}{N} \sum_{i=1}^{N} \left(\|\mathbf{y}_i - \mathbf{D}\boldsymbol{\gamma}_i\|_2^2 + \lambda \|\boldsymbol{\gamma}_i\|_1 \right).\end{aligned} \tag{6.8}$$

This is not jointly convex over $\mathbf{D}$ and $\mathbf{\Gamma}$. In order to find the optimized solution, a stochastic gradient descent approach was utilized in Mairal et al. (2009) to update $\mathbf{D}$ sequentially. Instead of using the full training set at each iteration as in the K-SVD algorithm, the online DL algorithm updates the dictionary atoms by accessing one training signal at a time. Assuming that the set of training signals are independent and identically distributed (i.i.d.), one signal is drawn for updating $\mathbf{D}$ at each iteration as in the stochastic gradient descent. The online optimization process is summarized in Algorithm 6.2. It follows classic DL algorithms and alternates the sparse coding step with the DL step. However, at the current iteration, the new dictionary $\mathbf{D}_t$ uses the previous dictionary $\mathbf{D}_{t-1}$ as a warm restart, which is different from other DL algorithms. The new dictionary $\mathbf{D}_t$ is updated by minimizing the following function (Mairal et al., 2009):

$$\mathbf{D}_t = \underset{\mathbf{D}}{\arg\min} \frac{1}{t} \sum_{i=1}^{t} \left(\|\mathbf{y}_i - \mathbf{D}\boldsymbol{\gamma}_i\|_2^2 + \lambda \|\boldsymbol{\gamma}_i\|_1 \right). \tag{6.9}$$

The coding coefficients $\hat{\boldsymbol{\gamma}}_i$ computed during the previous iterations aggregate past information. The information from past coefficients $\hat{\boldsymbol{\gamma}}_1, \hat{\boldsymbol{\gamma}}_2, \ldots, \hat{\boldsymbol{\gamma}}_t$ is carried forward in matrices:

$$\mathbf{A}_t \leftarrow \mathbf{A}_{t-1} + \hat{\boldsymbol{\gamma}}_t \hat{\boldsymbol{\gamma}}_t^T \quad \text{and} \quad \mathbf{B}_T \leftarrow \mathbf{B}_{t-1} + \mathbf{y}_t \hat{\boldsymbol{\gamma}}_t^T. \tag{6.10}$$

This enables updating dictionaries based on past information without accessing the past training samples again. The new dictionary $\mathbf{D}_t$ can then be optimized by using these matrices and the previous dictionary $\mathbf{D}_{t-1}$ as initialization. This optimization strategy leads to faster convergence performance and better dictionaries than classical batch algorithms, scaling up gracefully to large datasets even with millions of training samples (Mairal et al., 2009).

6.3 PATCH-BASED DICTIONARY SPARSE CODING

In this section we look at practical considerations of sparse recovery problems using dictionaries. Specifically, we present the implications for sparse coding brought on by overcompleteness, redundancy and adaptability. The experiments below use the Batch-OMP implementation described in Rubinstein et al. (2008).

6.3.1 OVERCOMPLETENESS

Consider the sparse recovery problem

$$\min_{\boldsymbol{\gamma}} \|\mathbf{y} - \mathbf{D}\boldsymbol{\gamma}\|_2^2 \quad \text{s.t.} \quad \|\boldsymbol{\gamma}\|_0 \leq S, \tag{6.11}$$

where $\mathbf{D} \in \mathbb{C}^{M\times K}$, $M \leq K$, is a dictionary. The solution to this problem is trivial in the complete, orthonormal case ($M = K$, $\mathbf{D}^H\mathbf{D} = \mathbf{I}$), given that the generalized Parseval theorem holds between $\mathbf{y}$ in the signal domain and $\boldsymbol{\gamma}$ in the sparsity domain. Therefore energy is preserved upon basis transformation, implying that $\|\mathbf{y} - \mathbf{D}\boldsymbol{\gamma}\|_2^2 = \|\mathbf{D}^H\mathbf{y} - \boldsymbol{\gamma}\|_2^2$, and so the best S sparse representation is trivially given by the S largest coefficients of the transform $\mathbf{D}^H\mathbf{y}$. Energy preservation between domains is, however, violated as soon as $M < K$, which adds considerable flexibility to the sparse recovery problem given that it brings about a regime where one signal can have multiple dictionary representations.

To illustrate this, let us assume we extract an $M = 8 \times 8$ patch from Fig. 6.2A. We then use OMP to solve Eq. (6.11) with a sparsity index $S = 6$. Lastly, we look at the mean squared error (MSE) $\frac{1}{M}\|\mathbf{y} - \mathbf{D}\boldsymbol{\gamma}\|_2^2$ of the result produced using DCT dictionaries of different sizes. This experiment is repeated with 10^4 different

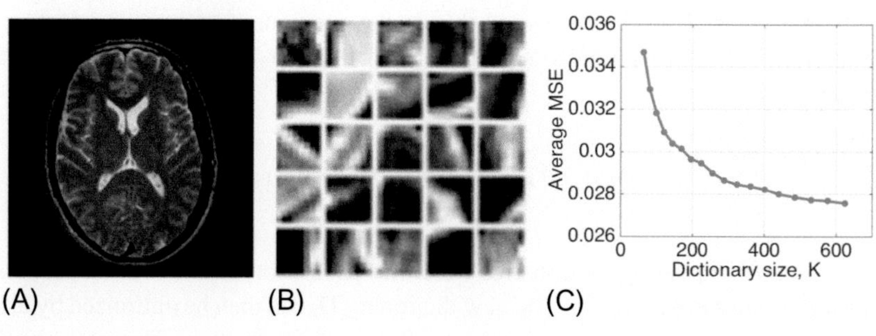

FIG. 6.2

Dictionary overcompleteness translates into increased representation sparsity. The plot in (C) shows the average representation error of 10^4 patches of size 8×8 from image (A) using a DCT dictionary of different sizes and a sparsity index $S = 6$. (A) Brain MR image. (B) Sample patches. (C) Reconstruction error.

patches and average results are plotted in Fig. 6.2C. Examples of the test patches are displayed in Fig. 6.2B. Despite the advantage of domain transformation that an orthonormal dictionary provides, the dictionary representation becomes more accurate with increasing overcompleteness.

6.3.2 REDUNDANCY

Redundant dictionaries are typically highly coherent, which immediately poses a problem for sparse recovery given that multiple representations are plausible for the same signal. If the focus, however, is not on the recovery of the particular sparse code that generated the signal but on any sparse code that will approximate it, redundant dictionaries can be useful. We analyze this statement by considering an overcomplete DCT dictionary with $K = 225$ atoms of size $M = 64$. Given a sparsity degree $S = 5$, we synthesize an 8×8 patch by linearly combining S randomly chosen atoms with random weights, and we then try to recover this sparse code with OMP solving

$$\min_{\boldsymbol{\gamma}} \|\boldsymbol{\gamma}\|_0 \quad \text{s.t.} \quad \|\mathbf{y} - \mathbf{D}\boldsymbol{\gamma}\|_2 \leq \epsilon, \tag{6.12}$$

for $\epsilon = 10^{-5}\sqrt{M}$.

The greedy approach of OMP to sparse recovery is sometimes able to perfectly find the sequence of sparse coefficients that make up the signal. However, the coherence of the dictionary can sometimes make OMP fail dramatically in finding the original support of the sparse code, but the redundancy in the dictionary makes it possible to find an alternative sparse configuration that still achieves the data consistency level required in the signal domain. This behavior is analyzed for different degrees of sparsity S in Fig. 6.3. The plots show average results for the same experiment with 10^4 different patches. Despite the correct support recovery decreasing quickly for $S \geq 2$ (Fig. 6.3A), the signal domain reconstruction accuracy can be maintained below the predefined threshold (Fig. 6.3B) at the expense of a denser representation $\boldsymbol{\gamma}$ (Fig. 6.3C).[1]

6.3.3 ADAPTABILITY

One of the main advantages of adaptive dictionaries over structured dictionaries is a sparser representation for a predefined set of signals. This comes at the cost of a computationally intensive training process and the loss of structure, meaning that implicit and efficient dictionary transforms are not available and theoretical guarantees of the dictionary are more difficult to derive.

In this section we compare the sparse representations of a structured dictionary with one that is trained using the K-SVD algorithm from Aharon et al. (2006). For the

[1] The sparsity of the result was measured as the number of nonzero coefficients accounting for 99.9% of the energy of the sparse code $\boldsymbol{\gamma}$.

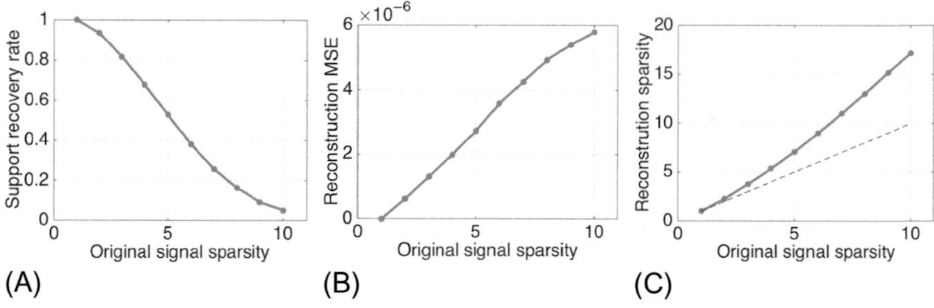

FIG. 6.3

Average empirical OMP recovery performance using an overcomplete DCT dictionary. The support recovery rate falls almost to zero for $S/M = 10/64 = 0.15$ sparsity (A). This is due to the high degree of redundancy in the dictionary, which makes OMP choose sparse coding configurations that are not the ones used to generate the original patches. Nevertheless, OMP is able to maintain the predefined data consistency tolerance (B) by using a few additional atoms relative to the original signal (C). (A) Support recovery rate. (B) Average MSE. (C) Average sparsity.

comparison we use the magnitude brain magnetic resonance (MR) image shown in Fig. 6.2A, of size $N = 256 \times 256$. Breaking the image down into $M = 8 \times 8$ overlapping patches and assuming patches wrap around the boundaries of the image, we have a total of N signals to be coded arranged as column vectors in $\mathbf{Y} \in \mathbb{R}^{M \times N}$. Extracting a subset of 2×10^4 training patches from a regular grid on the image, we analyze the first 30 iterations of the K-SVD algorithm with $S = 5$ for the training of a $K = 196$ atom dictionary. The initial dictionary is chosen to be a DCT dictionary.

The effects of the DL algorithm are shown in Fig. 6.4. It is clear how the MSE of the representation cost function decreases through the iterations of the K-SVD algorithm. This empirically confirms that the alternating strategy between a sparse coding stage and a dictionary update stage is effectively converging toward at least a local optimum. Furthermore, in order to achieve this, the dictionary is changing the shape of its atoms, and moves from the initially structured DCT dictionary toward one that incorporates new patterns.

We now focus on the implications that this adaptability has for the sparse approximation of the entire image. Assuming that the training set of patches, which is approximately a third of the full set, is a representative collection of the patches in the image, we should see the same improvement in the representation error when comparing the coding errors of both dictionaries. To recover a coded image from patches we average the contribution of overlapping patches and plot the MSE obtained from DCT coding and K-SVD coding with 30 iterations. Results are plotted in Fig. 6.5 for a range of sparsity indices $1 \leq S \leq 8$. We also show the accuracy of

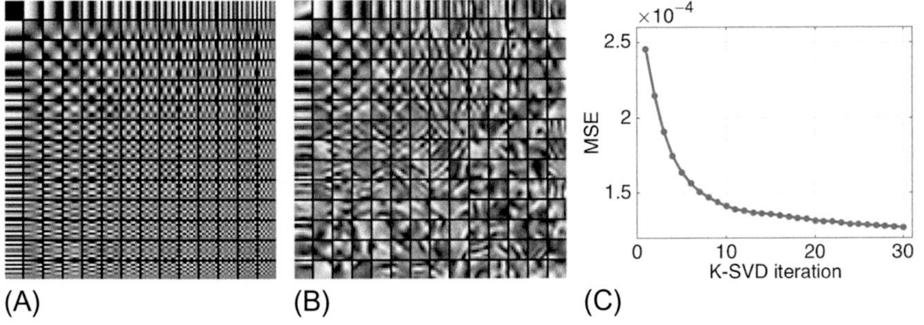

FIG. 6.4

Effect of K-SVD training on dictionary and on training dataset. (A) Initial dictionary. (B) Trained dictionary. (C) Cost function MSE.

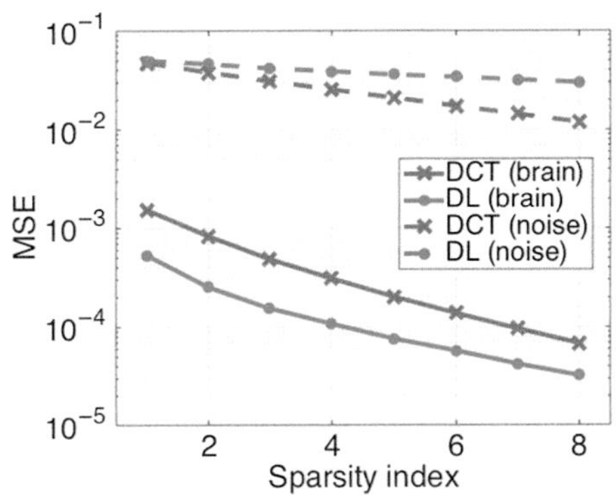

FIG. 6.5

Sparse coding reconstruction MSE of brain MR image. Adapting a dictionary to the brain image reduces the representation error with respect to it, while increasing the error with respect to data of different nature, such as a Gaussian noise image.

recovering an image of only random noise with both dictionaries, to highlight how the lack of structure in random features cannot be well captured by sparse coding.

The gap in representation accuracy represents the gain that can be achieved through DL relative to the initial structured dictionary. This gap is also visible in the error maps of the approximated images for $S = 3$, shown in Fig. 6.6. Notice how most of the representation error concentrates on edges and fine details of the image. This is expected given that those features are precisely the ones that will not conform to the sparsity criterion, and are therefore the first ones to be penalized with the assumption of sparsity.

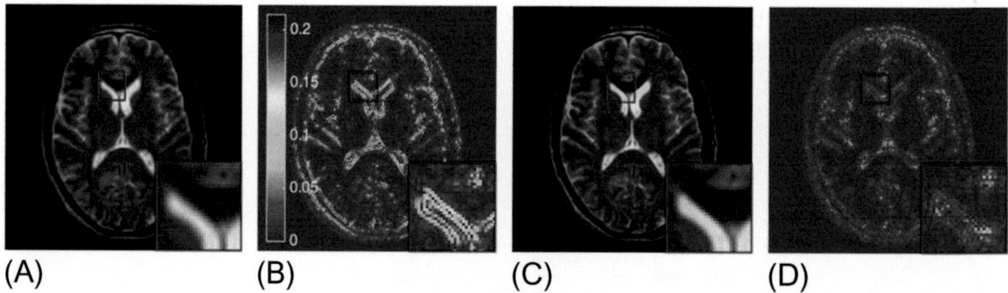

FIG. 6.6

Sparse coded approximations of a brain MR image using a DCT dictionary (A, B) and a trained dictionary (C, D) with a sparsity index $S = 3$. Error maps show absolute value differences with respect to the original image. The region of interest (ROI) in the red rectangle (dark gray in print versions) is enlarged for better visualization. (A) DCT result. (B) DCT error map. (C) DL result. (D) DL error map.

6.4 APPLICATION OF DICTIONARY LEARNING IN MEDICAL IMAGING

6.4.1 DENOISING

The quality of medical images plays an important role in the accuracy of clinical diagnosis. Denoising is a crucial step for improving image quality, which also enhances the ability of subsequent imaging analysis such as detection, segmentation and registration. The main challenge in image denoising is to remove noise corruption while keeping the integrity of relevant image information. One of the most representative methods is a nonlocal means (NLM) filter (Coupé et al., 2008). This method exploits the redundancy of image patterns for noise removal. In particular, similar patches are selected within an image and the restored intensities are the weighted average of the selected patches. Another well-known category of denoising methods utilizes sparse representations of image patterns (Bao et al., 2013). In such methods, a clean signal is assumed to be sparsely represented by a few bases such as sine or cosine functions in FFT or DCT transformations. The noise is then reduced by removing the noise-related coefficients in the transform domain.

More recently, DL has been used for image denoising (Elad and Aharon, 2006). Rather than using standard bases of DCT or FFT transforms, these techniques learn a set of bases from the image to form a dictionary. The learnt dictionary can be used to reduce the noise and restore the clean image. Given a noise image $\mathbf{y}$, the denoising process tries to remove the noise from $\mathbf{y}$ to provide an approximation of the original image $\mathbf{x}$, which can be formulated as

$$\min_{\mathbf{x},\boldsymbol{\Gamma},\mathbf{D}} \quad \|\mathbf{y}-\mathbf{x}\|_2^2 + \lambda\|\mathbf{R}_i\mathbf{x} - \mathbf{D}\boldsymbol{\gamma}_i\|_2^2 \quad \text{s.t.} \quad \|\boldsymbol{\gamma}_i\|_0 \leq S \quad \forall i. \tag{6.13}$$

Here $\mathbf{R}_i$ represents a patch extractor at pixel location i and $\boldsymbol{\Gamma}$ collects as columns $\boldsymbol{\gamma}_i$ the sparse coding of image patches $\mathbf{R}_i\mathbf{x}$. The first term in Eq. (6.13) enforces the consistency between image $\mathbf{y}$ and the denoised image $\mathbf{x}$. The second and third terms ensure that every patch $\mathbf{R}_i\mathbf{x}$ in the denoised image can be represented by a linear combination of a few atoms in $\mathbf{D}$. The solution to this equation is nonconvex and can be approximated using an iterative alternating strategy (Elad and Aharon, 2006) by repeating three steps—sparse coding, dictionary update, and the estimation of $\mathbf{x}$. The sparse coding and the dictionary update steps are the same as in the K-SVD algorithms. The denoised image $\mathbf{x}$ is estimated in the third step as

$$\hat{\mathbf{x}} = \frac{\lambda \mathbf{y} + \mathbf{R}_i^T \mathbf{D} \boldsymbol{\gamma}_i}{\lambda \mathbf{I} + \mathbf{R}_i^T \mathbf{R}_i}. \tag{6.14}$$

This denoising process has been successfully applied to remove the Gaussian noise from natural images (Elad and Aharon, 2006) and recently from medical images (Li et al., 2012; Bao et al., 2013). For example, the K-SVD method was used in Patel et al. (2011) to remove the noise from high-angular resolution diffusion images. Another study (Li et al., 2012) performed image denoising by exploiting the local geometrical structure of atoms of the learnt dictionaries, which showed improved performance compared to the K-SVD method. Overall, the advantage of DL methods over standard transforms such as DCT or FFT is that the learnt bases can be better adapted to the images, enabling a sparser representation and a better separation between clean signals and noise. However, it should be mentioned that noise in MR images follows a Rician distribution rather than a Gaussian distribution (Manjón et al., 2010c) and may have spatially varying noise levels (Manjón et al., 2010c). Thus it would be interesting to develop more advanced denoising algorithms based on DL to solve these problems in future work.

6.4.2 RECONSTRUCTION

Reconstruction refers to the processing required to turn raw acquisition measurements into an image. Magnetic resonance imaging (MRI) acquisition is performed in k-space, which is the Fourier description of the MR image. Traditionally, MR acquisition and reconstruction have been governed by Shannon's sampling criterion (Shannon, 1949), where samples acquired at a sampling frequency at least twice as large as the maximum signal frequency content are linearly reconstructed. MR physics impose that samples need to be drawn sequentially and at a limited maximum rate, and trying to satisfy Shannon's criterion will often lead to lengthy acquisition times. This constraint is particularly problematic in dynamic MR such as cardiac cine imaging, where spatial and temporal sampling need to be traded off against each other. This has motivated the exploration of sampling techniques that undersample k-space, introducing aliasing in the image domain, as illustrated in Fig. 6.7.

Assuming $\mathbf{x} \in \mathbb{C}^N$ to be the image of interest and $\mathbf{y} \in \mathbb{C}^M$ its k-space acquisition, an undersampled acquisition can be described as $\mathbf{y} = \mathbf{MFx}$, where

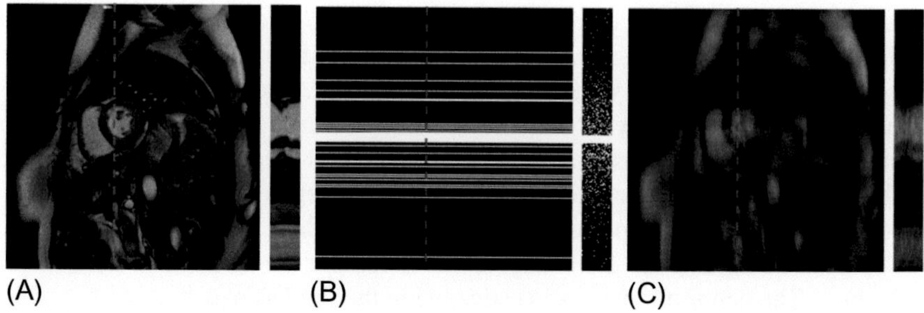

FIG. 6.7

Example of a magnitude temporal frame from a cardiac cine dataset (A). The undersampling mask (B) applied in k-space reduces acquisition time but introduces aliasing in image space (C). All figures show a 2D frame on the left-hand side and the temporal profile across the dashed line on the right-hand side. (A) Fully sampled. (B) Undersampling mask. (C) Zero-filled result.

$\mathbf{M} \in \mathbb{R}^{M \times N}$, $M < N$, is an undersampling mask equivalent to an identity matrix whose rows corresponding to nonacquired k-space locations are missing, and $\mathbf{F}$ is a discrete Fourier transform matrix. This underdetermined system of equations makes it impossible to reconstruct $\mathbf{x}$ from $\mathbf{y}$ unless some prior knowledge about the image is introduced. Compressed sensing (CS) reconstruction methods impose the condition of sparsity in image $\mathbf{x}$, and assuming that image information is redundant they look for a reconstructed image that can be sparsely represented in a transform domain. Assuming $\mathcal{S}(\mathbf{x})$ to be a sparse representation of $\mathbf{x}$, where we enforce a sparsity of S, the global CS MRI problem can be posed as the following combination of a data consistency and a sparsity term:

$$\min_{\mathbf{x}} \quad \|\mathbf{y} - \mathbf{MFx}\|_2^2 \quad \text{s.t.} \quad \|\mathcal{S}(\mathbf{x})\|_0 \leq S. \tag{6.15}$$

CS theory states that the recovery of $\mathbf{x}$ from $\mathbf{y}$ becomes theoretically possible if (a) the undersampling strategy complies with some incoherence criteria, (b) a nonlinear reconstruction method is employed to solve the nonconvex optimization problem, and (c) the image is sufficiently sparse in the transform domain (Lustig et al., 2007). A random subsampling of the Fourier domain sampling of MR can be shown to provide sufficient incoherence for CS reconstruction, while the solution to the optimization problem can be approximated either by greedy methods (Tropp, 2004) or by linear programming after convex relaxation (Candès and Tao, 2005).

A key observation in choosing the sparsity transform is to note that sparser descriptions of the data can better condition the CS problem, hence having better chances of providing an accurate reconstruction. The use of sparsifying transforms usually exploited in compression such as wavelet transforms or transforms assuming piece-wise constant structures like total variation have been proposed many times for CS reconstruction. Some examples of these approaches are presented in Lustig et al. (2007) and Ma and Yin (2008). A common sparse transform for cardiac cine data has been the Fourier transform along the temporal dimension studied, for instance, in Jung et al. (2007) and Gamper et al. (2008), given that it collapses information from slowly varying structures in a few low-frequency coefficients.

Global complete and nonadaptive transforms sometimes do not adhere well to the data and can become crude sparsity models. More recently, the use of redundant and adaptive dictionaries has been proposed to seek better, more flexible sparse models. Assuming patches overlap and the extraction operation wraps around image boundaries, the patch-based dictionary CS problem can be cast as

$$\min_{\mathbf{x},\boldsymbol{\Gamma},\mathbf{D}} \quad \|\mathbf{y}-\mathbf{MFx}\|_2^2+\lambda\|\mathbf{R}_i\mathbf{x}-\mathbf{D}\boldsymbol{\gamma}_i\|_2^2 \quad \text{s.t.} \quad \|\boldsymbol{\gamma}_i\|_0 \leq S \quad \forall i. \tag{6.16}$$

The solution to this problem looks for an image $\mathbf{x}$ weighting consistency with the original acquisitions $\mathbf{y}$ with a sparse and accurate representation of the image provided by the patch-based dictionary $\mathbf{D}$.

The content of dictionary atoms can take different forms depending on the data they relate to. They can describe spatial information in the form of 2D or 3D patches, as proposed respectively in Ravishankar and Bresler (2011a) and Song et al. (2013). Spatio-temporal dictionaries have also been explored in Caballero et al. (2014) and Wang and Ying (2014) for the representation of cardiac cine data. One-dimensional atoms have been proposed for the description of temporal magnitude data in Awate and Dibella (2012) as well as quantitative parameter mapping data as discussed in Doneva et al. (2010).

The solution to Eq. (6.16) is nontrivial because it is highly nonconvex. It can nevertheless be approximated by keeping two variables fixed at a time and alternating the solution to the dictionary $\mathbf{D}$, the sparse representation $\boldsymbol{\Gamma}$, and the image $\mathbf{x}$. As shown in Ravishankar and Bresler (2011a); Caballero et al. (2014), solving for $\mathbf{D}$ boils down to the global DL problem, and the solution to $\boldsymbol{\Gamma}$ is given as the solution to the sparse coding problem. Finally, solving for $\mathbf{x}$ can be posed as a least-squares problem that weights in k-space the original acquisition and the sparse coding image approximation.

Results are reported in Ravishankar and Bresler (2011a) where the dictionary-based approach produces more accurate reconstructions from MR undersampled acquisitions for 2D structural images than the pioneering wavelet-based method from Lustig et al. (2007). Similarly, Fig. 6.8 compares results on cardiac cine data from the dictionary-based method in Caballero et al. (2014) to the method k-t FOCUSS presented in Jung et al. (2007), which exploits sparsity in the temporal Fourier space.

6.4.3 SUPER-RESOLUTION

Resolution enhancement, or super-resolution, is an important tool in medical imaging, particularly in dynamic imaging where movement precludes the acquisition of high-resolution images. For example, in MR imaging of the lungs or heart,

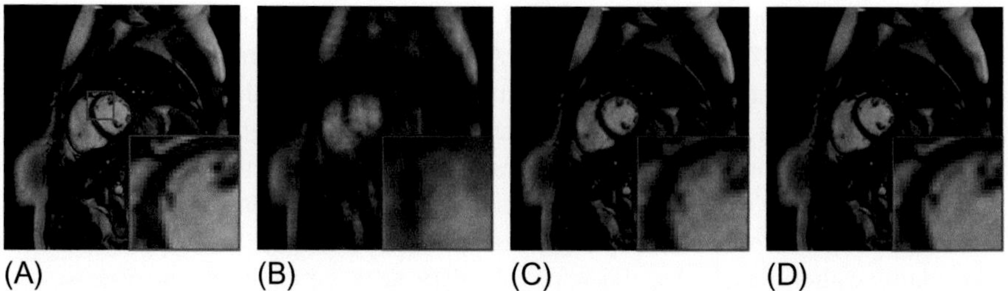

FIG. 6.8

Reconstruction from an eightfold undersampled MRI acquisition. A fully sampled magnitude frame (A), its undersampled by 8 zero-filled version (B), and reconstructions using k-t FOCUSS (C), and the dictionary learning method in Caballero et al. (2014) (D). The ROI in the red rectangle (dark gray in print versions) is enlarged for better visualization. (A) Original. (B) Zero-filled. (C) Temporal Fourier. (D) Dictionary learning.

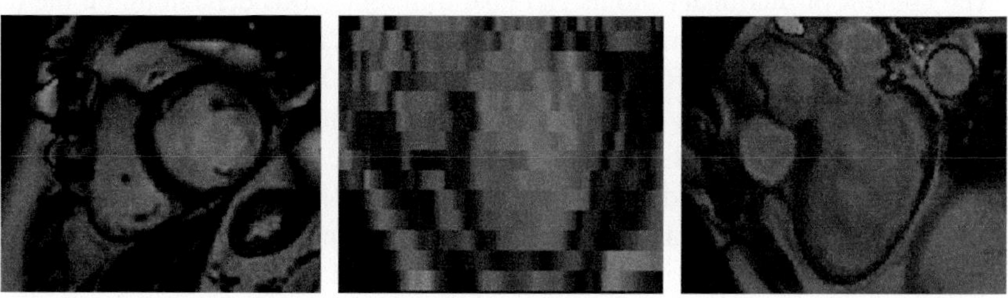

FIG. 6.9

Typical cardiac image acquisitions. Left to right: High-resolution short-axis slice, long-axis stack of short-axis slices showing low resolution in slice-select direction, frame of high-resolution long-axis slice sequence.

motion due to the respiratory or cardiac cycles forces a compromise between spatial resolution, temporal resolution, and volumetric coverage. A typical cardiac acquisition consists of imaging a number of anisotropic 2D slabs over time (Fig. 6.9). Improving the resolution of these acquired sequences is strongly motivated both by clinical visualization and analysis.

The super-resolution problem can be formulated as an inverse problem where it is assumed that the low-resolution (LR) image is a blurred and downsampled version of the original high-resolution (HR) image. Given an underlying unknown HR image $\mathbf{y}_H$, the acquired LR image $\mathbf{y}_L$ is modeled as

$$\mathbf{y}_L = (\mathbf{y}_H * \mathbf{B}) \downarrow s + \eta, \tag{6.17}$$

where $\mathbf{B}$ represents a blur operator, $\downarrow s$ is a downsampling operator that decreases the resolution by a factor of s, and η represents an additive noise term. Recovering the high-resolution image $\mathbf{y}_H$ from $\mathbf{y}_L$ is underdetermined and requires regularization or prior information on the nature of $\mathbf{y}_H$. Conventional super-resolution techniques provide this by fusing several views of the same object aligned with subpixel (or voxel in 3D) accuracy to constrain the solution. While this works well for objects which are deformed only by rigid or affine motion (such as the fetal brain (Gholipour et al., 2010)), and where accurate deformation can therefore be easily recovered, when the motion is nonrigid, such as that due to cardiac or respiratory dynamics, obtaining subpixel alignment becomes difficult.

This motivates the use of *example-based* or *hallucination* super-resolution methods. These avoid the need for accurate alignment by moving the analysis to a small patch scale. The aim is to upsample LR image patches using knowledge of the relationship between HR and LR features gained from example training data as in Manjón et al. (2010a,b), and Rousseau (2010). Such a relationship can be encapsulated using a pair of *correlated* (or *coupled*) HR and LR dictionaries. Central to this is how to train these dictionaries such that corresponding HR and LR patches have the same sparse representation with respect to their dictionary in both cases. The use of DL effectively enforces the prior on the reconstruction that signals such as image patches can be represented by a sparse combination of dictionary atoms.

The construction of correlated dictionaries for single-image super-resolution was proposed by Zeyde et al. (2012) and applied directly to single-image brain MRI enhancement in Rueda et al. (2013). However, a real benefit in medical applications is to improve resolution by combining several low-resolution views of a structure. This has been developed in Bhatia et al. (2014) using the same dictionary construction method. It is described in the following, with particular adaptation to cardiac image sequence enhancement.

Stack sequences such as those in Fig. 6.1 can be acquired; while HR is available only in two dimensions, orthogonal HR planes can be acquired and the information combined for super-resolution. For simplicity, we describe an upsampling of a stack which is HR in the cardiac short-axis (SA) only, using training data from a stack which is HR in the cardiac long-axis (LA) only. Training data consists of frames from an HR LA sequence $Y_H = \{\mathbf{y}_H\}$, and corresponding LR sequences obtained through blurring and downsampling according to Eq. (6.17). A Gaussian kernel, with full width at half maximum (FWHM) equal to the slice thickness, is used to blur in the slice-select direction only, in order to simulate the acquisition process (Greenspan, 2009). The resulting LR images are then upsampled using bicubic interpolation back to the original size, to avoid complications arising from differently sized patches, giving $Y_L = \{\mathbf{y}_L\}$. Patches at corresponding locations $P = \{\mathbf{p}_k^H, \mathbf{p}_k^L\}_k$ are extracted from these images: $\mathbf{p}_k = \mathbf{R}_k\mathbf{y}$, where $\mathbf{R}_k$ is an operator to extract a patch of size of size $n \times m$ from location k. These are used to co-train the HR and LR dictionaries, as shown diagrammatically in Fig. 6.10.

The aim of constructing correlated dictionaries is to ensure that an HR patch and its LR counterpart have the same sparse representations with respect to their individual dictionaries. It is also necessary to ensure that the LR patches can

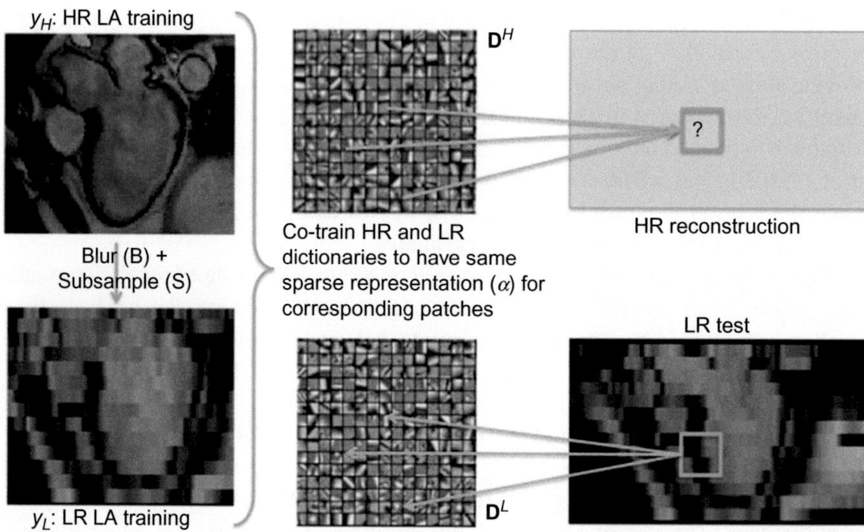

FIG. 6.10

High-resolution and low-resolution dictionaries can be co-trained such that corresponding high-resolution and low-resolution patches have the same sparse representation in each of their respective dictionaries.

be encoded sparsely *and in the same way* for both train and test data. When reconstructing an LR test patch, we find the sparse representation of that patch in terms of the LR dictionary only. The LR dictionary $\mathbf{D}^L$ is therefore also constructed using LR patches only:

$$\mathbf{D}^L, \boldsymbol{\gamma} = \arg\min_{\mathbf{D}^L, \gamma_k} \sum_k \|\mathbf{p}_k^L - \mathbf{D}^L \gamma_k\|_2^2 \text{ s.t.} \|\boldsymbol{\gamma}\|_0 < S, \tag{6.18}$$

where S denotes the desired sparsity of the reconstruction weights vector $\boldsymbol{\gamma}$. This is a standard DL equation which is solved sequentially for $\mathbf{D}^L$ and $\boldsymbol{\gamma}$ using K-SVD and OMP respectively. The resulting sparse code $\boldsymbol{\gamma}_k$ for each patch k is then used to solve for the HR dictionary by minimizing the reconstruction error of the HR training patches:

$$\mathbf{D}^H = \arg\min_{D_H} \sum_k \|\mathbf{p}_k^H - \mathbf{D}^H \gamma_k\|_2^2 = \arg\min_{\mathbf{D}^H} \|\mathbf{P}^H - \mathbf{D}^H \mathbf{A}\|_F^2, \tag{6.19}$$

where the columns of $\mathbf{P}$ are formed by the high-resolution training patches, $\mathbf{p}_k^H$ and the columns of $\mathbf{A}$ are formed by the atoms $\boldsymbol{\gamma}$. The solution is given by $\mathbf{D}^H = \mathbf{P}^H A^+ = \mathbf{P}^H \mathbf{A}^T (\mathbf{A}\mathbf{A}^T)^{-1}$ where $\mathbf{A}^+$ is the Pseudo-Inverse, noting that $\mathbf{A}$ has full row rank. This results in two correlated dictionaries, $\mathbf{D}^H$ and $\mathbf{D}^L$, which can be used to reconstruct HR patches from LR input.

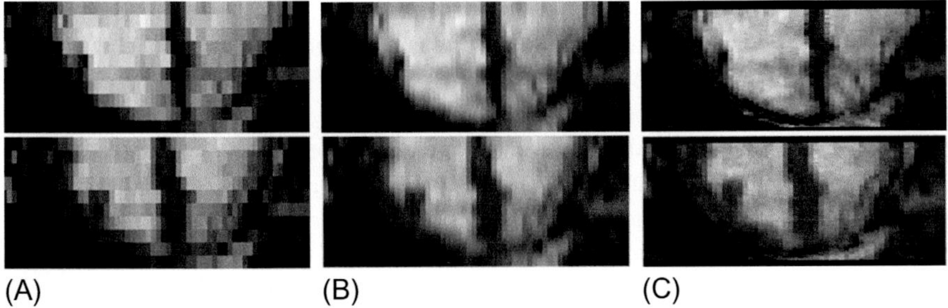

FIG. 6.11

Adult cardiac data; reconstruction using 30 high-resolution long-axis frames. (A) Original. (B) Bicubic. (C) Dictionary learning.

At the reconstruction stage, patches from an LR test image are extracted in the same way as for the LR training images. The sparse code for each patch with respect to the LR dictionary, $\mathbf{D}_L$, is found by solving:

$$\gamma_k = \arg\min_{\gamma_k} \sum_k \|p_k^L - \mathbf{D}^L \gamma_k\|_2^2 \text{ s.t.} \|\gamma_k\|_0 < S \tag{6.20}$$

using OMP. Crucially, this is the same sparse coding equation as in the training phase. The reconstruction weights vectors $\boldsymbol{\gamma}_k$ for each test patch are used to approximate high-resolution patches by $\{\tilde{\mathbf{p}}_k^H\}_k = \{\mathbf{D}^H \boldsymbol{\gamma}_k\}_k$. To create a smooth overall reconstruction, overlapping patches can be used, and the upsampled image given by the average of the overlapping reconstructed patches.

Example reconstructions for two adult subjects are shown in Fig. 6.11. Here, 30-frame sSA stack acquisitions are reconstructed to isotropic using an orthogonal HR LA stack sequence to train the dictionaries. This represents an upsampling factor of 8. Patch sizes of 12 × 24 pixels (6 × 12 pixel overlap) were used in both training and reconstruction. Training took less than 8],s, while reconstruction of all SA stack frames took under 2 s, using MATLAB on a 2.4-GHz Intel Core i5 machine in both cases. It can be clearly seen that in both cases the patch-based reconstruction algorithm produces sharper images than the standard interpolation.

6.4.4 SEGMENTATION

In previous sections, we have shown that DL is well suited for restoration tasks such as reconstruction and super-resolution. However, they cannot be directly used for image classification tasks because the learnt dictionaries only have the reconstructive power. There have been multiple attempts to learn discriminative dictionaries by using the class labels of training samples. One straightforward way is to use the training samples themselves as the dictionary without learning, in which case each atom in the dictionary has a training label. A testing sample can be represented by the training samples of all classes and be classified to the group that leads to the minimal reconstruction error. The so-called sparse representation-based classification (SRC) scheme has been successfully applied to face recognition (Wright et al., 2009) and also in medical image segmentation (Tong et al., 2013; Wang et al., 2014). For example, in Wang et al. (2014), the class labels of training patches were propagated to test patches via SRC. Specifically, all the training patches were first formed as a dictionary. The test patches can then be represented by this predefined dictionary. Since the class labels of atoms in the predefined dictionary are available, the class labels of test patches are estimated by using the coding coefficients as weights in a label fusion process. However, the noisy information in the training patches may make the segmentation less effective and the complexity of sparse coding can be very high when there are a large number of training patches. This problem can be addressed by properly learning a discriminative dictionary from the original training patches.

There are two main categories of discriminative DL methods. In the first category, dictionaries are learnt for different classes separately. These class-specific dictionaries are used to represent the test patches. The representation residual associated with each class can then be used to do classification. For example, Huang et al. (2014) proposed to learn a pair of dictionaries for the foreground and background training patches and used the representation residuals as new features for segmentation. Another category of discriminative DL methods learns a shared dictionary by all classes while training a classifier over the representation coefficients at the same time. In Tong et al. (2013) this type of discriminative DL methods has been successfully applied to the segmentation of the hippocampus in brain MR images. In the following we will describe this segmentation method, which is known as discriminative dictionary learning for segmentation (DDLS) and is illustrated in Fig. 6.12.

Given a set of training patches $\boldsymbol{P_L} = [p_1, p_2, \ldots, p_n] \in \mathbb{R}^{M \times N}$ with class labels, the segmentation process is to assign a class label for each test patch p_t. A reconstructive dictionary $\boldsymbol{D}$ with K atoms can be learnt from the training patches by solving the following problem:

$$\min_{\boldsymbol{D},\boldsymbol{\Gamma}} \|\boldsymbol{P_L} - \boldsymbol{D\Gamma}\|_F^2 \quad \text{s.t.} \quad \|\boldsymbol{\gamma}_i\|_0 \le S \;\; \forall i. \tag{6.21}$$

Here the objective function includes the reconstruction error term and the sparsity constraint term without considering the discriminative power. Thus the learnt dictionary is not suitable for the segmentation task. To address this problem, a linear classifier was added to the objective function for learning dictionaries with both reconstructive and discriminative power. The objective function is then formulated as

$$\min_{\boldsymbol{D},\boldsymbol{W},\boldsymbol{\Gamma}} \|\boldsymbol{P_L} - \boldsymbol{D\Gamma}\|_F^2 + \beta \|\boldsymbol{H} - \boldsymbol{W\Gamma}\|_F^2 \quad \text{s.t.} \quad \|\boldsymbol{\gamma}_i\|_0 \le S \;\; \forall i. \tag{6.22}$$

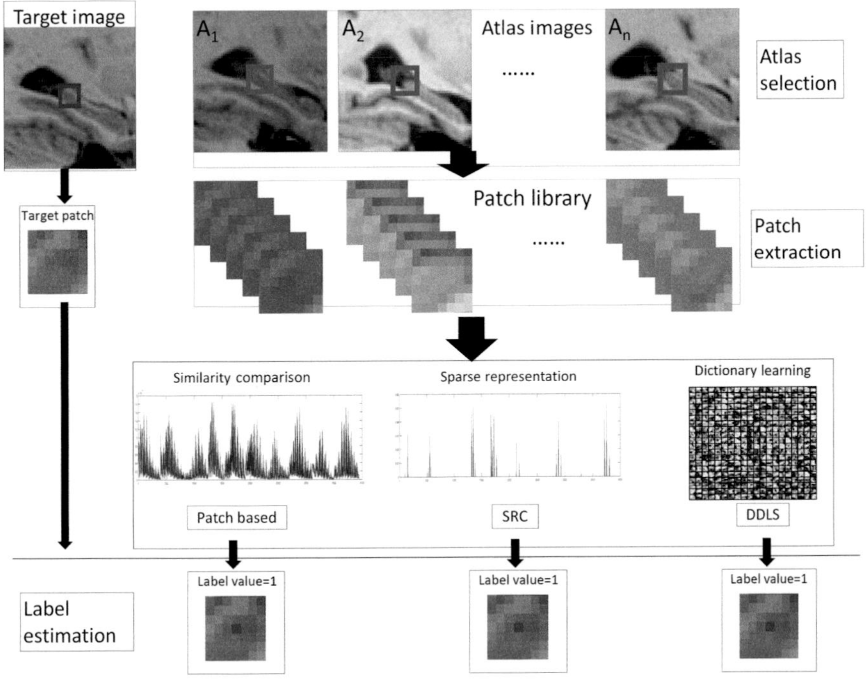

FIG. 6.12

Flow chart of labeling one target voxel by three different methods: Patch-based Labeling, Sparse Representation Classification (SRC) and Discriminative Dictionary Learning for Segmentation (DDLS). The red box (dark gray in print versions) in the target image represents the target patch. The blue boxes in atlas images represent the search volume area for extracting template patches.

Here the classification error term $\|\boldsymbol{H} - \boldsymbol{W\Gamma}\|_F^2$ is added to Eq. (6.21). $\boldsymbol{H}$ represents the labels of the central voxels of the patches in $\boldsymbol{P}_L$. Each column of $\boldsymbol{H}$ is a label vector corresponding to a training patch. Each label vector is defined as $h_i = [0, 0 \ldots 1 \ldots 0, 0]$, where the nonzero entry position indicates the label of the center voxel of the corresponding patch. $\boldsymbol{W}$ denotes the linear classifier parameters and β controls the trade-off between the reconstruction error term and the classification error term. In this testing stage, the learnt dictionary $\boldsymbol{D}$ can be used to represent the test patch while the learnt classifier $\boldsymbol{W}$ can be used for estimating a class label for the test patch.

The DDLS segmentation method has been successfully applied to MR hippocampal segmentation and multiorgan segmentation in abdominal CT images (Tong et al., 2015). Overall, the DDLS segmentation approach not only yields competitive segmentation accuracy but also can be implemented very efficiently.

However, the method was only evaluated on the segmentation of healthy structures. Another recent study in Weiss et al. (2013) utilized DL for the segmentation of multiple sclerosis lesions. The dictionary was learnt using patches from healthy regions, which is expected to reconstruct badly in lesion regions and yield large reconstruction errors. The final segmentation of the lesion was achieved by thresholding the reconstruction error map.

6.5 FUTURE DIRECTIONS

Although DL has been successfully applied in a large number of image processing tasks, there is still room for many improvements. Some immediate open questions are how to choose from the available DL algorithms and the selection of hyperparameters such as the number of atoms, the sparsity level, and the number of iterations for guaranteeing convergence. In general, there is no definite answer for choosing the optimal number of atoms or the best sparsity level in DL. The choices of these parameters are often made heuristically. In addition to learning the best dictionary for a specific application, there are several other aspects which have not yet been fully exploited for applications in medical imaging.

First, most of the DL techniques use patches at a single scale for learning dictionaries. However, in some applications, it is helpful to use patches at multiple scales for learning more efficient dictionaries. Local patches at coarser scales can provide global structural information (also called anatomical patterns in medical imaging) while patches at finer scales present important local appearance information such as intensity patterns. Learning dictionaries at multiple scales can fully utilize both the global anatomical information and local appearance information, which has been shown to be useful in image segmentations (Huang et al., 2014). However, multiscale DL approaches proposed, such as those by Mairal et al. (2007) and Ophir et al. (2011) have not been fully investigated in medical imaging. One example showing that multiscale DL could be beneficial is shown in Ravishankar and Bresler (2011b), where it is used for MR image reconstruction.

Another interesting research avenue is multimodal DL. The applications described in this chapter are based on images from a single modality, but it is possible to learn information about medical images across modalities using DL, as shown in Cao et al. (2013). For instance, learning a dictionary in a given modality to improve the reconstruction quality of a different modality, where data acquisition could be challenging, can be viable, as demonstrated in Tosic et al. (2010).

Exploiting structures in data is crucial for the success of many machine learning techniques. By adopting l_0-norm or l_1-norm for regularization, sparsity is achieved by treating each atom individually in DL techniques, regardless of its position in the dictionary or relations with other atoms. Therefore the relations and the structures between the atoms are ignored by just using the l_0-norm or l_1-norm constraint. However, this type of information could be helpful in many applications. Structured sparse coding (Elhamifar and Vidal, 2011) and structured DL (Szabó et al., 2011) have recently been introduced to encode these specific relations and structures in machine learning, which could be useful in many applications of medical image analysis.

To conclude, the description of DL throughout the chapter has been based on the sparse synthesis model, although there exist other types of signal modeling encompassed by DL such as analysis modeling (Rubinstein et al., 2013). Moreover, much research is still needed to better understand the low-dimensional spaces where medical images could potentially be efficiently represented. Although sparse coding and DL have been successfully applied in many applications, the performance of a linear model to represent medical images is limited for signals where nonlinearities can occur, such as in medical acquisition mechanisms.

6.6 CONCLUSION

In this chapter, we have provided an overview of DL and its application in medical imaging including image denoising, reconstruction, super-resolution, and segmentation. DL has been demonstrated to be very effective in these applications, showing its powerful ability in medical image analysis. In future, it would be very interesting to investigate the use of DL in new applications in medical imaging. However, it should be mentioned that identifying the appropriate applications is important as DL might not be able to provide a good solution to some specific medical image analysis problems.

REFERENCES

Aharon, M., Elad, M., Bruckstein, A., 2006. K-SVD: an algorithm for designing overcomplete dictionaries for sparse representation. IEEE Trans. Signal Process. 54 (11), 4311–4322.

Awate, S.P., Dibella, E.V.R., 2012. Spatiotemporal dictionary learning for undersampled dynamic MRI reconstruction via joint frame-based and dictionary-based sparsity. In: Proceedings of the IEEE International Symposium on Biomedical Imaging, pp. 318–321.

Bao, L., Robini, M., Liu, W., Zhu, Y., 2013. Structure-adaptive sparse denoising for diffusion-tensor MRI. Med. Image Anal. 17 (4), 442–457.

Beylkin, G., 1992. On the representation of operators in bases of compactly supported wavelets. SIAM J. Numer. Anal. 29 (6), 1716–1740.

Bhatia, K.K., Price, A.N., Shi, W., Hajnal, J.V., Rueckert, D., 2014. Super-resolution reconstruction of cardiac MRI using coupled dictionary learning. In: IEEE International Symposium on Biomedical Imaging. IEEE, Piscataway, NJ, pp. 947–950.

Bioucas-Dias, J.M., Figueiredo, M.A., 2008. An iterative algorithm for linear inverse problems with compound regularizers. In: International Conference on Image Processing, IEEE, Piscataway, NJ, pp. 685–688.

Caballero, J., Price, A.N., Rueckert, D., Hajnal, J.V., 2014. Dictionary learning and time sparsity for dynamic MR data reconstruction. IEEE Trans. Med. Imaging 33 (4), 979–994.

Candès, E.J., Donoho, D.L., 1999. Curvelets: a surprisingly effective nonadaptive representation for objects with edges. In: International Conference on Curves and Surfaces, vol. 2, pp. 105–120.

Candès, E.J., Tao, T., 2005. Decoding by linear programming. IEEE Trans. Inform. Theory 51 (12), 4203–4215.

Cao, T., Jojic, V., Modla, S., Powell, D., Czymmek, K., Niethammer, M., 2013. Robust multimodal dictionary learning. In: International Conference on Medical Image Computing and Computer-Assisted Intervention. Springer, New York, pp. 259–266.

Chen, S., Donoho, D., Saunders, M., 1998. Atomic decomposition by basis pursuit. SIAM J. Sci. Comput. 20 (1), 33–61.

Coifman, R.R., Meyer, Y., Wickerhauser, V., 1992. Wavelet analysis and signal processing. Wavelets Their Appl. 153–178.

Cooley, J.W., Tukey, J.W., 1965. An algorithm for the machine calculation of complex Fourier series. Math. Comput. 19 (90), 297–301.

Coupé, P., Yger, P., Prima, S., Hellier, P., Kervrann, C., Barillot, C., 2008. An optimized blockwise nonlocal means denoising filter for 3D magnetic resonance images. IEEE Trans. Med. Imaging 27 (4), 425–441.

Do, M.N., Vetterli, M., 2002. Contourlets : a directional multiresolution image representation. Image Process. 1, 357–360.

Doneva, M., Börnert, P., Eggers, H., Stehning, C., Sénégas, J., Mertins, A., 2010. Compressed sensing reconstruction for magnetic resonance parameter mapping. Magn. Reson. Med. 64 (4), 1114–1120.

Donoho, D.L., Tsaig, Y., Drori, I., Starck, J.L., 2012. Sparse solution of underdetermined systems of linear equations by stagewise orthogonal matching pursuit. IEEE Trans. Inform. Theory 58 (2), 1094–1121.

Elad, M., 2010. Sparse and Redundant Representations: From Theory to Applications in Signal and Image Processing. Springer, New York.

Elad, M., Aharon, M., 2006. Image denoising via sparse and redundant representations over learned dictionaries. IEEE Trans. Image Process. 15 (12), 3736–3745.

Elhamifar, E., Vidal, R., 2011. Robust classification using structured sparse representation. In: IEEE Conference on Computer Vision and Pattern Recognition, pp. 1873–1879.

Engan, K., Aase, S.O., Husoy, J., 1999. Frame based signal compression using method of optimal directions (MOD). In: IEEE International Symposium on Circuits and Systems, vol. 4, pp. 1–4.

Friedman, J., Hastie, T., Höfling, H., Tibshirani, R., et al., 2007. Pathwise coordinate optimization. Ann. Appl. Stat. 1 (2), 302–332.

Gabor, D., 1946. Theory of communication. J. Inst. Elect. Eng. 93 (26), 429–441.

Gamper, U., Boesiger, P., Kozerke, S., 2008. Compressed sensing in dynamic MRI. Magn. Reson. Med. 59 (2), 365–373.

Gholipour, A., Estroff, J.A., Warfield, S.K., 2010. Robust super-resolution volume reconstruction from slice-acquisitions: application to fetal brain MRI. IEEE Trans. Med. Imaging 29 (10), 1739–1758.

Greenspan, H., 2009. Super-resolution in medical imaging. Comput. J. 52(1), 43–63.

Grossmann, A., Morlet, J., 1984. Decomposition of Hardy functions into square integrable wavelets of constant shape. SIAM J. Math. Anal. 15 (4), 723–736.

Huang, X., Dione, D.P., Compas, C.B., Papademetris, X., Lin, B.A., Bregasi, A., Sinusas, A.J., Staib, L.H., Duncan, J.S., 2014. Contour tracking in echocardiographic sequences via sparse representation and dictionary learning. Med. Image Anal. 18 (2), 253–271.

Jolliffe, I., 2005. Principal Component Analysis. Wiley, Chichester.

Jung, H., Ye, J.C., Kim, E.Y., 2007. Improved k-t BLAST and k-t SENSE using FOCUSS. Phys. Med. Biol. 52 (11), 3201–3226.

Le Pennec, E., Mallat, S., 2005. Sparse geometric image representations with bandelets. IEEE Trans. Image Process. 14 (4), 423–438.

Li, S., Yin, H., Fang, L., 2012. Group-sparse representation with dictionary learning for medical image denoising and fusion. IEEE Trans. Biomed. Eng. 59 (12), 3450–3459.

Lustig, M., Donoho, D.L., Pauly, J.M., 2007. Sparse MRI: the application of compressed sensing for rapid MR imaging. Magn. Reson. Med. 58 (6), 1182–1195.

Ma, S., Yin, W., 2008. An efficient algorithm for compressed MR imaging using total variation and wavelets. In: IEEE Conference on Computer Vision and Pattern Recognition 2008, pp. 1–8.

Mairal, J., Sapiro, G., Elad, M., 2007. Learning multiscale sparse representations for image and video restoration. Technical Report, DTIC Document.

Mairal, J., Bach, F., Ponce, J., Sapiro, G., 2009. Online dictionary learning for sparse coding. In: Proceedings of the 26th Annual International Conference on Machine Learning. ACM, pp. 689–696.

Mallat, S.G., 1999. A Wavelet Tour of Signal Processing. Academic Press, San Diego, CA.

Mallat, S.G., Zhang, Z., 1993. Matching pursuits with time-frequency dictionaries. IEEE Trans. Signal Process. 41 (12), 3397–3415.

Manjón, J.V., Coupé, P., Buades, A., Collins, D.L., Robles, M., 2010a. MRI superresolution using self-similarity and image priors. J. Biomed. Imaging 2010 (17), 1–11.

Manjón, J.V., Coupé, P., Buades, A., Fonov, V., Collins, D.L., Robles, M., 2010b. Non-local MRI upsampling. Med. Image Anal. 14 (6), 784–792.

Manjón, J.V., Coupé, P., Martí-Bonmatí, L., Collins, D.L., Robles, M., 2010c. Adaptive non-local means denoising of MR images with spatially varying noise levels. J. Mag. Reson. Imaging 31 (1), 192–203.

Needell, D., Tropp, J.A., 2009. CoSaMP: iterative signal recovery from incomplete and inaccurate samples. Appl. Comput. Harmon. Anal. 26 (3), 301–321.

Needell, D., Vershynin, R., 2010. Signal recovery from incomplete and inaccurate measurements via regularized orthogonal matching pursuit. IEEE J. Sel. Top. Signal Process. 4 (2), 310–316.

Ophir, B., Lustig, M., Elad, M., 2011. Multi-scale dictionary learning using wavelets. IEEE J. Sel. Top. Signal Process. 5 (5), 1014–1024.

Osher, S., Burger, M., Goldfarb, D., Xu, J., Yin, W., 2005. An iterative regularization method for total variation-based image restoration. Multiscale Model. Simul. 4 (2), 460–489.

Patel, V., Shi, Y., Thompson, P.M., Toga, A.W., 2011. K-SVD for Hardi denoising. In: IEEE International Symposium on Biomedical Imaging: From Nano to Macro, pp. 1805–1808.

Ravishankar, S., Bresler, Y., 2011a. MR image reconstruction from highly undersampled k-space data by dictionary learning. IEEE Trans. Med. Imaging 30 (5), 1028–41.

Ravishankar, S., Bresler, Y., 2011b. Multiscale dictionary learning for MRI. In: Proceedings of ISMRM, p. 2830.

Rousseau, F., 2010. A non-local approach for image super-resolution using intermodality priors. Med. Image Anal. 14, 594–605.

Rubinstein, R., Zibulevsky, M., Elad, M., 2008. Efficient implementation of the K-SVD algorithm using batch orthogonal matching pursuit. CS Technion 40 (8), 1–15.

Rubinstein, R., Peleg, T., Elad, M., 2013. Analysis K-SVD: a dictionary-learning algorithm for the analysis sparse model. IEEE Trans. Signal Process. 61 (3), 661–677.

Rueda, A., Malpica, N., Romero, E., 2013. Single-image super-resolution of brain MR images using overcomplete dictionaries. Med. Image Anal. 17(1),113–132.

Shannon, C.E., 1949. Communication in the presence noise. Proc. IRE 37 (1), 10–21.

Simoncelli, E.P., Freeman, W.T., Adelson, E.H., Heeger, D.J., 1992. Shiftable multiscale transforms. IEEE Trans. Inform. Theory 38 (2), 587–607.

Skodras, A., Christopoulos, C., Ebrahimi, T., 2001. The JPEG 2000 still image compression standard. IEEE Signal Process. Mag. 18 (5), 36–58.

Song, Y., Zhu, Z., Lu, Y., Liu, Q., Zhao, J., 2013. Reconstruction of magnetic resonance imaging by three-dimensional dual-dictionary learning. Magn. Reson. Med. 71 (3), 1285–1298.

Szabó, Z., Póczos, B., Lorincz, A., 2011. Online group-structured dictionary learning. In: IEEE Conference on Computer Vision and Pattern Recognition. IEEE, Piscataway, NJ, pp. 2865–2872.

Tibshirani, R., 1996. Regression shrinkage and selection via the lasso. J. R. Stat. Soc. B 267–288.

Tong, T., Wolz, R., Coupé, P., Hajnal, J.V., Rueckert, D., Initiative, A.D.N., et al., 2013. Segmentation of MR images via discriminative dictionary learning and sparse coding: application to hippocampus labeling. NeuroImage 76, 11–23.

Tong, T., Wolz, R., Wang, Z., Gao, Q., Misawa, K., Fujiwara, M., Mori, K., Hajnal, J.V., Rueckert, D., 2015. Discriminative dictionary learning for abdominal multi-organ segmentation. Med. Image Anal. 23 (1), 92–104.

Tosic, I., Frossard, P., 2011. Dictionary learning. IEEE Signal Process. Mag. 28 (2), 27–38.

Tosic, I., Jovanovic, I., Frossard, P., Vetterli, M., Duric, N., 2010. Ultrasound tomography with learned dictionaries. In: IEEE International Conference on Acoustics, Speech, and Signal Processing, pp. 5502–5505.

Tropp, J., 2004. Greed is good: algorithmic results for sparse approximation. IEEE Trans. Inform. Theory 50 (10), 2231–2242.

Wallace, G., 1992. The JPEG still picture compression standard. IEEE Trans. Consumer Electron. 38 (1), xviii–xxxiv.

Wang, Y., Ying, L., 2014. Compressed sensing dynamic cardiac cine MRI using learned spatiotemporal dictionary. IEEE Trans. Biomed. Eng. 61 (4), 1109–1120.

Wang, L., Shi, F., Gao, Y., Li, G., Gilmore, J.H., Lin, W., Shen, D., 2014. Integration of sparse multi-modality representation and anatomical constraint for isointense infant brain MR image segmentation. NeuroImage 89, 152–164.

Weiss, N., Rueckert, D., Rao, A., 2013. Multiple sclerosis lesion segmentation using dictionary learning and sparse coding. In: International Conference on Medical Image Computing and Computer-Assisted Intervention. Springer, New York, pp. 735–742.

Wright, J., Yang, A., Ganesh, A., Sastry, S., Ma, Y., 2009. Robust face recognition via sparse representation. IEEE Trans. Pattern Anal. Mach. Intell. 31 (2), 210–227.

Zeyde, R., Elad, M., Protter, M., 2012. On single image scale-up using sparse-representations. In: International Conference on Curves and Survaces. Springer, New York, pp. 711–730.

GLOSSARY

CS compressed sensing
DCT discrete cosine transform
DL dictionary learning
MR magnetic resonance
MRI magnetic resonance imaging
MSE mean squared error
OMP orthogonal matching pursuit

CHAPTER

Advanced sparsity techniques in magnetic resonance imaging

J. Huang, Y. Li

University of Texas at Arlington, Arlington, TX, United States

CHAPTER OUTLINE

7.1 INTRODUCTION

Magnetic resonance imaging (MRI) has been widely used in medical diagnosis because of its noninvasive manner and excellent depiction of soft tissue changes. Recent developments in compressive sensing (CS) theory (Candès et al., 2006c; Donoho, 2006) show that it is possible to accurately reconstruct the magnetic

Machine Learning and Medical Imaging. http://dx.doi.org/10.1016/B978-0-12-804076-8.00007-4

resonance (MR) images from highly undersampled K-space data and therefore significantly reduce the scanning duration.

Suppose x is an MR image and R is a partial Fourier transform, the sampling measurement b of x in K-space is defined as $b = Ax$. The compressed MR image reconstruction problem is to reconstruct x given the measurement b and the sampling matrix A. Sometimes the data is not sparse but compressible under some base Φ such as wavelet, and the corresponding problem is $A\Phi^{-1}\theta = b$, where θ denotes the set of wavelet coefficients. Although the problem is underdetermined, the data can be perfectly reconstructed if the sampling matrix satisfies the restricted isometry property (RIP) (Candès, 2006) and the number of measurements is larger than $\mathcal{O}(k + k\log(N/k))$ for k-sparse data[1] (Candès and Romberg, 2007; Candès et al., 2006b).

To solve the underdetermined problem, we may find the sparsest solution via ℓ_0-norm regularization. However, because the problem is NP-hard (Natarajan, 1995) and impractical for most applications, ℓ_1-norm regularization methods such as the lasso (Tibshirani, 1996) and basis pursuit (BP) (Chen et al., 1998) are first used to pursue the sparse solution. It has been proved that the ℓ_1-norm regularization can exactly recover the sparse data for the CS inverse problem under mild conditions (Candès et al., 2006a; Donoho and Elad, 2002). Therefore a lot of efficient algorithms have been proposed for standard sparse recovery. Generally speaking, those algorithms can be classified into three groups: greedy algorithms (Needell and Tropp, 2009; Tropp, 2004), convex programming (Beck and Teboulle, 2009b; Figueiredo et al., 2007; Koh et al., 2007), and probability-based methods (Donoho et al., 2009; Ji et al., 2008).

Beyond standard sparsity, the nonzero components of x often tend to be in some structures. This leads to the concept of *structured sparsity* or model-based compressed sensing (Baraniuk et al., 2010; Huang et al., 2011c; Huang, 2011). In contrast to standard sparsity that only relies on the sparseness of the data, structured sparsity models exploit both the nonzero values and the corresponding locations. For example, in the multiple measurement vector (MMV) problem, the data consists of several vectors that share the same support.[2] This is called *joint sparsity* that is common in cognitive radio networks (Meng et al., 2011), direction-of-arrival estimation in radar (Krim and Viberg, 1996), multichannel compressed sensing (Baron et al., 2005; Majumdar and Ward, 2010), remote sensing (Chen et al., 2014b), and medical imaging (Bilgic et al., 2011; Huang et al., 2012). If the data $X \in \mathbb{R}^{TN\times 1}$ consists of T k-sparse vectors, the measurement bound could be substantially reduced to $\mathcal{O}(Tk + k\log(N/q))$ instead of $\mathcal{O}(Tk + Tk\log(N/q))$ for standard sparsity (Baraniuk et al., 2010; Huang et al., 2011c, 2009, 2010).

A common way to implement joint sparsity in convex programming is to replace the ℓ_1-norm with $\ell_{2,1}$-norm, which is the summation of ℓ_2-norms of the correlated entries (Bach, 2008; Yuan and Lin, 2005). $\ell_{2,1}$-norm for joint sparsity has been

[1] The term "k-sparse data" means there are at most k nonzero components in the data.

[2] The set of indices corresponding to the nonzero entries is often called the support.

used in many convex solvers and algorithms (Cotter et al., 2005; Van Den Berg and Friedlander, 2008; Huang et al., 2012; Deng et al., 2011). In Bayesian sparse learning or approximate message passing (Ji et al., 2009; Wipf and Rao, 2007; Ziniel and Schniter, 2011), data from all channels contribute to the estimation of parameters or hidden variables in the sparse prior model.

Another common structure would be the hierarchical tree structure, which has already been successfully utilized in image compression (Manduca, 1996), compressed imaging (Chen and Huang, 2012b; He and Carin, 2009; Som and Schniter, 2012; Rao et al., 2011), and machine learning (Kim and Xing, 2012). Most nature signals/images are approximately tree-sparse under the wavelet basis. A typical relationship with *tree sparsity* is that, if a node on the tree is nonzero, all of its ancestors leading to the root should be nonzero. For multichannel data $X = [x_1; x_2; \ldots; x_T] \in \mathbb{R}^{NT \times 1}$,[3] $\mathcal{O}(Tk + T\log(N/k))$ measurements are required if each channel x_t is tree-sparse.

Due to the overlapping and intricate structure of tree sparsity, it is much harder to implement. For greedy algorithms, structured orthogonal matching pursuit (StructOMP) (Huang et al., 2011c) and tree-based orthogonal matching pursuit (TOMP) (La and Do, 2006) have been developed for exploiting tree structure where the coefficients are updated by only searching the subtree blocks instead of all subspace. In statistical models (He and Carin, 2009; Som and Schniter, 2012), hierarchical inference is used to model the tree structure, where the value of a node is not independent but relies on the distribution or state of its parent. In convex programming (Chen and Huang, 2014a; Rao et al., 2011), due to the trade-off between the recovery accuracy and computational complexity, this is often approximated as overlapping group sparsity (Jacob et al., 2009), where each node and its parent are assigned into one group.

Although both joint sparsity and tree sparsity have been widely studied, unfortunately there is no work that studies the benefit of their combinations so far. Actually, in many multichannel compressed sensing or MMV problems, the data has joint sparsity across different channels and each channel itself is tree-sparse. Note that this differs from C-HiLasso (Sprechmann et al., 2011), where sparsity is assumed inside the groups. No method has fully exploited both priors and no theory guarantees the performance. In practical applications, researchers and engineers have to choose either joint sparsity algorithms by giving up their intra tree-sparse prior, or tree sparsity algorithms by ignoring their intercorrelations.

A new sparsity model called *forest sparsity* is proposed to bridge this gap (Chen et al., 2014a). It is a natural extension of existing structured sparsity models by assuming that the data can be represented by a forest of mutually connected trees. Based on compressed sensing theory, it is proved that for a forest of T k-sparse trees, only $\mathcal{O}(Tk + \log(N/k))$ measurements are required for successful recovery with high probability (Chen et al., 2014a). That is much less than the bounds of

[3] In this chapter, the notation [;] denotes concatenating the data vertically.

joint sparsity $\mathcal{O}(Tk + k\log(N/k))$ and tree sparsity $\mathcal{O}(Tk + T\log(N/k))$ on the same data. The theory is further extended to the case of MMV problems, which is ignored in existing structured sparsity theories (Baraniuk et al., 2010; Huang et al., 2011c; Huang, 2011). In this chapter we will also show an efficient algorithm to optimize the forest sparsity model. The algorithm is applied to medical imaging applications such as multicontrast MRI and parallel MRI (pMRI).

The rest of this chapter is organized as follows. In Section 7.2 we discuss the application of the standard sparsity to MR image reconstruction. Then, Section 7.3 introduces the benefits of the group sparsity in multicontrast MRI reconstruction, while Section 7.4 discusses the benefit of tree sparsity in accelerated MRI. An extension of the tree sparsity named forest sparsity is discussed for multichannel CS-MRI in Section 7.5. We conclude this chapter in Section 7.6.

7.2 STANDARD SPARSITY IN CS-MRI

In this section we present an efficient algorithm for MR image reconstruction. The algorithm minimizes a linear combination of three terms corresponding to a least square data fitting, total variation (TV), and $L1$-norm regularization. This has been shown to be very powerful for the MR image reconstruction.

Suppose x is an MR image and R is a partial Fourier transform, the sampling measurement b of x in K-space is defined as $b = Rx$. The compressed MR image reconstruction problem is to reconstruct x given the measurement b and the sampling matrix R. Motivated by the compressive sensing theory, Lustig et al. (2007) proposed their pioneering work for MR image reconstruction. Their method can effectively reconstruct MR images with only 20% sampling. Improved results were obtained by having both a wavelet transform and a discrete gradient in the objective, which is formulated as follows:

$$\hat{x} = \arg\min_{x} \left\{ \frac{1}{2}\|Rx - b\|^2 + \alpha\|x\|_{TV} + \beta\|\Phi, x\|_1 \right\} \tag{7.1}$$

where α and β are two positive parameters, b is the undersampled measurements of K-space data, R is a partial Fourier transform, and Φ is a wavelet transform. It is based on the fact that the piecewise smooth MR images of organs can be sparsely represented by the wavelet basis and should have small total variations. The TV was defined discretely as $\|x\|_{TV} = \sum_i \sum_j \sqrt{(\nabla_1 x_{ij})^2 + (\nabla_2 x_{ij})^2}$, where ∇_1 and ∇_2 denote the forward finite difference operators on the first and second coordinates respectively. Since both $L1$- and TV-norm regularization terms are nonsmooth, this problem is very difficult to solve. The conjugate gradient (CG) (Lustig et al., 2007) and partial differential equation (He et al., 2006) methods have been used to do this. However, they are very slow and impractical for real MR images. Computation became the bottleneck that prevented this good model (7.1) from being used in practical MR image reconstruction. Therefore the key problem in compressed MR image reconstruction is to develop efficient algorithms to solve problem (7.1) with nearly optimal reconstruction accuracy.

7.2.1 MODEL AND ALGORITHM

7.2.1.1 Related acceleration algorithm

We first briefly review the fast iterative shrinkage-thresholding algorithm (FISTA) from (Beck and Teboulle, 2009b), since our methods are motivated by this. FISTA considers minimizing the following problem:

$$\min\{F(x) \equiv f(x) + g(x), x \in \mathbf{R^p},\} \tag{7.2}$$

where f is a smooth convex function with Lipschitz constant L_f, and g is a convex function which may be nonsmooth.

ϵ-Optimal solution: Suppose x^* is an optimal solution to (7.2). $x \in \mathbf{R}^p$ is called an ϵ-optimal solution to (7.2) if $F(x) - F(x^*) \leq \epsilon$ holds.

Gradient: $\nabla f(x)$ denotes the gradient of the function f at the point x.

The proximal map: Given a continuous convex function $g(x)$ and any scalar $\rho > 0$, the proximal map associated with function g is defined as follows (Beck and Teboulle, 2009a,b):

$$prox_\rho(g)(x) := \arg\min_u \left\{ g(u) + \frac{1}{2\rho}\|u - x\|^2 \right\}, \tag{7.3}$$

Algorithm 7.1 outlines the FISTA. It can obtain an ϵ-optimal solution in $\mathcal{O}(1/\sqrt{\epsilon})$ iterations.

Theorem 7.1 (Theorem 4.1 in Beck and Teboulle (2009b)). *Suppose* $\{x^k\}$ *and* $\{r^k\}$ *are iteratively obtained by the FISTA, then we have*

$$F(x^k) - F(x^*) \leq \frac{2L_f\|x^0 - x^*\|^2}{(k+1)^2}, \forall x^* \in X_*.$$

ALGORITHM 7.1 FISTA Beck and Teboulle (2009b)

series Input: $\rho = 1/L_f, r^1 = x^0, t^1 = 1$
for $k = 1$ **series to** K **do**
  $x_g = r^k - \rho\nabla f(r^k)$
  $x^k = prox_\rho(g)(x_g)$
  $t^{k+1} = \frac{1+\sqrt{1+4(t^k)^2}}{2}$
  $r^{k+1} = x^k + \frac{t^k-1}{t^{k+1}}(x^k - x^{k-1})$
end for

ALGORITHM 7.2 CSD

series Input: $\rho = 1/L, \alpha, \beta, z_1^0 = z_2^0 = x_g$
for $j = 1$ **series to** J **do**
$x_1 = prox_\rho(2\alpha\|x\|_{TV})(z_1^{j-1})$
$x_2 = prox_\rho(2\beta\|\Phi x\|_1)(z_2^{j-1})$
$x^j = (x_1 + x_2)/2$
$z_1^j = z_1^{j-1} + x^j - x_1$
$z_2^j = z_2^{j-1} + x^j - x_2$
end for

The efficiency of the FISTA depends on being able to quickly solve its second step $x^k = prox_\rho(g)(x_g)$. For simpler regularization problems, this is possible, that is, the FISTA can rapidly solve the l_1 regularization problem with cost $\mathcal{O}(p\log(p))$ (Beck and Teboulle, 2009b) (where p is the dimension of x), since the second step $x^k = prox_\rho(\beta\|\Phi x\|_1)(x_g)$ has a closed-form solution; it can also quickly solve the TV regularization problem, since the step $x^k = prox_\rho(\alpha\|x\|_{TV})(x_g)$ can be computed with cost $\mathcal{O}(p)$ (Beck and Teboulle, 2009a). However, the FISTA cannot efficiently solve the composite l_1 and TV regularization problem (7.1), since no efficient algorithm exists to solve the step

$$x^k = prox_\rho(\alpha\|x\|_{TV} + \beta\|\Phi x\|_1)(x_g). \tag{7.4}$$

To solve problem (7.1), the key problem is thus to develop an efficient algorithm to solve problem (7.4). In the following section, we will show that a scheme based on composite splitting techniques can be used to do this.

7.2.1.2 CSA and FCSA

From the above introduction, we know that, if we can develop a fast algorithm to solve problem (7.4), the MR image reconstruction problem can then be efficiently solved by the FISTA, which obtains an ϵ-optimal solution in $\mathcal{O}(1/\sqrt{\epsilon})$ iterations. Actually, problem (7.4) can be considered as a denoising problem:

$$x^k = \arg\min_x \left\{\frac{1}{2}\|x - x_g\|^2 + \rho\alpha\|x\|_{TV} + \rho\beta\|\Phi x\|_1\right\}. \tag{7.5}$$

We use composite splitting techniques to solve this problem: (1) splitting variable x into two variables $\{x_i\}_{i=1,2}$; (2) performing operator splitting over each of $\{x_i\}_{i=1,2}$ independently; and (3) obtaining the solution x by linear combination of $\{x_i\}_{i=1,2}$. We call this the composite splitting denoising (CSD) method, which is outlined in Algorithm 7.2. Its validity is guaranteed by the following theorem:

Theorem 7.2. *Suppose* $\{x^j\}$ *is the sequence generated by the CSD. Then,* x^j *will converge to* $prox_\rho(\alpha\|x\|_{TV} + \beta\|\Phi x\|_1)(x_g)$*, which means that we have* $x^j \to prox_\rho(\alpha\|x\|_{TV} + \beta\|\Phi x\|_1)(x_g)$.

Sketch Proof of Theorem 7.2
Consider a more general formulation:

$$\min_{x \in \mathbf{R^p}} F(x) \equiv f(x) + \sum_{i=1}^{m} g_i(B_i x), \tag{7.6}$$

where f is the loss function and $\{g_i\}_{i=1,\ldots,m}$ are the prior models, both of which are convex functions; $\{B_i\}_{i=1,\ldots,m}$ are orthogonal matrices.

Proposition 7.1 (Theorem 3.4 in Combettes and Pesquet (2008)). *Let $\mathcal{H}$ be a real Hilbert space, and let $g = \sum_{i=1}^{m} g_i$ in $\Gamma_0(\mathcal{H})$ such that $domg_i \cap domg_j \neq \emptyset$. Let $r \in \mathcal{H}$ and $\{x_j\}$ be generated by Algorithm 7.3. Then, x_j will converge to $prox(g)(r)$.*

The detailed proof for this proposition can be found in Combettes and Pesquet (2008) and Combettes (2009).

ALGORITHM 7.3 ALGORITHM 3.1 IN Combettes and Pesquet (2008)

series Input: ρ, $\{z_i\}_{i=1,\ldots,m} = r$, $\{w_i\}_{i=1,\ldots,m} = 1/m$,
for $j = 1$ **series to** J **do**
  for $i = 1$ **series to** m **do**
    $p_{i,j} = prox_\rho(g_i/w_i)(z_j)$
  end for
  $p_j = \sum_{i=1}^{m} w_i p_{i,j}$
  $\lambda_j \in [0, 2]$
  for $i = 1$ **series to** m **do**
    $z_{i,j+1} = z_{i,j+1} + \lambda_j(2p_j - x_j - p_{i,j})$
  end for
  $x_{j+1} = x_j + \lambda_j(p_j - x_j)$
end for

Suppose that $y_i = B_i x$, $s_i = B_i^T r$, and $h_i(y_i) = m\rho g_i(B_i x)$. Because the operators $\{B_i\}_{i=1,\ldots,m}$ are orthogonal, we can easily obtain that $\frac{1}{2\rho}\|x - r\|^2 = \sum_{i=1}^{m} \frac{1}{2m\rho} \|y_i - s_i\|^2$. The above problem is transferred to:

$$\hat{y}_i = \arg\min_{y_i} \sum_{i=1}^{m} \left[\frac{1}{2}\|y_i - s_i\|^2 + h_i(y_i) \right], \; x = B_i^T y_i, \; i = 1, \ldots, m. \tag{7.7}$$

Obviously, this problem can be solved by Algorithm 7.3. According to Proposition 7.1, we know that x will converge to $prox(g)(r)$. Assuming $g_1(x) = \alpha\|x\|_{TV}$, $g_2(x) = \beta\|x\|_1$, $m = 2$, $w_1 = w_2 = 1/2$ and $\lambda_j = 1$, we obtain the proposed CSD algorithm. x will converge to $prox(g)(r)$, where $g = g_1 + g_2 = \alpha\|x\|_{TV} + \beta\|\Phi x\|_1$.

End of Proof

Combining the CSD with FISTA, a new algorithm, the fast composite splitting algorithm (FCSA), is proposed for MR image reconstruction problem (7.1). In practice, we found that a small iteration number J in the CSD is enough for the FCSA to obtain good reconstruction results. In particular, it is set as 1 in our algorithm. Numerous experimental results in the next section will show that it is good enough for real MR image reconstruction.

Algorithm 7.5 outlines the proposed FCSA. In this algorithm, if we remove the acceleration step by setting $t^{k+1} \equiv 1$ in each iteration, we will obtain the composite splitting algorithm (CSA), which is outlined in Algorithm 7.4. A key feature of the FCSA is its fast convergence performance borrowed from the FISTA. From Theorem 7.1, we know that the FISTA can obtain an ϵ-optimal solution in $\mathcal{O}(1/\sqrt{\epsilon})$ iterations.

Another key feature of the FCSA is that the cost of each iteration is $\mathcal{O}(p\log(p))$, as confirmed by the following observations. Steps 4, 6, and 7 only involve adding vectors or scalars, thus cost only $\mathcal{O}(p)$ or $\mathcal{O}(1)$. In step 1, $\nabla f(r^k = R^T(Rr^k - b)$ since $f(r^k) = \frac{1}{2}\|Rr^k - b\|^2$ in this case. Thus this step only costs $\mathcal{O}(p\log(p))$. As introduced above, the step $x^k = prox_\rho(2\alpha\|x\|_{TV})(x_g)$ can be computed quickly with cost $\mathcal{O}(p)$ (Beck and Teboulle, 2009a); the step $x^k = prox_\rho(2\beta\|\Phi x\|_1)(x_g)$ has a closed-form solution and can be computed with cost $\mathcal{O}(p\log(p))$. In the step $x^k = project(x^k, [l, u])$, the function $x = project(x, [l, u])$ is defined as: (1) $x = x$ if $l \leq x \leq u$; (2) $x = l$ if $x < u$; and (3) $x = u$ if $x > u$, where $[l, u]$ is the range of x. For example, in the case of MR image reconstruction, we can let $l = 0$ and $u = 255$ for 8-bit gray MR images. This step costs $\mathcal{O}(p)$. Thus the total cost of each iteration in the FCSA is $\mathcal{O}(p\log(p))$.

With these two key features, the FCSA efficiently solves the MR image reconstruction problem (7.1) and obtains better reconstruction results in terms of both the reconstruction accuracy and computation complexity. The experimental results in the next section demonstrate its superior performance compared with all previous methods for compressed MR image reconstruction.

ALGORITHM 7.4 CSA

series Input: $\rho = 1/L, \alpha, \beta, t^1 = 1\ x^0 = r^1$
for $k = 1$ **series to** K **do**
$x_g = r^k - \rho\nabla f(r^k)$
$x_1 = prox_\rho(2\alpha\|x\|_{TV})(x_g)$
$x_2 = prox_\rho(2\beta\|\Phi x\|_1)(x_g)$
$x^k = (x_1 + x_2)/2$
x^k=project(x^k, [l, u])
$r^{k+1} = x^k$
end for

ALGORITHM 7.5 FCSA

series Input: $\rho = 1/L, \alpha, \beta, t^1 = 1\ x^0 = r^1$
for $k = 1$ **series to** K **do**
$x_g = r^k - \rho \nabla f(r^k)$
$x_1 = prox_\rho(2\alpha\|x\|_{TV})(x_g)$
$x_2 = prox_\rho(2\beta\|\Phi x\|_1)(x_g)$
$x^k = (x_1 + x_2)/2$; x^k=project(x^k, [l, u])
$t^{k+1} = (1 + \sqrt{1 + 4(t^k)^2})/2$
$r^{k+1} = x^k + ((t^k - 1)/t^{k+1})(x^k - x^{k-1})$
end for

7.2.2 EVALUATION

7.2.2.1 Experimental setup

Suppose an MR image x has n pixels, the partial Fourier transform R in problem (7.1) consists of m rows of an $n \times n$ matrix corresponding to the full 2D discrete Fourier transform. The m selected rows correspond to the acquired b. The sampling ratio is defined as m/n. The scanning duration is shorter if the sampling ratio is smaller. In MR imaging, we have certain freedom to select rows, which correspond to certain frequencies. In the following experiments, we select the corresponding frequencies in the following manner. In the K-space, we randomly obtain more samples at low frequencies and fewer samples at higher frequencies. This sampling scheme has been widely used for compressed MR image reconstruction (Lustig et al., 2007; Ma et al., 2008; Yang et al., 2010). Practically, the sampling scheme and speed in MR imaging also depend on the physical and physiological limitations (Lustig et al., 2007).

We implement our CSA and FCSA for problem (7.1) and apply them on 2D real MR images. The code that was used for the experiment is available for download.[4] All experiments were conducted on a 2.4-GHz PC in a Matlab environment. We compare the CSA and FCSA with the classic MR image reconstruction method based on the CG (Lustig et al., 2007). We also compare them with two of the fastest MR image reconstruction methods, total variation-based compressed MRI (TVCMRI)[5] (Ma et al., 2008) and reconstruction from partial Fourier data (RecPF)[6] (Yang et al., 2010). For fair comparisons, we download the codes from their websites and carefully follow their experiment setup. For example, the observation measurement b is synthesized as $b = Rx + \mathbf{n}$, where $\mathbf{n}$ is the Gaussian white noise with standard deviation $\sigma = 0.01$. The regularization parameters α and β are set as 0.001 and 0.035. R and b are given as inputs, and x is the unknown target. For quantitative evaluation, the signal-to-noise ratio (SNR) is computed for each reconstruction result. Let x_0 be the original image and x a reconstructed image, the SNR is computed as:

[4]http://ranger.uta.edu/~huang/R_FCSAMRI.htm.
[5]http://www1.se.cuhk.edu.hk/~sqma/TVCMRI.html.
[6]http://www.caam.rice.edu/~optimization/L1/RecPF/.

SNR $= 10\log_{10}(V_s/V_n)$, where V_n is the mean square error between the original image x_0 and the reconstructed image x; $V_s = var(x_0)$ denotes the power level of the original image where $var(x_0)$ denotes the variance of the values in x_0.

7.2.2.2 Visual comparisons

We apply all methods on four 2D MR images: cardiac, brain, chest, and artery. Fig. 7.1 shows these images. For convenience, they have the same size of 256 × 256. The sample ratio is set to be approximately 20%. To perform fair comparisons, all methods run 50 iterations except that the CG runs only 8 iterations due to its higher computational complexity.

Figs. 7.2–7.5 show the visual comparisons of the reconstructed results by different methods. The FCSA always obtains the best visual effects on all MR images in less CPU time. The CSA is always inferior to the FCSA, which shows the effectiveness of acceleration steps in the FCSA for MR image reconstruction. The classical CG (Lustig et al., 2007) is far worse than the others because of its higher cost in each iteration, and the RecPF is slightly better than the TVCMRI, which is consistent with the observations in Ma et al. (2008) and Yang et al. (2010).

In our experiments, these methods have also been applied to the test images with the sample ratio set to 100%. We observe that all methods obtain almost the same reconstruction results, with SNR 64.8, after sufficient iterations. This was to be expected, since all methods are essentially solving the same formulation "Model (7.1)."

7.2.2.3 CPU time and SNRs

Fig. 7.6 gives the performance comparisons between different methods in terms of the CPU time over the SNR. Tables 7.1 and 7.2 tabulate the SNR and CPU time respectively by different methods, averaged over 100 runs for each experiment. The FCSA always obtains the best reconstruction results on all MR images by achieving the highest SNR in less CPU time. The CSA is always inferior to the FCSA, which shows the effectiveness of acceleration steps in the FCSA for MR image reconstruction. While the classical CG (Lustig et al., 2007) is far worse than

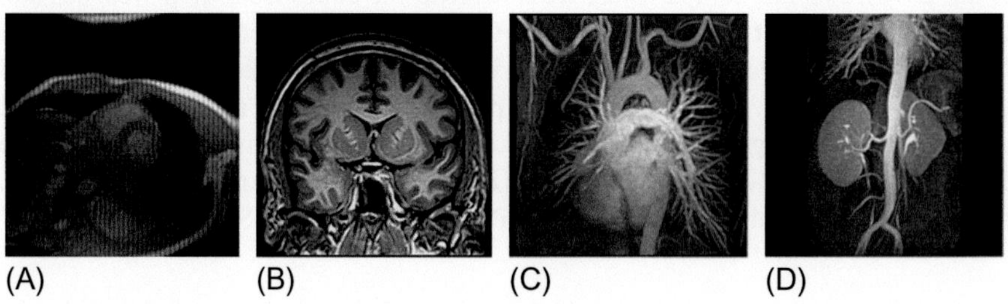

FIG. 7.1

MR images: (A) cardiac; (B) brain; (C) chest; (D) artery.

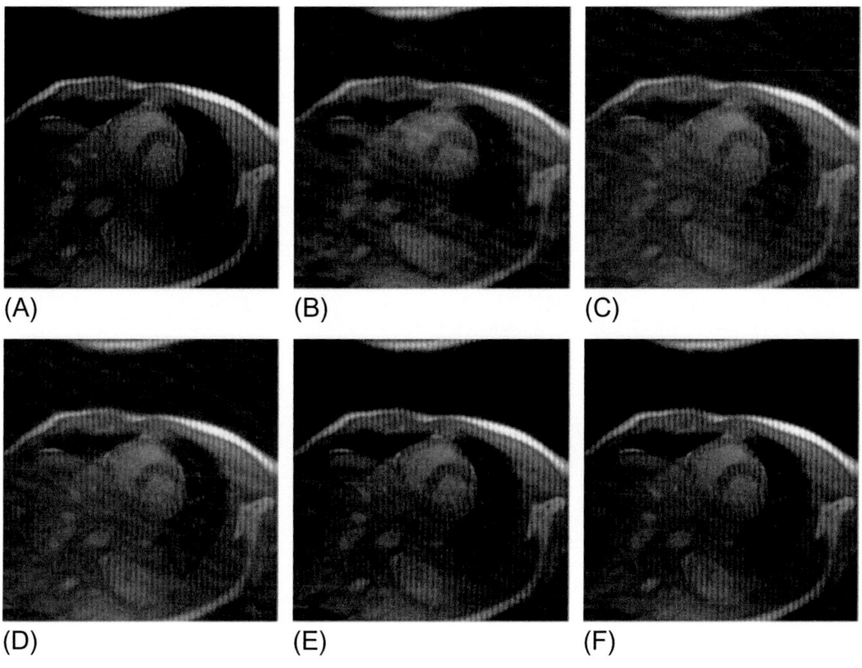

FIG. 7.2

Cardiac MR image reconstruction from 20% sampling: (A) Original image; (B)–(F) are the reconstructed images by the CG (Lustig et al., 2007), TVCMRI (Ma et al., 2008), RecPF (Yang et al., 2010), CSA, and FCSA. Their SNRs are 9.86, 14.43, 15.20, 16.46, and 17.57 (dB). Their CPU times are 2.87, 3.14, 3.07, 2.22, and 2.29 (s).

the others because of its higher cost in each iteration, the RecPF is slightly better than the TVCMRI, which is consistent with the observations in Ma et al. (2008) and Yang et al. (2010).

7.2.2.4 Sample ratios

To test the efficiency of the discussed method, we further perform experiments on a full-body MR image with size of 924×208. Each algorithm runs 50 iterations. Since we have shown that the CG method is far less efficient than the other methods, we will not include it in this experiment. The sample ratio is set to be approximately 25%. To reduce the randomness, we run each experiment 100 times for each parameter setting of each method. Examples of the original and recovered images by different algorithms are shown in Fig. 7.7. From there, we can observe that the results obtained by the FCSA are not only visibly better, but also superior in terms of both the SNR and CPU time.

To evaluate the reconstruction performance with different sampling ratios, we use sampling ratios of 36%, 25%, and 20% to obtain the measurement b. Different

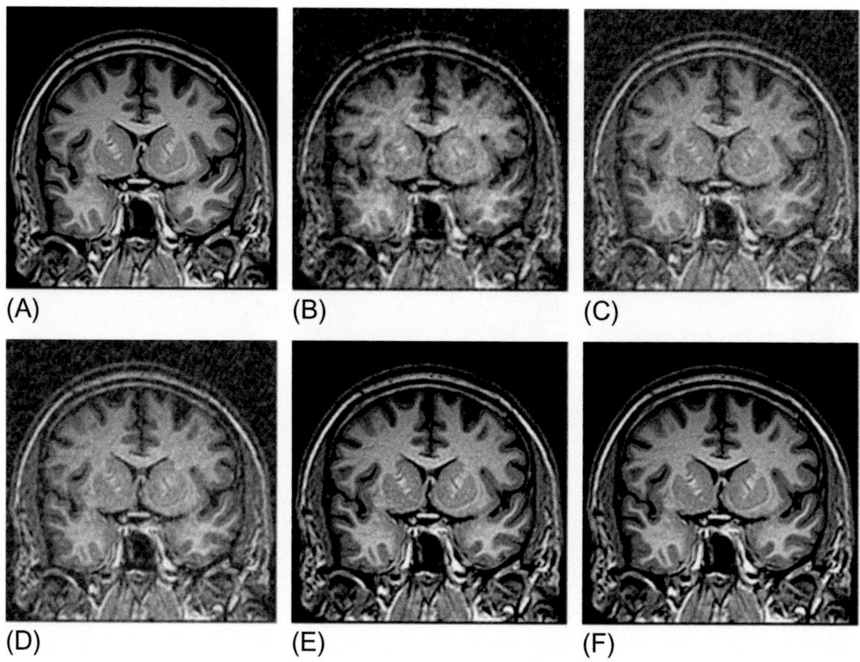

FIG. 7.3

Brain MR image reconstruction from 20% sampling: (A) Original image; (B)–(F) are the reconstructed images by the CG (Lustig et al., 2007), TVCMRI (Ma et al., 2008), RecPF (Yang et al., 2010), CSA, and FCSA. Their SNRs are 8.71, 12.12, 12.40, 18.68, and 20.35 (dB). Their CPU times are 2.75, 3.03, 3.00, 2.22, and 2.20 (s).

methods are then used to perform reconstruction. To reduce the randomness, we run each experiment 100 times for each parameter setting of each method. The SNR and CPU time are traced in each iteration for each method.

Fig. 7.8 gives the performance comparisons between different methods in terms of the CPU time and SNR for sampling ratios of 36%, 25%, and 20%. The reconstruction results produced by the FCSA are far better than those produced by the CG, TVCMRI, and RecPF. The reconstruction performance of the FCSA is always the best in terms of both the reconstruction accuracy and the computational complexity, which further demonstrates the effectiveness and efficiency of the FCSA for compressed MR image construction.

7.2.3 SUMMARY

We have discussed an efficient algorithm for compressed MR image reconstruction. This work has the following benefits. First, the FCSA can efficiently solve a composite regularization problem including both TV term and l_1-norm term,

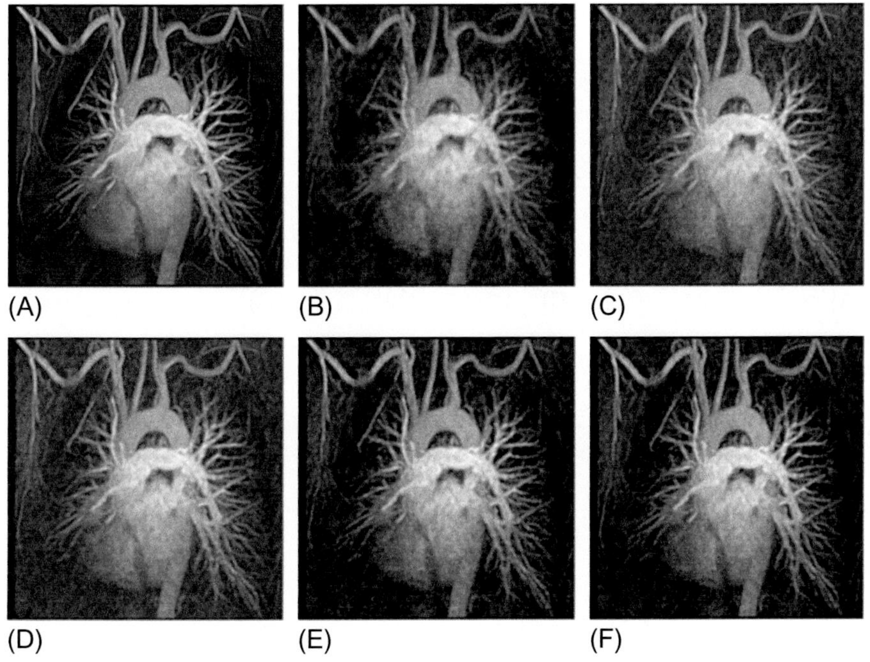

FIG. 7.4

Chest MR image reconstruction from 20% sampling: (A) Original image; (B)–(F) are the reconstructed images by the CG (Lustig et al., 2007), TVCMRI (Ma et al., 2008), RecPF (Yang et al., 2010), CSA, and FCSA. Their SNRs are 11.80, 15.06, 15.37, 16.53, and 16.07 (dB). Their CPU times are 2.95, 3.03, 3.00, 2.29, and 2.234 (s).

which can be easily extended to other medical image applications. Second, the computational complexity of the FCSA is only $\mathcal{O}(p\log(p))$ in each iteration where p is the pixel number of the reconstructed image. It also has strong convergence properties. Experimental results further validate these benefits.

7.3 GROUP SPARSITY IN MULTICONTRAST MRI

This section investigates the benefits of the group sparsity in multicontrast MRI reconstruction. Multicontrast MRI is a useful technique to aid clinical diagnosis since it can achieve superior performance for clinical diagnosis over individual T1, T2, or proton-density weighted images. The superiority of multicontrast MRI lies in the fact that different contrasts emphasize different kinds of materials, which gives richer information for diagnosis. For group sparsity models, this section discusses

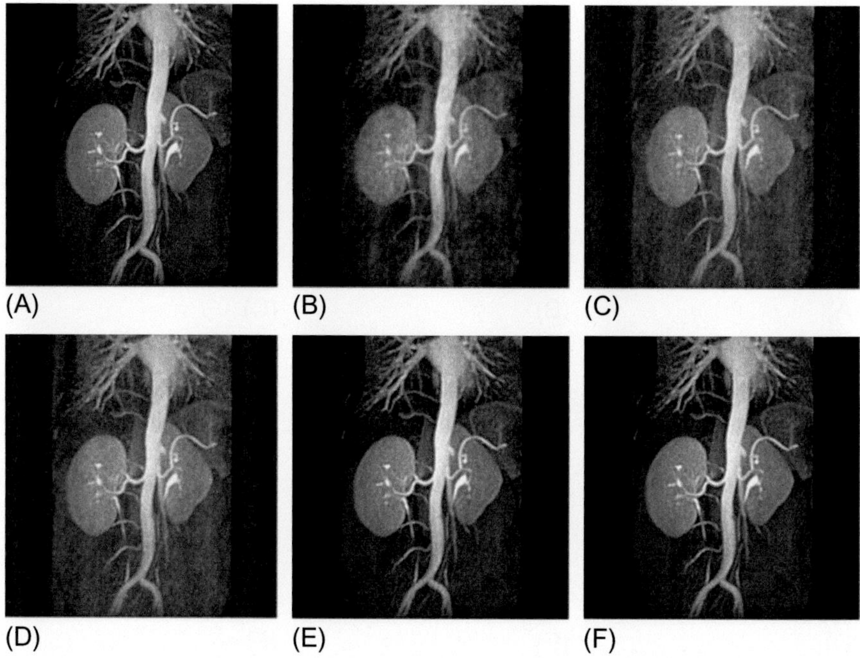

FIG. 7.5

Artery MR image reconstruction from 20% sampling: (A) Original image; (B)–(F) are the reconstructed images by the CG (Lustig et al., 2007), TVCMRI (Ma et al., 2008), RecPF (Yang et al., 2010), CSA, and FCSA. Their SNRs are 11.73, 15.49, 16.05, 22.27, and 23.70 (dB). Their CPU times are 2.78, 3.06, 3.20, 2.22, and 2.20 (s).

an efficient algorithm to jointly reconstruct multiple T1/T2-weighted images of the same anatomical cross-section from partially sampled K-space data.

7.3.1 MODEL AND ALGORITHM

7.3.1.1 Proposed fast multicontrast reconstruction

In the multicontrast imaging setting, the different MR images denote MRI scans with different imaging weights. We make two observations about them: (1) the relative magnitudes of the gradients of images should be similar for the same spatial positions across multiple contrasts; (2) the wavelet coefficients of all MR images from the same spatial positions should have similar sparse modes. Intuitively, better performance can be achieved by fully exploiting group sparsity in both the wavelet and gradient domains. Motivated by these considerations, the joint reconstruction problem can be formulated as follows:

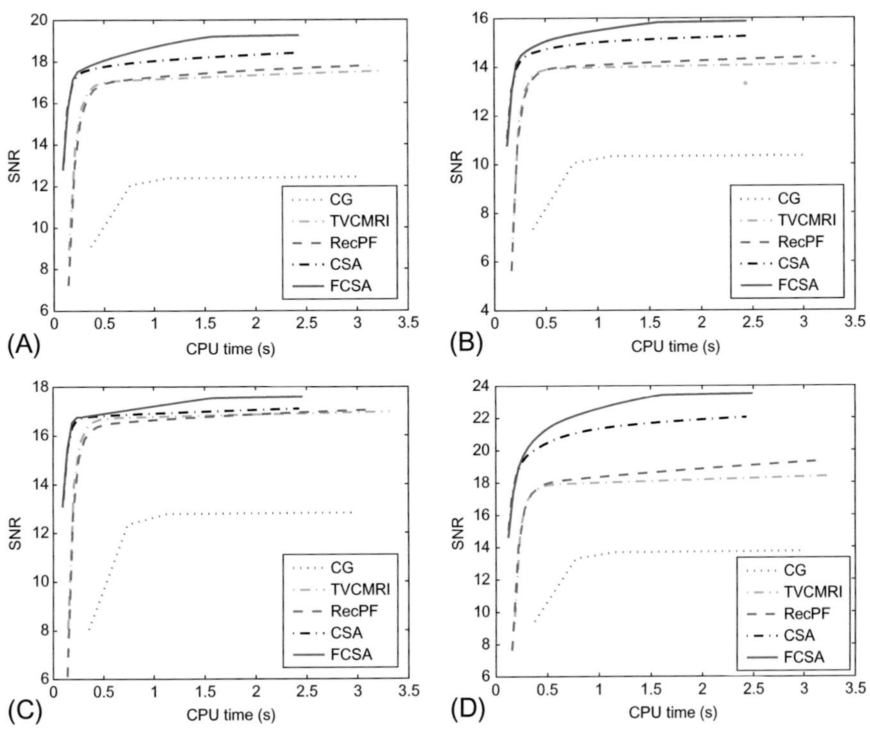

FIG. 7.6

Performance comparisons (CPU time vs. SNR) on different MR images: (A) cardiac image; (B) brain image; (C) chest image; (D) artery image.

Table 7.1 Comparisons of the SNR (dB) Over 100 Runs

	CG	TVCMRI	RecPF	CSA	FCSA
Cardiac	12.43±1.53	17.54±0.94	17.79±2.33	18.41±0.73	19.26±0.78
Brain	10.33±1.63	14.11±0.34	14.39±2.17	15.25±0.23	15.86±0.22
Chest	12.83±2.05	16.97±0.32	17.03±2.36	17.10±0.31	17.58±0.32
Artery	13.74±2.28	18.39±0.47	19.30±2.55	22.03±0.18	23.50±0.20

Table 7.2 Comparisons of the CPU Time (s) Over 100 Runs

	CG	TVCMRI	RecPF	CSA	FCSA
Cardiac	2.82±0.16	3.16±0.10	2.97±0.12	2.27±0.08	2.30±0.08
Brain	2.81±0.15	3.12±0.15	2.95±0.10	2.27±0.12	2.31±0.13
Chest	2.79±0.16	3.00±0.11	2.89±0.07	2.21±0.06	2.26±0.07
Artery	2.81±0.17	3.04±0.13	2.94±0.09	2.22±0.07	2.27±0.13

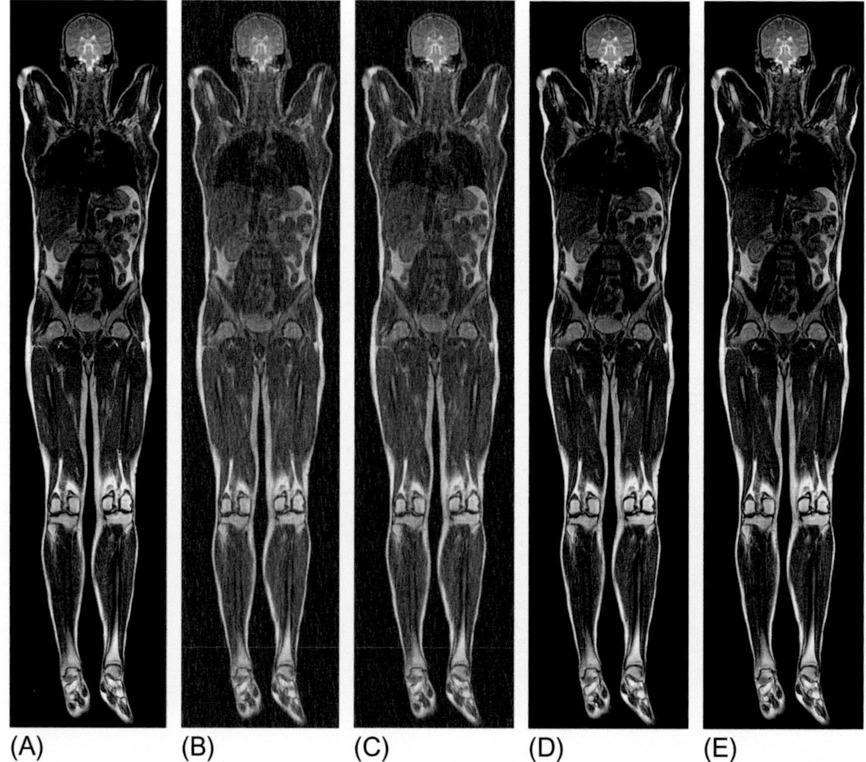

FIG. 7.7

Full-Body MR image reconstruction from 25% sampling: (A) Original image; (B)–(E) are the reconstructed images by the TVCMRI (Ma et al., 2008), RecPF (Yang et al., 2010), CSA, and FCSA. Their SNRs are 12.56, 13.06, 18.21, and 19.45 (dB). Their CPU times are 12.57, 11.14, 10.20, and 10.64 (s).

$$\hat{X} = \arg\min_{X} \left\{ F(x) = \frac{1}{2}\sum_{s=1}^{T} \|F_s X(:,s) - y_s\|^2 + \alpha\|X\|_{JTV} + \beta\|\Phi X\|_{2,1} \right\}, \tag{7.8}$$

where $X = [x_1, x_2, x_T]$ is the set of all multicontrast images, and F_s and y_s are the sampling matrix and K-space measurements for the sth image respectively. The $\ell_{2,1}$-norm is defined as $||X||_{2,1} = \sum_{i=1}^{N}\left(\sqrt{\sum_{s=1}^{T}(\Phi X_{is})^2}\right)$, which is the summation of the ℓ_2-norm for each row. The JTV of X is defined as

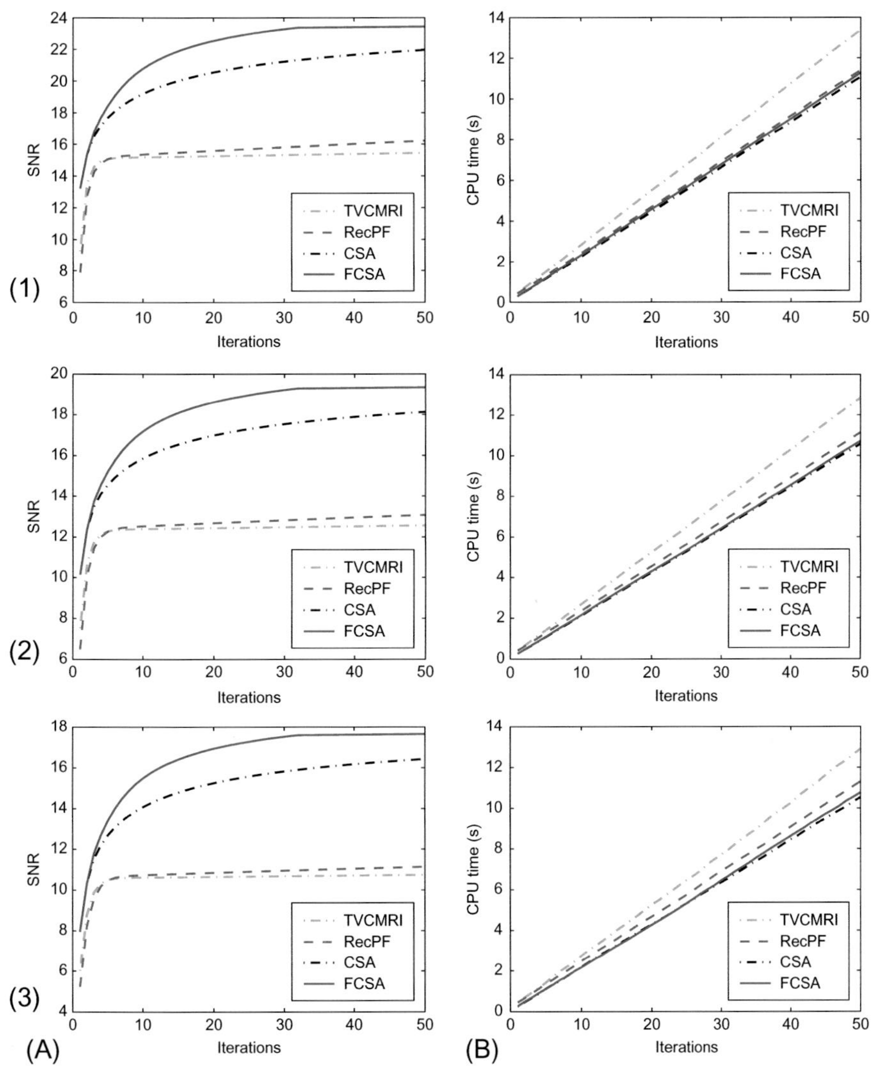

FIG. 7.8

Performance comparisons on the full-body MR image with different sampling ratios. The sample ratios are: (1) 36%; (2) 25%, and (3) 20%. (A) Iterations vs. SNR (dB). (B) Iterations vs. CPU time (s).

$$\|X\|_{JTV} = \sum_{i=1}^{N} \sqrt{\sum_{s=1}^{T}((\nabla_1 X_{is})^2 + (\nabla_2 X_{is})^2)}.$$

The JTV is also known as color total variation for color image denoising (Blomgren and Chan, 1998; Saito et al., 2011). To solve problem (7.8), we follow the FCSA (Huang et al., 2011b) scheme to decompose it into two subproblems. Let $f(X) = \frac{1}{2}\sum_{s=1}^{T} ||F_s X_s - y_s||_2^2$, $g_1(X) = \alpha||X||_{JTV}$, and $g_2(X) = \beta\|\Phi X\|_{2,1}$. Algorithm 7.6 outlines the whole algorithm for this problem. Here, $\nabla f(X)$ denotes the gradient of $f(X)$ and L_f denotes its Lipschitz constant.

ALGORITHM 7.6 PROPOSED FCSA-MT

series Input: $\rho = \frac{1}{L_f}, \alpha, \beta, t^1 = 1\ Z = X^0$
for $k = 1$ **series to** K **do**
  $Y = Z - \rho\nabla f(Z), s = 1, \ldots, T$
  $X_1 = \arg\min_X\{\frac{1}{4\rho}\|X - Y\|^2 + \alpha\|X\|_{JTV}\}$
  $X_2 = \arg\min_X\{\frac{1}{4\rho}\|X - Y\|^2 + \beta\|\Phi X\|_{2,1}\}$
  $X^k = \frac{X_1+X_2}{2}; t^{k+1} = \frac{1+\sqrt{1+4(t^k)^2}}{2}$
  $Z = X^k + \frac{t^k-1}{t^{k+1}}[X^k - X^{k-1}]$
end for

For the subproblem of X_2:

$$X_2 = \arg\min_X \left\{\frac{1}{4\rho}\|X - Y\|^2 + \beta\|\Phi X\|_{2,1}\right\}, \tag{7.9}$$

there is a closed-form solution, by soft thresholding:

$$(X_2)_i = \Phi^T\left(\max\left(1 - \frac{2\rho\beta}{\|(\Phi X)_i\|_2}, 0\right)(\Phi X)_i\right), \tag{7.10}$$

where $(\cdot)_i$ denotes the ith row of the matrix. The efficiency of the whole algorithm is highly dependent on how quickly we can solve the X_1 subproblem in each iteration.

For the subproblem of X_1:

$$X_1 = \arg\min_X \left\{\frac{1}{4\rho}\|X - Y\|^2 + \alpha\|X\|_{JTV}\right\}. \tag{7.11}$$

There is no closed-form solution. The fast gradient projection (FGP) algorithm for TV (Beck and Teboulle, 2009a) previously proposed cannot directly solve it, due to the different formulation. Fortunately, we have developed a new method, called the fast joint-gradient projection (FJGP) algorithm, for this JTV problem. Following the FGP (Beck and Teboulle, 2009a), we consider a dual method for problem (7.11). Supposing the size of each image is m by n with $m \times n = N$, we reshape the image matrices X, Y to $m \times n \times T$ for convenience. Let P be an $(m-1) \times n \times T$ matrix, Q be an $m \times (n-1) \times T$ matrix, and that they satisfy:

$$\begin{cases} \sum_{s=1}^{T} (P_{i,j,s}^2 + Q_{i,j,s}^2) \le 1 & i = 1, 2, \ldots, m-1; j = 1, 2, \ldots, n-1, \\ |P_{i,n,s}| \le 1 & i = 1, 2, \ldots, m-1; s = 1, 2, \ldots, T, \\ |Q_{m,j,s}| \le 1 & j = 1, 2, \ldots, n-1; s = 1, 2, \ldots, T. \end{cases} \tag{7.12}$$

A linear operator is defined as $\mathcal{L}(P, Q)_{i,j,s} = P_{i,j,s} - P_{i-1,j,s} + Q_{i,j,s} - Q_{i,j-1,s}$, where $i = 1, \ldots, n_1, j = 1, \ldots, n_2$, and $s = 1, \ldots, T$. The $\mathcal{L}^T$ is defined as $\mathcal{L}^T(X) = (P, Q)$, where $P \in \mathbf{R}^{(n_1-1)\times n_2 \times T}$ and $Q \in \mathbf{R}^{n_1 \times (n_2-1)\times T}$ are defined as

$$P_{i,j,s} = x_{i,j,s} - x_{i+1,j,s}, \quad i = 1, \ldots, n_1 - 1, j = 1, \ldots, n_2, s = 1, \ldots, T,$$
$$Q_{i,j,s} = x_{i,j,s} - x_{i,j+1,s}, \quad i = 1, \ldots, n_1, j = 1, \ldots, n_2 - 1, s = 1, \ldots, T.$$

Therefore the optimal solution for problem (7.11) is $X^* = Y - 2\alpha\rho\mathcal{L}(P^*, Q^*)$, where (P^*, Q^*) is the optimal solution for

$$\min_{P,Q} \left\{ h(P, Q) = Y - 2\alpha\rho\mathcal{L}(P, Q)_F^2 \right\}, \tag{7.13}$$

where $||\cdot||_F$ denotes the Frobenius norm. Note that problem (7.13) could be accelerated by FISTA (Beck and Teboulle, 2009a). The whole algorithm for problem (7.11) is summarized in Algorithm 7.7. Our recent study shows that the JTV can also improve parallel MRI reconstruction (Chen et al., 2013).

ALGORITHM 7.7 PROPOSED FJGP FOR JOINT TOTAL VARIATION

series Input: $\rho, \alpha, Y, P^0, Q^0, U, V$
for $k = 1$ **series to** K **do**
  $t^{k+1} = \frac{1+\sqrt{1+4(t^k)^2}}{2}$
  for $s = 1$ **series to** T **do**
    $(P^k, Q^k) = \text{Proj}[(U, V) + \frac{1}{16\rho\alpha}\mathcal{L}^T[Y - 2\rho\alpha\mathcal{L}(U, V)]]$
    $(U, V) = (P^k, Q^k) + \frac{t^k - 1}{t^{k+1}}(P^k - P^{k-1}, Q^k - Q^{k-1})$
  end for
end for
$X = Y - 2\rho\alpha\mathcal{L}(P^K, Q^K)$

The projection operator $\text{Proj}(P, Q) = (U, V)$ is used to force (P, Q) to satisfy the conditions (7.12):

$$U_{i,j,s} = \begin{cases} \dfrac{P_{i,j,s}}{\max(1, \sqrt{\sum_{s=1}^{T} P_{i,j,s}^2 + Q_{i,j,s}^2})} & i = 1, 2, \ldots, m-1; j = 1, 2, \ldots, n-1, \\ \dfrac{P_{i,n,s}}{\max(1, \sqrt{\sum_{s=1}^{T} P_{i,n,s}^2})} & i = 1, 2, \ldots, m-1 \end{cases} \tag{7.14}$$

and

$$V_{i,j,s} = \begin{cases} \dfrac{Q_{i,j,s}}{\max(1,\sqrt{\sum_{s=1}^{T} P_{i,j,s}^2 + Q_{i,j,s}^2})} & i = 1,2,\ldots,m-1; j = 1,2,\ldots,n-1, \\ \dfrac{Q_{m,j,s}}{\max(1,\sqrt{\sum_{s=1}^{T} Q_{m,j,s}^2})} & j = 1,2,\ldots,n-1. \end{cases} \tag{7.15}$$

It can be observed that all operations in Algorithm 7.2 are linear. Therefore the total computational complexity is $\mathcal{O}(TN)$. It can be easily proved that Algorithm 7.2 achieves the optimal convergence rate $F(X^k)-F(X^*) \leq \mathcal{O}(1/k^2)$ (Beck and Teboulle, 2009a,b). Due to the trade-off between efficiency and effectiveness, the FJGP algorithm only runs for one iteration in our implementation. For the entire Algorithm 7.1, step 1 takes $\mathcal{O}(TN \log N)$ if the fast Fourier transform (FFT) is applied. Steps 2 and 3 for the two subproblems take $\mathcal{O}(TN)$. Therefore the computational complexity for the whole algorithm is $\mathcal{O}(TN \log N)$. In addition, it is accelerated by FISTA, which has very fast convergence speed.

7.3.2 EVALUATION

The proposed method was evaluated on three datasets: the SRI24 brain atlas (Rohlfing et al., 2009), the complex-valued Shepp-Logan phantom data, and in vivo brain data. All reconstruction methods were implemented in Matlab (MathWorks, Natick, MA) on a desktop computer with an Intel i7-3770 central processing unit and 12-GB random-access memory. We compared our algorithm with the conventional CS-MRI methods SparseMRI (Lustig et al., 2007), TVCMRI (Ma et al., 2008), RecPF (Yang et al., 2010), FCSA (Huang et al., 2011b), and multicontrast MRI methods Bayesian CS (BCS) (Bilgic et al., 2011) and GroupSparseMRI (Majumdar and Ward, 2011) (GSMRI for short). For fair comparison, all codes were downloaded from the authors' websites and we carefully followed their experimental setup. All the convex relaxation methods were run for 100 iterations and the parameters were set the same for them. Due to the slow convergence speed of BCS (eg, 26.4 h for the SRI24 brain atlas (Rohlfing et al., 2009)), we only ran it for 6000 iterations as we are interested in fast reconstruction. The BCS method has been accelerated in a recent work (Cauley et al., 2013), but it still requires 9 min to reconstruct the images. The accelerated version is much less accurate than the original version and is therefore not compared. We added Gaussian white noise with 0.01 standard deviation for the simulated data. The images with Nyquist rate sampling were used as reference images. SNR and relative error (RE) were used for result evaluation. $SNR = 10 \log 10(V_s/V_n)$, where V_n is the mean square error between the reference image x_0 and the reconstructed image x; $V_s = var(x_0)$ denotes the power level of the original image, where $var(x_0)$ denotes the variance of the values in x_0. $RE = 100\% \times ||x - x_0||_2/||x_0||_2$.

7.3.2.1 SRI24 multichannel brain atlas data

This experiment was conducted on an MR image extracted from the SRI24 atlas (Rohlfing et al., 2009). Structural scans of the atlas features were obtained on a 3.0T GE scanner with an eight-channel head coil with three different contrast settings:

i. For T1-weighted structural images: 3D axial IR-prep SPoiled Gradient Recalled (SPGR), TR = 6.5 ms, TE = 1.54 ms, number of slices = 124, slice thickness = 1.25 mm.
ii. For proton density-weighted (early-echo) and T2-weighted (late-echo) images: 2D axial dual-echo fast spin echo (FSE), TR = 10,000 ms, TE = 14/98 ms, number of slices = 62, slice thickness = 2.5 mm.

The field-of-view covers a region of 240 mm ×240 mm with resolution of 256 × 256 pixels. The sampling mask is Gaussian random with variable density and reduction factor $R = 4$. More samples were acquired at low frequencies and fewer samples were acquired at higher frequencies (Huang et al., 2011b; Lustig et al., 2007). Reconstructions were performed using conventional CS-MRI methods (Huang et al., 2011b; Lustig et al., 2007; Ma et al., 2008; Yang et al., 2010), multicontrast MRI methods (Bilgic et al., 2011; Majumdar and Ward, 2011), and the proposed method. All the convex relaxation methods ran for 100 iterations and BCS ran for 6000 iterations, due to its higher computational complexity.

7.3.2.2 Complex-valued Shepp-Logan phantoms data

To validate the proposed method on complex-valued data, we conducted experiments on two complex-valued numerical phantoms (Bilgic et al., 2011). Each of these has a resolution of 128 × 128 pixels. The real parts and imaginary parts tend to be similar but not exactly the same, which was used to validate the stability of the proposed method. The phantom images are piecewise smooth, where the TV or JTV would significantly increase the reconstruction accuracy. The sampling mask is single-slice (Bilgic et al., 2011; Lustig et al., 2007) with variable density and reduction factor $R = 4$. Reconstructions were performed with BCS (Bilgic et al., 2011), GSMRI (Majumdar and Ward, 2011), and the proposed method.

7.3.2.3 Complex-valued turbo spin echo slices with early and late TEs data

In vivo images were used to further validate the performance of the proposed method. These images were obtained with two different TE settings, using a TSE sequence (1 mm ×1 mm in-plane spatial resolution with 3-mm-thick contiguous slices, TR = 6000 ms, TE1 = 27 ms, TE2= 94 ms) (Bilgic et al., 2011). This dataset contains 38 slices with 256 × 256 resolution. They are much less compressible than the previous two datasets. The sampling mask consisted of multiple radial lines (Block et al., 2007; Chang et al., 2006; Yang et al., 2010; Ye et al., 2007) with reduction factor $R = 4$. Reconstructions were performed using BCS (Bilgic et al., 2011), GSMRI (Majumdar and Ward, 2011), and the proposed method.

7.3.2.4 The benefit of group sparsity on both wavelet and gradient domains

Fig. 7.9 demonstrates the benefit of the proposed method when utilizing JTV and group wavelet sparsity on the SRI24 dataset. Comparing the proposed FJGP and conventional FGP (Beck and Teboulle, 2009a), the JTV denoising is much more powerful on multicontrast MR data. With more iterations, our algorithm achieves higher accuracy and less computational time. The results produced by the combination of group wavelet sparsity and JTV are always better than those produced by the group wavelet sparsity or JTV only. This confirms the benefit of the combination of group wavelet sparsity and JTV for multicontrast CS-MRI.

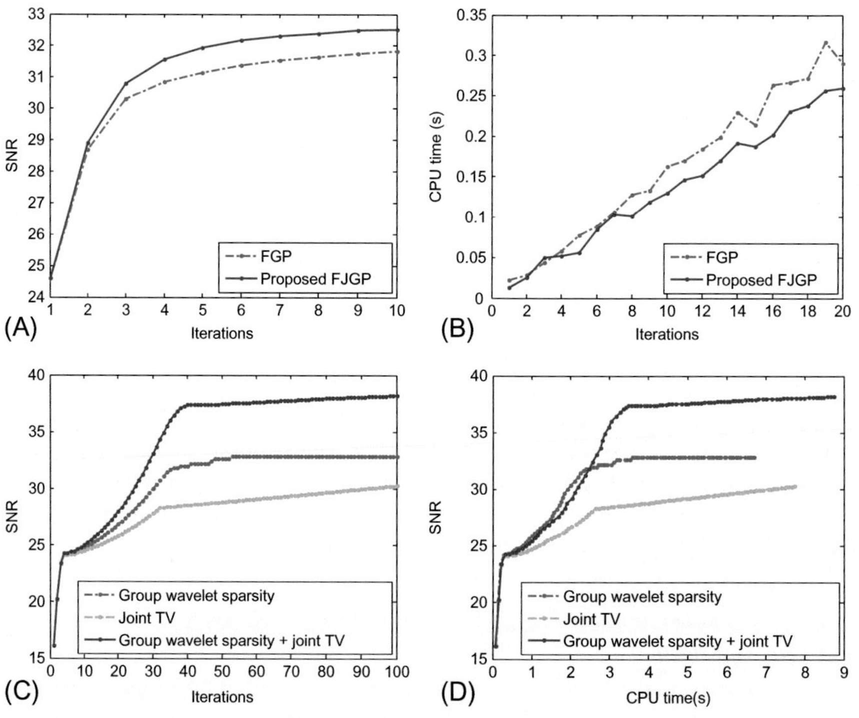

FIG. 7.9

Demonstration of the benefit of the proposed method for SRI24 dataset. (A) The SNRs of the denoising result with FGP (18) and the proposed FJGP. (B) Comparison of the computational time between FPG (18) and the proposed FJGP. (C) Comparison of the SNR in terms of iterations among only group wavelet sparsity, only joint TV, and their combination. (D) Comparison of the SNR in terms of CPU time among only group wavelet sparsity, only joint TV, and their combination.

7.3.2.5 Results on SRI24 multichannel brain atlas data

Fig. 7.10 depicts the SRI24 reconstruction results obtained by the conventional CS-MRI methods (Huang et al., 2011b; Lustig et al., 2007; Ma et al., 2008; Yang et al., 2010), multicontrast MRI methods (Bilgic et al., 2011; Majumdar and Ward, 2011), and the proposed method. These images are very compressible. It is difficult to observe visible artifacts for all reconstruction results. For better visualization, the reconstruction errors are also shown on the same scale. We found the reconstruction with our method had the smallest error. Table 7.3 shows all the SNRs, REs as well as the reconstruction time for each algorithm. Our algorithm had the highest accuracy with comparable computational cost to the fastest CS-MRI algorithms. Due to the inherent shortcoming of the Bayesian CS framework (Ji et al., 2009), it has a huge computational cost that may make it practically impossible to use in most applications. GSMRI applies SPGL1 (Van Den Berg and Friedlander, 2008) to solve the group wavelet sparsity problem. It had similar accuracy with BCS on this data, while the computational cost was much less.

Fig. 7.11 presents the convergence speed of each algorithm with convex relaxation in terms of iterations and CPU time. Inherent from the fast convergence rate of FISTA, FCSA, and the proposed algorithm outperformed all other algorithms on this dataset. However, FCSA cannot reconstruct the multicontrast MR images jointly, but only individually. That is why it is always inferior to the proposed method. In general, the conventional CS-MRI methods are not good as joint reconstruction methods, which has also been validated in previous works (Bilgic et al., 2011; Majumdar and Ward, 2011). In later experiments, we only compared the proposed method with multicontrast reconstruction methods (Bilgic et al., 2011; Majumdar and Ward, 2011).

7.3.2.6 Results on Complex-valued Shepp-Logan phantoms data

Absolute values of the reconstruction results after undersampling with $R = 4$ are presented in Fig. 7.12. The proposed method was better than BCS in terms of SNR and RE on this dataset, and far better than GSMRI. At first glance, it seems that BCS achieved similar reconstruction results from visual observation. This is because such images are ideal examples for BCS, where the images are extremely piecewise smooth. However, for in vivo data, as we will show later, our method will be much more stable and robust. In addition, BCS converges the fastest on these ideal images among all those used in the experiments, but it is still substantially slower than our method. Visible artifacts can be found in the results reconstructed by GSMRI. This is because these images are piecewise smooth, which is perfect for using gradient-based methods. However, it is still unknown how to combine TV or JTV in SPGL1 for GSMRI. Across different contrasts, the gradients and wavelet coefficients of these phantom images are not exactly the same, which validates the stability of the proposed method. When the sparsity patterns of different images tend to be the same, our algorithm can obtain very good results.

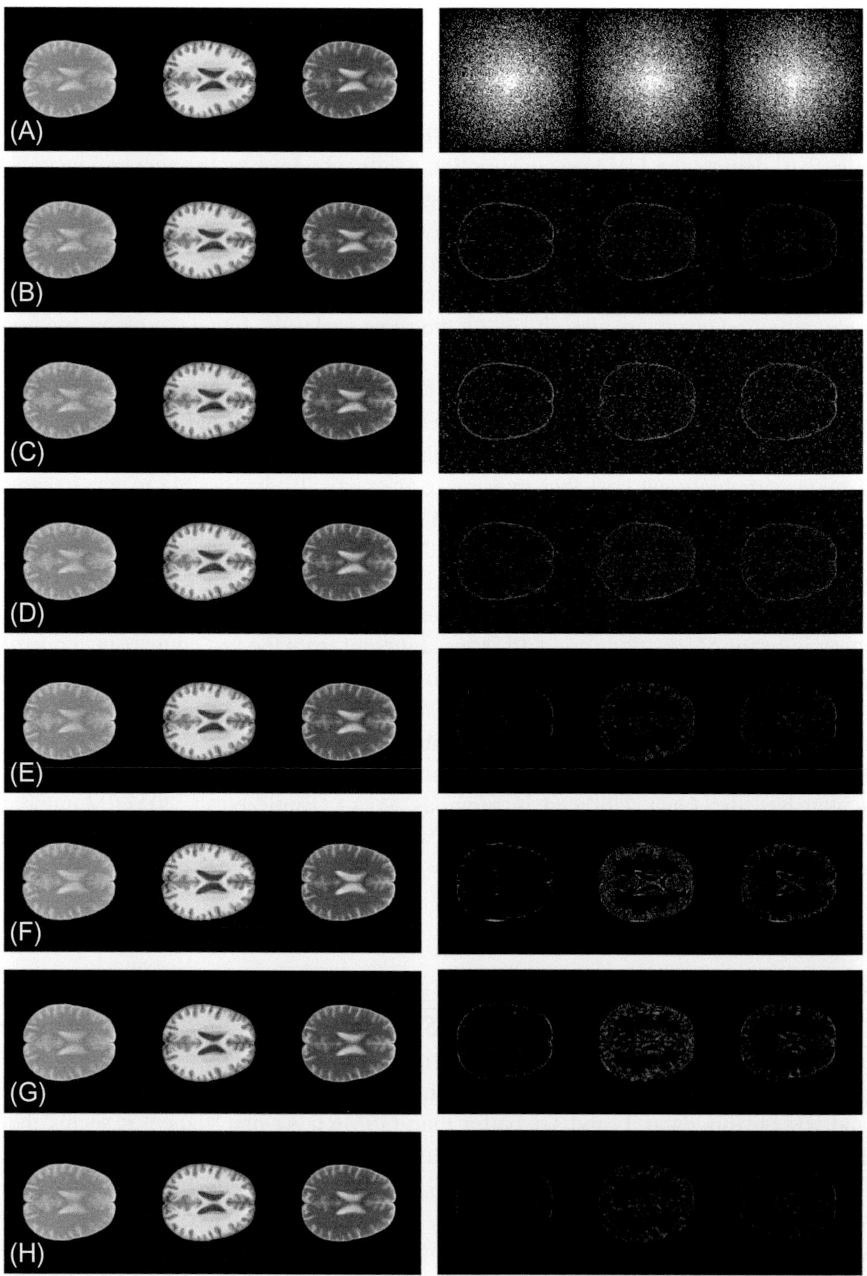

FIG. 7.10

Representative reconstruction results for SRI24 atlas images after undersampling with $R = 4$. (A) Atlas images at Nyquist rate sampling (left) and the sampling masks (right). (B)–(H) The reconstruction results (left) and the absolute errors (right) with SparseMRI (Lustig et al., 2007), TVCMRI (Ma et al., 2008), RecPF (Yang et al., 2010), FCSA (Huang et al., 2011b), BCS (Bilgic et al., 2011), GSMRI (Majumdar and Ward, 2011), and the proposed method respectively.

Table 7.3 The SNRs, Computational Times, and REs for the Reconstructions in Fig. 7.10

	SparseMRI	TVCMRI	RecPF	FCSA	BCS	GSMRI	Proposed
SNR (dB)	25.13	24.03	25.80	35.64	31.49	30.40	**37.97**
RE (%)	4.3	4.9	4.0	1.3	2.3	2.1	**0.9**
Time (s)	44.96	7.60	8.53	8.36	7134.1	11.81	8.78

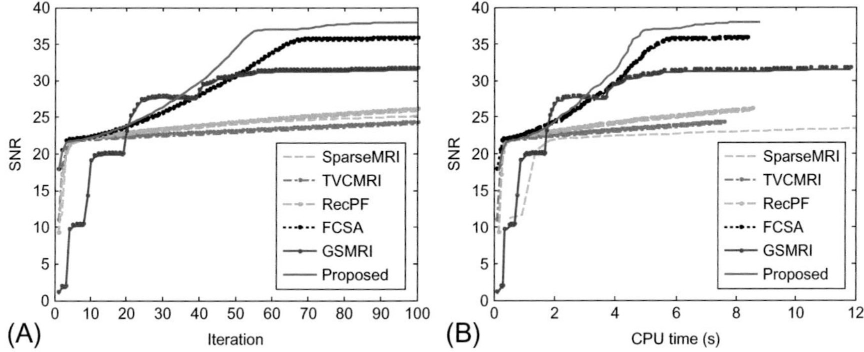

FIG. 7.11

Performance comparisons of reconstruction results for SRI24 atlas among algorithms with convex relaxation. (A) SNR vs. iteration. (B) SNR vs. CPU time (s). Due to the high computational cost of SparseMRI, we only show its SNRs for the first 12 s.

7.3.2.7 Results on Complex-valued turbo spin echo slices with early and late TEs

The reconstruction results on the in vivo human brain MRI data are shown in Fig. 7.13. These images are less compressible than the previous two datasets. With the same reduction factor, all algorithms performed worse on this dataset than they performed on the previous datasets. The result of BCS is significantly blurred at the edges, while the result of GSMRI contains obvious artifacts. By contrast, the result of our method is closest to the original one. It takes more than 10,000 s for BCS for this reconstruction (it was still not converged), while our algorithm obtained an acceptable result within only around 12 s.

With the same datasets, additional reconstructions were performed to quantify the performance of the three multicontrast MRI methods. The efficiency and effectiveness of the proposed method were validated under different sampling schemes with various reduction factors. The comparisons with GSMRI (Majumdar and Ward, 2011) are tabulated in Table 7.4, and the comparisons with BCS (Bilgic et al., 2011) are shown in Table 7.5. For speed, we used 128 × 128 low-resolution images when comparing the results with BCS. For the same data, the results on the non-Cartesian

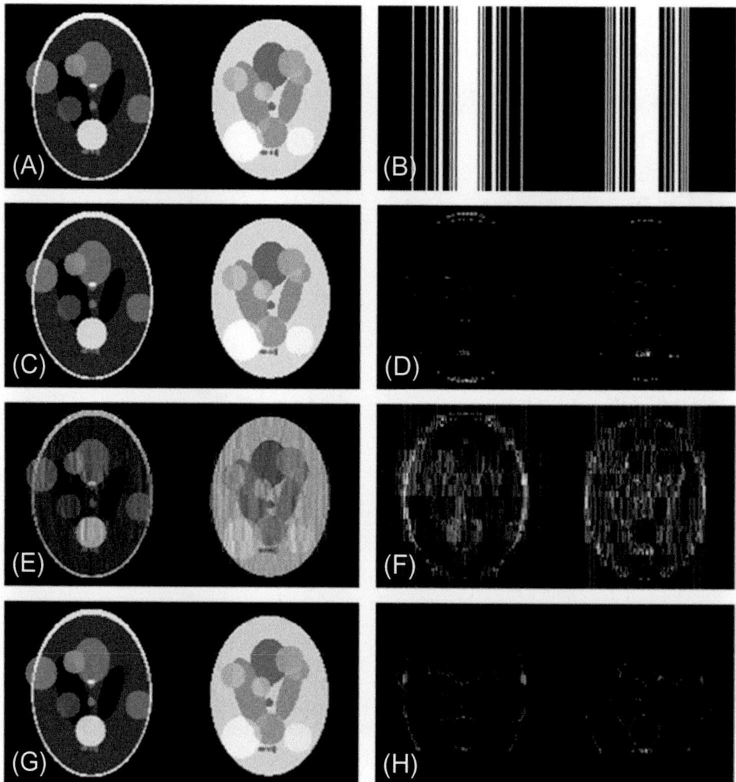

FIG. 7.12

Reconstruction results for the complex-valued Shepp-Logan phantoms after undersampling with reduction factor $R = 4$. (A) Magnitudes of phantoms at Nyquist rate sampling. (B) The sampling masks. (C) The reconstruction result with Bayesian CS. Its SNR is 26.43. (D) Absolute error of Bayesian CS reconstruction. $RE = 4.3\%$. (E) The reconstruction result with GSMRI. Its SNR is 14.32. (F) Absolute error of GSMRI reconstruction. $RE = 17.7\%$. (G) The reconstruction result with the proposed method. Its SNR is 27.15. (H) Absolute error of the proposed reconstruction. $RE = 4.2\%$. The reconstruction times were 271.2, 10.37, and 8.64 s, respectively.

masks were much better than those on the Cartesian mask. This is because the Cartesian sampling matrix is less incoherent. Also, with the same sampling scheme, the accuracy of each algorithm is higher with more measurements. When compared with different algorithms in the same condition (same sampling scheme and same reduction factor), our algorithm always outperformed the other two in terms of both reconstruction accuracy and computational complexity.

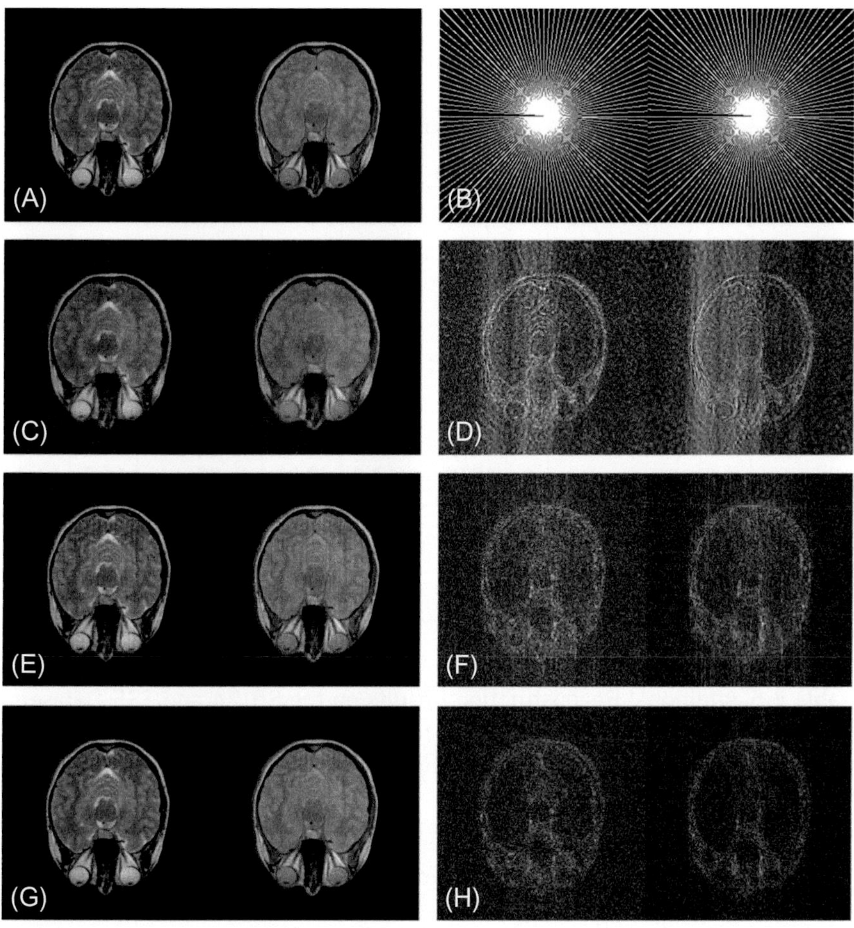

FIG. 7.13

Reconstruction results for the complex-valued in vivo TSE images after undersampling with reduction factor $R = 4$. (A) Magnitude images at Nyquist rate sampling. (B) The sampling masks. (C) The reconstruction result with Bayesian CS. Its SNR is 12.34. (D) Absolute error of Bayesian CS reconstruction. $RE = 27.4\%$. (E) The reconstruction result with GSMRI. Its SNR is 15.07. (F) Absolute error of GSMRI reconstruction. $RE = 18.9\%$. (G) The reconstruction result with the proposed method. Its SNR is 17.07. (H) Absolute error of the proposed reconstruction. $RE = 14.7\%$. The reconstruction times were: 10,074.3, 18.21, and 12.04 s, respectively.

7.3.2.8 Discussion

One may note that there are two parameters that need to be tuned for most convex relaxation methods including the proposed method, while no parameters are required for BCS (Bilgic et al., 2011) and GSMRI (Majumdar and Ward, 2011). Fortunately,

Table 7.4 Additional Reconstruction Results on the SRI24 and TSE Datasets, With Comparisons Between GSMRI (Majumdar and Ward, 2011) and the Proposed Method

			GSMRI			Proposed		
Dataset	**Schemes**	***R***	**SNR (dB)**	**RE (%)**	**Time (s)**	**SNR (dB)**	**RE (%)**	**Time (s)**
SRI24	Random (Fig. 7.10)	4	30.40	2.1	11.81	37.97	0.9	8.78
	Random	5	29.19	2.7	11.40	34.33	1.5	8.31
	Slice (Fig. 7.12)	3	22.16	6.1	11.65	26.54	3.7	8.68
	Slice	4	20.08	7.8	11.54	24.77	4.5	9.39
	Radial (Fig. 7.13)	4	28.98	2.8	11.58	33.63	1.6	9.06
	Radial	5	27.09	3.4	11.20	32.33	1.9	9.25
TSE	Random	3	17.86	14.0	18.88	19.08	12.0	12.85
	Random	4	16.26	16.4	17.86	17.08	13.7	11.47
	Slice	2	16.34	16.2	20.19	17.28	14.2	12.39
	Slice	3	13.33	22.3	18.30	14.38	19.4	11.51
	Radial	3	17.18	15.1	20.19	18.88	12.3	13.17
	Radial	4	15.07	18.9	18.21	17.07	14.7	12.04

Table 7.5 Additional Reconstruction Results on the SRI24 and TSE Datasets, With Comparisons Between BCS (Bilgic et al., 2011) and the Proposed Method (for Speed, We Used 128 × 128 Low-Resolution Images for These Comparisons)

			BCS			Proposed		
Dataset	**Schemes**	***R***	**SNR (dB)**	**RE (%)**	**Time (s)**	**SNR (dB)**	**RE (%)**	**Time (s)**
SRI24	Random	3.5	32.29	2.0	2706.8	36.21	0.4	2.48
	Random	4	30.13	1.1	2129.9	33.53	0.6	2.46
	Slice	3	23.04	5.8	1919.8	24.68	4.8	2.02
	Slice	4	19.78	8.5	1838.4	20.70	7.6	2.09
	Radial	4	30.65	2.4	2384.4	32.72	1.9	2.15
	Radial	5	27.73	3.3	2335.0	29.82	2.6	2.09
TSE	Random	3	15.63	17.2	9262.7	23.87	6.7	3.25
	Random	4	12.95	23.1	5306.9	21.39	8.5	3.24
	Slice	3	15.44	16.0	177.6	15.72	15.1	2.65
	Slice	3.5	13.86	18.7	378.3	14.02	18.2	2.97
	Radial	3	15.91	16.5	3384.1	22.73	7.6	3.25
	Radial	4	13.52	21.3	4759.3	19.67	10.5	3.18

the parameters are easy to tune when the sampling scheme is fixed. The parameters determine the relative weights between the least squares fitting and the sparsity terms. The user could manually set how important the sparseness of the data is relative to the least squares fitting. However, BCS and GSMRI always seek the sparsest solution under a fixed tolerance of the least squares fitting, which may result in oversmoothing, especially for those less compressible images. The parameters to obtain the experimental results are set as $\alpha = 0.01$ or 0.1, $\beta = 0.035$. Although they are not necessarily the optimal values, they were good enough to demonstrate the superiority of the proposed method. Interested readers may refer to existing methods (Liang et al., 2009; Weller et al., 2010) for more information on how to combine parallel imaging techniques.

7.3.3 SUMMARY

We have discussed an efficient algorithm for multicontrast CS-MRI. In the discussion, we demonstrated the model that applied group sparsity in both wavelet and gradient domains and the algorithm to efficiently solve the optimization problem. The presented FCSA algorithm has fast convergence speed borrowed from FISTA, and each iteration only costs $\mathcal{O}(TN \log N)$.

7.4 TREE SPARSITY IN ACCELERATED MRI

This section investigates the benefits of tree sparsity in accelerated MRI. In contrast to conventional CS-MRI that only relies on the sparsity of MR images in the wavelet or gradient domain, the tree sparsity exploits the wavelet tree structure to improve CS-MRI, which can further improve the reconstruction quality.

7.4.1 MODEL AND ALGORITHM

To validate the benefit of tree sparsity in accelerated MRI, we demonstrate two algorithms that efficiently solve the constrained and unconstrained tree-based CSMRI problems. The tree structure in MR images is approximated as overlapping groups (Jacob et al., 2009; Rao et al., 2011). The unconstrained problem is solved in a FISTA (Beck and Teboulle, 2009a) framework and the constrained problem is solved in a NEsterov Shrinkage Thresholding Algorithm (Becker et al., 2011) framework. Both FISTA and NESTA have the optimal convergence rate for first-order methods, that is, $\mathcal{O}(1/k^2)$ in function value where k is the iteration number (Nesterov, 1983).

7.4.1.1 Unconstrained tree-based MRI

Following overlapping group sparsity algorithms (Jacob et al., 2009; Rao et al., 2011), the unconstrained MRI problem with tree sparsity can be formulated as

$$\hat{x} = \min_{x} \left\{ \frac{1}{2} \|Rx - b\|_2^2 + \beta \sum_{g \in \mathcal{G}} ||(\Phi x)_g||_2 \right\}, \tag{7.16}$$

where x is the MR image to be reconstructed, R is the partial Fourier transform, b is the measurement vector, Φ denotes the wavelet transform, and β is a positive parameter that needs to be tuned. Here, g denotes one of the groups that encourages tree sparsity (eg, one node and its parent) and $\mathcal{G}$ denotes the set of all such groups. Due to the nonsmoothness and nonseparability of the overlapping group penalty, it is not easy to solve the problem directly. Instead, we introduce a variable z to constrain the problem:

$$\hat{x} = \arg\min_{x,z} \left\{ \frac{1}{2} \|Rx - b\|_2^2 + \beta \sum_{g \in \mathcal{G}} ||z_g||_2 + \frac{\lambda}{2} ||z - G\Phi x||_2^2 \right\}, \tag{7.17}$$

where λ is another positive parameter and G is a binary matrix to duplicate the overlapped entries. z is the extended vector of wavelet coefficients x without overlapping.

All terms in our model are convex. For the z subproblem:

$$z_g = \arg\min_{z_g} \left\{ \beta ||z_g||_2 + \frac{\lambda}{2} ||z_g - (G\Phi x)_g||_2^2 \right\}, \qquad g \in \mathcal{G}. \tag{7.18}$$

It has closed-form solution by soft thresholding:

$$z_g = \max\left(||r||_2 - \frac{\beta}{\lambda}, 0 \right) \frac{r}{||r||_2}, \qquad g \in \mathcal{G}, \tag{7.19}$$

where $r = (G\Phi x)_g$. We denote this step by $z = \mathit{shrinkgroup}(G\Phi x, \frac{\beta}{\lambda})$ for convenience. For the x subproblem:

$$x = \arg\min_{x} \left\{ \frac{1}{2} \|Rx - b\|_2^2 + \frac{\lambda}{2} ||z - G\Phi x||_2^2 \right\}. \tag{7.20}$$

This is a combination of two quadratic terms and has closed-form solution: $x = (R^T R + \lambda \Phi^T G^T G \Phi)^{-1} (R^T b + \Phi^T G^T z)$. However, the inverse of $R^T R + \lambda \Phi^T G^T G \Phi$ is not easily obtained. In order to validate the benefit tree structure, we apply FISTA to solve the x subproblem, which can match the convergence rate of FCSA. Let $f(x) = \frac{1}{2}\|Rx - b\|_2^2 + \frac{\lambda}{2}||z - G\Phi x||_2^2$, which is a convex and smooth function with Lipschitz constant L_f, and $g(x) = 0$. Then our algorithm can be summarized in Algorithm 7.8, which is called FISTA_Tree. Here $\nabla f(r^k) = R^T(Rr^k - b) + \lambda \Phi^T G^T (G\Phi r^k - z)$. R^T and Φ^T denote the inverse partial Fourier transform and the inverse wavelet transform.

ALGORITHM 7.8 FISTA_TREE

series Input: $\rho = 1/L_f, r^1 = x^0, t^1 = 1, \beta, \lambda, N$
for $k = 1$ **series to** N **do**
  $z = shrinkgroup(G\Phi x^{k-1}, \beta/\lambda)$
  $x^k = r^k - \rho \nabla f(r^k)$
  $t^{k+1} = [1 + \sqrt{1 + 4(t^k)^2}]/2$
  $r^{k+1} = x^k + \frac{t^k - 1}{t^{k+1}}(x^k - x^{k-1})$
end for

Computational complexity. Note that $G \in \mathbb{R}^{N' \times N}$ is a sparse matrix with each row containing only one nonzero element 1. Therefore the multiplication by G only cost $\mathcal{O}(N') = \mathcal{O}(N)$ with our group configuration. Suppose x is an image with N pixels. The *shrinkgroup* step can be implemented in only $\mathcal{O}(N \log N)$ time and the gradient step also takes $\mathcal{O}(N \log N)$. We can find that the total time complexity in each iteration is still $\mathcal{O}(N \log N)$, the same as that of TVCMRI, RecPF, and FCSA. This good feature guarantees that the proposed algorithm is comparable with the fastest MRI algorithms in terms of execution speed.

7.4.1.2 Constrained tree-based MRI

NESTA (Becker et al., 2011) solves the constrained problem of standard sparsity:

$$\min_{\theta} ||\theta||_1, \quad s.t. \quad ||b - A\theta||_2 \leq \epsilon, \tag{7.21}$$

where θ denotes the set of wavelet coefficients with $\theta = \Phi x$, $A = R\Phi^{\mathrm{T}}$, Φ^{T} denotes the inverse wavelet transform, and ϵ is a small constant. It reaches the optimal convergence rate for first-order methods. Similar to the previous section, we extend it to solve the tree-based MRI problem:

$$\min_{\theta} ||G\theta||_{2,1}, \quad s.t. \quad ||b - A\theta||_2 \leq \epsilon, \tag{7.22}$$

where $||G\theta||_{2,1} = \sum_{g \in \mathcal{G}} ||(G\theta)_g||_2$, and $g, \mathcal{G}$ are the same as those in Algorithm 7.8. Recall that the $\ell_{2,1}$-norm also has the form:

$$||G\theta||_{2,1} = \max_{u \in \mathcal{Q}} < u, Gx >, \tag{7.23}$$

where the dual feasible set is

$$\mathcal{Q} = \{u : ||u||_{2,\infty} \leqslant 1\} = \left\{u : \max_{g \in \mathcal{G}} ||u_g||_2 \leqslant 1\right\}. \tag{7.24}$$

We relax the nonsmooth $\ell_{2,1}$-norm to a smooth function with:

$$f_\mu(\theta) = \max_{u \in \mathcal{Q}} \left(< u, G\theta > - \frac{\mu}{2} ||u||_2^2\right), \tag{7.25}$$

where μ is a small fixed number.

Note that $(G\theta)_g = G_g\theta$, where G_g is the row of G, to group g. The first-order gradient of $f_\mu(\theta)$ with Lipschitz constant L_μ is given by

$$\nabla f_\mu(\theta)_g = \begin{cases} \mu^{-1}G_g^{\mathrm{T}}G_g\theta, & ||G_g\theta||_2 < \mu \\ G_g^{\mathrm{T}}G_g\theta/||G_g\theta||_2, & \texttt{otherwise}, \end{cases} \tag{7.26}$$

NESTA assumes the rows of the sampling matrix A are orthogonal, that is, $AA^{\mathrm{T}} = I$, where I denotes the identical matrix. Fortunately, the partial Fourier transform in compressed sensing MRI satisfies this assumption: $AA^{\mathrm{T}} = R\Phi^{\mathrm{T}}\Phi R^{\mathrm{T}} = RR^{\mathrm{T}} = I$, where R^{T} denotes the inverse operator of R. The whole algorithm based on the NESTA (Becker et al., 2011) framework is given in Algorithm 7.9.

ALGORITHM 7.9 NESTA_TREE

series Input: $\theta_0, \epsilon, k = 1, L_\mu, \mu$
while not meeting the stopping criterion **do**
1. Compute $\nabla f_\mu(\theta)$
2. Compute y^k
   $q = \theta^k - L_\mu^{-1}\nabla f_\mu(\theta)$
   $\lambda_\epsilon = \max(0, \epsilon^{-1}||b - Aq||_2 - L_\mu)$
   $y^k = (I - \frac{\lambda_\epsilon}{\lambda_\epsilon + L_\mu}A^{\mathrm{T}}A)(\frac{\lambda_\epsilon}{L_\mu}A^{\mathrm{T}}b + q)$
3. Compute z^k
   $\alpha^k = 1/2(k+1)$
   $q = x_0 - L_\mu^{-1}\sum_{i\le k}\nabla\alpha_i f_\mu(\theta)$
   $\lambda_\epsilon = \max(0, \epsilon^{-1}||b - Aq||_2 - L_\mu)$
   $z^k = (I - \frac{\lambda_\epsilon}{\lambda_\epsilon + L_\mu}A^{\mathrm{T}}A)(\frac{\lambda_\epsilon}{L_\mu}A^{\mathrm{T}}b + q)$
4. Update θ^k
   $\tau^k = 2(k+3)$
   $\theta^k = \tau^k z^k + (1 - \tau^k)y^k$
5. $k = k + 1$

end while

Computational complexity. As shown in Algorithm 7.9, the complexity of the proposed algorithm is the same as the original NESTA algorithm (Becker et al., 2011). It is $6\mathcal{C} + \mathcal{O}(N)$, where $\mathcal{C}$ denotes the complexity of applying A or A^{T}. In CSMRI, $\mathcal{C} = \mathcal{O}(N\log N)$ if FFT is applied. Therefore the total computational complexity is $\mathcal{O}(N\log N)$ for each iteration, the same as that of Algorithm 7.8.

If we compare the two types of algorithms, the parameters can be manually set in the unconstrained algorithm to determine how sparse the data is. Or the weights between sparseness and the least squares fitting can be controlled. However, the constrained algorithm always seeks the sparsest solution that satisfies the constraint. In the application of MRI, we find that if good parameters can be tuned, the unconstrained algorithm (Algorithm 7.8) performs better, or vice versa. In contrast, the constrained algorithm (Algorithm 7.9) has the convenience of not requiring tuning of the parameter.

7.4.2 EVALUATION

7.4.2.1 Experimental setup

We compare the unconstrained algorithm FISTA_Tree with CG (Lustig et al., 2007), TVCMRI (Ma et al., 2008), RecPF (Yang et al., 2010), and FCSA (Huang et al., 2011b), and compare the constrained algorithm NESTA_Tree with the original NESTA (Becker et al., 2011) algorithm for CSMRI. For fair comparisons, all code were downloaded from the authors' websites and we carefully followed their experimental setup. We applied all these methods on four real-valued MR images: cardiac, brain, chest, and shoulder (shown in Fig. 7.14). In addition, a complex-valued MR brain image[7] is added to validate the benefit of tree sparsity on complex-valued data. Suppose R is a partial Fourier transform with M rows and N columns. The sampling ratio is defined as M/N. For simulations with real-valued images, we follow the sampling strategy of previous works (Ma et al., 2008; Huang

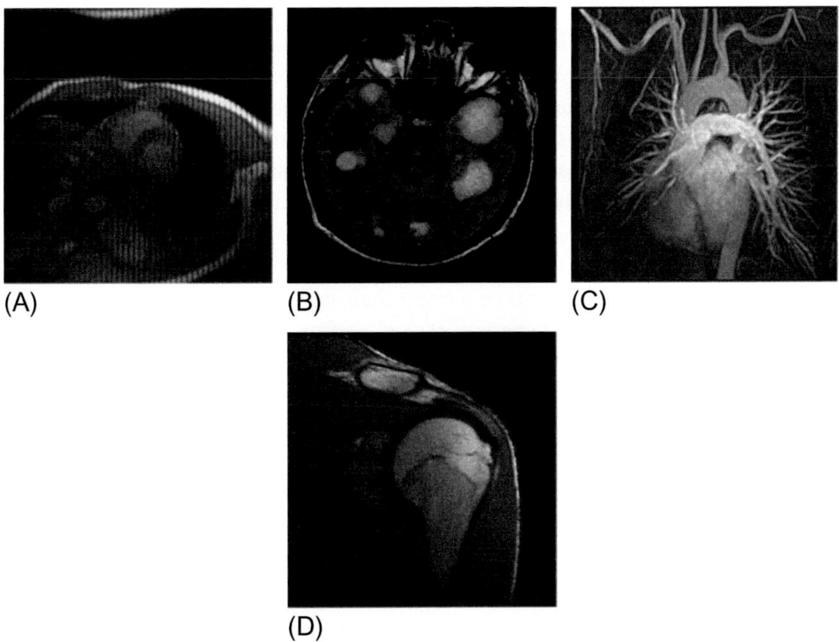

FIG. 7.14

MR images: (A) cardiac; (B) brain; (C) chest; (D) shoulder.

[7] http://www.eecs.berkeley.edu/~mlustig/CS.html.

et al., 2011b), which randomly choose more Fourier coefficients at low frequency and less at high frequency. For complex-valued data, the radial sampling mask is used (Yang et al., 2010), which is more feasible in practice.

In order to study the benefit of the tree structure in CSMRI, we remove the TV term in all algorithms. The parameters for real-valued images and complex-valued images are tuned separately. There is no continuation step (Becker et al., 2011) in the NESTA_Tree algorithm. All experiments are on a desktop with 3.4-GHz Intel core i7 3770 CPU. Matlab version is 7.8 (2009a). Measurements are added by Gaussian white noise with 0.01 standard deviation. The SNR is used for result evaluation:

$$SNR = 10\log_{10}(V_s/V_n), \tag{7.27}$$

where V_n is the mean square error between the original image x_0 and the solution x; V_s denotes the variance of the values in x_0.

7.4.2.2 Group configuration for tree sparsity

In all previous works, the tree structures are approximated as overlapping groups (Chen and Huang, 2012a; Rao et al., 2011). In addition, for each of them the wavelet coefficient and its parent are assigned into one group. However, the relationship between a coefficient and its grandparent is not exploited. We first conduct an experiment to validate the influence of the group size on the reconstruction result. Four group sizes are compared: (a) each group contains one coefficient, which is the same case as standard sparsity; (b) each group contains a coefficient and its parent, which is the same as previous works (Chen and Huang, 2012a; Rao et al., 2011); (c) each group contains a coefficient, its parent, and its grandparent; (d) each group contains four coefficients, where the grandparent's parent is also assigned in the same group.

With these group configurations, we test their performance on the FISTA _Tree algorithm, except the standard sparsity case is performed on FISTA. The parameter β determines how strong the tree sparsity assumption is. Tables 7.6 and 7.7 show the average SNRs and CPU time respectively for the four MR images with various parameter settings. With smaller parameters, the third group configuration performs the best, while the second group configuration is the best with larger parameters.

Table 7.6 Comparisons of SNR (dB) for Different Group Sizes With Tree Sparsity

β\Group Size	1	2	3	4
5×10^{-2}	17.30	**17.94**	16.45	15.33
10^{-2}	16.49	**16.99**	16.95	16.53
5×10^{-3}	16.36	16.62	**16.66**	16.48
10^{-3}	16.21	16.27	**16.29**	16.27

Table 7.7 Comparisons of Computational Cost (s) for Different Group Sizes With Tree Sparsity

β\Group Size	1	2	3	4
5×10^{-2}	0.69	0.99	1.11	1.17
10^{-2}	0.70	0.95	1.09	1.15
5×10^{-3}	0.72	0.97	1.07	1.11
10^{-3}	0.70	0.97	1.07	1.11

The computational time increases monotonously as the size of the group becomes larger. Due to the above two reasons, we encourage the use of the second group configuration on CS-MRI.

7.4.2.3 Visual comparisons

We compare the proposed tree-based algorithms with the fastest MRI algorithms to validate how much the tree structure can improve existing results. To perform fair comparisons, all methods run 50 iterations except that the CG runs only eight iterations due to its higher computational complexity. Total variation terms are removed in all algorithms, as we only want to validate how much benefit the wavelet tree sparsity can bring compared to standard wavelet sparsity. In this case, FCSA (Huang et al., 2011b) is similar to FISTA (Beck and Teboulle, 2009a). Figs. 7.15–7.18 show the visual results for the four MR images with 20% sampling. It can be seen that the proposed unconstrained algorithm FISTA_Tree is always better than CG (Lustig et al., 2007), TVCMRI (Ma et al., 2008), RecPF (Yang et al., 2010), and FCSA (Huang et al., 2011b). These results are consistent with previous observations (Huang et al., 2011b). Compared to the proposed NESTA_Tree with NESTA, our method is still much better. These results are reasonable because no structured prior information has been exploited in previous algorithms other than sparsity, while the tree structure in our algorithms is utilized. Any coefficient that disobeys the tree structure will be penalized in our algorithms, which makes the results closer to the original ones.

7.4.2.4 SNRs and CPU time

Fig. 7.19 gives the performance comparisons between different methods in terms of SNR with 50 iterations. Due to the faster convergence rate of FISTA and NESTA, they always outperform CG, TVCMRI, and RecPF. Moreover, the tree-based algorithms approximated by overlapping group sparsity are always better than those with standard sparsity. Table 7.8 shows all computational costs for the different algorithms. CG has the highest computational complexity. TVCMRI and RecPF are much faster than CG and slower than FCSA. It is to be expected that the tree-based algorithms FISTA_Tree and NESTA_Tree are slower than FISTA and NESTA respectively, since the overlapping structure requires more time for computing than

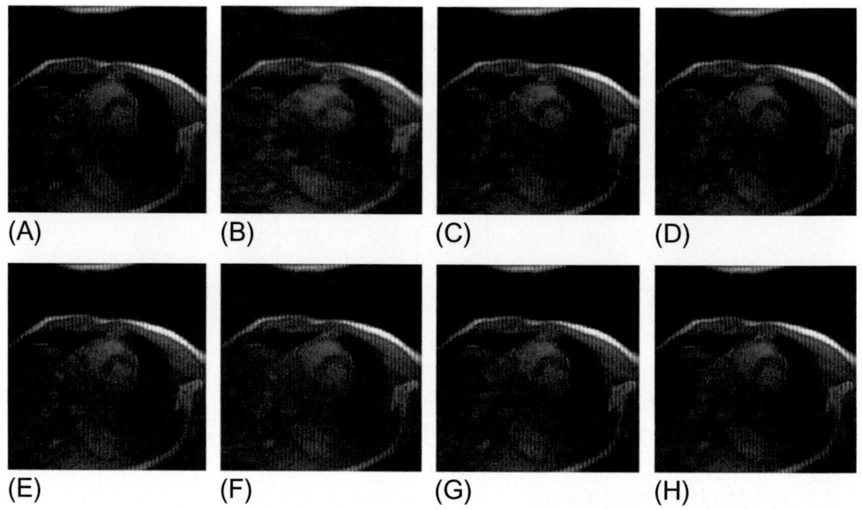

FIG. 7.15

Cardiac MR image reconstruction from 20% sampling. (A) The original image. Also shown are images recovered by: (B) CG (Lustig et al., 2007); (C) TVCMRI (Ma et al., 2008); (D) RecPF (Yang et al., 2010); (E) FCSA (Huang et al., 2011b); (F) FISTA_Tree; (G) NESTA (Becker et al., 2011); (H) NESTA_Tree. All algorithms are without total variation regularization. Their SNRs are 9.86, 14.70, 15.14, 17.31, 17.93, 16.31, and 16.96 respectively. Their computational time costs are 1.34, 1.12, 1.25, 0.67, 0.85, 0.88, and 1.05 s respectively.

the nonoverlapping structure. However, applying the wavelet transform and the Fourier transform is still the dominant cost, which is the same for all algorithms. As a result, FISTA_Tree and NESTA_Tree are comparable to the corresponding standard sparsity algorithms in terms of reconstruction speed, and provide greater improvement in terms of accuracy.

7.4.2.5 Sampling ratios

All algorithms are compared under different sampling ratios for the four MR images. Since we have shown that the CG method is far less efficient than the other methods, we do not include it in this experiment. To reduce the randomness, we run each experiments 100 times to obtain the average results for each method. The sampling ratio ranges from 17% to 25%. Fig. 7.20 shows these results for the four images. We can observe that TVCMRI and RecPF are not comparable to recent algorithms with fast convergence rates. Under the same framework and with similar convergence rate, the tree-based algorithms (ie, FISTA_Tree and NESTA_Tree) are always better than the corresponding standard sparsity algorithms (ie, FCSA and NESTA). These results further demonstrate the benefit of tree sparsity in accelerated MRI.

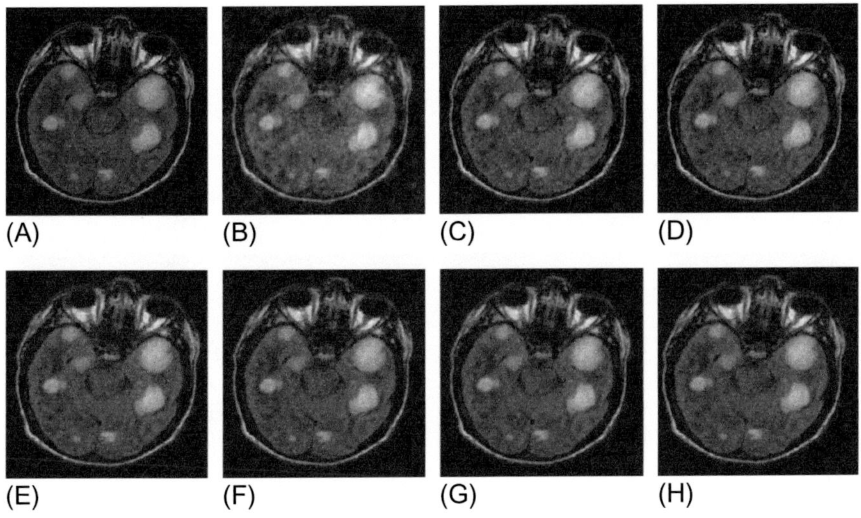

FIG. 7.16

Brain MR image reconstruction from 20% sampling. (A) The original image. Also shown are images recovered by: (B) CG (Lustig et al., 2007); (C) TVCMRI (Ma et al., 2008); (D) RecPF (Yang et al., 2010); (E) FCSA (Huang et al., 2011b); (F) FISTA_Tree; (G) NESTA (Becker et al., 2011); (H) NESTA_Tree. None of the algorithms have total variation regularization. Their SNRs are 10.25, 13.81, 14.22, 15.65, 16.13, 15.05, and 15.52 respectively. Their computational time costs are 1.36, 1.11, 1.17, 0.71, 1.02, 0.91, and 1.03 s respectively.

7.4.2.6 Complex-valued image with radial sampling mask

We have observed superior performance for tree-based MRI algorithms from numerical simulations. In this section, we validate their performance on a complex-valued MR image with 512 × 512 pixels. The sampling mask is a radial mask, which is more feasible than the random sampling mask in practice. Here, we only compare the classical method CG (Lustig et al., 2007) and the fastest algorithm FCSA (Huang et al., 2011b) with the proposed tree-based algorithms.

Fig. 7.21 presents the visual results reconstructed by difference methods. The image with full sampling is used as referent image. We can observe that the tree-based algorithms FISTA_Tree and NESTA_Tree achieve higher SNRs than the standard sparsity algorithms CG and FCSA. Due to the relatively slow convergence rate of CG, it has still not converged after 50 iterations. That is why it has inferior performance to FCSA. This data is scanned with noise. Therefore we also compare image quality besides SNR. From the zoomed-in areas, image details are lost in the image reconstructed by CG and blurred in that reconstructed by FCSA. However, both tree-based algorithms can preserve significant features on the MR image even with a low sampling ratio.

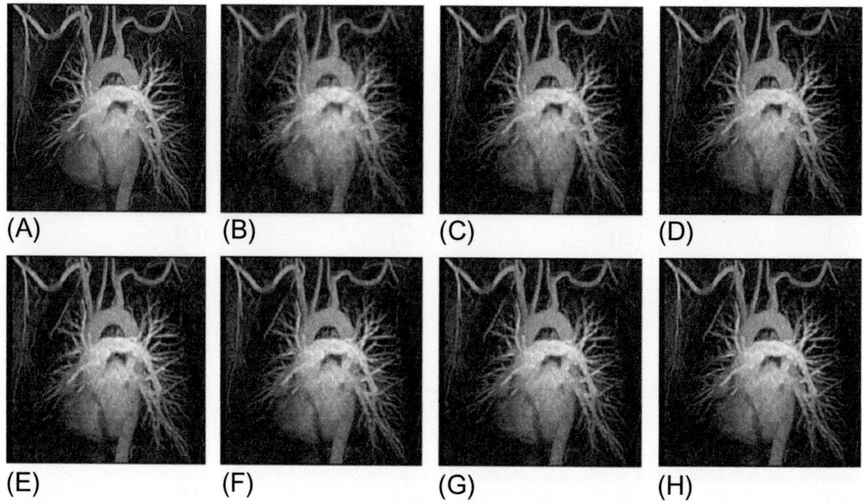

FIG. 7.17

Chest MR image reconstruction from 20% sampling. (A) The original image. Also shown are images recovered by: (B) CG (Lustig et al., 2007); (C) TVCMRI (Ma et al., 2008); (D) RecPF (Yang et al., 2010); (E) FCSA (Huang et al., 2011b); (F) FISTA_Tree; (G) NESTA (Becker et al., 2011); (H) NESTA_Tree. None of the algorithms have total variation regularization. Their SNRs are 11.82, 15.09, 15.36, 15.98, 16.35, 15.91, and 16.30 respectively. Their computational time costs are 1.28, 1.12, 1.23, 0.67, 0.96, 0.84, and 1.07 s respectively.

7.4.3 SUMMARY

We discussed two tree-sparsity-based algorithms for CS-MRI and compared them with the state-of-the-art algorithms based on standard sparsity. In order to observe the benefit of tree sparsity more clearly, total variation terms were removed in all algorithms. Evaluation results demonstrated the practical improvement of the tree-sparsity-based algorithm on MR images. The results show that the benefit of the presented algorithm is greater than predicted by structured sparsity theory. That is because the tree structure is not adhered to as strictly as the structured sparsity theories assumed for practical data.

7.5 FOREST SPARSITY IN MULTICHANNEL CS-MRI

This section discusses the application of forest sparsity (Chen et al., 2014a) in CS-MRI. In practical applications, it is usually observed that multichannel images, such as color images, multispectral images and MR images, have joint sparsity and tree sparsity simultaneously. Therefore the support of such data consists of several connected trees and is like a forest. Fig. 7.22 shows the forest structure in

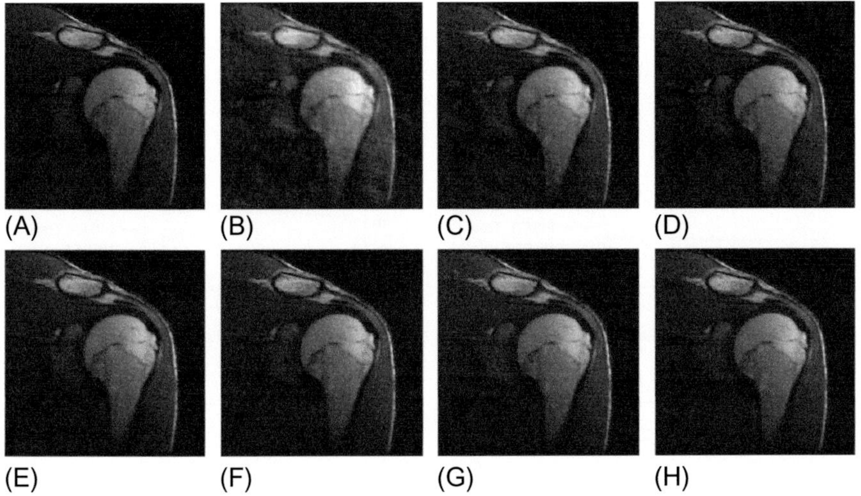

FIG. 7.18

Shoulder MR image reconstruction from 20% sampling. (A) The original image. Also shown are images recovered by: (B) CG (Lustig et al., 2007); (C) TVCMRI (Ma et al., 2008); (D) RecPF (Yang et al., 2010); (E) FCSA (Huang et al., 2011b); (F) FISTA_Tree; (G) NESTA (Becker et al., 2011); (H) NESTA_Tree. None of the algorithms have total variation regularization. Their SNR are 12.31, 16.80, 17.90, 20.77, 21.04, 20.17, and 20.62 respectively. Their computational time costs are 1.36, 1.07, 1.25, 0.67, 0.95, 0.82, and 1.07 s respectively.

multicontrast MR images. We can see that the nonzero coefficients are not randomly distributed but form a connected forest. Therefore forest sparsity is a favorable model for these kinds of problems.

7.5.1 MODEL AND ALGORITHM

In this section, the forest structure is approximated as overlapping group sparsity (Jacob et al., 2009) with mixed $\ell_{2,1}$-norm. Although it may not be the best approximation, it is enough to demonstrate the benefit of forest sparsity. To evaluate the forest sparsity model, we need to compare different models via a similar framework. From the definition of forest-sparse data, we can find that if a coefficient is large/small, its parent and "neighbors"[8] also tend to be large/small. All parent-child pairs in the same position across different channels are assigned into one group, and the problem becomes overlapping group sparsity regularization. A similar scheme has been used

[8]Parent denotes the parent node on the same channel while neighbors mean coefficients at the same position on other channels.

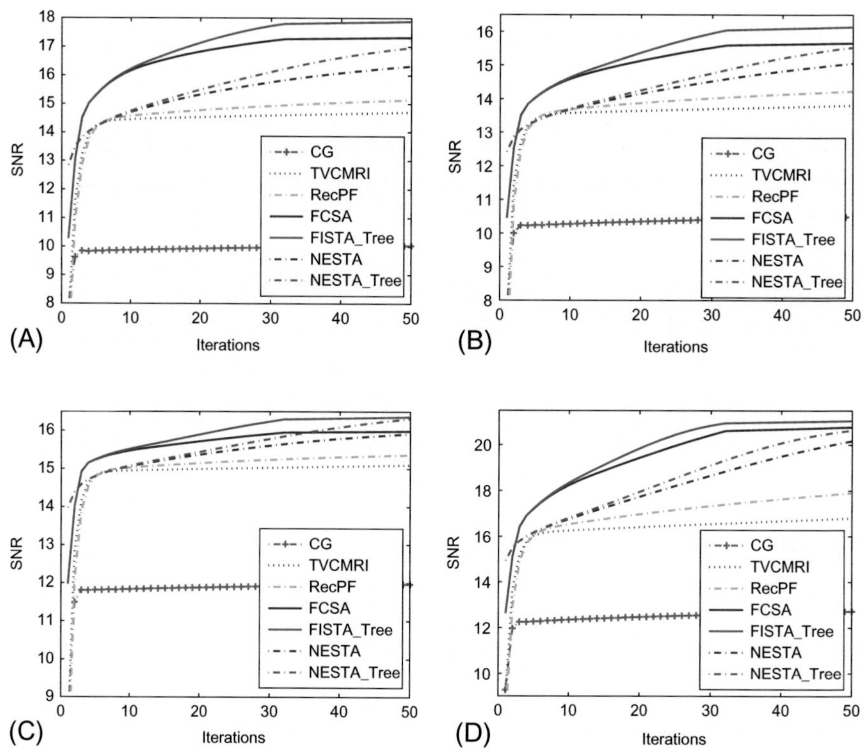

FIG. 7.19

Performance comparisons (SNRs) for different MR images: (A) cardiac image; (B) brain image; (C) chest image; (D) shoulder image.

Table 7.8 Comparison of Average Computational Costs (s) for Different MR Images With 20% Sampling

	Cardiac	Brain	Chest	Shoulder
CG	9.14	8.87	9.21	9.17
TVCMRI	1.12	1.11	1.12	1.07
RecPF	1.25	1.17	1.23	1.25
FCSA	0.67	0.71	0.67	0.67
FISTA_Tree	0.85	1.02	0.96	0.95
NESTA	0.88	0.91	0.84	0.82
NESTA_Tree	1.05	1.03	1.07	1.07

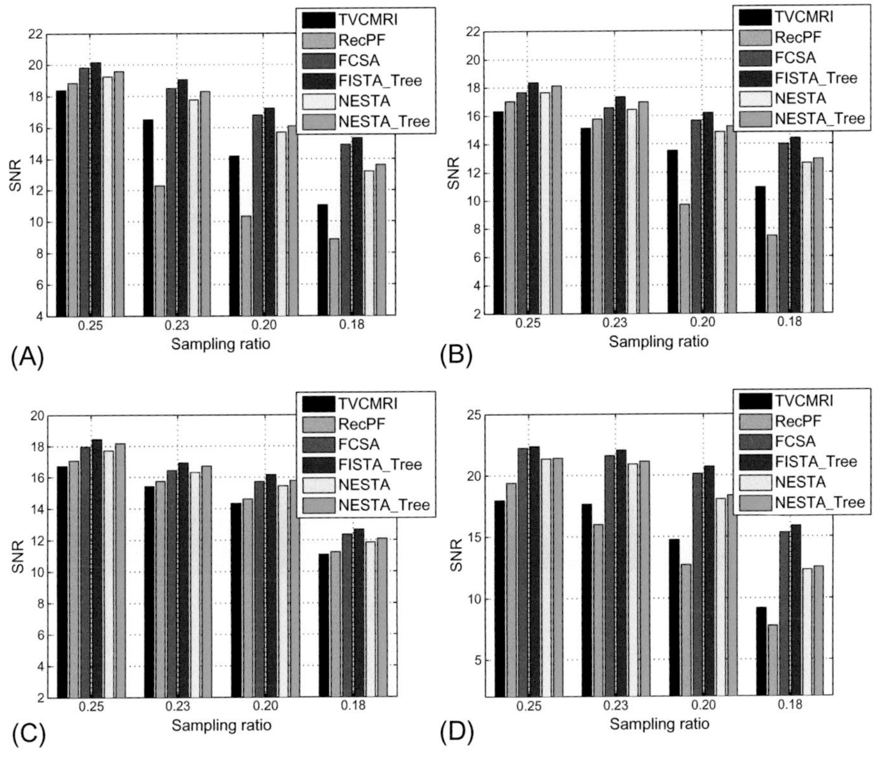

FIG. 7.20

Performance comparisons for the four MR images with different sampling ratios: (A) cardiac image; (B) brain image; (C) chest image; (D) shoulder image.

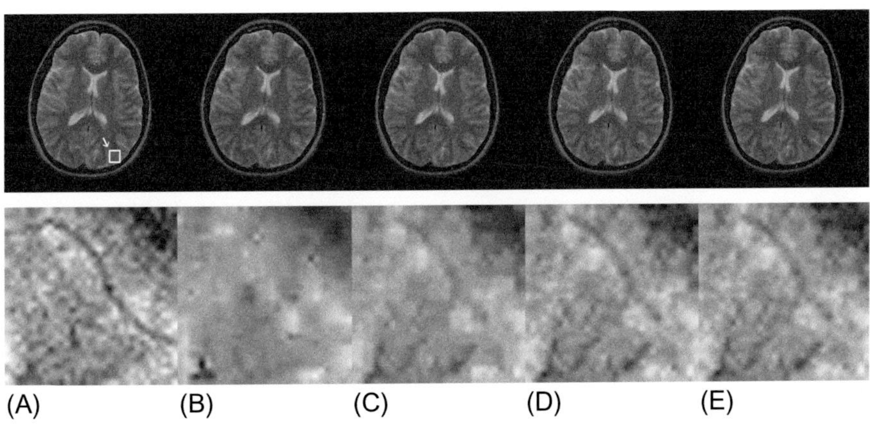

FIG. 7.21

Reconstruction on complex-valued MR image with 20% sampling. The first shows the visual results of by different algorithm. The second row shows the zoomed-in areas indicated by the white box. (A) Inverse FFT with full sampling. (B) CG. (C) FCSA. (D) FISTA_Tree. (E) NESTA_Tree. Their SNRs are 15.31, 16.32, 16.65, and 16.68, respectively.

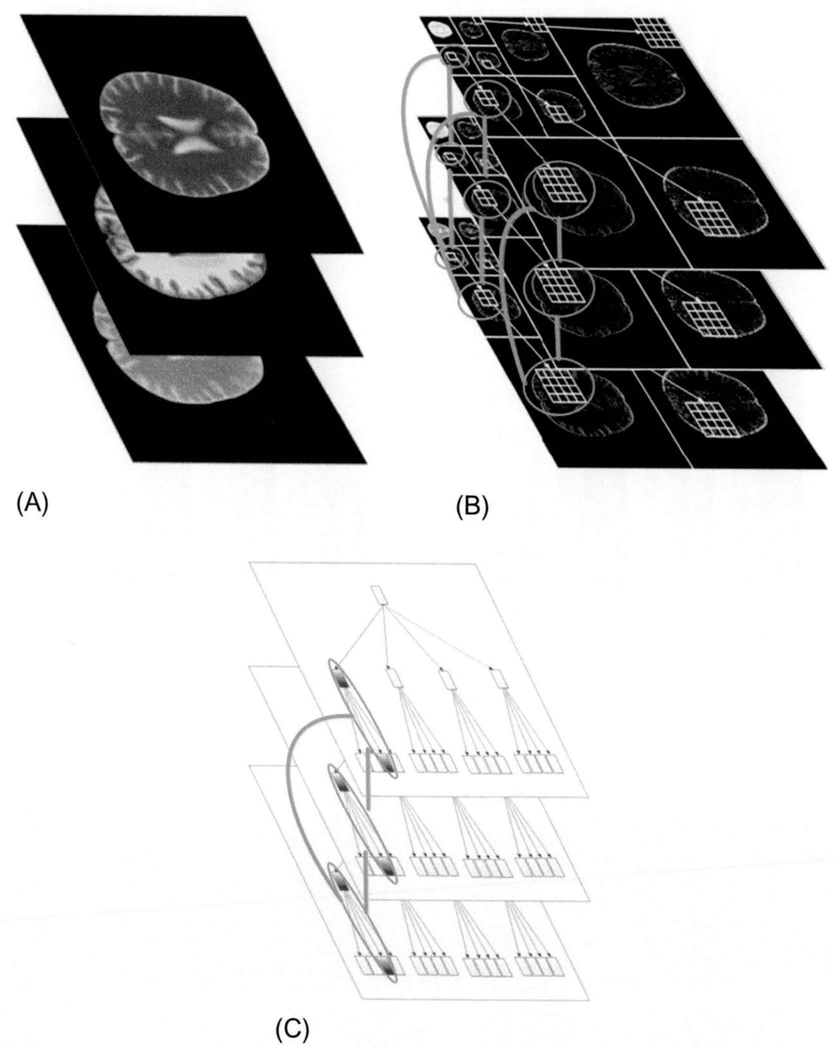

(A) (B)

(C)

FIG. 7.22

Forest structure on multicontrast MR images. (A) Three multicontrast MR images. (B) The wavelet coefficients of the images. Each coefficient tends to be consistent with its parent and children, and the coefficients across different trees at the same position. (C) One joint parent-child group across different trees that was used in our algorithm.

in approximating tree sparsity (Chen and Huang, 2012b; Rao et al., 2011), where each node and its parent are assigned into one group. We write the approximated problem as

$$\min_x \frac{1}{2}||Ax - b||_2^2 + \lambda \sum_{g \in \mathcal{G}} ||(\Phi x)_g||_2, \tag{7.28}$$

where g denotes one of the coefficient groups discussed above (an example is demonstrated in Fig. 7.22C), $(\cdot)_g$ denotes the coefficients in group g, and $\mathcal{G}$ is the set of all groups.

The mixed $\ell_{2,1}$-norm encourages all the components in the same group g to be zeros or nonzeros simultaneously. With our group configuration, this encourages forest sparsity. We present an efficient implementation based on the fast iterative shrinkage-thresholding algorithm (FISTA) (Beck and Teboulle, 2009b) framework for this problem. This is because FISTA can be easily applied for standard sparsity and joint sparsity, which makes the validation of the benefit of the proposed model more convenient. In addition, the formulation (7.28) can be easily extended to the combination of total variation (TV) via the fast composite splitting algorithm (FCSA) scheme (Huang et al., 2011b). Note that other algorithms may be used to solve the forest sparsity problems (eg, (Deng et al., 2011; Jacob et al., 2009; Kowalski et al., 2013)), but determining the optimal algorithm for forest sparsity is beyond the scope of this chapter.

FISTA (Beck and Teboulle, 2009b) is an accelerated version of the proximal method which minimizes the object function with the following form:

$$\min\{F(x) = f(x) + g(x)\}, \tag{7.29}$$

where $f(x)$ is a convex smooth function with Lipschitz constant L_f and $g(x)$ is a convex but usually nonsmooth function. It reverts to the original FISTA when $f(x) = \frac{1}{2}||Ax - b||_2^2$ and $g(x) = \lambda\|\Phi x\|_1$, which is summarized in Algorithm 7.10, where A^{T} denotes the transpose of A.

ALGORITHM 7.10 FISTA Beck and Teboulle (2009b)

series Input: $\rho = 1/L_f, \lambda, n = 1, t^1 = 1\ r^1 = x^0$
while not meeting the stopping criterion **do**
$y = r^n - \rho A^{\mathrm{T}}(Ar^n - b)$
$x = \arg\min_x\{\frac{1}{2\rho}\|x - y\|^2 + \lambda\|\Phi x\|_1\}$
$t^{n+1} = 1 + \sqrt{1 + 4(t^n)^2}/2$
$r^{n+1} = x^n + \frac{t^n - 1}{t^{n+1}}(x^n - x^{n-1})$
$n = n + 1$
end while

For the second step, there is a closed-form solution by soft-thresholding. For a joint sparsity problem where $g(x) = \lambda\|\Phi x\|_{2,1}$, the second step also has a closed-form solution. We call this version FISTA_Joint for joint sparsity. However, for the problem (7.28) with overlapping groups, we cannot directly apply FISTA to solve it.

In order to transfer the problem (7.28) to the nonoverlapping version, we introduce a binary matrix $G \in \mathbb{R}^{D\times TN}$ $(D > TN)$ to duplicate the overlapping coefficients. Each row of G only contains one 1 and all others are 0s. The 1 appears in the ith column corresponds to the ith coefficient of Φx. Intuitively, if the coefficient is included in j groups, G will contains j such rows. An auxiliary variable z is used to constrain $G\Phi x$. This scheme is widely utilized in the alternating direction method (ADM) (Deng et al., 2011). The alternating formulation becomes:

$$\min_{x,z}\left\{\frac{1}{2}\|Ax-b\|_2^2+\lambda\sum_{g\in\mathcal{G}}\|z_g\|_2+\frac{\gamma}{2}\|z-G\Phi x\|_2^2\right\}, \tag{7.30}$$

where γ is another positive parameter. We iteratively solve this alternative formulation by minimizing x and z subproblems respectively. For the z subproblem:

$$\hat{z_g} = \arg\min_{z_g}\left\{\lambda\|z_g\|_2+\frac{\gamma}{2}\|z_g-(G\Phi x)_g\|_2^2\right\},\ g\in\mathcal{G} \tag{7.31}$$

which has the closed-form solution:

$$\hat{z_g} = \max(\|(G\Phi x)_g\|_2-\frac{\lambda}{\gamma},0)\frac{(G\Phi x)_g}{\|(G\Phi x)_g\|_2},\ g\in\mathcal{G}. \tag{7.32}$$

We denote it as a shrinkgroup operation. For the x-subproblem:

$$\hat{x} = \arg\min_{x}\left\{\frac{1}{2}\|Ax-b\|_2^2+\frac{\gamma}{2}\|z-G\Phi x\|_2^2\right\} \tag{7.33}$$

The optimal solution is $x = (A^{\mathrm{T}}A+\lambda\Phi^{\mathrm{T}}G^{\mathrm{T}}G\Phi)^{-1}(A^{\mathrm{T}}b+\lambda\Phi^{\mathrm{T}}G^{\mathrm{T}}z)$, which contains a large-scale inverse problem. Actually, this problem can be efficiently solved by various methods. In order to compare with FISTA and FISTA_Joint, we apply FISTA to solve (7.33). This will demonstrate the benefit of forest sparsity more clearly. Let $f(x) = \frac{1}{2}\|Ax-b\|_2^2+\frac{\lambda}{2}\|z-G\Phi x\|_2^2$ and $g(x) = 0$. Supposing its Lipschitz constant is L_f, the whole algorithm is summarized in Algorithm 7.11.

ALGORITHM 7.11 FISTA_FOREST

series Input: $\rho = 1/L_f, r^1 = x^0, t^1 = 1, \lambda, \gamma, n = 1$
while not meeting the stopping criterion **do**
  $z = shrinkgroup(G\Phi x^{n-1}, \lambda/\gamma)$
  $x^n = r^n - \rho[A^{\mathrm{T}}(Ar^n - b) + \gamma\Phi^{\mathrm{T}}G^{\mathrm{T}}(G\Phi r^n - z)]$
  $t^{n+1} = [1+\sqrt{1+4(t^n)^2}]/2$
  $r^{n+1} = x^n + \frac{t^n-1}{t^{n+1}}(x^n - x^{n-1})$
  $n = n+1$
end while

For the first step, we solve (7.31) while $\frac{1}{2}\|Ax - b\|_2^2$ stays the same. The object function value in (7.30) decreases. For the second step, (7.33) is solved by FISTA iteratively while $\lambda \sum_{g\in\mathcal{G}} ||z_g||_2$ stays the same. Therefore the object function value in (7.30) decreases in each iteration and the algorithm is convergent. Algorithm 7.11 is also used to implement tree sparsity by recovering the data channel by channel separately. We call it FISTA_Tree.

In some practical applications, the data tends to be forest-sparse but not strictly so. We can soften and complement the forest assumption with other penalties, such as joint $\ell_{2,1}$-norm or TV. For example, after combining TV, problem (7.30) becomes:

$$\min_{x,z} \left\{ \frac{1}{2}\|Ax - b\|_2^2 + \lambda \sum_{g\in\mathcal{G}} ||z_g||_2 + \frac{\gamma}{2}||z - G\Phi x||_2^2 \right.$$
$$\left. + \mu||x||_{TV} \right\}, \quad (7.34)$$

where $||x||_{TV} = \sum_{i=1}^{TN} \sqrt{(\nabla_1 x_i)^2 + (\nabla_2 x_i)^2}$; ∇_1 and ∇_2 denote the forward finite difference operators on the first and second coordinates respectively; μ is a positive parameter. Compared with Algorithm 7.11, we only need to set $g(x) = \mu||x||_{TV}$ and the corresponding subproblem has already been solved (Beck and Teboulle, 2009b; Huang et al., 2011a,b). This TV combined algorithm is called FCSA_Forest, which will be used in the experiments. To avoid repetition, it is not listed.

7.5.2 EVALUATION

We conduct experiments on RGB color images, multicontrast MR images, and MR images of multichannel coils to validate the benefit of forest sparsity. All experiments are conducted on a desktop with 3.4-GHz Intel core i7 3770 CPU. Matlab version is 7.8 (2009a). If the sampling matrix A is M by N, the sampling ratio is defined as M/N. All measurements are mixed with Gaussian white noise of 0.01 standard deviation. The SNR is used as the metric for evaluations:

$$SNR = 10 \log_{10}(V_s/V_n) \quad (7.35)$$

where V_n is the Mean Square Error between the original data x_0 and the reconstructed x; $V_s = var(x_0)$ denotes the power level of the original data where $var(x_0)$ denotes the variance of the values in x_0.

7.5.2.1 Multicontrast MRI

Multicontrast MRI is a popular technique to aid clinical diagnosis. For example T1 weighted MR images can distinguish fat from water, with water appearing darker and fat brighter. In T2 weighted images fat is darker and water is lighter, which is better suited to imaging edema. Although having different intensities, T1/T2 or proton-density weighted MR images are scanned at the same anatomical position. Therefore they are not independent but highly correlated. Multicontrast MR images are typically forest-sparse under the wavelet basis. Suppose $\{x_t\}_{t=1}^{T} \in \mathbb{R}^N$ are the multicontrast images for the same anatomical cross-section and $\{b_t\}_{t=1}^{T}$ are the corresponding undersampled data in the Fourier domain, the forest-sparse reconstruction can be formulated as

$$\hat{x} = \arg\min_{x} \|\Phi x\|_{\mathcal{F},T} + \lambda \sum_{s=1}^{T} \|R_t x_t - b_t\|^2, \tag{7.36}$$

where x is the vertorized data of $[x_1, \ldots, x_T]$ and R_t is the measurement matrix for the image x_t. This is an extension of conventional CS-MRI (Lustig et al., 2007). Fig. 7.22 shows an example of the forest structure in multicontrast MR images.

The data is extracted from the SRI24 Multichannel Brain Atlas Dataset (Rohlfing et al., 2009). In the Fourier domain, we randomly obtain more samples at low frequencies and fewer samples at higher frequencies. This sampling scheme has been widely used for CS-MRI (Huang et al., 2011b; Lustig et al., 2007; Ma et al., 2008). Fig. 7.23 shows the original multicontrast MR images and the sampling mask.

We compare four algorithms on this dataset: FISTA, FISTA_Joint, FISTA_Tree, and FISTA_Forest. The parameter λ is set to 0.035 and γ is set to 0.5λ. We run each algorithm for 400 iterations. Fig. 7.24A demonstrates the performance comparisons among different algorithms. From the figure, we can observe that modeling with forest sparsity achieves the highest SNR after convergence. Although the algorithm for forest sparsity takes more time due to the overlapping structure, it always outperforms all other methods in terms of accuracy.

In addition, as total variation is very popular in CS-MRI (Huang et al., 2012, 2011b; Lustig et al., 2007), we compare our FCSA_Forest algorithm with FCSA (Huang et al., 2011b) (TV is combined in FISTA), FCSA_Joint (Huang et al., 2012) (TV is combined in FISTA_Joint), and FCSA_Tree. The parameter μ for TV is set to 0.001, the same as in previous works (Huang et al., 2011b; Ma et al., 2008). Fig. 7.24B demonstrates the performance comparison including TV regularization. Compared with Fig. 7.24A, all algorithms improve at different degrees. However,

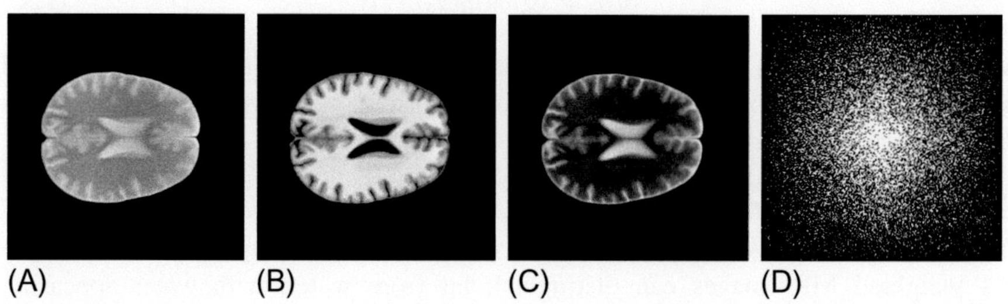

FIG. 7.23

(A)–(C) The original multiconstrast images. (D) The sampling mask.

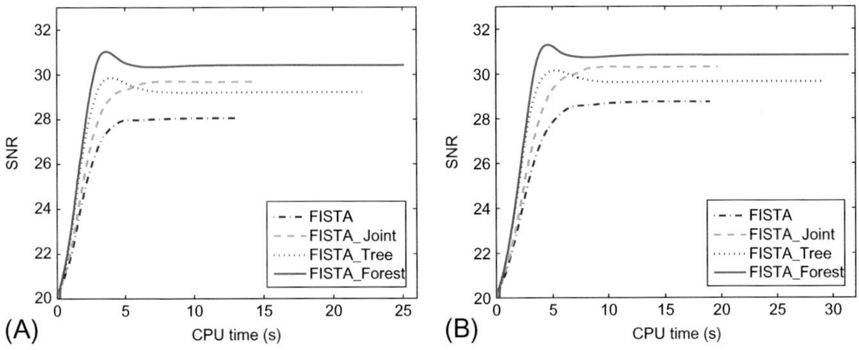

FIG. 7.24

Performance comparisons among different algorithms. (A) Multicontrast MR image reconstruction with 20% sampling. Their final SNRs are 28.05, 29.69, 29.22, and 30.42 respectively. The time costs are 13.11, 14.43, 22.08, and 25.11 s respectively. (B) Multicontrast MR images reconstruction with 20% sampling by both wavelet sparsity and TV regularization. Their final SNRs are 28.75, 30.30, 29.65, and 30.83 respectively. The time costs are 19.00, 19.68, 29.11, and 31.41 s, respectively.

the ranking does not change, which validates the superiority of forest sparsity. As FCSA has been proved to be better than other algorithms for general compressed sensing MRI (CS-MRI) (Lustig et al., 2007; Ma et al., 2008; Yang et al., 2010) and FCSA_Joint (Huang et al., 2012) is better (Bilgic et al., 2011; Majumdar and Ward, 2011) in multicontrast MRI, the proposed method further improves CS-MRI and make it more feasible than before.

In order to validate the benefit of forest sparsity in terms of measurement number, we conducted an experiment to reconstruct multicontrast MR images from different sampling ratios. Fig. 7.25 demonstrates the final results of four algorithms with sampling ratio from 16% to 26%. With more sampling, all algorithms show better performance. However, the forest sparsity algorithm always achieves the best reconstruction. For the same reconstruction accuracy, the FISTA_Forest algorithm only requires about 16% measurements to achieve SNR 28, which is approximately 2%, 3%, and 5% less than those of FISTA_Joint, FISTA_Tree, and FISTA respectively. More results of forest sparsity on multicontrast MRI can be found in Chen and Huang (2014b).

7.5.2.2 Parallel MRI

To improve the scanning speed of MRI, an efficient and feasible way is to acquire the data in parallel with multichannel coils. The scanning time depends on the number of measurements in the Fourier domain, and it will be significantly reduced when each coil only acquires a small fraction of the whole measurements. The bottleneck is how to reconstruct the original MR image efficiently and precisely. This is called

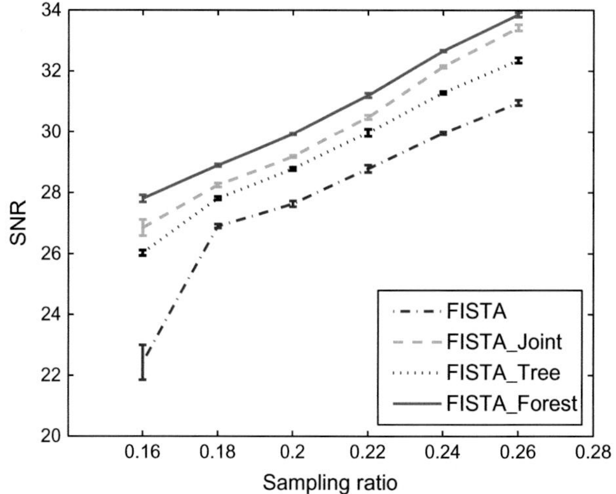

FIG. 7.25

Reconstruction performance with different sampling ratios.

parallel MRI (pMRI) in the literature. Sparsity techniques have been used to improve the classical method SENSE (Pruessmann et al., 1999). However, when the coil sensitivity cannot be estimated precisely, the final image will contain visual artifacts. Unlike the previous CS-SENSE (Liang et al., 2009), which reconstructs the images of multicoils individually, calibrationless parallel MRI (Chen et al., 2013; Majumdar and Ward, 2012) recovers the aliased images of all coils jointly by assuming the data is jointly sparse.

Let T equal the number of coils and b_t be the measurement vector from coil t. It is therefore the same CS problem as (7.36). The final result of CaLM-MRI is obtained by a sum-of-square (SoS) approach without coil sensitivity and SENSE encoding. It shows comparable results with those methods which need precise coil configuration. As shown in Fig. 7.26, the appearances of different images obtained from multicoils are very similar. This method can be improved with forest sparsity, since the images follow the forest sparsity assumption.

There are two steps for compressed sensing pMRI reconstruction in CaLM-MRI (Majumdar and Ward, 2012): (1) the aliased images are recovered from the undersampled Fourier signals of different coil channels by CS methods; (2) the final image for clinical diagnosis is synthesized by the recovered aliased images using the SoS approach. As discussed above, these aliased images should be forest-sparse under the wavelet basis. We compare our algorithm with FISTA_Joint and SPGL1 (Van Den Berg and Friedlander, 2008), which solves the joint $\ell_{2,1}$-norm problem in CaLM-MRI. For the second step, all methods use the SoS approach from the aliased images that they recovered. All algorithms run for sufficient time until it has converged.

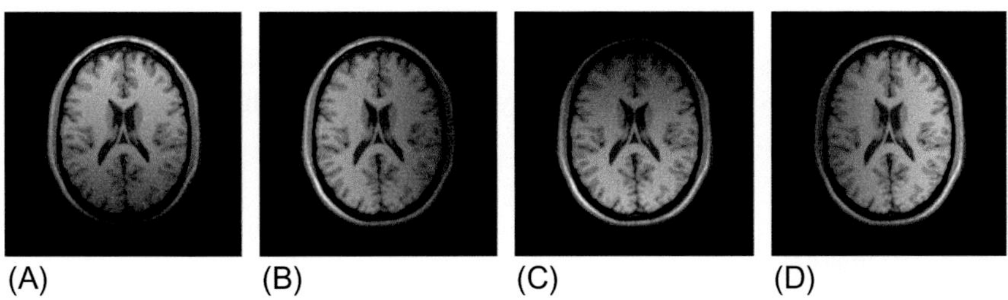

FIG. 7.26

The aliased MR images of multicoils. Due to the different locations of the coils, they have different sensitivities to the same image.

Tables 7.9 and 7.10 show all the comprehensive comparisons among these algorithms. For the same algorithm, more measurements or a greater number of coils tend to increase the SNRs of aliased images, although this does not result in linear improvement for the final image reconstruction. Another observation is that FISTA_Joint and SPGL1 show similar performance in terms of SNR on this data. This is because both solve the same joint sparsity problem, even with different

Table 7.9 Comparison of SNRs (dB) at Different Sampling Ratios With Four Coils

	Sampling Ratios	25%	20%	17%	15%
SNR of aliased images	SPGL1	26.72	24.59	23.08	22.31
	FISTA_Joint	26.95	24.73	23.06	22.21
	FISTA_Forest	**27.47**	**25.22**	**23.37**	**22.59**
SNR of final image	SPGL1	20.64	20.35	19.12	18.64
	FISTA_Joint	20.79	20.41	19.75	18.49
	FISTA_Forest	**22.62**	**22.29**	**21.03**	**20.47**

Table 7.10 Comparison of SNRs (dB) for Different Number of Coils With 20% Sampling Ratio

	Number of Coils	2	4	6	8
SNR of aliased images	SPGL1	23.33	24.61	24.74	25.16
	FISTA_Joint	23.41	24.71	24.89	25.23
	FISTA_Forest	**24.25**	**25.12**	**25.29**	**25.52**
SNR of final image	SPGL1	21.76	18.95	21.05	21.32
	FISTA_Joint	21.90	18.94	21.15	21.87
	FISTA_Forest	**22.44**	**22.22**	**22.52**	**22.52**

schemes. Upgrading the model to forest sparsity, significant improvement can be gained. Finally, it is unknown how to combine TV in SPGL1. However, both FISTA_Joint and FISTA_Forest can easily combine TV, which can further enhance the results (Huang et al., 2012).

7.5.3 SUMMARY

In this section, we have discussed the *forest sparsity* model for sparse learning and compressed sensing. This model enriches the family of structured sparsity and can be widely applied to numerous fields of sparse regularization problems. The benefit of the forest sparsity model has been theoretically proved and empirically validated in practical applications. Under compressed sensing assumptions, significant reduction of measurements is achieved with forest sparsity compared with standard sparsity, joint sparsity or independent tree sparsity. We also discuss a fast algorithm for efficiently solving the forest sparsity problem.

7.6 CONCLUSION

In this chapter, we have presented several sparsity models for MRI reconstruction. A set of efficient algorithms is discussed to solve sparsity regularization MRI reconstruction problems. First, we introduce standard sparsity for MR image reconstruction. Second, we discussed an efficient algorithm for fast multicontrast MRI reconstruction based on group sparsity constraints. Third, we introduce wavelet tree sparsity in accelerated MRI. Finally, we discuss forest sparsity on multichannel data, which is a natural extension of tree sparsity.

REFERENCES

Bach, F., 2008. Consistency of the group lasso and multiple kernel learning. J. Mach. Learn. Res. 9, 1179–1225.

Baraniuk, R., Cevher, V., Duarte, M., Hegde, C., 2010. Model-based compressive sensing. IEEE Trans. Inform. Theory 56 (4), 1982–2001.

Baron, D., Wakin, M., Duarte, M., Sarvotham, S., Baraniuk, R., 2005. Distributed compressed sensing. arXiv 0901.3403

Beck, A., Teboulle, M., 2009a. Fast gradient-based algorithms for constrained total variation image denoising and deblurring problems. IEEE Trans. Image Process. 18 (113), 2419–2434.

Beck, A., Teboulle, M., 2009b. A fast iterative shrinkage-thresholding algorithm for linear inverse problems. SIAM J. Imaging Sci. 2 (1), 183–202.

Becker, S., Bobin, J., Candès, E.J., 2011. Nesta: a fast and accurate first-order method for sparse recovery. SIAM J. Imaging Sci. 4 (1), 1–39.

Bilgic, B., Goyal, V., Adalsteinsson, E., 2011. Multi-contrast reconstruction with Bayesian compressed sensing. Magn. Reson. Med. 66 (6), 1601–1615.

Block, K.T., Uecker, M., Frahm, J., 2007. Undersampled radial MRI with multiple coils. iterative image reconstruction using a total variation constraint. Magn. Reson. Med. 57 (6), 1086–1098.

Blomgren, P., Chan, T.F., 1998. Color TV: total variation methods for restoration of vector-valued images. IEEE Trans. Image Process. 7 (3), 304–309.

Candès, E., 2006. Compressive sampling. In: Proceedings of the International Congress of Mathematicians, pp. 1433–1452.

Candès, E., Romberg, J., 2007. Sparsity and incoherence in compressive sampling. Inverse Problems 23 (3), 969.

Candès, E., Romberg, J., Tao, T., 2006a. Robust uncertainty principles: Exact signal reconstruction from highly incomplete frequency information. IEEE Trans. Inform. Theory 52 (2), 489–509.

Candès, E., Romberg, J., Tao, T., 2006b. Stable signal recovery from incomplete and inaccurate measurements. Commun. Pure Appl. Math. 59 (8), 1207–1223.

Candès, E.J., Romberg, J., Tao, T., 2006c. Robust uncertainty principles: exact signal reconstruction from highly incomplete frequency information. IEEE Trans. Inform. Theory 52, 489–509.

Cauley, S.F., Xi, Y., Bilgic, B., Setsompop, K., Xia, J., Adalsteinsson, E., Balakrishnan, V.R., Wald, L.L., 2013. Scalable and accurate variance estimation (save) for joint Bayesian compressed sensing. In: 21st Annual Meeting of ISMRM, p. 2603.

Chang, T.-C., He, L., Fang, T., 2006. MR image reconstruction from sparse radial samples using Bregman iteration. In: Proceedings of the 13th Annual Meeting of ISMRM, Seattle, p. 696.

Chen, C., Huang, J., 2012a. The benefit of tree sparsity in accelerated MRI. In: MICCAI Workshop on Sparsity Techniques in Medical Imaging. IEEE, Piscataway, NJ.

Chen, C., Huang, J., 2012b. Compressive sensing MRI with wavelet tree sparsity. In: Proceedings of the Annual Conference on Advances in Neural Information Processing Systems (NIPS), pp. 1124–1132.

Chen, C., Huang, J., 2014a. The benefit of tree sparsity in accelerated MRI. Med. Image Anal. 18 (6), 834–842.

Chen, C., Huang, J., 2014b. Exploiting both intra-quadtree and inter-spatial structures for multi-contrast MRI. In: Proceedings of the International Symposium on Biomedical Imaging (ISBI).

Chen, S., Donoho, D., Saunders, M., 1998. Atomic decomposition by basis pursuit. SIAM J. Sci. Comput. 20 (1), 33–61.

Chen, C., Li, Y., Huang, J., 2013. Calibrationless parallel MRI with joint total variation regularization. In: Proceedings of the Annual International Conference on Medical Image Computing and Computer-Assisted Intervention (MICCAI), pp. 106–114.

Chen, C., Li, Y., Huang, J., 2014a. Forest sparsity for multi-channel compressive sensing. IEEE Trans. Signal Process. 62 (11), 2803–2813.

Chen, C., Li, Y., Liu, W., Huang, J., 2014b, Image fusion with local spectral consistency and dynamic gradient sparsity. In: Proceedings of the IEEE Conference on Computer Vision and Pattern Recognition (CVPR), pp. 2760–2765.

Combettes, P.L., 2009. Iterative construction of the resolvent of a sum of maximal monotone operators. J. Convex Anal. 16, 727–748.

Combettes, P.L., Pesquet, J.C., 2008. A proximal decomposition method for solving convex variational inverse problems. Inverse Problems 24, 1–27.

Cotter, S., Rao, B., Engan, K., Kreutz-Delgado, K., 2005. Sparse solutions to linear inverse problems with multiple measurement vectors. IEEE Trans. Signal Process. 53 (7), 2477–2488.

Deng, W., Yin, W., Zhang, Y., 2011. Group sparse optimization by alternating direction method, TR11-06. Department of Computational and Applied Mathematics, Rice University.

Donoho, D., 2006. Compressed sensing. IEEE Trans. Inform. Theory 52 (4), 1289–1306.

Donoho, D., Elad, M., 2002. Optimally sparse representation in general (nonorthogonal) dictionaries via ℓ_1 minimization. Optimally 100 (5), 2197–2202.

Donoho, D., Maleki, A., Montanari, A., 2009. Message-passing algorithms for compressed sensing, Proc. Natl. Acad. Sci. 106, (45), 18914–18919.

Figueiredo, M., Nowak, R., Wright, S., 2007. Gradient projection for sparse reconstruction: application to compressed sensing and other inverse problems. IEEE J. Sel. Top. Signal Process. 1 (4), 586–597.

He, L., Carin, L., 2009. Exploiting structure in wavelet-based Bayesian compressive sensing. IEEE Trans. Signal Process. 57 (9), 3488–3497.

He, L., Chang, T.C., Osher, S., Fang, T., Speier, P., 2006. MR image reconstruction by using the iterative refinement method and nonlinear inverse scale space methods. Technical Report UCLA CAM 06-35. ftp://ftp.math.ucla.edu/pub/camreport/cam06-35.pdf.

Huang, J., 2011. Structured sparsity: theorems, algorithms and applications. Ph.D. thesis, Rutgers University.

Huang, J., Huang, X., Metaxas, D., 2009. Learning with dynamic group sparsity. In: Proceedings of IEEE International Conference on Computer Vision (ICCV), pp. 64–71.

Huang, J., Zhang, T., et al., 2010. The benefit of group sparsity. Ann. Stat. 38 (4), 1978–2004.

Huang, J., Zhang, S., Li, H., Metaxas, D., 2011a. Composite splitting algorithms for convex optimization. Comput. Vis. Image Understand. 115 (12), 1610–1622.

Huang, J., Zhang, S., Metaxas, D., 2011b. Efficient MR image reconstruction for compressed MR imaging. Med. Image Anal. 15 (5), 670–679.

Huang, J., Zhang, T., Metaxas, D., 2011c. Learning with structured sparsity. J. Mach. Learn. Res. 12, 3371–3412.

Huang, J., Chen, C., Axel, L., 2012. Fast Multi-contrast MRI Reconstruction. In: Proceedings of Medical Image Computing and Computer-Assisted Intervention, pp. 281–288.

Jacob, L., Obozinski, G., Vert, J., 2009. Group lasso with overlap and graph lasso. In: Proceedings of the International Conference on Machine Learning (ICML), pp. 433–440.

Ji, S., Xue, Y., Carin, L., 2008. Bayesian compressive sensing. IEEE Trans. Signal Process. 56 (6), 2346–2356.

Ji, S., Dunson, D., Carin, L., 2009. Multitask compressive sensing. IEEE Trans. Signal Process. 57 (1), 92–106.

Kim, S., Xing, E., 2012. Tree-guided group lasso for multi-response regression with structured sparsity, with an application to eqtl mapping. Ann. Appl. Stat. 6 (3), 1095–1117.

Koh, K., Kim, S., Boyd, S., 2007. An interior-point method for large-scale l1-regularized logistic regression. J. Mach. Learn. Res. 8 (8), 1519–1555.

Kowalski, M., Siedenburg, K., Dörfler, M., 2013. Social sparsity! Neighborhood systems enrich structured shrinkage operators. IEEE Trans. Signal Process. 61 (10), 2498–2511.

Krim, H., Viberg, M., 1996. Two decades of array signal processing research: the parametric approach. IEEE Signal Process. Mag. 13 (4), 67–94.

La, C., Do, M., 2006. Tree-based orthogonal matching pursuit algorithm for signal reconstruction. In: Proceedings of the IEEE International Conference on Image Processing (ICIP), pp. 1277–1280.

Liang, D., Liu, B., Wang, J., Ying, L., 2009. Accelerating sense using compressed sensing. Magn. Reson. Med. 62 (6), 1574–1584.

Lustig, M., Donoho, D., Pauly, J., 2007. Sparse MRI: the application of compressed sensing for rapid MR imaging. Magn. Reson. Med. 58, 1182–1195.

Ma, S., Yin, W., Zhang, Y., Chakraborty, A., 2008. An efficient algorithm for compressed MR imaging using total variation and wavelets. In: Proceedings of the IEEE Conference on Computer Vision and Pattern Recognition (CVPR), pp. 1–8.

Majumdar, A., Ward, R., 2010. Compressive color imaging with group-sparsity on analysis prior. In: Proceedings of the 17th IEEE International Conference on Image Processing (ICIP), pp. 1337–1340.

Majumdar, A., Ward, R., 2011. Joint reconstruction of multiecho MR images using correlated sparsity. Mag. Reson. Imaging 29 (7), 899–906.

Majumdar, A., Ward, R., 2012. Calibration-less multi-coil MR image reconstruction. Mag. Reson. Imaging.

Manduca, A., 1996. Wavelet compression of medical images with set partitioning in hierarchical trees. In: Proceedings of the Annual International Conference of the IEEE Engineering in Medicine and Biology Society, vol. 3, pp. 1224–1225.

Meng, J., Yin, W., Li, H., Hossain, E., Han, Z., 2011. Collaborative spectrum sensing from sparse observations in cognitive radio networks. IEEE J. Sel. Areas Commun. 29 (2), 327–337.

Natarajan, B., 1995. Sparse approximate solutions to linear systems. SIAM J. Imaging Sci. 24 (2), 227–234.

Needell, D., Tropp, J., 2009. CoSaMP: iterative signal recovery from incomplete and inaccurate samples. Appl. Comput. Harmon. Anal. 26 (3), 301–321.

Nesterov, Y., 1983. A method for unconstrained convex minimization problem with the rate of convergence $\mathcal{O}(1/k^2)$. Sov. Math. Dokl. 269, 543–547.

Pruessmann, K., Weiger, M., Scheidegger, M., Boesiger, P., et al., 1999. SENSE: sensitivity encoding for fast MRI. Magn. Reson. Med. 42 (5), 952–962.

Rao, N., Nowak, R., Wright, S., Kingsbury, N., 2011. Convex approaches to model wavelet sparsity patterns. In: Proceedings of the IEEE International Conference on Image Processing (ICIP), pp. 1917–1920.

Rohlfing, T., Zahr, N., Sullivan, E., Pfefferbaum, A., 2009. The SRI24 multichannel atlas of normal adult human brain structure. Hum. Brain Map. 31 (5), 798–819.

Saito, T., Takagaki, Y., Komatsu, T., 2011. Three kinds of color total-variation semi-norms and its application to color-image denoising. In: Proceedings of the IEEE International Conference on Image Processing (ICIP), pp. 1457–1460.

Som, S., Schniter, P., 2012. Compressive imaging using approximate message passing and a Markov-tree prior. IEEE Trans. Signal Process. 60 (7), 3439–3448.

Sprechmann, P., Ramirez, I., Sapiro, G., Eldar, Y.C., 2011. C-HiLasso: a collaborative hierarchical sparse modeling framework. IEEE Trans. Signal Process. 59 (9), 4183–4198.

Tibshirani, R., 1996. Regression shrinkage and selection via the lasso. J. R. Stat. Soc. B 267–288.

Tropp, J., 2004. Greed is good: algorithmic results for sparse approximation. IEEE Trans. Inform. Theory 50 (10), 2231–2242.

Van Den Berg, E., Friedlander, M., 2008. Probing the Pareto frontier for basis pursuit solutions. SIAM J. Imaging Sci. 31 (2), 890–912.

Weller, D., Polimeni, J., Grady, L., Wald, L., Adalsteinsson, E., Goyal, V., 2010. Combining nonconvex compressed sensing and grappa using the nullspace method. In: 18th Annual Meeting of ISMRM, p. 4880.

Wipf, D., Rao, B., 2007. An empirical Bayesian strategy for solving the simultaneous sparse approximation problem. IEEE Trans. Signal Process. 55 (7), 3704–3716.

Yang, J., Zhang, Y., Yin, W., 2010. A fast alternating direction method for TVL1-L2 signal reconstruction from partial fourier data. IEEE J. Sel. Top. Signal Process. 4 (2), 288–297.

Ye, J., Tak, S., Han, Y., Park, H., 2007. Projection reconstruction MR imaging using FOCUSS. Magn. Reson. Med. 57 (4), 764–775.

Yuan, M., Lin, Y., 2005. Model selection and estimation in regression with grouped variables. J. R. Stat. Soc. B 68 (1), 49–67.

Ziniel, J., Schniter, P., 2011. Efficient high-dimensional inference in the multiple measurement vector problem. arXiv 1111.5272.

CHAPTER

Hashing-based large-scale medical image retrieval for computer-aided diagnosis

X. Zhang, S. Zhang

University of North Carolina at Charlotte, Charlotte, NC, United States

CONTENTS

8.1 INTRODUCTION

An important goal in medical imaging informatics is transforming raw images into a quantifiable symbolic form for *indexing, retrieval, and reasoning*. In particular, content-based image retrieval (CBIR) has become important in medical informatics by providing doctors with diagnostic aid in the form of visualizing existing and relevant cases, along with diagnosis information. Therefore computer-aided diagnosis techniques such as case-based reasoning or evidence-based medicine have a strong need to retrieve images that can be valuable for diagnosis. However, this task, of interest to both academia and the healthcare industry, is challenging because medical image content is essentially unstructured. Take tissue images as an example: "accidental characteristics" can be caused by incompletely controlled illumination, variability in tissue texture, image noise introduced in the staining process and normal variations among subjects, impeding the success of existing systems. Most importantly, few systems are able to analyze large-scale (ie, ever-increasing amount and complexity of) medical image databases in real-time. When applied to

Machine Learning and Medical Imaging. http://dx.doi.org/10.1016/B978-0-12-804076-8.00008-6

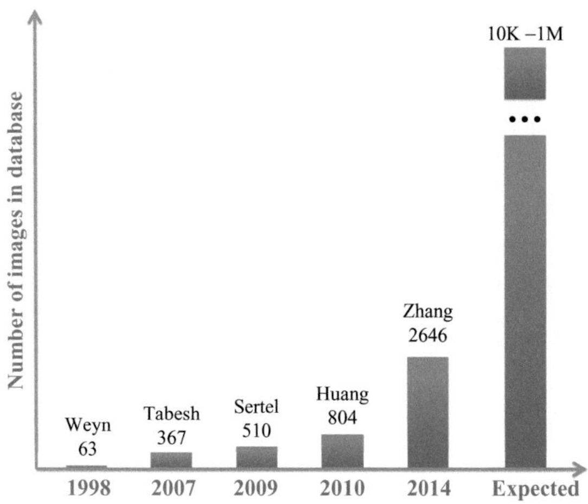

FIG. 8.1

The number of histopathological images analyzed by Weyn et al. (1998), Tabesh et al. (2007), Sertel et al. (2009), Huang and Lai (2010), and in our preliminary work (Zhang et al., 2014a). The expected size is 10K to one million. (The *x*-axis represents the years, and the *y*-axis denotes the number of images.)

full-resolution, real-world data, most methods in this area can only handle relatively small data sets (eg, several hundred images). For the problem of histopathological image analysis, Fig. 8.1 shows the capacity, in terms of number of database images processed, for recent approaches compared to the expected size for a clinically relevant database. The development of large-scale medical image analysis algorithms has lagged greatly behind the increasing quality (and complexity) of medical images and the imaging modalities themselves. These drawbacks limit the effectiveness of current image retrieval systems in research and industrial settings for mining the ever-growing number of medical images stored digitally.

Considering these factors, there is an urgent need to develop an innovative, integrated framework enabling robust and timely measurement, analysis, and characterization of such databases. Efforts in web-scale computer vision and multimedia databases have hinted at the promise of large-scale, data-driven methods for robust tagging, fine-grained object classification. In medical imaging informatics, the ever-increasing amount of medical images also provides a foundation for novel methods of semantic analysis. Transforming these raw images into a quantifiable, symbolic form will facilitate indexing and retrieval, and potentially lead to new avenues of knowledge discovery and decision support. The goal of this chapter is to introduce recent progress in large-scale visual data mining and information retrieval methods for knowledge discovery in potentially massive databases of medical images, particularly in the domain of histopathological image analysis.

8.2 RELATED WORK

Information retrieval in medical images has been widely investigated in this community. For example, Comaniciu et al. (1999) proposed a content-based image-retrieval system that supports decision making in clinical pathology, in which a central module and fast color segmenter are used to extract features such as shape, area, and texture of the nucleus. System performance was assessed through a 10-fold cross-validated classification and compared with that of a human expert on a database containing 261 digitized specimens. Dy et al. (2003) described a new hierarchical approach of CBIR based on multiple feature sets and a two-step approach. The query image is classified into different classes with best discriminative features between the classes. Then similar images are searched in the predicted class with the features customized to distinguish subclasses. El-Naqa et al. (2004) proposed a hierarchical learning approach that consists of a cascade of a binary classifier and a regression module to optimize retrieval effectiveness and efficiency. They applied this to retrieve digital mammograms and evaluated it on 76 mammograms. Greenspan and Pinhas (2007) proposed a CBIR system that consists of a continuous and probabilistic image-representation scheme. It uses information-theoretic image matching via the Kullback-Leibler (KL) measure to match and categorize X-ray images by body region. Song et al. (2011) designed a hierarchical spatial matching-based image-retrieval method using spatial pyramid matching to effectively extract and represent the spatial context of pathological tissues. CBIR has also been employed for histopathological image analysis. For example, Schnorrenberg et al. (2000) extended the biopsy analysis support system to include indexing and content-based retrieval of biopsy slide images. A database containing 57 breast cancer cases was used for evaluation. Zheng et al. (2003) designed a CBIR system to retrieve images and their associated annotations from a networked microscopic pathology image database based on four types of image features. Akakin and Gurcan (2012) proposed a CBIR system using the multitiered approach to classify and retrieve microscopic images. It enables both multi-image query and slide-level image retrieval in order to protect the semantic consistency among the retrieved images. Foran et al. (2011) designed a CBIR system named ImageMiner for comparative analysis of tissue microarrays by harnessing the benefits of high-performance computing and grid technology.

As emphasized in Zhou et al. (2008), scalability is the key factor in CBIR for medical image analysis. In fact, with the ever-increasing amount of annotated medical data, large-scale, data-driven methods provide the promise of bridging the semantic gap between images and diagnoses. However, the development of large-scale medical image analysis algorithms has lagged greatly behind the increasing quality and complexity of medical images. Specifically, owing to the difficulties in developing scalable CBIR systems for large-scale data sets, most previous systems have been tested on a relatively small number of cases. With the goal of comparing CBIR methods on a larger scale, ImageCLEF and VISCERAL provide benchmarks for medical image retrieval tasks (Müller et al., 2005; Langs et al., 2013; Hanbury et al., 2013). Recently, hashing methods have been intensively investigated in the machine learning and computer vision community for large-scale image retrieval (Wang et al., 2015). They enable fast approximated nearest neighbors (ANN) search to deal with the scalability issue. For example, locality sensitive hashing (LSH) (Andoni and Indyk, 2006) uses random projections to map data to binary codes, resulting in highly compact binary codes and enabling efficient comparison within a large database using the Hamming distance. Anchor graph hashing (AGH) (Liu et al., 2011) has been proposed to use neighborhood graphs which reveal the underlying manifold of features, leading to a high search accuracy. Recent research has focused on data-dependent hash functions, such as spectral graph partitioning and hashing (Weiss et al., 2009) and supervised hashing with kernels (Liu et al., 2012) incorporating the pairwise semantic similarity and dissimilarity constraints from labeled data. These hashing methods have also been employed to solve the dimensionality problem in medical image analysis. In particular, Zhang et al. (2014a, 2015c) built a scalable image-retrieval framework based on the supervised hashing technique and validated its performance on several thousand histopathological images acquired from breast microscopic tissues. It leverages a small amount of supervised information in learning to compress high-dimensional image feature vectors into only tens of binary bits with the informative signatures preserved. The supervised information is employed to bridge the semantic gap between low-level image features and high-level diagnostic information, which is critical to medical image analysis. Instead of hashing and searching the whole image, another approach is to segment all cells from histopathological images and conduct large-scale retrieval among cell images (Zhang et al., 2015d,e,f). This enables cell-level and fine-grained analysis, achieving high accuracy. In addition to using a single feature, it is also possible to fuse multiple types of features in a hashing framework to improve the accuracy of medical image retrieval. Specifically, a composite AGH algorithm (Liu et al., 2011) has been employed for retrieving medical images (Zhang et al., 2014b; Liu et al., 2014), for example, retrieving lung microscopic tissue images for the differentiation of adenocarcinoma and squamous carcinoma. Besides hashing-based methods, vocabulary trees have also been intensively investigated (Nister and Stewenius, 2006) and employed for medical image analysis (Jiang et al., 2015b, 2014).

In this chapter, we introduce technical details of hashing-based large-scale image retrieval for computer-aided diagnosis, with the use case of histopathological image analysis.

8.3 SUPERVISED HASHING FOR LARGE-SCALE RETRIEVAL

8.3.1 OVERVIEW OF SCALABLE IMAGE RETRIEVAL FRAMEWORK

Fig. 8.2 shows a framework for the scalable image retrieval-based diagnosis system. It includes offline learning and run-time search. During the offline learning, we first extract high-dimensional visual features from digitized histopathological images.

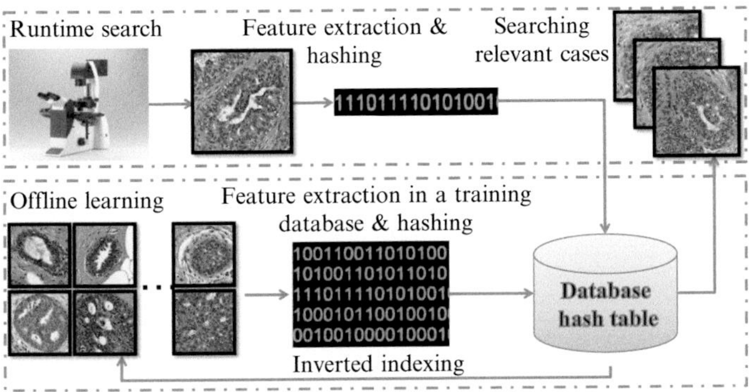

FIG. 8.2

Framework of our large-scale image retrieval system (Zhang et al., 2014a).

These features model texture and appearance information based on SIFT (Lowe, 2004) and are quantized with a bag-of-words (Sivic and Zisserman, 2003). The SIFT descriptor is an effective local texture feature that uses the difference of Gaussian (DoG) detection result and considers the gradient of pixels around the detected region. It can provide an informative description of cell appearance and is robust to subtle changes in staining color. It has been used in both general computer vision tasks and histopathological image analysis.

Although these features can be used directly to measure the similarity among images, computational efficiency is an issue, especially when searching in a large database (eg, exhaustively searching k-nearest neighbors (kNN)). Therefore we employ a hashing method to compress these features into binary codes with tens of bits. Such short binary features allow easy mapping into a hash table for real-time search. Each feature is then linked to the corresponding training images using an inverted index. During a run-time query, high-dimensional features are extracted from the query image and then projected to the binary codes. With a hash table, searching for nearest neighbors can be achieved in a constant time, irrespective of the number of images. The retrieved images (via inverted indices of nearest neighbors) can be used to interpret this new case or for decision support based on majority voting.

8.3.2 KERNELIZED AND SUPERVISED HASHING

In this section, we introduce the key module for histopathological image retrieval, a kernelized and supervised hashing method.

8.3.2.1 Hashing method

Given a set of image feature vectors $\mathcal{X} = \{\boldsymbol{x}_1, \ldots, \boldsymbol{x}_n\} \subset \mathbb{R}^d$ (in our case, $\boldsymbol{x}_i$ is the high-dimensional texture feature extracted from the ith histopathological image), a hashing method aims to find a group of proper hash functions $h: \mathbb{R}^d \mapsto \{1, -1\}^1$, each

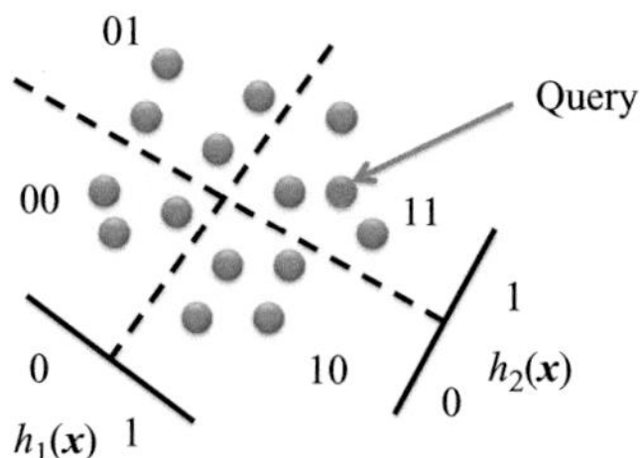

FIG. 8.3

Visualization of desirable hash functions as a hyperplane.

of which generates a single hash bit to preserve the similarity of original features. Searching kNN using tens of bits is significantly faster than traditional methods (eg, Euclidean distance-based brute-force search), owing to constant-time hash-table lookups and/or efficient Hamming distance computation. Note that hashing methods are different from dimensionality-reduction techniques, since a fundamental requirement of hashing is to map similar feature vectors into the same bucket with high probability. Fig. 8.3 visualizes desirable hash functions as a hyperplane to separate higher-dimensional features. Therefore hashing methods need to ensure that the generated hash bits have balanced and uncorrelated bit distributions, which leads to maximum information at each single bit and minimum redundancy among all bits.

8.3.2.2 Kernelized hashing

Kernel methods can handle practical data that are mostly linearly inseparable. For histopathological images, linear inseparability is an important constraint that needs to be taken into account when building hashing methods. Therefore kernel functions should be considered in hashing methods, $h = sgn(f(x))$ (Kulis and Grauman, 2012), to map the feature vectors into higher-dimensional space. A kernel function is denoted as $\kappa: \mathbb{R}^d \times \mathbb{R}^d \mapsto \mathbb{R}$. The prediction function $f: \mathbb{R}^d \mapsto \mathbb{R}$ with kernel κ plugged in is defined as

$$f(\boldsymbol{x}) = \sum_{j=1}^{m} \kappa(\boldsymbol{x}_{(j)}, \boldsymbol{x}) a_j - b, \tag{8.1}$$

where $\boldsymbol{x}_{(1)}, \ldots, \boldsymbol{x}_{(m)}$ are m ($m \ll n$) feature vectors randomly selected from $\mathcal{X}$, $a_j \in \mathbb{R}$ is the coefficient, and $b \in \mathbb{R}$ is the bias.

The bits generated from hash functions h using f aim to keep as much information as possible, so the hash functions should produce a balanced distribution of bits, that is, $\sum_{i=1}^{n} h(\boldsymbol{x}_i) = 0$. Therefore b is set as the median of $\{\sum_{j=1}^{m} \kappa(\boldsymbol{x}_{(j)}, \boldsymbol{x}_i) a_j\}_{i=1}^{n}$, which is usually approximated by the mean. Adding this constraint into Eq. 8.1, we obtain

$$f(\boldsymbol{x}) = \sum_{j=1}^{m} \left(\kappa(\boldsymbol{x}_{(j)}, \boldsymbol{x}) - \frac{1}{n} \sum_{i=1}^{n} \kappa(\boldsymbol{x}_{(j)}, \boldsymbol{x}_i) \right) a_j = \boldsymbol{a}^{\mathrm{T}} \bar{\boldsymbol{k}}(\boldsymbol{x}), \tag{8.2}$$

where $\boldsymbol{a} = [a_1, a_2, \ldots, a_m]^{\mathrm{T}}$. $\bar{\boldsymbol{k}} : \mathbb{R}^d \mapsto \mathbb{R}^m$ is $\bar{\boldsymbol{k}}(\boldsymbol{x}) = \left[\kappa(\boldsymbol{x}_{(1)}, \boldsymbol{x}) - \mu_1, \ldots, \kappa(\boldsymbol{x}_{(m)}, \boldsymbol{x}) - \mu_m\right]^{\mathrm{T}}$, in which $\mu_j = \sum_{i=1}^{n} \kappa(\boldsymbol{x}_{(j)}, \boldsymbol{x}_i)/n$.

The vector $\boldsymbol{a}$ is the most important factor that determines hash functions. In traditional kernelized hashing methods, $\boldsymbol{a}$ is defined as a random direction drawn from a Gaussian distribution (Kulis and Grauman, 2012), without using any other prior knowledge (ie, no semantic information). This scheme works well for natural images, especially scenes, because of large differences in their appearance. However, such differences are very subtle in histopathological images. For example, identifying subtle differences between benign and actionable categories may require characterizing cytoplasmic texture or nuclear appearance. This subtlety motivates us to leverage supervised information to design discriminative hash functions that are suitable for histopathological image retrieval.

8.3.2.3 Supervised hashing

Intuitively, hashing methods minimize the Hamming distance of "neighboring" image pairs (eg, close in terms of the Euclidean distance in the raw feature space). "Neighboring" in our case is defined by its semantic meaning, that is, whether the two images belong to the same category or not. Therefore supervised information can be naturally encoded as similar and dissimilar pairs. Specifically, we assign the label 1 to image pairs when both are benign or actionable, and -1 to pairs when one is benign and the other is actionable (as shown in Fig. 8.4). Then, l ($l \ll n$) feature vectors are randomly selected from $\mathcal{X}$ to build the label matrix S. Note that we need to provide labels for only a small number of image pairs. Therefore labeled data are explicitly constrained by both semantic information and visual similarities, whereas unlabeled data are mainly constrained by visual similarities and implicitly affected by labeled data.

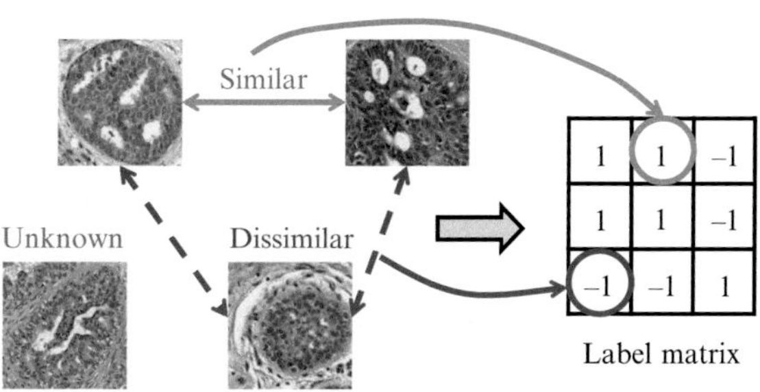

FIG. 8.4

Supervised information is encoded in the label matrix S.

Using this supervision scheme to bridge the semantic gap, r hash functions $h_k(\boldsymbol{x})_{k=1}^r$ are then designed to generate r discriminative hash bits based on Hamming distances. However, direct optimization of the following Hamming distances $\mathcal{D}_h(\boldsymbol{x}_i, \boldsymbol{x}_j) = |\{k|h_k(\boldsymbol{x}_i) \neq h_k(\boldsymbol{x}_j), 1 \leq k \leq r\}|$ is nontrivial. Therefore code inner products can be used to simplify the optimization process. As shown in Liu et al. (2012), a Hamming distance and a code inner product are actually equivalent:

$$code_r(\boldsymbol{x}_i) \circ code_r(\boldsymbol{x}_j) = r - 2\mathcal{D}_h(\boldsymbol{x}_i, \boldsymbol{x}_j), \tag{8.3}$$

where $code_r(\boldsymbol{x})$ are r-bit hash codes and the symbol $\circ$ is the code inner product.

Therefore the objective function $\mathcal{Q}$ to the binary codes H_l is defined as

$$\min_{H_l \in \{1,-1\}^{l\times r}} \mathcal{Q} = \left\| \frac{1}{r} H_l H_l^{\mathrm{T}} - S \right\|_F^2, \tag{8.4}$$

where $H_l = \begin{bmatrix} h_1(\boldsymbol{x}_1), \ldots, h_r(\boldsymbol{x}_1) \\ \cdots\cdots \\ h_1(\boldsymbol{x}_l), \ldots, h_r(\boldsymbol{x}_l) \end{bmatrix}$ is the code matrix of the labeled data $\mathcal{X}_l$ and S is a label matrix with 1 for similar pairs and -1 for dissimilar pairs. $\|.\|_F$ denotes the Frobenius norm. Define $\bar{K}_l$ as $[\bar{\boldsymbol{k}}(\boldsymbol{x}_1), \ldots, \bar{\boldsymbol{k}}(\boldsymbol{x}_l)]^{\mathrm{T}} \in \mathbb{R}^{l\times m}$, $\bar{\boldsymbol{k}}(\boldsymbol{x}_i)$. The inner product of code matrix H_l can be represented as $H_l H_l^{\mathrm{T}} = \sum_{k=1}^r sgn(\bar{K}_l \boldsymbol{a}_k)(sgn(\bar{K}_l \boldsymbol{a}_k))^{\mathrm{T}}$ for binarization. Therefore the new objective function $\mathcal{Q}$ that offers a clearer connection and easier access to the model parameter $\boldsymbol{a}_k$ is

$$\min_{\boldsymbol{a}_k} \mathcal{Q}(\boldsymbol{a}_k) = \left\| \sum_{k=1}^{r} sgn(\bar{K}_l \boldsymbol{a}_k)(sgn(\bar{K}_l \boldsymbol{a}_k))^{\mathrm{T}} - rS \right\|_F^2. \tag{8.5}$$

This can be optimized using (1) spectral relaxation (Weiss et al., 2009) to drop the sign functions and hence convexify the object function, or (2) sigmoid smoothing to replace *sgn*() with the sigmoid-shaped function. In our implementation, we employ the first strategy to efficiently obtain a solution as the initialization, and use the second strategy to produce an accurate solution.

8.4 RESULTS

Breast-tissue specimens available for this study were collected on a retrospective basis from the IU Health Pathology Lab (IUHPL) according to the protocol approved by the Institutional Review Board (IRB) for this study. All the slides were imaged using a ScanScope digitizer (Aperio, Vista, CA) available in the tissue archival service at IUHPL. 3121 images (around 2250K pixels) were sampled from 657 larger region-of-interest images (eg, 5K×7K) of microscopic breast tissue, which were gathered from 116 patients. Fifty-three of these patients were labeled as benign (usual ductal hyperplasia (UDH)) and 63 as actionable (atypical ductal hyperplasia

(ADH) and ductal carcinoma in situ (DCIS)), based on the majority diagnosis of nine board-certified pathologists. To demonstrate the efficiency of our method, one-fourth of all patients in each category were randomly selected as the test set and the remainder used for training. Note that each patient may have a different number of images. Therefore the number of testing images is not fixed. The approximate number is about 700–900 in each testing process. All the experiments were conducted on a 3.40-GHz CPU with four cores and 16GB RAM, in a MATLAB implementation.

In each image, 1500–2000 SIFT descriptors were extracted from key points detected by DoG (Lowe, 2004). These descriptors were quantized into sets of cluster centers using bag-of-words, in which the feature dimension equals the number of clusters. Specifically, we quantize them into high-dimensional feature vectors of length 10,000, to maximally utilize these millions of cell-level texture features. We provide both qualitative and quantitative evaluations for our proposed framework on two tasks, image classification (ie, benign vs. actionable category) and image retrieval, in terms of accuracy and computational efficiency.

In our system, classification is achieved using the majority vote of the top images retrieved by hashing. We compare our approach with various classifiers that have been widely used in systems for histopathological image analysis. Specifically, kNN has often been used as the baseline in analyzing histopathological images (Tabesh et al., 2007; Yang et al., 2009), owing to its simplicity and proved lower bound, despite the inefficiency in large-scale databases. The Bayesian method is another solution to ensemble statistics of all extracted features and minimize the classification metric, which shows its efficacy in classifying histopathological images (Comaniciu et al., 1999). Boosting methods are always employed to combine multiple weak classifiers for higher accuracy (Yang et al., 2009; Doyle et al., 2012). A support vector machine (SVM) with a nonlinear kernel is commonly used in histopathological images because of its efficiency and the ability to handle linearly inseparable cases (Tuzel et al., 2007; Caicedo et al., 2009; Nguyen et al., 2010; Huang and Lai, 2010). For fair comparison, all parameters of these compared methods were optimized by cross-validation.

In addition, we also compared our proposed method with several dimensionality-reduction algorithms in terms of classification accuracy. Principal component analysis (PCA) has been widely used in this area to preserve variance of original features (Sertel et al., 2009). Graph embedding is a nonlinear dimensionality-reduction algorithm that performs well in grading of lymphocytic infiltration in HER2+ breast cancer histopathology (Basavanhally et al., 2010). Since we use supervised information in generating hash functions, a supervised dimensionality-reduction algorithm, neighborhood components analysis (NCA) (Goldberger et al., 2004), was also chosen for our experimental comparisons.

Fig. 8.5 shows the quantitative results for the classification accuracy. Most methods achieve better accuracy with higher-dimensional features. This is very intuitive, as finer quantization of SIFT features usually provides richer information. In particular, since the SIFT interest points cover most nuclear regions in images, fine quantization (ie, high-dimensional features) indicates analysis on a small scale.

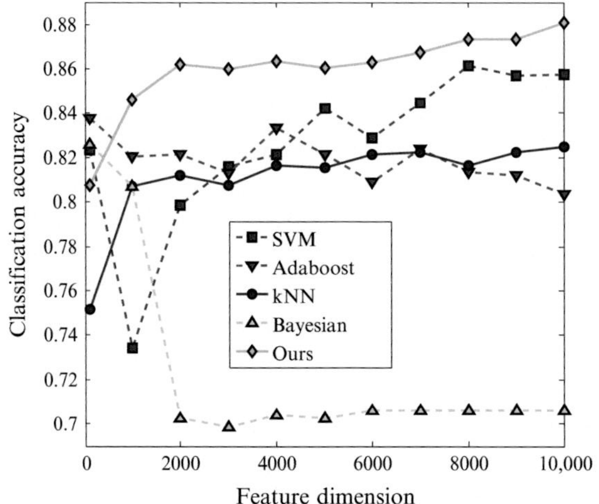

FIG. 8.5

Comparison of classification accuracy with different dimensions of features (from 100 to 10,000).

Exceptions are the Adaboost and Bayesian methods, whose accuracy drops when the feature dimensions increase. This indicates that high-dimensional features do not guarantee the improvement of accuracy. An important factor is the proper utilization of such information. For example, Adaboost is essentially a feature-selection method that chooses only an effective subset of features for the classification. Therefore it may lose important information, especially in high-dimensional space, resulting in accuracy worse than that of our hashing method. Our method is also generally better than kNN and its variations, owing to the semantic information (ie, labels of similar and dissimilar pairs in hashing) that bridges the semantic gap between images and diagnoses. Note that our hashing method needs only a small amount of supervision—in this case, similar or dissimilar pairs of 40% images. This is generally less than the supervised information required by SVM in the training stage. It compares favorably to all other methods when the feature dimension is larger than 1000. The overall classification accuracy is 88.1% for 10,000-dimensional features, 2–18% better than the other methods.

Fig. 8.6 compares the computational efficiency of these methods. With increasing dimensionality the running time of some compared methods increases dramatically. When feature dimensionality reaches 10,000, kNN needs 16 s to classify all query images, and Adaboost needs 5 s. SVM, dimensionality-reduction methods, and the proposed method are much faster. However, the running time for SVM increases with the feature dimensionality, as shown in the expanded view of Fig. 8.6. In contrast, PCA, graph embedding, NCA, and our method achieve constant running time in this data set owing to the fixed size of features after compression. Compared to other

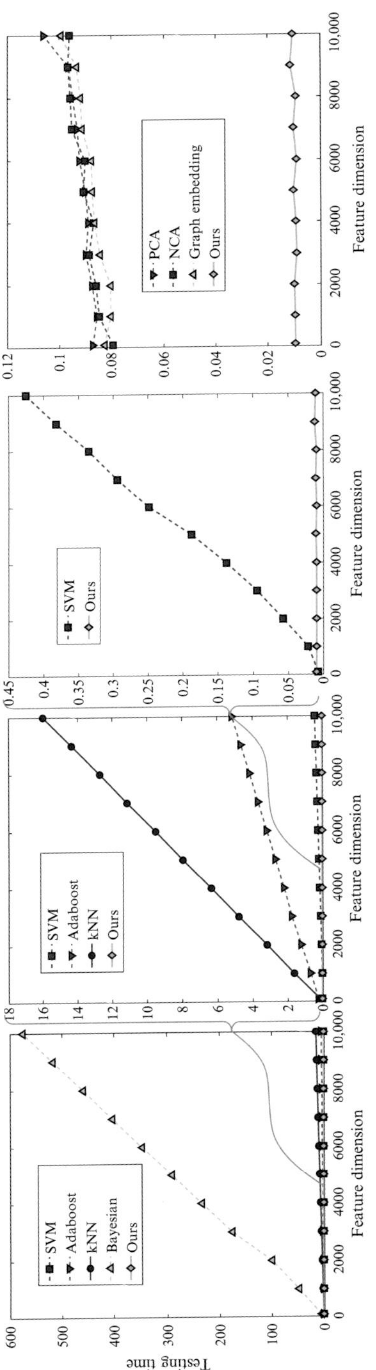

FIG. 8.6

Comparison of the classification running time (seconds) with different dimensions of features, which means the average time of classifying hundreds of test images.

dimensionality-reduction methods, our approach is about 10 times faster because of the efficient comparison among binary codes. In addition, the running time of all kNN-based methods increases with the number of images in a data set, as exhaustive search is needed, while hashing-based methods can achieve $\mathcal{O}(1)$ efficiency using a hash table. To summarize, the average running time of our method is only 0.01 s for all testing images, which is 40 times faster than SVM and 1500 times faster than kNN.

We have also conducted experiments on image retrieval using 10,000-dimensional features. The retrieval precision, evaluated at a given cut-off rank and considering only the topmost results, is reported in Table 8.1, along with the query time and memory cost. The results are quite consistent with the image classification. The mean precision of the hashing method is around 83%, and the standard deviation is 1.1%, which is much better than PCA (Sertel et al., 2009), graph embedding (Basavanhally et al., 2010), and NCA (Goldberger et al., 2004). In most cases, the precision of our method is at least 6% better than the others, except for NCA. Our method is around 3.5% better than NCA on benign cases. To demonstrate statistical significance, we perform t-test for the precision obtained by NCA and by the proposed method on benign cases, under the null hypothesis using a significance level of 0.05. The p-values are found to be 3.6×10^{-6}, 3.2×10^{-6}, and 5.7×10^{-6} at the range of the top 10, 20, and 30 retrievals, respectively, demonstrating that precision values achieved by the proposed technique are indeed significantly better than NCA for the benign cases. In addition, our method is around 14% better than NCA in the actionable cases, resulting in much higher average precision. In fact, most traditional methods produce results as highly unbalanced NCA, that is, the retrieval precision of the benign category is much higher than that of the actionable one. In contrast, our method does not have this problem, owing to the supervised information and the optimization for balanced hash bits. Our framework is also computationally more efficient than traditional methods. The query time of our hashing method is 1000 times faster than kNN and 10 times faster than other dimensionality-reduction methods. Note that our method takes a constant time when using the hash table, independent of the number of feature dimensions and the number of samples. Furthermore, the memory cost is also considerably reduced (10,000 times less than that of kNN). Therefore this method is more applicable to large-scale databases (millions of images) than are other methods.

Fig. 8.7 shows our image-retrieval results. The top five relevant images are listed for each query image. The differences between certain images in different categories are very subtle. Our accurate results demonstrate the efficacy of the proposed method. Specifically, the features capturing local texture and appearance are very robust to various image sizes, cell distributions, and occlusions by the blood. The supervised information also improves the retrieval precision by correlating binary code with diagnosis information. These retrieved images are clinically relevant in potential (ie, retrieved images belong to the same category as the query image) and thus can be useful for decision support.

Table 8.1 Comparison of Retrieval Precision for the Top 10, 20, and 30 Results (Denoted as P@10, P@20, and P@30, Respectively), Along With the Memory Cost of Training Data and Query Time of All Test Images

	kNN		PCA		NCA		Graph Embedding		Ours	
	Benign	**Actionable**	**Benign**	**Actionable**	**Benign**	**Actionable**	**Benign**	**Actionable**	**Benign**	**Actionable**
P@10	0.779	0.687	0.762	0.705	0.799	0.697	0.672	0.487	**0.836**	**0.830**
P@20	0.773	0.653	0.758	0.681	0.800	0.689	0.673	0.486	**0.839**	**0.829**
P@30	0.770	0.631	0.755	0.667	0.800	0.685	0.670	0.480	**0.837**	**0.833**
STD	0.024		0.028		0.020		0.012		0.011	
Time (s)	15.77		10.07		10.04		10.03		<0.01	
Memory	134.58MB		0.65MB		0.65MB		0.65MB		0.01MB	

Both mean values and the standard deviation (STD) of 20 experiments are reported. The best precision in each row for benign and actionable categories are highlighted in bold.

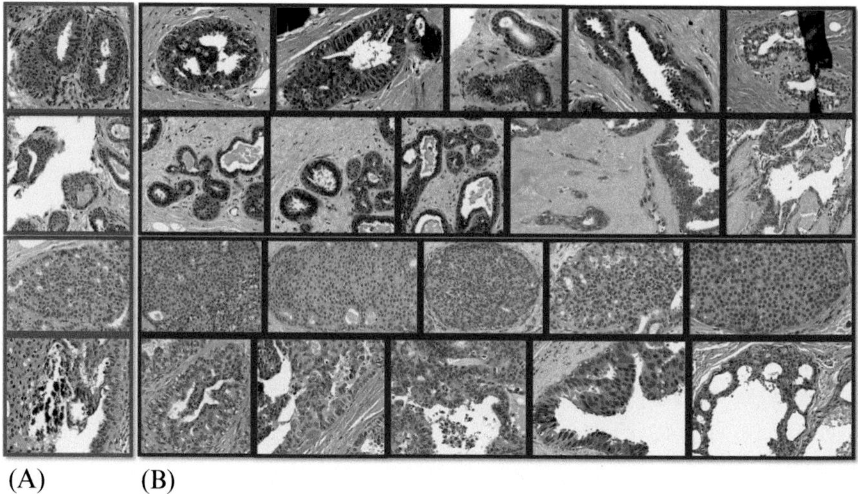

FIG. 8.7

Four examples of our image retrieval (query marked in red and in the first column, and retrieved images marked in blue). The first two rows are benign; the last two rows are actionable. (A) Query. (B) Retrieved Images

8.5 DISCUSSION AND FUTURE WORK

In this chapter, we introduce a *large-scale image-retrieval framework* for medical image analysis. We employ hashing to achieve efficient image retrieval and present an improved kernelized and supervised hashing approach for real-time image retrieval. Note that even though we use histopathological image analysis as a use case, this framework is applicable to other problems in this area. The potential applications of our framework include image-guided diagnosis, decision support, education, and efficient data management. For example, the efficient retrieval of relevant cases from medical databases will provide usable tools to assist clinicians' diagnoses and support efficient medical image data management, such as picture archiving and communication systems (PACS). More specifically, it provides efficient reasoning in large-scale medical image databases using techniques for scalable and accurate medical image retrieval in potentially massive databases to provide real-time querying for the most relevant and consistent instances (eg, similar morphological profiles) for decision support. In addition to the resulting tools for medical image processing, disease detection and information retrieval, their use will allow for the exploration of structured image databases, in medical education and training.

An important extension to the existing framework is to integrate multiple features for accurate retrieval and diagnosis. For example, accurate analysis of histopathological images requires examination of cell-level information for accurate diagnosis, including individual cells (eg, appearance (Caicedo et al., 2009; Zhang et al., 2015c) and shapes (Dundar et al., 2011)) and architecture of tissue (eg, topology and layout of all cells (Basavanhally et al., 2010)). These features cover both local and holistic information, all benefitting the diagnostic accuracy of histopathological images. Therefore the complementary descriptive capability of local and holistic features motivates us to integrate their strengths to yield more satisfactory results. However, their characteristics, algorithmic procedures, and representations can be dramatically different, making them nontrivial to fuse. For example, architecture features (Basavanhally et al., 2010) are represented as a low-dimensional vector of statistics, while local features can be represented as high-dimensional bag-of-words (BoW) (Sivic and Zisserman, 2003) and compressed as binary codes to improve the efficiency (Liu et al., 2012; Zhang et al., 2015c). To tackle this feature fusion problem, we have focused on the rank-level fusion of local and holistic features for the image-guided diagnosis of breast cancer, that is, differentiation of the benign and actionable cases. In particular, we conduct image retrieval to discover clinically relevant instances from an image database, which can be used to infer and classify the new data. Given image ranks (ie, retrieval results) obtained from different features, a data-driven and graph-based method (Zhang et al., 2012, 2015a,b) is employed for accurate, robust, and efficient fusion, by evaluating the quality of each rank online. Fig. 8.8 shows the overview of this graph-based feature fusion for image retrieval. This provides an effective solution for the fusion of heterogeneous information in the domain of histopathological image analysis, and the preliminary experimental results demonstrate the accuracy and efficiency of our framework. In addition to rank-based fusion, it is also possible to integrate multiple features in the hashing framework, either at the distance level or at the kernel level (Jiang et al., 2015a).

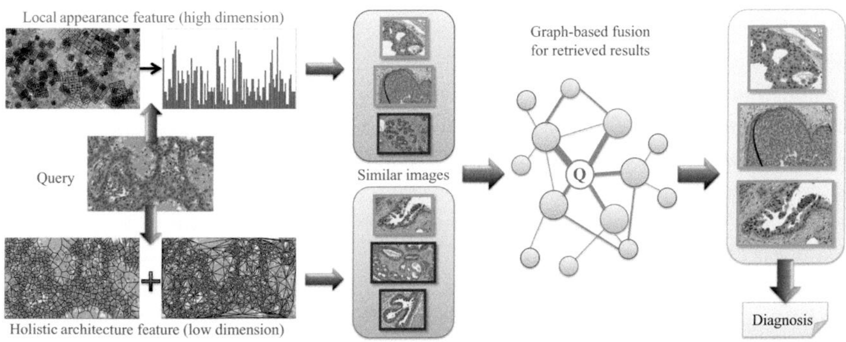

FIG. 8.8

Overview of the graph-based feature fusion for image retrieval (Zhang et al., 2015b). Both holistic architecture feature and local appearance feature are extracted and employed for image retrieval. The retrieval results are fused via the graph-based framework to improve the accuracy. Note that majority voting does not work in this example, since two ranks have no intersection.

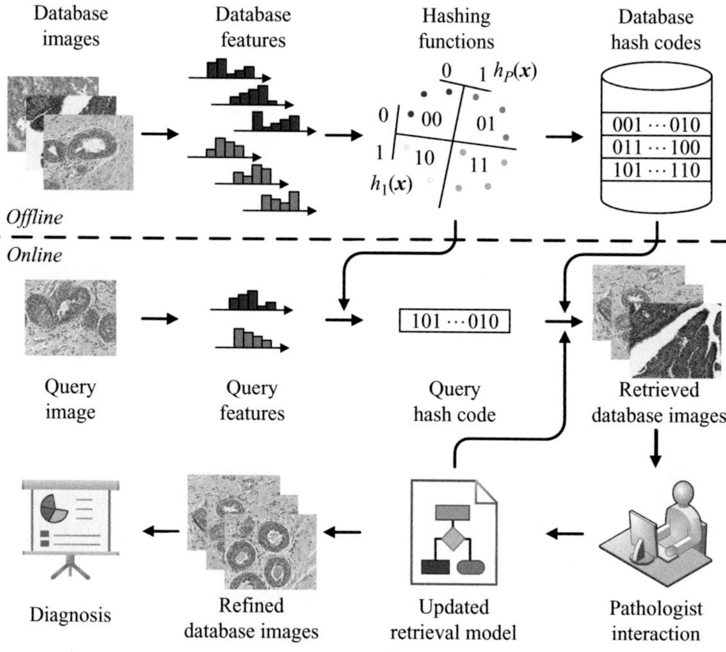

FIG. 8.9

Example of incorporating domain knowledge from pathologists into the loop of hashing model updating.

In the future, we will focus on the intelligent interaction and visualization that integrates expert feedback and automated algorithms for efficient decision making and provides a comprehensive understanding of the query results and supports semantic interaction functions. Interaction and visualization is another important yet challenging tool for effective computer-aided diagnosis and medical data mining. To achieve our ultimate goal of assisting efficient decision making and reasoning using medical image databases, we plan to incorporate users in the loop to incorporate the domain knowledge of experts, as shown in Fig. 8.9. While the automated methods are designed to process millions of images, human users can only reasonably work with much fewer images at a time. The main challenge will be bridging the gap between the large-scale automated algorithms and the knowledge that domain experts can provide, but at much smaller scales. We plan to design a visual analysis system with a set of feature-based query, visualization, comparison, and learning methods for revealing the relevant image features and relationships. This system will support the analysis of the retrieved relevant image sets, extracted image features, and feature similarities among the retrieved image sets, and will provide efficient interaction methods to enhance the query algorithms and obtain finer-tuned results. To summarize, the components of large-scale retrieval and intelligent interaction will be coordinated for the purpose of scalable and interactive mining to provide a semantic interface between users and data through the language of feature similarities. The overall framework will be designed to address the challenges of both *scalable and interactive* mining within medical imaging informatics, and each aspect of the design and development will be driven by the goals of efficiency, robustness, and effective integration of user input.

REFERENCES

Akakin, H.C., Gurcan, M.N., 2012. Content-based microscopic image retrieval system for multi-image queries. IEEE Trans. Inform. Technol. Biomed. 16 (4), 758–769.

Andoni, A., Indyk, P., 2006. Near-optimal hashing algorithms for approximate nearest neighbor in high dimensions. In: IEEE Symposium on Foundations of Computer Science (FOCS), Berkeley, CA.

Basavanhally, A.N., Ganesan, S., Agner, S., Monaco, J.P., Feldman, M.D., Tomaszewski, J.E., Bhanot, G., Madabhushi, A., 2010. Computerized image-based detection and grading of lymphocytic infiltration in HER2+ breast cancer histopathology. IEEE Trans. Biomed. Eng. 57 (3), 642–653.

Caicedo, J.C., Cruz, A., Gonzalez, F.A., 2009. Histopathology image classification using bag of features and kernel functions. In: Artificial Intelligence in Medicine. Springer, New York, pp. 126–135.

Comaniciu, D., Meer, P., Foran, D.J., 1999. Image-guided decision support system for pathology. Mach. Vis. Appl. 11 (4), 213–224.

Doyle, S., Feldman, M., Tomaszewski, J., Madabhushi, A., 2012. A boosted Bayesian multiresolution classifier for prostate cancer detection from digitized needle biopsies. IEEE Trans. Biomed. Eng. 59 (5), 1205–1218.

Dundar, M.M., Badve, S., Bilgin, G., Raykar, V., Jain, R., Sertel, O., Gurcan, M.N., 2011. Computerized classification of intraductal breast lesions using histopathological images. IEEE Trans. Biomed. Eng. 58 (7), 1977–1984.

Dy, J.G., Brodley, C.E., Kak, A., Broderick, L.S., Aisen, A.M., 2003. Unsupervised feature selection applied to content-based retrieval of lung images. IEEE Trans. Pattern Anal. Mach. Intell. 25 (3), 373–378.

El-Naqa, I., Yang, Y., Galatsanos, N.P., Nishikawa, R.M., Wernick, M.N., 2004. A similarity learning approach to content-based image retrieval: application to digital mammography. IEEE Trans. Med. Imaging 23 (10), 1233–1244.

Foran, D.J., Yang, L., et al., 2011. Imageminer: a software system for comparative analysis of tissue microarrays using content-based image retrieval, high-performance computing, and grid technology. J. Am. Med. Inform. Assoc. 18 (4), 403–415.

Goldberger, J., Hinton, G.E., Roweis, S.T., Salakhutdinov, R., 2004. Neighbourhood components analysis. In: Advances in Neural Information Processing Systems, pp. 513–520.

Greenspan, H., Pinhas, A.T., 2007. Medical image categorization and retrieval for PACS using the GMM-KL framework. IEEE Trans. Inform. Technol. Biomed. 11 (2), 190–202.

Hanbury, A., Müller, H., Langs, G., Menze, B.H., 2013. Cloud-based evaluation framework for big data. In: FIA Book 2013, LNCS volume. Springer, New York.

Huang, P.W., Lai, Y.H., 2010. Effective segmentation and classification for HCC biopsy images. Pattern Recogn. 43 (4), 1550–1563.

Jiang, M., Zhang, S., Liu, J., Shen, T., Metaxas, D.N., 2014. Computer-aided diagnosis of mammographic masses using vocabulary tree-based image retrieval. In: IEEE International Symposium on Biomedical Imaging. ISBI, pp. 1123–1126.

Jiang, M., Zhang, S., Huang, J., Yang, L., Metaxas, D.N., 2015a, Joint kernel-based supervised hashing for scalable histopathological image analysis. In: Medical Image Computing and Computer-Assisted Intervention. Springer, New York, pp. 366–373.

Jiang, M., Zhang, S., Li, H., Metaxas, D.N., 2015b. Computer-aided diagnosis of mammographic masses using scalable image retrieval. IEEE Trans. Biomed. Eng. 62 (2), 783–792.

Kulis, B., Grauman, K., 2012. Kernelized locality-sensitive hashing. IEEE Trans. Pattern Anal. Mach. Intell. 34 (6), 1092–1104.

Langs, G., Müller, H., Menze, B.H., Hanbury, A., 2013. VISCERAL: towards large data in medical imaging—challenges and directions. In: MCBR-CDS MICCAI workshop, LNCS vol. 7723. Springer, New York.

Liu, W., Wang, J., Kumar, S., Chang, S.F., 2011. Hashing with graphs. In: Proceedings of the 28th International Conference on Machine Learning (ICML-11), pp. 1–8.

Liu, W., Wang, J., Ji, R., Jiang, Y.G., Chang, S.F., 2012. Supervised hashing with kernels. In: IEEE International Conference on Computer Vision and Pattern Recognition, pp. 2074–2081.

Liu, J., Zhang, S., Liu, W., Zhang, X., Metaxas, D.N., 2014. Scalable mammogram retrieval using anchor graph hashing. In: IEEE International Symposium on Biomedical Imaging. ISBI, pp. 898–901.

Lowe, D.G., 2004. Distinctive image features from scale-invariant keypoints. Int. J. Comput. Vis. 60 (2), 91–110.

Müller, H., Geissbühler, A., Ruch, P., 2005. ImageCLEF 2004: combining image and multi-lingual search for medical image retrieval. In: Multilingual Information Access for Text, Speech and Images. Springer, New York, pp. 718–727.

Nguyen, K., Jain, A.K., Allen, R.L., 2010. Automated gland segmentation and classification for Gleason grading of prostate tissue images. In: IEEE International Conference on Pattern Recognition, pp. 1497–1500.

Nister, D., Stewenius, H., 2006. Scalable recognition with a vocabulary tree. In: IEEE International Conference on Computer Vision and Pattern Recognition, vol. 2, pp. 2161–2168.

Schnorrenberg, F., Pattichis, C., Schizas, C., Kyriacou, K., 2000. Content-based retrieval of breast cancer biopsy slides. Tech. Health Care 8 (5), 291–297.

Sertel, O., Kong, J., Catalyurek, U.V., Lozanski, G., Saltz, J.H., Gurcan, M.N., 2009. Histopathological image analysis using model-based intermediate representations and color texture: follicular lymphoma grading. J. Signal Process. Syst. 55 (1–3), 169–183.

Sivic, J., Zisserman, A., 2003. Video google: a text retrieval approach to object matching in videos. In: IEEE International Conference on Computer Vision, pp. 1470–1477.

Song, Y., Cai, W., Feng, D., 2011. Hierarchical spatial matching for medical image retrieval. In: ACM International Workshop on Medical Multimedia Analysis and Retrieval, pp. 1–6.

Tabesh, A., Teverovskiy, M., Pang, H.Y., Kumar, V.P., Verbel, D., Kotsianti, A., Saidi, O., 2007. Multifeature prostate cancer diagnosis and Gleason grading of histological images. IEEE Trans. Med. Imaging 26 (10), 1366–1378.

Tuzel, O., Yang, L., Meer, P., Foran, D.J., 2007. Classification of hematologic malignancies using texton signatures. Pattern Anal. Appl. 10 (4), 277–290.

Wang, J., Liu, W., Kumar, S., Chang, S.F., 2015. Learning to hash for indexing big data—a survey. arXiv 1509.05472

Weiss, Y., Torralba, A., Fergus, R., 2009. Spectral hashing. In: Advances in Neural Information Processing Systems, pp. 1753–1760.

Weyn, B., van de Wouwer, G., van Daele, A., Scheunders, P., van Dyck, D., van Marck, E., Jacob, W., 1998. Automated breast tumor diagnosis and grading based on wavelet chromatin texture description. Cytometry 33 (1), 32–40.

Yang, L., Chen, W., Meer, P., Salaru, G., Goodell, L.A., Berstis, V., Foran, D.J., 2009. Virtual microscopy and grid-enabled decision support for large-scale analysis of imaged pathology specimens. IEEE Trans. Inform. Technol. Biomed. 13 (4), 636–644.

Zhang, S., Yang, M., Cour, T., Yu, K., Metaxas, D., 2012. Query specific fusion for image retrieval. In: Fitzgibbon, A., Lazebnik, S., Perona, P., Sato, Y., Schmid, C. (Eds.), European Conference on Computer Vision, Lecture Notes in Computer Science. Springer, Berlin, Heidelberg, pp. 660–673.

Zhang, X., Liu, W., Zhang, S., 2014a, Mining histopathological images via hashing-based scalable image retrieval. In: IEEE International Symposium on Biomedical Imaging, pp. 1111–1114.

Zhang, X., Yang, L., Liu, W., Su, H., Zhang, S., 2014b, Mining histopathological images via composite hashing and online learning. In: International Conference on Medical Image Computing and Computer Assisted Intervention. Springer, New York, pp. 479–486.

Zhang, S., Yang, M., Cour, T., Yu, K., Metaxas, D.N., 2015a. Query specific rank fusion for image retrieval. IEEE Trans. Pattern Anal. Mach. Intell. 37 (4), 803–815.

Zhang, X., Dou, H., Ju, T., Xu, J., Zhang, S., 2015b. Fusing heterogeneous features from stacked sparse autoencoder for histopathological image analysis. IEEE J. Biomed. Health Inform. PP (99), 1–1.

Zhang, X., Liu, W., Dundar, M., Badve, S., Zhang, S., 2015c, Towards large-scale histopathological image analysis: hashing-based image retrieval. IEEE Trans. Med. Imaging 34 (2), 496–506.

Zhang, X., Su, H., Yang, L., Zhang, S., 2015d, Fine-grained histopathological image analysis via robust segmentation and large-scale retrieval. In: IEEE International Conference on Computer Vision and Pattern Recognition, pp. 5361–5368.

Zhang, X., Su, H., Yang, L., Zhang, S., 2015e, Weighted hashing with multiple cues for cell-level analysis of histopathological images. In: Information Processing in Medical Imaging, pp. 303–314.

Zhang, X., Xing, F., Su, H., Yang, L., Zhang, S., 2015f. High-throughput histopathological image analysis via robust cell segmentation and hashing. Med. Image Anal. 26 (1), 306–315.

Zheng, L., Wetzel, A.W., Gilbertson, J., Becich, M.J., 2003. Design and analysis of a content-based pathology image retrieval system. IEEE Trans. Inform. Technol. Biomed. 7 (4), 249–255.

Zhou, X.S., Zillner, S., Moeller, M., Sintek, M., Zhan, Y., Krishnan, A., Gupta, A., 2008. Semantics and CBIR: a medical imaging perspective. In: ACM International Conference on Content-Based Image and Video Retrieval, pp. 571–580.

PART

2

Successful applications in medical imaging

CHAPTER

9 Multitemplate-based multiview learning for Alzheimer's disease diagnosis

M. Liu[1,2,3], R. Min[1], Y. Gao[4], D. Zhang[2], D. Shen[1]

University of North Carolina at Chapel Hill, Chapel Hill, NC, United States[1] Nanjing University of Aeronautics and Astronautics, Nanjing, China[2] Taishan University, Taian, China[3] School of Software, Tsinghua University, Beijing, China[4]

CHAPTER OUTLINE

Machine Learning and Medical Imaging. http://dx.doi.org/10.1016/B978-0-12-804076-8.00009-8

9.1 BACKGROUND

Alzheimer's disease (AD), characterized by progressive impairment of cognitive and memory function, is the sixth leading cause of death in the United States for Americans aged 65 years or older. According to a recent report from the Alzheimer's Association (2013), the total estimated prevalence of AD is expected to be 13.8 million in the United States by 2050. As there is no cure for AD to reverse its progression, early diagnosis and monitoring of AD at its early prodromal stage, that is, mild cognitive impairment (MCI), is of vital importance.

Over the past decade, advances in magnetic resonance imaging (MRI) have enabled significant progress in understanding neural changes that are related to AD (Chan et al., 2003; Davatzikos et al., 2001; Fan et al., 2008; Fox et al., 1996; Hinrichs et al., 2009; Magnin et al., 2009; Mueller et al., 2005). By directly accessing the structures provided by MRI, brain morphometry can identify the anatomical differences between populations of AD patients and normal controls (NCs) for assisting diagnosis and also evaluating the progression of MCI (Fox et al., 1996; Dickerson et al., 2001; Jack et al., 2008; Wang et al., 2014; Liu et al., 2015). In general, MRI-based classification methods can be roughly divided into two categories, that is, (1) methods using single-template-based morphometric representation of brain structures (Cuingnet et al., 2011; Liu et al., 2012; Argyriou et al., 2008; Zhang et al., 2011) and (2) methods using multitemplate-based representation of brain structures (Liu et al., 2015; Koikkalainen et al., 2011; Leporé et al., 2008; Min et al., 2014a,b).

In the first category of methods, researchers mainly utilize a single template as a benchmark space to provide a representative basis for comparing the common anatomical structures of different brain images. More specifically, they first obtain a morphometric representation of each brain image by spatially normalizing it onto a common space (eg, a predefined template) via nonlinear registration, and thus the corresponding regions in different brain images can be compared (Sotiras et al., 2013; Tang et al., 2009; Yap et al., 2009). Usually, such a predefined template is an image of a single subject, a general average template, or a specific template generated from a particular data set under study (Leporé et al., 2008; Chung et al., 2001; Teipel et al., 2007). In the literature, many single-template-based morphometric pattern analysis methods, such as voxel-based morphometry (VBM) (Davatzikos et al., 2001; Ashburner and Friston, 2000; Davatzikos et al., 2008; Thompson et al., 2001), deformation-based morphometry (DBM) (Chung et al., 2001; Ashburner et al., 1998; Gaser et al., 2001; Joseph et al., 2014), and tensor-based morphometry (TBM) (Koikkalainen et al., 2011; Leporé et al., 2008; Kipps et al., 2005; Whitford et al., 2006; Leow et al., 2006; Hua et al., 2008), have been proposed and have demonstrated promising results in AD diagnosis with different classification techniques (Bozzali et al., 2006; Frisoni et al., 2002; Hua et al., 2013). Specifically, in these methods, after nonrigidly transforming each individual brain image onto a common template space, VBM measures the local tissue density of the original brain image directly, while DBM and TBM measure the local deformation and the Jacobian of local deformation, respectively. For example, researchers in Fan et al. (2007) proposed a classification of morphological patterns using adaptive regional elements (COMPARE) algorithm to extract volumetric features from self-organized and spatial-adaptive local regions based on a single template. However, due to the potential bias associated with the use of a particular template, the feature representation extracted from a single (particular) template may not be sufficient to reveal the underlying complicated differences between populations of disease-affected patients and NCs.

In the second category of methods, researchers attempt to use multiple templates to minimize the bias associated with the use of a single template. Although requiring higher computational cost, this kind of method can help reduce the negative impact of registration errors in morphometric analysis of brain images. Recently, several studies (Liu et al., 2015; Koikkalainen et al., 2011; Leporé et al., 2008; Min et al., 2014a,b) have shown that the multitemplate-based methods can often offer more accurate diagnosis results than the single-template-based methods. For example, researchers in Leporé et al. (2008) registered each brain image onto multiple templates (which had already been nonlinearly aligned to a new common template), and then averaged their respective Jacobian maps of the estimated deformation fields to improve the TBM-based monozygotic/dizygotic twin classification. In order to reduce errors caused by registration in the TBM-based classification, researchers in Koikkalainen et al. (2011) investigated the effects of utilizing mean deformation fields, mean volumetric features, and mean predicted responses of regression-based classifiers from multiple templates, and obtained improved results for AD analysis. However, one main disadvantage of the above-mentioned methods is that, after averaging the features from multiple templates, morphometric representations for a subject (although generated from different templates) could become less powerful in revealing the underlying complicated differences between AD patients and NCs, because they ignore the characteristics of each template.

It is worth noting that, due to the fact that anatomical structures among different templates can be very different from each other, a subject's corresponding representations generated from different templates (also named as views later) will also be distinctive, as shown in Fig. 9.1. Fig. 9.1 illustrates (1) how different morphometric patterns can be generated from different templates via nonlinear transformation, where we show an example of the tissue density map of white matter (WM) calculated from the registration by HAMMER (Shen and Davatzikos, 2002), and (2) also the amplified differences in comparison of two subjects when different templates are jointly considered. Actually, a similar philosophy is widely applied in other domains. For example, a side-view camera can capture the profile of an object, which is able to provide supplemental information for object recognition in addition to the frontal shot of the same object. In brain morphometry, multiple templates can be similarly regarded as different "cameras" in such measurements for the same "object" brain MRI.

On the other hand, in machine learning and pattern recognition domains, multiview-based learning methods have been well studied to make full use of features from multiple views to represent an object (Li et al., 2002; Liu et al., 2014; Thomas et al., 2006). For example, in multiview face recognition, a human

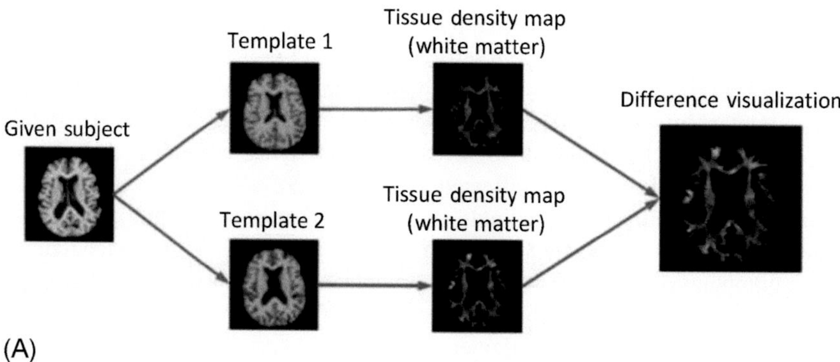

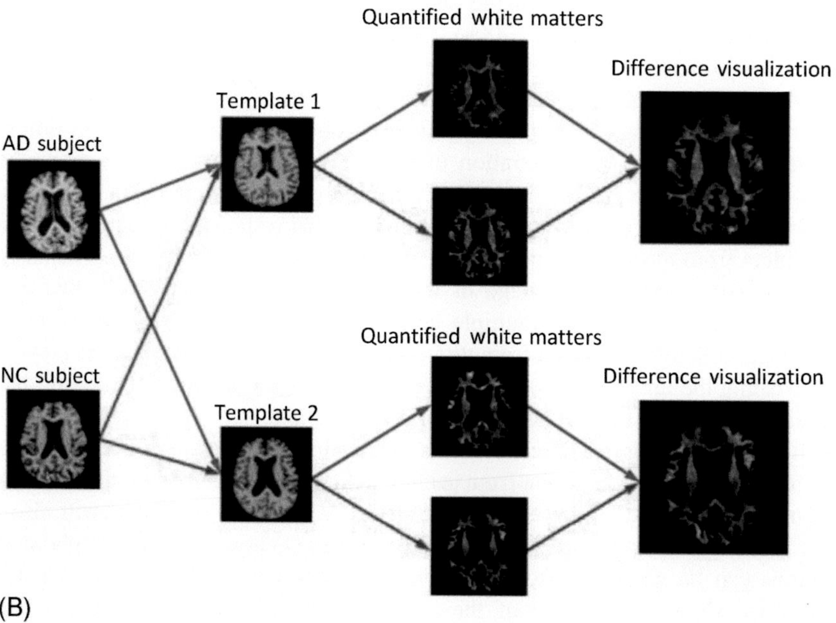

FIG. 9.1

Illustration of different morphometric patterns generated from different templates. (A) Registration of an image to different templates leads to different representations. It can be seen that the geometrical structures of white matter (WM) represented in different templates are different. In addition, tissue density distributions within each tissue are also different from the two different templates. (B) Registration of different images (eg, an AD subject and an NC subject) to different templates: the differences between their representations from individual templates are different (implying the amplified discriminative power when jointly considered in classification).

face can be represented by both the frontal- and side-view images. Since these images can provide different information for the same person, the use of multiple sets of features from different views can largely enrich the representation of the person and achieve significantly enhanced discriminative power, in comparison to features from only one view, as suggested in various studies (Li et al., 2002; Thomas et al., 2006; Basha et al., 2013; Gong et al., 2014; Xu et al., 2014). Similarly, in brain morphometry, multiple templates can also be regarded as multiple views for representing the same brain. Thus a representation generated from a specific template can be regarded as a profile for the brain, and can be used to provide supplementary or side information for other representations generated from the other templates (ie, views).

The rest of this chapter is organized as follows. In Section 9.2 we first introduce a multiview feature representation method for AD diagnosis, by using multiple templates selected from data. In Section 9.3 we present four multiview learning methods for the automatic diagnosis of AD and MCI. Section 9.4 introduces experiments and corresponding analysis. In Section 9.5 we conclude this chapter.

9.2 MULTIVIEW FEATURE REPRESENTATION WITH MR IMAGING

Min et al. (2014b) and Liu et al. (2015) propose a multiview feature representation method by using MRI data. Specifically, they propose to measure brain morphometry via multiple templates, in order to generate a rich representation of anatomical structures, which will be more discriminative to separate different groups of subjects. Unlike previous multitemplate-based works (Koikkalainen et al., 2011; Leporé et al., 2008), which register their templates to a common space via deformable registration, they retain the selected templates in their original (linearly aligned) spaces without nonlinearly registering them to the common space, in order to consider different information provided by different templates. In their method, affinity propagation (Frey and Dueck, 2007) is first applied to select the most distinctive and representative templates. Then, subjects from different groups are registered to different templates by using HAMMER (Shen and Davatzikos, 2002). By adopting a feature extraction method used in COMPARE (Fan et al., 2007), the most discriminative regional features can be extracted in different template spaces.

9.2.1 PREPROCESSING

A standard preprocessing procedure is applied to the T1-weighted MR brain images. First of all, nonparametric nonuniform bias correction (N3) (Sled et al., 1998) is applied to correct intensity inhomogeneity. Then, a skull-stripping method (Wang et al., 2011, 2014) is performed, followed by manual review or correction to ensure clean skull and dura removal. Cerebellum removal is subsequently conducted by warping a labeled template to each skull-stripping image. Afterwards, each brain image is segmented into three tissues (gray matter (GM), WM, and cerebrospinal fluid (CSF)) by using FAST (Zhang et al., 2001), and finally all brain images are affine aligned by FLIRT (Jenkinson and Smith, 2001; Jenkinson et al., 2002).

9.2.2 TEMPLATE SELECTION

For obtaining a multitemplate-based human brain representation, the first question to address is how to select those multiple templates. In Koikkalainen et al. (2011), 30 templates are randomly selected from different categories (10 for AD, 10 for MCI, and 10 for NC). However, these randomly selected templates cannot guarantee to appropriately reflect the distribution of the whole population. Also, redundant information could be introduced with this random selection; moreover, the selection of unrepresentative images as templates could further cause large registration errors. To overcome these limitations, we propose a data-driven template selection scheme to obtain the most distinctive and representative templates.

In order to select templates that can yield discriminative morphometric representations, differences among the selected templates should be maximized. On the other hand, to reduce registration errors, selected templates should be representative enough to cover the entire population. To this end, the affinity propagation (Frey and Dueck, 2007) algorithm is used to partition the entire population (of AD and NC images) into K (eg, $K = 10$ in this chapter) nonoverlapping clusters. Note that, by performing affinity propagation, an exemplar image will be automatically selected for each cluster, which can then be used as a representative image or template for this cluster. Finally, by combining all exemplar images from all different clusters, we can obtain a set of templates to form the template pool. In the clustering process, a bisection method (Frey and Dueck, 2007) is applied to find the appropriate preference value, and the image similarity is computed as normalized mutual information. The clustering results and the respective selected templates are shown in Fig. 9.2. It should be noted that, although it is possible to add more templates to the set of selected templates, those additional templates could introduce just the redundant information and thus affect the optimal representation of each subject. Here, only templates from the AD and NC subjects are selected, but not from the MCI subjects. This is because MCI can be considered as an intermediate stage between AD and NC and is associated with both AD and NC characteristics.

9.2.3 REGISTRATION AND QUANTIFICATION

The core steps in morphometric pattern analysis (eg, VBM, DBM, or TBM) include (1) a registration step for spatial normalization of different images into a common space and (2) a quantification step for morphometric measurement. Similar to Fan et al. (2007), a mass-preserving shape transformation framework

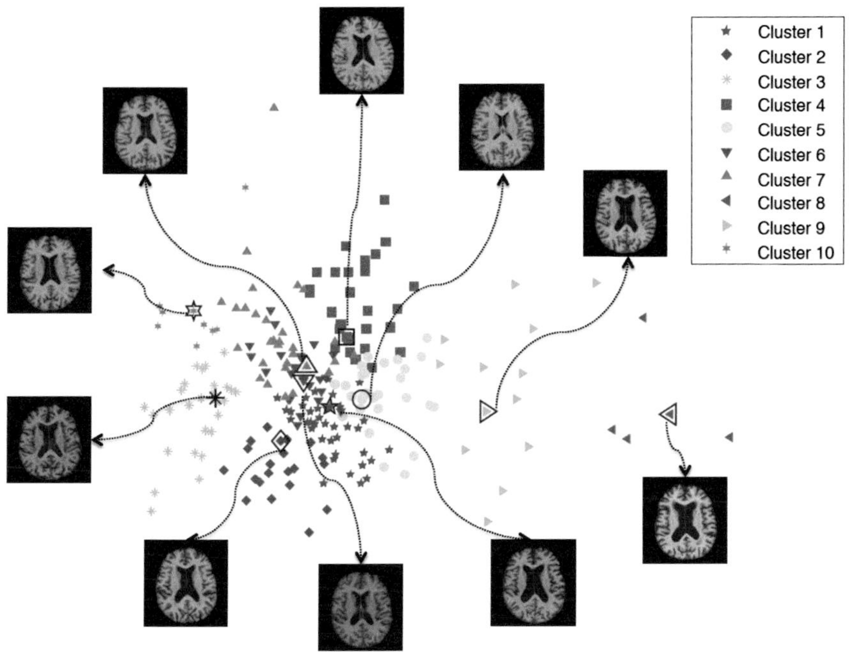

FIG. 9.2

The clustering result of AD/NC subjects, using affinity propagation with normalized mutual information. Each selected template corresponds to an exemplar image of the respective cluster. Points are visualized by multidimensional scaling (Kruskal, 1964).

(Shen and Davatzikos, 2003) is adopted to capture the morphometric patterns of any given subject on the spaces of different templates.

Fig. 9.3 illustrates the multitemplate registration and quantification steps. First, for a given subject with three segmented tissues (ie, GM, WM, and CSF), it will be registered onto multiple (ie, K) selected templates by using a high-dimensional elastic warping tool (ie, HAMMER (Shen and Davatzikos, 2002)). Then, based on those K estimated deformation fields, for each tissue one can quantify its voxel-wise tissue density map in each of the K different template spaces. All these quantified tissue density maps (Davatzikos et al., 2001; Goldszal et al., 1998; Davatzikos, 1998) can thus reflect the unique deformation behavior of the given subject with respect to each different template. In Fig. 9.3 it is clear that the K generated tissue density maps are different in terms of both their density values and tissue structures, which lead to different feature representations as introduced below. Since the GM is mostly affected by AD and thus widely investigated in the literature (Liu et al., 2012; Zhang et al., 2011; Zhang and Shen, 2012), only the GM density map is used for subsequent feature extraction and classification.

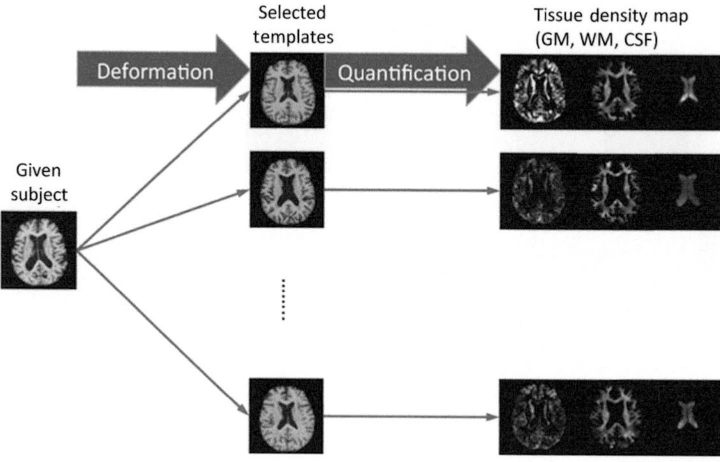

FIG. 9.3

Registration and quantification of a subject registered to multiple templates using HAMMER. Registration to different templates leads to different quantification results. In the figure, the generated tissue density maps (GM, WM, and CSF) are different from registration via different templates.

9.2.4 FEATURE EXTRACTION

Features are first extracted from each individual template space, and then integrated together for a more complete representation. In Section 9.2.4.1 a set of regions-of-interest (ROIs) in each template space is first adaptively determined by performing watershed segmentation (Vincent and Soille, 1991; Grau et al., 2004) on the correlation map obtained between the voxel-wise tissue density values and the class labels from all training subjects. Then, to improve both discrimination and robustness of the volumetric feature computed from each ROI, in Section 9.2.4.2 each ROI is further refined by picking only voxels with reasonable representation power. Finally, to show the consistency and difference of ROIs obtained in all templates, in Section 9.2.4.3 some analysis is provided to demonstrate the capability of the feature extraction method in extracting the complementary features from multiple templates for representing each subject brain.

9.2.4.1 Watershed segmentation

For robust feature extraction, it is important to group voxel-wise morphometric features into regional features. Voxel-wise morphometric features (such as the Jacobian determinants, voxel-wise displacement fields, and tissue density maps) usually have very high feature dimensionality, which includes a large amount of redundant/irrelevant information as well as noises that are due to registration errors.

On the other hand, using regional features can alleviate the above issues and thus provide more robust features in classification.

A traditional way to obtain regional features is to use prior knowledge, that is, predefined ROIs, which summarizes all voxel-wise features in each predefined ROI. However, this method is inappropriate in the case of using multiple templates for complementary representation of brain images, since in this way ROI features from multiple templates will be very similar (we use the volume-preserving measurement to calculate the template-specific morphometric pattern of tissue density change within the same ROI w.r.t. each different template). To capture different sets of distinctive brain features from different templates, a clustering method (Fan et al., 2007) is adopted for adaptive feature grouping. Since clustering will be performed on each template space separately, the complementary information from different templates can be preserved for the same subject image. As indicated in Fan et al. (2007), the clustering algorithm can improve the discriminative power of the obtained regional features, and reduce the negative impacts from registration errors.

Let $I_i^k(u)$ denote a voxel-wise tissue density value at voxel u in the kth template for the ith training subject, $i \in [1, N]$. The ROI partition for the kth template is based on the combined discrimination and robustness measure, $\mathrm{DRM}^k(u)$, computed from all N training subjects, which takes into account both feature relevance and spatial consistency as defined below:

$$\mathrm{DRM}^k(u) = P^k(u)C^k(u), \tag{9.1}$$

where $P^k(u)$ is the voxel-wise Pearson correlation (PC) between tissue density set $\{I_i^k(u), i \in [1, N]\}$ and label set $\{y_i \in [-1, 1], i \in [1, N]\}$ (1 for AD and -1 for NC) from all N training subjects, and $C^k(u)$ denotes the spatial consistency among all features in the spatial neighborhood (Fan et al., 2007).

Watershed segmentation is then performed on each calculated DRM^k map for obtaining the ROI partitions for the kth template. Note that, before applying watershed segmentation, we use a Gaussian kernel to smooth each map DRM^k, to avoid any possible oversegmentation, as also suggested in Fan et al. (2007). As a result, for example, we can partition the kth template into totally R^k nonoverlapping regions, $\{r_l^k, l \in [1, R^k]\}$, with each region r_l^k owning U_l^k voxels. It is worth noting that each template will yield its own unique ROI partition, since different tissue density maps (of same subject) are generated in different template spaces.

Fig. 9.4 shows the partition results obtained from the same group of images registered to the two different templates. It is clear that the obtained ROIs are very different, in terms of both their structures and discriminative powers (as indicated by different colors). Those differences will naturally guide the subsequent steps of feature extraction and selection, and thus provide the complementary information to represent each subject and also improve its classification.

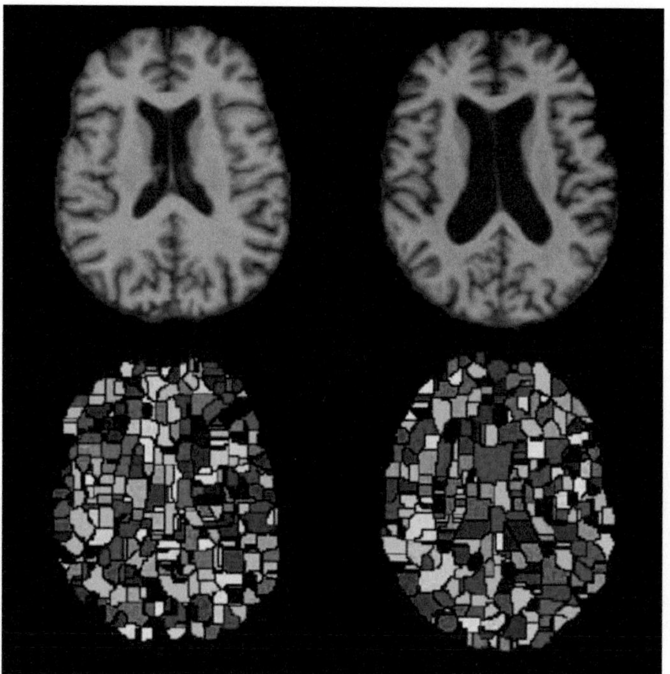

FIG. 9.4

Watershed segmentation of the same group of subjects on two different templates. Color indicates the discriminative power learned from the group of subjects (with the hotter color denoting more discriminative regions). Upper row: two different templates. Lower row: the corresponding partition results.

9.2.4.2 Regional feature aggregation

Instead of using all U_l^k voxels in each region r_l^k for total regional volumetric measurement, only a subregion $\tilde{r}_l^k$ in each region r_l^k is aggregated to further optimize the discriminative power of the obtained regional feature, by employing an iterative voxel selection algorithm. Specifically, one first selects a most relevant voxel, according to the PC calculated between this voxel's tissue density values and class labels from all N training subjects. Then the neighboring voxels are iteratively included to increase the discriminative power of all selected voxels, until no increase is found when adding new voxels. Note that this iterative voxel selection process will finally lead to a voxel set (called the optimal subregion) $\tilde{r}_l^k$ with $\tilde{U}_l^k$ voxels, which are selected from the region r_l^k. In this way, for a given subject i, its lth regional feature $V_{i,l}^k$ in the region $\tilde{r}_l^k$ of the kth template can be computed as

$$V_{i,l}^k = \sum_{\forall u \in \tilde{r}_l^k} \frac{I_i^k(u)}{\tilde{U}_l^k}. \tag{9.2}$$

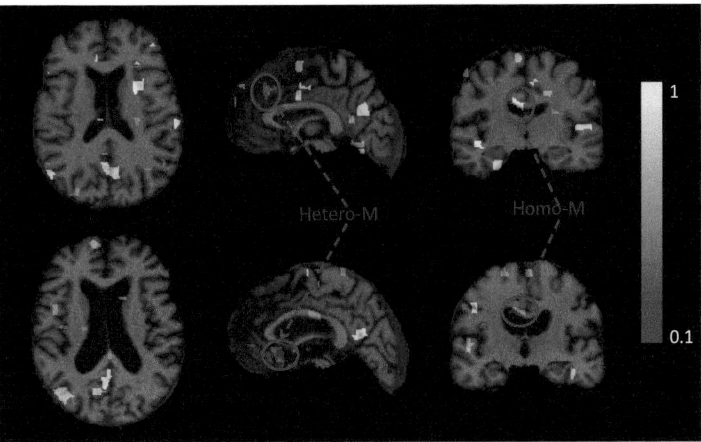

FIG. 9.5

Illustration of the top 100 regions identified using the regional feature aggregation scheme, where the same subject is registered to two different templates. The axial, sagittal, and coronal views of the original MR image of the subject after warping to each of the two different templates are displayed. Color indicates the discriminative power of the identified region (with the hotter color denoting more discriminative region). Upper row: image registered to template 1. Lower row: image registered to template 2. (For the definitions of both hetero-M and homo-M, please refer to Section 9.2.4.3.)

Each regional feature is then normalized to have zero mean and unit variance, across all N training subjects. Finally, from each template, M (out of R^k) most discriminative features are selected using their PC. Thus for each subject, its feature representation from all K templates consists of $M \times K$ features, which will be further selected for classification. Fig. 9.5 shows the top 100 regions selected using the regional feature aggregation scheme, for the same image registered to two templates (as shown in Fig. 9.4). It clearly shows the structural and discriminative differences of regional features from different templates.

9.2.4.3 Anatomical analysis

It is important to understand how the identified regions (ROIs) from different templates are correlated with the target brain abnormality (ie, AD), in order to better reveal the advantages of using multiple templates for morphometric pattern analysis in comparison to using only a single template. Accordingly, we categorize the identified regions (ROIs) into two classes: (1) the class with homogeneous measurements (*homo-M*) and (2) the class with heterogeneous measurements (*hetero-M*) (see Fig. 9.5). The homo-M refers to the regions that are simultaneously identified from different templates, whereas the hetero-M refers to the regions identified in a certain template but not in other templates. In Fig. 9.5, it can be observed that a region within the left corpus callosum is identified in both templates 1 and 2 (see the coronal view). On the other hand, a region within the frontal lobe is only identified in template 1, and a region within the temporal lobe is only identified in template 2 (see the sagittal view). When jointly considering all identified regions from different templates in the classification, the integration of homo-M features is helpful to improve both robustness and generalization of feature extraction for the unseen subjects, while the combination of hetero-M features can provide complementary information for distinguishing subjects during the classification.

9.3 MULTIVIEW LEARNING METHODS FOR AD DIAGNOSIS

9.3.1 FEATURE FILTERING-BASED MULTIVIEW LEARNING

Although the most representative regional features are selected from each template, many regional features, after combination with other features from other templates, could be redundant or even deteriorate the classification of unseen subjects. Therefore selecting a subset of robust regional features (from all templates) is an essential step to achieve good classification performance. Min et al. (2014a) develop a feature filtering-based method to make use of those multiview feature representations.

It has been demonstrated via Fig. 9.5 that the regional features identified from different templates could be heterogeneous. Therefore selecting features jointly from multiple templates can potentially aggregate complementary information that is helpful for the classification. Specifically, for the N training images that have been registered to K templates, all features extracted from K templates can be denoted as $V = \{v_{n,m}^k, m \in [1, M], k \in [1, K], n \in [1, N]\}$, where M top selected features are extracted *independently* from each template by using the method described in Section 9.2.4.2. For each subject, that is, the nth subject, its feature vector $V_n = \{v_{n,m}^k, m \in [1, M], k \in [1, K]\}$ has in total $\xi = M \times K$ features. The goal is to select the top T features out of ξ features to gather the most discriminative and robust information jointly from all templates. The detail of selecting the top M features is provided in the following paragraph.

Because the regional features extracted from different templates are finally used for the same classification task, a "good" feature should be agreed not only by one template, but also by the other templates. In other words, a "good" feature selected from one template should strongly correlate to the "good" features selected from the other templates. Meanwhile, features that are helpful for classification should also strongly correlate with the training labels. To this end, for feature selection (FS), we propose to maximize both the feature relevance w.r.t. labels (ie, according to the PC), and the correlation with features from other templates. This can be done by introducing the "intertemplate" correlation ψ, and combining it with the PC ω by imposing a balancing factor λ as follows:

$$\Delta_m^k = \omega_m^k + \lambda \psi_m^k, \tag{9.3}$$

where Δ_m^k indicates the importance of the mth feature computed from the kth template. The FS can then be achieved by ranking this feature importance for all $\xi = M \times K$ features, $\{\Delta_m^k, m \in [1, M], k \in [1, K]\}$. In Eq. (9.3), ω_m^k denotes the PC between the mth feature from the tth template and the class label from all training subjects. Similarly, the "intertemplate" correlation ψ_m^k can be obtained by first computing the correlation between this mth feature in the kth template and each feature in *other* templates, and then integrating all these correlation coefficients (via summation and normalization) as the final measure. By using the above scheme, one can select a total of M top features with the highest feature importance values.

9.3.2 MAXIMUM-MARGIN-BASED REPRESENTATION LEARNING

The high-dimensional representations generated from multiple templates in their original spaces can form a low-dimensional manifold, in which the optimal representation for classification might be neither a representation generated from one of the existing templates nor the average representation located at the manifold centroid. Instead, the optimal representation could lie somewhere within the manifold of representations from multiple templates, which is most discriminative for the classification. Accordingly, Min et al. (2014b) propose a maximum-margin-based representation learning (MMRL) method to learn the optimal representation from multiple templates for AD classification, which can not only reduce the negative impact due to registration errors but also aggregate the complementary information captured from different template spaces. First, multiple templates are selected to serve as unique common spaces based on affinity propagation (Frey and Dueck, 2007). Then each studied subject is nonlinearly registered to the selected templates, and multiple representations from different template spaces are further generated by an autonomous feature extraction algorithm (Fan et al., 2007). Afterwards, the optimal representation from multiple representations (of multiple templates) in conjunction with the learning of a support vector machine (SVM) (Cortes and Vapnik, 1995) is learned based on the maximum-margin criteria. Finally, the learned representation and SVM are used for classification. Unlike traditional methods enforcing a prior in the representation learning (eg, variance maximization in PCA-based dimensionality reduction (DR) (Jolliffe, 2002), or the locality-preserving property in Laplacian score (LS)-based FS (He et al., 2005), which is independent from the classification stage), the MMRL method learns both the optimal representation and the classifier jointly, in order to make the two different tasks consistently conform to the same classification objective.

Fig. 9.6 illustrates the main idea, where a subject is first nonlinearly registered to multiple templates. Volumetric features are then extracted within each template space, so that multiple representations are generated from different templates. Based on the representations obtained, an optimal representation is finally learned to maximize the classification accuracy. To this end, an MMRL method is introduced to jointly learn both the optimal representation and the classifier for AD classification.

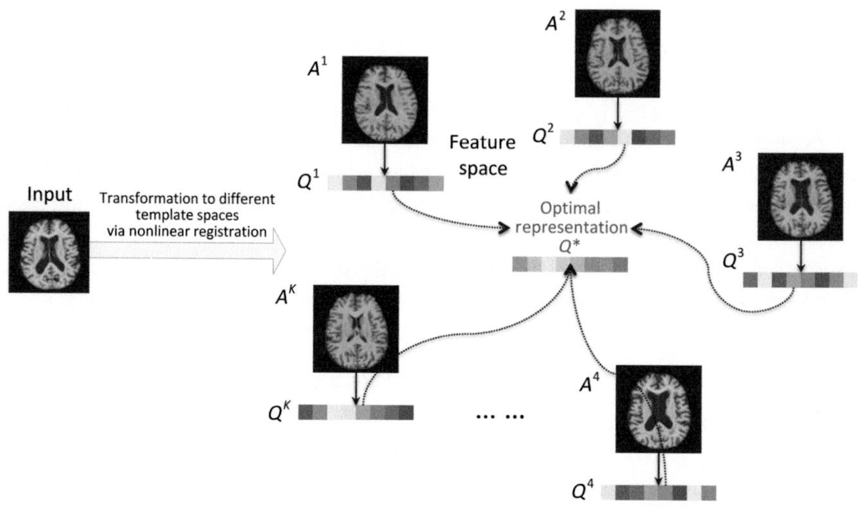

FIG. 9.6

Framework of the MMRL method: learning an optimal representation (Q^*) from the representations ($Q^1 - Q^K$) generated in multiple template spaces ($A^1 - A^K$).

In the multiview feature extraction method introduced in Section 9.2, it is assumed that different templates can then be used to capture complementary information for the same subject, by performing feature extraction in each individual template space. Given the set of representations of a subject generated from K different templates $\boldsymbol{X} = \{\boldsymbol{x}^k \in \mathbb{R}^M, k \in [1, K]\}$, we want to find a new representation $\boldsymbol{x}^* \in \mathbb{R}^L$, which can yield the best classification result. Suppose that the new representation can be generated by applying a mapping to the set of original representations as

$$\boldsymbol{x}^* = f(\boldsymbol{X}). \tag{9.4}$$

The goal is to learn the optimal mapping function $f(\cdot)$ which can yield the best representation $\boldsymbol{x}^*$ for classification. To achieve this goal, we propose an MMRL method to learn $f(\cdot)$ in conjunction with the learning of an SVM classifier, where the jointly learned mapping and classifier are both optimal for the targeted classification task.

Given a training set $\{(\boldsymbol{x}_i, y_i),\ i \in [1, N]\}$, where $\boldsymbol{x}_i \in \mathbb{R}^M$ and $y_i \in \{-1, 1\}$ denote the feature vector and label of the ith subject, respectively, a soft-margin SVM tries to find a hyperplane that maximizes the margin between two classes of samples and also minimizes the cost of misclassification:

$$\arg\min_{\boldsymbol{w},b} \frac{1}{2}||\boldsymbol{w}||_2^2 + C\sum_{i=1}\left[1 - y_i\left(\boldsymbol{w}^{\mathrm{T}}\boldsymbol{x}_i + b\right)\right]_+, \tag{9.5}$$

where $\{\boldsymbol{w}, b\}$ defines the SVM hyperplane, $[\cdot]_+$ denotes the hinge loss function, and C is the balancing factor between the hinge loss and the margin regularization.

When the feature vector $\tilde{x}$ of an unknown subject is input, its associated label $\tilde{y}$ can be predicted using the learned hyperplane $\{\boldsymbol{w}, b\}$ as

$$\tilde{y} = \text{sign}\left(\boldsymbol{w}^{\mathrm{T}}\tilde{\boldsymbol{x}} + b\right). \tag{9.6}$$

Given a training set $\{(\boldsymbol{X}_i, y_i), i \in [1, N]\}$, where $\boldsymbol{X}_i = \left\{x_i^k, k \in [1, K]\right\}$ is the set of representations generated from all K templates, and $\boldsymbol{x}_i^k \in \mathbb{R}^M$ denotes one representation extracted from the kth template for the ith subject. In order to learn the optimal representation $\boldsymbol{x}^*$ (ie, learning the optimal mapping function $f(\cdot)$) jointly with the classification model as defined in Eq. (9.5), the mapping to the new representation is first defined as a linear combination of the K existing representations generated from different templates:

$$f\left(X_i|P^k, \forall k\right) = \sum_{k=1}^{K} P^k \boldsymbol{x}_i^k, \tag{9.7}$$

where $P^k \in \mathbb{R}^{L\times L}$ is a diagonal coefficient matrix to assign different weights to different features of the kth representation (with all nondiagonal elements equal to zero). Then the goal is to find the optimal mapping $f\left(\cdot|P^k, \forall k\right)$ and hyperplane $\{\boldsymbol{w}, b\}$ that maximize the margin between different classes and also reduce the misclassification rate on the training set:

$$\begin{aligned} &\arg\min_{\boldsymbol{w},b,\{P^k,\forall k\}} \frac{1}{2}\|\boldsymbol{w}\|_2^2 + C\sum_{i=1}\left[1 - y_i\left(\boldsymbol{w}^{\mathrm{T}} f\left(X_i|P^k, \forall k\right) + b\right)\right]_+ \\ &\text{s.t.} \quad \forall j, P_j^k > 0 \quad \text{and} \quad \sum_{k=1}^{K} P_j^k = 1, \end{aligned} \tag{9.8}$$

where P_j^k denotes the jth diagonal element of the coefficient matrix P^k (ie, the weight for the jth feature from the kth template). The constraints in Eq. (9.8) confine the estimated weights into the first quadrant of a unit square, so that the generated representation lies within the polygon in the manifold of original representations.

To avoid overfitting, features are further partitioned into different groups, where features within the same group will be assigned to the same mapping weights (ie, $P_{j_1}^k = P_{j_2}^k$ if the j_1th and j_2th features are in the same group). Introducing this additional constraint can efficiently reduce the degree of freedom of the proposed model, thus achieving improved generalization with limited training samples. The feature grouping strategy used in this chapter is implemented by performing affinity propagation on the feature covariance matrix calculated from the training set.

To optimize Eq. (9.8), one can adopt the coordinate descent method to estimate the parameters. The mapping weights $\left\{P^k, \forall k\right\}$ and the hyperplane $\{\boldsymbol{w}, b\}$ are optimized in an iterative manner. In each iteration, one term is optimized while the other is fixed, and thus each optimization step is convex. With the learned mapping $f\left(\cdot|P^k, \forall k\right)$ and the learned decision boundary $\{\boldsymbol{w}, b\}$, given the feature vectors $\tilde{\boldsymbol{X}}$ of an unknown sample extracted from multiple templates, the associated label $\tilde{y}$ can be predicted as

$$\tilde{y} = \text{sign}\left(w^{\mathrm{T}} f\left(X_i | P^k, \forall k\right) + b\right). \tag{9.9}$$

9.3.3 VIEW-CENTRALIZED MULTIVIEW LEARNING

Given multiview feature representation, one can observe that a representation generated from a specific template can be regarded as a profile for the brain, and can be used to provide supplementary or side information for other representations generated from the other templates (ie, views). Accordingly, Liu et al. (2015) develop a view-centralized multitemplate (VCM) classification method, with flowchart illustrated in Fig. 9.7.

As can be seen from Fig. 9.7, brain images are first nonlinearly registered to multiple templates individually, and then their volumetric features are extracted within each template space. In this way, multiple feature representations can be generated

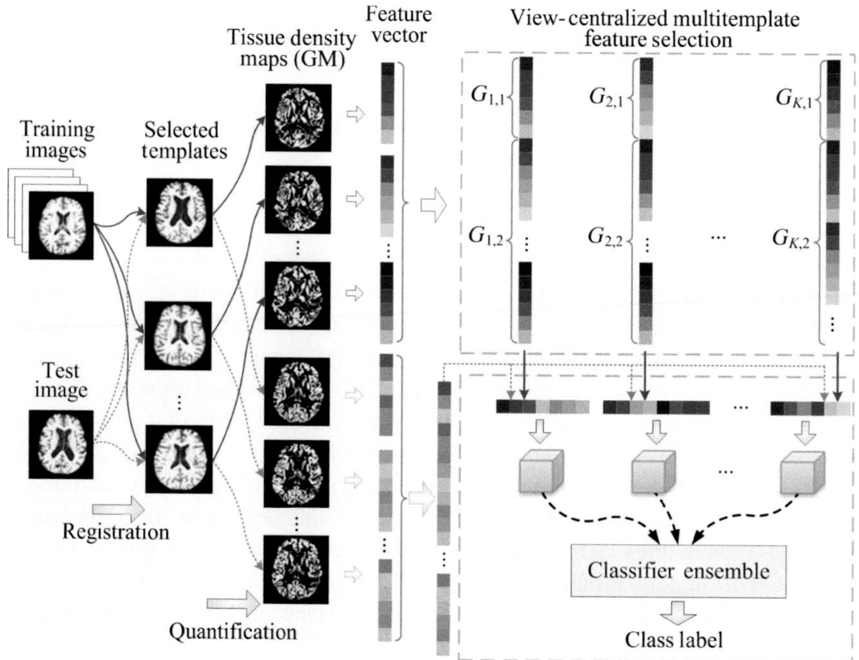

FIG. 9.7

The framework of the view-centralized multitemplate classification method, which includes four main steps: (1) preprocessing and template selection, (2) feature extraction, (3) feature selection, and (4) ensemble classification.

from different templates for each specific subject. Based on such representations, the proposed VCM FS method can be applied to select the most discriminative features, by focusing on the main-view template along with the extra guidance from side-view templates. Finally, multiple SVM classifiers are constructed based on multiple sets of selected features, followed by a classifier ensemble strategy to combine multiple outputs from all SVM classifiers for making the final decision.

Given N training images that have been registered to K templates, we denote $\boldsymbol{X} = \{\boldsymbol{x}_i\}_{i=1}^{N} \in \mathbb{R}^{D \times N}$ ($D = M \times K$ in this chapter) as the training data, where $\boldsymbol{x}_i \in \mathbb{R}^D$ is the feature representation generated from K templates for the ith training image. Let $\boldsymbol{Y} = \{y_i\}_{i=1}^{N} \in \mathbb{R}^N$ be the class labels of N training data, and $\boldsymbol{w} \in \mathbb{R}^D$ be the weight vector for the FS task. For clarity, we divide the feature representations from multiple templates into a main-view group and a side-view group, as illustrated in Fig. 9.8. As can be seen from Fig. 9.8, the main-view group (corresponding to the main template) contains features from a certain template, while the side-view group (corresponding to other supplementary templates) contains features from all other (supplementary) templates.

Denote $a^{(1)}$ as the weighting value for the main-view (ie, main template) group and $a^{(2)}$ as the weighting value for the side-view (ie, supplementary templates) group. By setting different weighting values for features from the main view and the side views, we can incorporate the prior information into the following learning model:

$$\min_{\boldsymbol{w}} \frac{1}{2N} \sum_{i=1}^{N} ||y_i - \boldsymbol{w}^{\mathrm{T}} \boldsymbol{x}_i||_2^2 + \lambda_1 ||\boldsymbol{w}||_1 + \lambda_2 \sum_{g=1}^{2} a^{(g)} ||\boldsymbol{w}^{(g)}||_2$$

$$\text{s.t.} \sum_{g=1}^{2} a^{(g)} = 1;\ a^{(g)} > 0,\ g = 1, 2, \tag{9.10}$$

where $\boldsymbol{w}^{(g)}$ represents the weight vector for the gth group. The first term in Eq. (9.10) is the empirical loss on the training data, and the second one is the l_1-norm regularization term that enforces some elements of $\boldsymbol{w}$ to be zero. It is worth noting

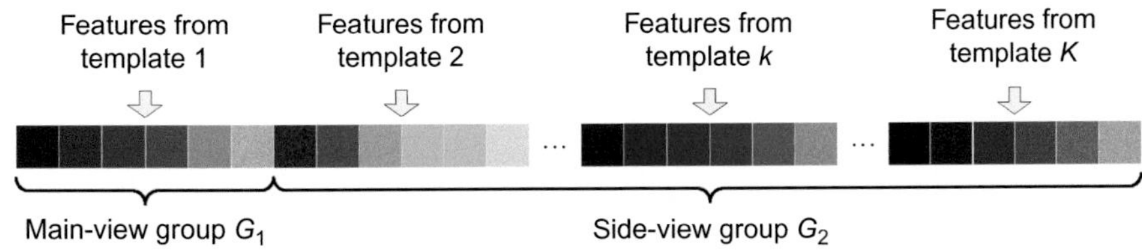

FIG. 9.8

Illustration of group information for feature representations generated from multiple templates. The first group G_1 (ie, the main-view group) consists of features from a certain template, while the second group G_2 (ie, the side-view group) contains features from all other (supplementary) templates.

that the last term in Eq. (9.10) is a view-centralized regularization term, which treats features in the main-view group and the side-view group differently by using different weighting values (ie, $a^{(1)}$ and $a^{(2)}$). For example, a small $a^{(1)}$ (as well as a large $a^{(2)}$) implies that the coefficients for features in the main-view group will be penalized lightly, while features in the side-view group will be penalized severely, because the goal of the model defined in Eq. (9.10) is to minimize the objective function. Accordingly, most elements in the weight vector corresponding to the side-view group will be zero, while those corresponding to the main-view group will not. In this way, the prior knowledge that one focuses on the representation from the main template (ie, main view) with extra guidance from other templates can be incorporated into the learning model naturally. In addition, two constraints in Eq. (9.10) are used to ensure that the weighting values for different groups are greater than 0 and not greater than 1. By introducing such constraints, one can efficiently reduce the degrees of freedom of the proposed model, and avoid overfitting with limited training samples.

Based on the VCM FS model defined in Eq. (9.10), one can obtain a feature subset by selecting features with nonzero coefficients in $\boldsymbol{w}$. Each time, one performs the above-mentioned FS procedure by focusing on one of multiple templates, with other templates used as extra guidance. Accordingly, given K templates, one can get K selected feature subsets, with each of them reflecting the information learned from a certain main template and corresponding supplementary templates.

Ensemble classification

After obtaining K feature subsets by using the view-centralized FS algorithm, one can then learn K base classifiers individually. In this study, a linear SVM classifier is used to identify AD patients from NCs, and progressive MCI patients from stable MCI patients, since the linear SVM model has good generalization capability across different training data, as shown in extensive studies (Zhang and Shen, 2012; Burges, 1998; Pereira et al., 2009). Finally, a classifier ensemble strategy is used to combine these K base classifiers to construct a more accurate and robust learning model, where the majority voting strategy is employed for the fusion of multiple classifiers. Thus the class label of an unseen test sample can be determined by majority voting for the outputs of base classifiers.

9.3.4 RELATIONSHIP-INDUCED MULTIVIEW LEARNING

The main limitation of existing multiview learning models is that only the relationship between samples and their corresponding class labels is considered. Actually, there exist some other important structure information in multiview feature representation using multitemplate MR imaging data, for example, (1) the relationship among multiple templates and (2) the relationship among different subjects. Accordingly, (Liu et al., 2016) develop a relationship-induced multitemplate learning (RIML) method to explicitly model the relationships among templates and among subjects. The flowchart of the RIML method is provided in Fig. 9.9. As can be seen, there are

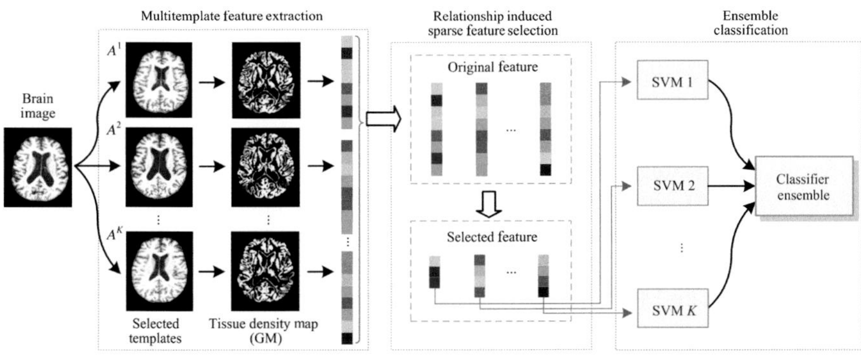

FIG. 9.9

The framework of the relationship-induced multitemplate learning (RIML) method, which consists of three main steps: (1) multitemplate feature extraction, (2) feature selection, and (3) ensemble classification.

three main steps in the RIML method: (1) feature extraction, (2) feature selection, and (3) ensemble classification.

To model the relationships among templates and among subjects, a *relationship-induced sparse* (RIS) FS method is proposed under the multitask learning framework (Argyriou et al., 2008; Zhang and Shen, 2012), by treating the classification in each template space as a specific task. In this chapter, we have K learning tasks corresponding to K templates. Denote $\boldsymbol{X}^k = [\boldsymbol{x}_1^k, \ldots, \boldsymbol{x}_n^k, \ldots, \boldsymbol{x}_N^k]^{\mathrm{T}} \in \mathbb{R}^{N\times M}$ as training data for the kth learning task (corresponding to the kth template) containing a total of N subjects, where $\boldsymbol{x}_n^k \in \mathbb{R}^M$ represents a feature vector of the nth subject in the kth template space ($n = [1, N]$). Similarly, denote $\boldsymbol{Y} = [y_1, \ldots, y_n, \ldots, y_N]^{\mathrm{T}} \in \mathbb{R}^N$ as the response vector for training data $\boldsymbol{X}^k$, where $y_n \in \{-1, 1\}$ is the class label (ie, NC or patient) for the nth subject. Let $\boldsymbol{W} = \left[\boldsymbol{w}^1, \ldots, \boldsymbol{w}^k, \ldots, \boldsymbol{w}^K\right] \in \mathbb{R}^{M\times K}$ represent the weight vector matrix, where $\boldsymbol{w}^k \in \mathbb{R}^M$ parameterizes a linear discriminant function for the kth task ($k = [1, K]$). Then, the multitask feature learning model can be formulated by solving the following objective function (Zhang and Shen, 2012; Caruana, 1997; Baxter, 1997):

$$\min_{\boldsymbol{W}} \sum_{k=1}^{K} ||\boldsymbol{Y} - \boldsymbol{X}^k \boldsymbol{w}^k||_2^2 + \lambda ||\boldsymbol{W}||_{2,1}. \tag{9.11}$$

The first term in Eq. (9.11) is the empirical loss on the training data. The second one is a group-sparsity regularizer to encourage the weight matrix $\boldsymbol{W}$ to have many zero rows, where $||\boldsymbol{W}||_{2,1}$ is the sum of the l_2-norm of the rows in matrix $\boldsymbol{W}$. For FS purposes, only those features corresponding to those rows with nonzero coefficients in $\boldsymbol{W}$ are selected, after solving Eq. (9.11). That is, the $l_{2,1}$-norm regularization term ensures that only a small number of common features are jointly selected across different tasks. The parameter λ is a regularization parameter that is used to balance the relative contributions of those two terms in Eq. (9.11). Specifically, a large λ means that a lower number of features will be selected, while a small λ denotes more features will be selected.

It is worth noting that, due to the anatomical differences across templates, different sets of features obtained for each brain image generally come from different ROIs. Thus the $l_{2,1}$-norm regularization in Eq. (9.11) is not appropriate for the case of using multiple templates, since it jointly selects features across different tasks (ie, templates). To encourage the sparsity of the weight matrix $\boldsymbol{W}$, as well as to select the most informative features corresponding to each template space, we propose the following multitask sparse feature learning model:

$$\min_{\boldsymbol{W}} \sum_{k=1}^{K} ||\boldsymbol{Y} - \boldsymbol{X}^k \boldsymbol{w}^k||_2^2 + \lambda ||\boldsymbol{W}||_{1,1}, \tag{9.12}$$

where $||\boldsymbol{W}||_{1,1}$ is the sum of l_1-norm of the rows in matrix $\boldsymbol{W}$. It is worth noting that the $l_{1,1}$-norm does not necessarily ensure many rows in $\boldsymbol{W}$ are zero, but can help select features that are discriminative for specific tasks.

In Eqs. (9.11) and (9.12), a linear mapping function (ie, $f(\boldsymbol{x}) = \boldsymbol{x}^{\mathrm{T}}\boldsymbol{w}$) is learned to transform data in the original high-dimensional feature space to a one-dimensional label space. The main limitation of these models is that only the relationship between samples and their corresponding class labels is considered. Actually, there exists some important structure information in the multitemplate data, for example, (1) the relationship among multiple templates (*template relationship*) and (2) the relationship among different subjects (*subject relationship*).

(1) As illustrated in Fig. 9.10A, a subject $\boldsymbol{x}_n$ is represented as $\boldsymbol{x}_n^{k_1}$ and $\boldsymbol{x}_n^{k_2}$ in the k_1th template space and in the k_2th template spaces, respectively. After being mapped to the label space, they should be close to each other (ie, $f(\boldsymbol{x}_n^{k_1})$ should be similar to $f(\boldsymbol{x}_n^{k_2})$), since they represent the same subject.

(2) Similarly, as shown in Fig. 9.10B, if two subjects $\boldsymbol{x}_{n_1}^k$ and $\boldsymbol{x}_{n_2}^k$ in the same kth template space are very similar, the distance between $f(\boldsymbol{x}_{n_1}^k)$ and $f(\boldsymbol{x}_{n_2}^k)$ should be small, implying that estimated labels of these two subjects are similar.

Accordingly, a novel *template relationship*-induced regularization term is defined as follows:

$$\begin{aligned} &\sum_{n=1}^{N} \sum_{k_1=1}^{K} \sum_{k_2=1}^{K} \left(f(\boldsymbol{x}_n^{k_1}) - f(\boldsymbol{x}_n^{k_2}) \right)^2 \\ &= \sum_{n=1}^{N} tr((\boldsymbol{B}_n \boldsymbol{W})^{\mathrm{T}} \boldsymbol{L}_n (\boldsymbol{B}_n \boldsymbol{W})), \end{aligned} \tag{9.13}$$

where $tr(\cdot)$ denotes the trace of a square matrix, $\boldsymbol{B}_n = [\boldsymbol{x}_n^1, \ldots, \boldsymbol{x}_n^k, \ldots, \boldsymbol{x}_n^K]^{\mathrm{T}} \in \mathbb{R}^{K \times M}$ represents multiple sets of features derived from K templates for the nth subject,

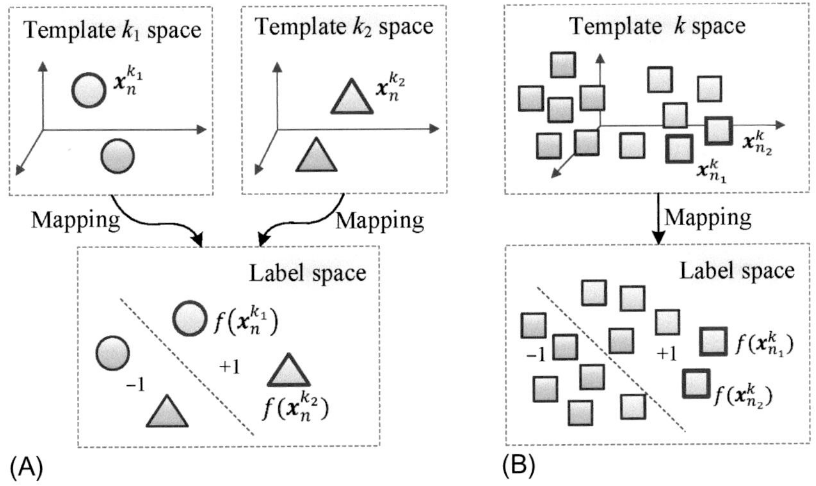

FIG. 9.10

Illustration of structure information, conveyed by: (A) relationship between features of two templates (ie, features of the nth subject in the k_1th template space and the k_2th template space, respectively), and (B) relationship between features of two subjects in the same template (ie, features of the n_1th subject and the n_2th subject in the kth template space). Here, yellow denotes positive training subjects, while blue denotes negative training subjects. Different shapes (circle, triangle, and square) denote samples in three different template spaces (ie, the k_1th template, k_2th template, and kth template).

and $\boldsymbol{L}_n \in \mathbb{R}^{K \times K}$ is a matrix with diagonal elements being $K-1$ and all other elements being -1. By using Eq. (9.13), we can model the relationship among multiple templates explicitly.

Similarly, the *subject relationship*-induced regularization term is defined as follows:

$$\sum_{k=1}^{K}\sum_{n_1=1}^{N}\sum_{n_2=1}^{N} S_{n_1,n_2}^k \left(f\left(\boldsymbol{x}_{n_1}^k\right) - f\left(\boldsymbol{x}_{n_2}^k\right)\right)^2 = \sum_{k=1}^{K}\left(\boldsymbol{X}^k\boldsymbol{w}^k\right)^{\mathrm{T}} \boldsymbol{L}^k \left(\boldsymbol{X}^k\boldsymbol{w}^k\right), \tag{9.14}$$

where $\boldsymbol{X}^k$ is the data matrix in the kth learning task (ie, kth template) as mentioned above, and $\boldsymbol{S}^k = \left\{S_{n_1,n_2}^k\right\}_{n_1,n_2=1}^N \in \mathbb{R}^{N\times N}$ denotes a similarity matrix with elements defining the similarity among N training subjects in the kth template space. Here, $\boldsymbol{L}^k = \boldsymbol{D}^k - \boldsymbol{S}^k$ represents the Laplacian matrix for task k, where $\boldsymbol{D}^k$ is a diagonal matrix with diagonal element $D_{n_1,n_1}^k = \sum_{n_2=1}^N S_{n_1,n_2}^k$, and S_{n_1,n_2}^k is defined as

$$S^k_{n_1,n_2} = \begin{cases} e^{-\frac{||x^k_{n_1} - x^k_{n_2}||^2}{\sigma}}, & \text{if } \boldsymbol{x}^k_{n_1} \text{ and } \boldsymbol{x}^k_{n_2} \text{ are } q \text{ neighbors,} \\ 0, & \text{otherwise,} \end{cases} \tag{9.15}$$

where σ is a constant to be set, and q is set as 3 empirically. It is easy to see that Eq. (9.14) aims to preserve the local neighboring structure of the original data during the mapping, through which one can capture the relationship among different subjects explicitly.

By incorporating two relationship-induced regularization terms defined in Eqs. (9.13) and (9.14) into Eq. (9.12), the objective function of the RIS FS model can be obtained as follows:

$$\begin{aligned} \min_{\boldsymbol{W}} \sum_{k=1}^{K} ||\boldsymbol{Y} - \boldsymbol{X}^k \boldsymbol{w}^k||_2^2 + \lambda_1 ||\boldsymbol{W}||_{1,1} + \lambda_2 \sum_{n=1}^{N} tr\left((\boldsymbol{B}_n \boldsymbol{W})^{\mathrm{T}} \boldsymbol{L}_n (\boldsymbol{B}_n \boldsymbol{W}) \right) \\ + \lambda_3 \sum_{k=1}^{K} \left(\boldsymbol{X}^k \boldsymbol{w}^k \right)^{\mathrm{T}} \boldsymbol{L}^k \left(\boldsymbol{X}^k \boldsymbol{w}^k \right), \end{aligned} \tag{9.16}$$

where λ_1, λ_2, and λ_3 are positive constants used to balance the relative contributions of the four terms in the proposed RIS model, and their values can be determined via inner cross-validation on training data. In Eq. (9.16), the $l_{1,1}$-norm regularization term (the second term) ensures only a small number of features to be selected for each task. The *template relationship*-induced regularization term (the third term) is used to capture the relationship among different templates, while the *subject relationship* regularization term (the fourth term) is employed to preserve the local neighboring structure of data in each template space. After FS using the proposed RIS FS algorithm, an ensemble classification process (similar to Section 9.3.3) is adopted for making a final decision for a test subject.

9.4 EXPERIMENTS

9.4.1 SUBJECTS

The Alzheimer's Disease Neuroimaging Initiative (ADNI) database (http://adni.loni.ucla.edu) (Jack et al., 2008) is employed to evaluate the performance of the proposed classification algorithm. The primary goal of ADNI has been to test whether serial MRI, positron emission tomography (PET), other biological markers, and clinical and neuropsychological assessment can be combined to measure the progression of MCI and early AD. Determination of sensitive and specific markers of very early AD progression is intended to aid researchers and clinicians to develop new treatments and monitor their effectiveness, as well as lessen the time and cost of clinical trials. Since we focus on the morphometric study of AD, T1-weighted MRI data from ADNI is used in the experiments. In total, 459 subjects, scanned with a 1.5T scanner,

Table 9.1 Demographic Information of the Studied Subjects From ADNI Database

Diagnosis	Number	Age	Gender (M/F)	MMSE
AD	97	75.90 ± 6.84	48/49	23.37 ± 1.84
NC	128	76.11 ± 5.10	63/65	29.13 ± 0.96
pMCI	117	75.18 ± 6.97	67/50	26.45 ± 1.66
sMCI	117	75.09 ± 7.65	79/38	27.42 ± 1.78

Note: Values are denoted as mean ± deviation; MMSE means mini-mental state examination; M and F represent male and female, respectively.

were randomly selected, comprising of 97 AD, 128 NC, and 234 MCI (117 pMCI and 117 sMCI) subjects. The demographic information of the used dataset is shown in Table 9.1.

As mentioned in Section 9.2, K ($K = 10$ in the experiments) representative templates are first selected from AD and NC subjects. In the multitemplate feature extraction stage, there are a total of M ($M = 1500$ in the experiments) features extracted in each template space. Thus for each subject, its feature representation from K templates consists of 1500×10 features, which will be further selected for classification.

9.4.2 EXPERIMENTAL SETTINGS

The evaluation of different methods is conducted on two different problems: (1) AD diagnosis such as AD versus NC classification and (2) progressive MCI diagnosis such as pMCI versus sMCI classification. Note that the second problem is considered more difficult than the first problem, but received relatively less attention in previous works. However, it is important to identify progressive MCI patients from stable MCI patients, in order to possibly prevent the progression of MCI to AD via timely therapeutic interventions.

In general, a 10-fold cross-validation strategy is adopted to evaluate the performances of different methods, which has been widely used in recent studies (Zhang and Shen, 2012; Burges, 1998; Pereira et al., 2009). Specifically, all samples are partitioned into 10 subsets (with each subset having a roughly equal size), and *each time* samples in one subset are successively selected as the test data, while those in all other nine subsets are used as the training data to perform FS and classifier construction. This process is repeated 10 times independently to avoid any bias introduced by the random partitioning of the original data in the cross-validation process. Finally, the mean values of corresponding classification results are recorded for comparison.

The performance of different methods is evaluated via four evaluation criteria, that is, classification accuracy (ACC), classification sensitivity (SEN), classification specificity (SPE), and the area under the receiver operating characteristic (ROC) curve (AUC). More specifically, the accuracy measures the proportion of subjects that are correctly predicted among all studied subjects, the sensitivity denotes the proportion of patients that are correctly predicted, and the specificity represents the proportion of NCs that are correctly predicted.

9.4.3 RESULTS OF FEATURE FILTERING-BASED METHOD FOR AD/MCI DIAGNOSIS

In this group of experiments, the balancing factor λ in Eq. (9.3) is set to 0.38. The SVM classifier used here is implemented by the LIBSVM library (Chang and Lin, 2011), using a linear kernel and $C = 1$ (the default cost). Finally, $M = 1{:}1500$ features are tested, and the best results are reported for quantitative comparison.

Table 9.2 first shows the results using a single template for AD/NC classification, to demonstrate the variability of classification results when using different templates even for the same classification task, where the best results are marked in boldface. Because the proposed FS method integrates not only the PC but also the "intertemplate" correlation from the multiple templates, two conventional FS methods are examined based on single templates. The first FS method is simply based on the ranking of PC, and the second method combines PC with SVM-RFE-based FS (Guyon et al., 2002) (as proposed in Fan et al. (2007)) for jointly considering multiple features in the selection. It should be noted that, in the single template case, the feature extraction performed in the proposed method is the same as COMPARE (Fan et al., 2007). Therefore in this chapter, the PC+SVM-RFE-based method using a single template is denoted as COMPARE.

Table 9.2 reports the best classification accuracies (ACC) for each of the 10 templates using PC and COMPARE, along with their respective sensitivities (SEN) and specificities (SPEC). Note that the sensitivity and the specificity refer to the portions of correctly identified AD patients and correctly classified NC subjects, respectively. From Table 9.2, it is clear that COMPARE outperforms PC when using their own best templates (ie, A^5 for PC and A^7 for COMPARE). However, for some templates (ie, A^1, A^2, A^5, A^9, and A^{10}), the use of additional SVM-RFE-based FS (in COMPARE) cannot further improve the simple PC-based classification (in terms of the best classification accuracy). That is, the result improvement brought by SVM-RFE is limited, but at a cost of increased computational burden.

Furthermore, the results of AD versus NC and pMCI versus sMCI classification using multiple templates are given in Table 9.3. The proposed (multitemplate-based) FS method (namely MA_Proposed) that considers both PC and "intertemplate" correlation is compared with both PC- and COMPARE-based FS methods using either a single template (namely SA_PC and SA_COMPARE) or multiple templates (namely MA_PC and MA_COMPARE). For fair comparison, the averaged results of single-template-based methods (SA_PC and SA_COMPARE) across all 10 templates are reported. In MA_PC, all regional features extracted from 10 different templates are used, thus resulting in a feature representation with $M \times K = 15{,}000$ dimensions for each subject; afterwards, the top 1500 features are selected out

Table 9.2 Results of AD Versus NC and pMCI Versus sMCI Classification Using Single Templates (A^1–A^{10})

	AD vs. NC Classification						pMCI vs. sMCI Classification					
	PC			COMPARE			PC			COMPARE		
Template	**ACC**	**SEN**	**SPE**	**ACC**	**SEN**	**SPE**	**ACC**	**SEN**	**SPE**	**ACC**	**SEN**	**SPE**
A^1	84.09	78.33	88.40	83.16	75.33	89.17	68.93	64.62	73.18	71.03	68.79	73.18
A^2	84.94	80.56	88.30	81.95	73.67	88.40	68.87	68.56	69.09	71.46	71.97	70.76
A^3	83.12	77.33	87.56	84.50	78.44	89.17	69.34	65.15	73.41	69.81	69.47	70.08
A^4	84.87	80.44	88.33	85.72	82.22	88.40	**72.71**	73.56	71.82	71.82	**72.58**	71.06
A^5	**85.85**	82.56	88.46	84.05	76.22	90.00	70.66	69.39	71.82	71.93	71.21	72.80
A^6	84.38	78.33	**89.04**	85.35	**83.56**	86.73	71.04	65.98	**75.98**	72.86	69.62	76.14
A^7	82.23	77.22	86.09	**87.07**	81.33	**91.54**	71.08	**73.94**	68.18	**74.56**	70.8	**78.64**
A^8	83.59	79.44	86.86	84.48	79.44	88.46	70.27	68.71	71.67	71.88	68.56	75.00
A^9	83.65	77.33	88.40	82.27	78.44	85.38	68.55	66.36	70.68	71.10	66.97	75.15
A^{10}	83.28	**83.78**	83.01	83.20	76.56	88.46	69.00	72.05	65.83	71.74	70.15	73.41

Table 9.3 Results of AD Versus NC and pMCI Versus sMCI Classification Using Single Templates (SA_PC, SA_COMPARE, SA_Proposed) and Multiple Templates (MA_PC, MA_COMPARE, MA_Proposed)

	AD vs. NC			pMCI vs. sMCI		
Method	**ACC (%)**	**SEN (%)**	**SPE (%)**	**ACC (%)**	**SEN (%)**	**SPE (%)**
SA_PC	82.01	75.88	86.76	68.49	67.80	69.10
SA_COMPARE	81.52	77.11	84.92	70.06	68.08	72.02
MA_PC	85.91	81.56	89.23	72.78	74.62	70.91
MA_COMPARE	87.19	80.56	92.31	73.35	**75.76**	70.83
MA_Proposed	**91.64**	**88.56**	**93.85**	**72.41**	72.12	**72.58**

of 15,000 features based on the PC, and $M = 1{:}1500$ features are subsequently selected and used for classification. In MA_COMPARE, the top 1500 features are first selected in the same way as MA_PC, but additionally using SVM-RFE to further refine the selected features, before inputting them to the SVM for classification.

For both AD versus NC and pMCI versus sMCI classification, the best classification accuracies (ACC) as well as the corresponding sensitivities (SEN) and specificities (SPEC) of all methods are illustrated in Table 9.3. The results clearly show that MA_Proposed is better than any other methods in terms of all metrics. It should be noted that the sensitivities of SA_PC, SA_COMPARE, MA_PC, and MA_COMPARE are much lower in comparison to their corresponding specificities. A low sensitivity value indicates low confidence on AD diagnosis, which will greatly limit their practical usage. On the other hand, MA_Proposed gives a significantly improved sensitivity value. Together with its high specificity (93.85% for AD vs. NC classification), the MA_Proposed method produces more confident AD diagnosis results.

In addition, Fig. 9.11 illustrates the results of SA_PC, SA_COMPARE, MA_PC, MA_COMPARE, and MA_Proposed in AD versus NC and pMCI versus sMCI classification with respect to different numbers of top selected features. From Fig. 9.11, it is clear that the results of multitemplate-based methods (MA_PC, MA_COMPARE, and MA_Proposed) outperform the results of single-template-based methods (SA_PC and SA_COMPARE) by a significant margin. Specifically, in Fig. 9.11 (left), SA_PC and SA_COMPARE reach their best classification accuracy with a small portion of top selected features, and their performances decline rapidly when more features are included in AD versus NC classification. This indicates that many of their selected features are noisy and redundant, if using only a single template. In contrast, multitemplate-based methods consistently increase or maintain their performance with the increase of the number of features used, which demonstrates that the complementary information from different templates is aggregated together to improve the classification. In addition, with the assistance of SVM-RFE, the COMPARE-based methods (SA_COMPARE and MA_COMPARE) achieve

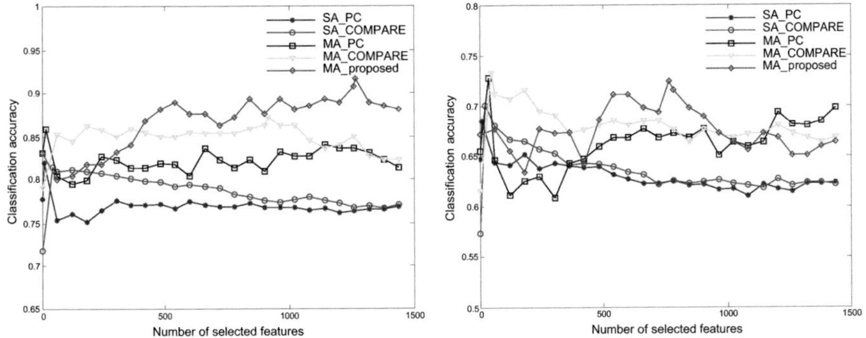

FIG. 9.11

Results of SA_PC, SA_COMPARE, MA_PC, MA_COMPARE, and MA_Proposed in (left) AD versus NC classification and (right) pMCI versus sMCI classification.

better performance than the PC-based methods (SA_PC and MA_PC) in both cases of using single template and multiple templates. Fig. 9.11 (left) also demonstrates that MA_Proposed significantly outperforms all other comparison methods. Although only a small portion of features can give good classification accuracy for the single-template-based methods, the performance of the MA_Proposed method is consistently improved with use of more features (ie, 91.64% when using 1268 features for AD versus NC classification). This phenomenon shows that the redundant features from a single template can be integrated with the features from other templates (in an effective way) to yield more robust and discriminative representations. From Fig. 9.11 (right), we can observe again that all three multitemplate-based methods (MA_PC, MA_COMPARE, and MA_Proposed) perform significantly better than the two single-template-based methods (SA_PC, SA_COMPARE) in pMCI versus sMCI classification, indicating the power of using multiple templates in aggregating more useful information for classification. Among all three multitemplate-based methods, MA_Proposed demonstrates comparable performance to both MA_PC and MA_COMPARE. When using the $M = 500{:}1000$ top selected features, the proposed method (MA_Proposed) gives the best overall classification results. On the other hand, MA_COMPARE gets its best results when using $M = 1{:}500$ features, and MA_PC achieves its best results when using $M = 1000{:}1500$ features.

9.4.4 RESULTS OF MAXIMUM-MARGIN-BASED LEARNING FOR AD/MCI DIAGNOSIS

In this group of experiments, the number of selected templates is $K = 10$ from affinity propagation, and the number of biomarkers identified on each template is $L = 20$. Table 9.4 compares the classification performance of the learned representation

Table 9.4 Comparison of MMRL to Representation Generated From Single Template (SA) and the Average Representation From Multiple Templates for AD/NC Classification and pMCI/sMCI Classification

	AD vs. NC			pMCI vs. sMCI		
Method	**ACC (%)**	**SEN (%)**	**SPE (%)**	**ACC (%)**	**SEN (%)**	**SPE (%)**
Mean_SA	83.23	82.28	84.06	67.56	70.30	67.71
Best_SA	85.35	82.33	87.69	71.08	72.88	69.02
Average	86.68	85.67	87.63	70.70	73.11	68.18
MMRL	**90.69**	**87.56**	**93.01**	**73.69**	**76.44**	**70.76**

(using MMRL from multiple templates) with single template (SA) representations and the average representation of multiple templates. The classification rate of the best template (Best SA) and the average result across the 10 templates (Mean SA) are also reported in Table 9.4. Additionally, the classification performance obtained by the average representation from multiple representations generated from all 10 templates is given in Table 9.4. From Table 9.4 it is clear that the representation learned by MMRL significantly outperforms both Best_SA and Average according to all evaluation metrics (accuracy, sensitivity, and specificity) for both AD/NC classification and pMCI/sMCI classification.

Table 9.5 shows the results comparing MMRL with four popular DR and FS methods when multiple templates are used. In Table 9.5, PCA (Jolliffe, 2002) and AutoEncoder (Bengio, 2009) are DR methods, whereas LS (He et al., 2005) and mRMR (Peng et al., 2005) are widely used FS techniques. For fair comparison, all techniques reduce the feature dimension to 20 (the same as the MMRL learned representation). For AutoEncoder, a widely used configuration with a three-layer architecture (Bengio, 2009) was adopted. These results demonstrate that the proposed joint learning method yields the best classification results (90.69% for AD vs. NC and

Table 9.5 Comparison of MMRL to Different Dimensionality Reduction and Feature Selection Methods for AD/NC Classification and pMCI/sMCI Classification

	AD vs. NC			pMCI vs. sMCI		
Method	**ACC (%)**	**SEN (%)**	**SPE (%)**	**ACC (%)**	**SEN (%)**	**SPE (%)**
PCA	83.08	81.22	84.55	68.96	72.12	68.68
AutoEncoder	87.60	86.56	88.40	66.42	70.45	62.20
LS	87.19	84.56	89.23	67.73	71.36	63.94
mRMR	85.31	83.33	86.79	69.42	69.62	69.17
MMRL	**90.69**	**87.56**	**93.01**	**73.69**	**76.44**	**70.76**

73.69% for pMCI vs. sMCI) in comparison to the others, whose representations are learned prior to the final classification.

9.4.5 RESULTS OF VIEW-CENTRALIZED LEARNING FOR AD/MCI DIAGNOSIS

To demonstrate the variability of classification results using different templates, we first report the results of AD versus NC and pMCI versus sMCI classification based on a single template. Note that, in the single template case, the VCM FS method (Liu et al., 2015) only uses features from the selected template space, while features from other template spaces are completely ignored (ie, $a^{(1)} = 1$ and $a^{(2)} = 0$); thus Eq. (9.1) is similar to the formulation of an elastic net (Zou and Hastie, 2005). Two conventional FS methods are employed for comparison. The first one is based on the ranking of PC coefficients, and the second one is the COMPARE method proposed in Fan et al. (2007) that combines PC and SVM-RFE (Guyon et al., 2002). For fair comparison, the linear SVM with default parameter ($C = 1$) is adopted as a classifier after FS using PC, COMPARE, and the VCM method (Liu et al., 2015), respectively. Fig. 9.12 reports the distribution of classification results achieved by PC, COMPARE, and VCM using 10 single templates in AD versus NC classification.

As can be seen from Fig. 9.12, classification results based on different single templates are very different, regardless of the use of any FS methods. The underlying reason may be that the anatomical structure of a certain template may be more representative of the entire population, compared with other templates. In this case, the overall registration errors to this template are smaller and thus the feature representation generated from this template includes less noise. Another possible reason could be that the AD-related patterns generated from a certain template may

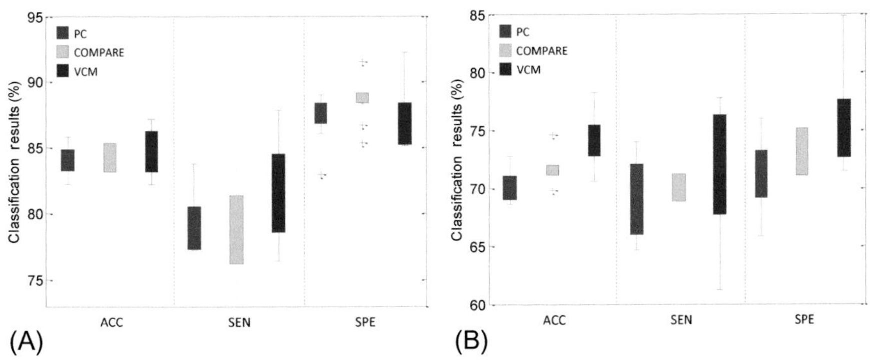

FIG. 9.12

Distribution of accuracy (ACC), sensitivity (SEN), and specificity (SPE) achieved by different single-template-based methods in (A) AD versus NC classification and (B) pMCI versus sMCI classification.

be more discriminative than those generated from other templates, thus having better generalization capability in identifying unseen test subjects.

The results for AD versus NC and pMCI versus sMCI classification by using multiple templates are shown in Table 9.6. The VCM method is compared with six FS methods, including (1) single-template-based PC (PC_SA); (2) single-template-based COMPARE (COMPARE_SA); (3) multiple-template-based PC (PC_MA); (4) multiple-template-based COMPARE (COMPARE_MA); (5) random subspace (RS) (Ho, 1998) that randomly selects features from the original feature space; and (6) Lasso (Tibshirani, 1996) that is a widely used FS method in neuroimaging analysis. Specifically, the averaged classification results of single-template-based methods (ie, PC and COMPARE) among all 10 templates are reported for PC_SA and COMPARE_SA. For both PC_MA and COMPARE_MA methods, we first concatenate all regional features (ie, 15,000-dimensional) extracted from multiple templates. Then, the top M ($M = \{1, 2, \ldots, 1500\}$) features are sequentially selected according to the PC (with respect to class labels) for PC_MA and PC+SVM-RFE for COMPAE_MA, and the best classification results are reported. It is worth noting that, in the proposed method, we learn 10 SVM classifiers based on different template-centralized feature subsets determined by the VCM FS method, and then construct a classifier ensemble with these learned base classifiers. For fair comparison, for the RS method, we randomly select M ($M = \{1, 2, \ldots, 1500\}$) features from each template for classification, and then record the best result. For the Lasso method, one first learns a Lasso model in a specific template space, and then selects features with nonzero coefficient in the learned weight vector. Given 10 templates, 10 classifiers can be constructed based on the features selected by RS and Lasso. Finally, for the RS and Lasso methods, these classifiers are combined using the same ensemble strategy as in the proposed method. The experimental results are summarized in Table 9.6.

From Table 9.6, it is clear to see that multitemplate-based methods (ie, PC_MA, COMPRE_MA, RS, Lasso, and VCM) generally achieve much better performance than single-template-based methods (ie, PC_SA and COMPARE_SA). Specifically, the best accuracies in AD versus NC classification achieved by PC_SA and COMPARE_SA are only 84.00% and 84.18%, respectively, which are much lower than those of COMPARE_MA, Lasso, and VCM. On the other hand, Table 9.6 shows that the VCM method consistently outperforms other methods in terms of classification accuracy, sensitivity, and AUC value. Obviously, by focusing on the representation from a certain template with other templates as extra guidance, the VCM method achieves better performance than the compared methods. In addition, from Table 9.6, one can observe that the sensitivities of PC_SA, COMPARE_SA, PC_MA, CPMPARE_MA, RS, and Lasso are much lower than their corresponding specificities. Here, low sensitivity values indicate low confidence in AD diagnosis, which will greatly limit practical usage in real-world applications. In contrast, VCM achieves a significantly improved sensitivity value in AD versus NC classification (ie, nearly 8% higher than the second best sensitivity achieved by Lasso).

Table 9.6 Results of AD Versus NC Classification Using Single Template and Multiple Templates

Method	AD vs. NC				pMCI vs. sMCI			
	ACC (%)	SEN (%)	SPE (%)	AUC	ACC (%)	SEN (%)	SPE (%)	AUC
PC_SA	84.00	79.53	87.45	0.7692	68.49	67.80	69.10	0.6285
COMPARE_SA	84.18	75.33	89.17	0.7870	70.06	68.08	72.02	0.6356
PC_MA	85.91	81.56	89.23	0.8191	72.78	74.62	70.91	0.7245
COMPARE_MA	87.19	80.56	92.31	0.8495	73.35	75.76	70.83	0.7405
RS	85.44	69.00	92.75	0.7688	69.05	68.10	72.94	0.6912
Lasso	87.27	84.78	89.23	0.9004	75.32	81.36	69.17	0.7602
VCM	**92.51**	**92.89**	88.33	**0.9583**	**78.88**	**85.45**	**76.06**	**0.8069**

9.4.6 RESULTS OF RELATIONSHIP-INDUCED LEARNING FOR AD/MCI DIAGNOSIS

To better make use of multiple sets of features generated from multiple templates, the following two strategies are used in this group of experiments, including (1) the feature concatenation method and (2) ensemble-based method. Specifically, in the feature concatenation method, features from multiple templates are *simply* concatenated into a long vector, and the corresponding classifier is constructed by using this feature vector. In the ensemble-based method, each feature set generated from a specific template space is treated individually, and multiple SVM classifiers based on these feature sets are constructed separately, followed by an ensemble strategy to combine the outputs of all SVMs for making a final classification decision.

In addition, the RIML method using the RIS FS algorithm (Liu et al., 2016) is compared with four methods, that is, (1) PC, (2) COMPARE method proposed in Fan et al. (2007) that combines PC and SVM-RFE, (3) statistical *t*-test method (Guyon et al., 2002), and (4) Lasso (Tibshirani, 1996) that is widely used for sparse FS in neuroimaging analysis. Here, we use PC<con>, COMPARE<con>, *t*-test<con>, and Lasso<con> to denote the four methods using four different FS algorithms (ie, PC, COMPARE, *t*-test, and Lasso) and the feature concatenation strategy (ie, <con>). Similarly, we use PC<ens>, COMPARE<ens>, *t*-test<ens>, and Lasso<ens> as another four methods using four different FS algorithms in each of the multiple template spaces during FS and then the proposed ensemble method (ie, <ens>) in the final classification step.

For comparison, the averaged classification results of single-template-based methods (including PC, COMPARE, *t*-test, and Lasso) are reported, with results given in Tables 9.7 and 9.8. Furthermore, the ROC curves achieved by five ensemble-based methods (RIML and four comparison methods) are plotted in Fig. 9.13.

Table 9.7 Performance of AD Versus NC Classification With Multiple Templates

	Method	ACC (%)	SEN (%)	SPE (%)	AUC
Single-template-based methods	PC	84.00	79.53	87.45	0.7692
	COMPARE	84.18	75.33	89.17	0.7870
	t-test	76.27	68.50	83.01	0.7496
	Lasso	84.32	81.66	86.36	0.8402
Multitemplate-based methods	PC<con>	84.01	81.56	89.23	0.8191
	COMPARE<con>	84.93	80.11	87.03	0.7907
	t-test<con>	81.87	70.77	**90.71**	0.8178
	Lasso<con>	86.62	84.78	89.80	0.8729
	PC<ens>	85.59	82.44	89.93	0.9151
	COMPARE<ens>	86.61	85.44	89.23	0.9085
	t-test<ens>	84.31	74.56	89.70	0.8878
	Lasso<ens>	87.27	84.78	89.23	0.9279
	RIML	**93.06**	**94.85**	90.49	**0.9579**

Table 9.8 Performance of pMCI Versus sMCI Classification With Multiple Templates

	Method	ACC (%)	SEN (%)	SPE (%)	AUC
Single-template-based methods	PC	68.49	67.80	69.10	0.6285
	COMPARE	70.06	68.08	72.02	0.6356
	t-test	61.99	64.93	73.11	0.6516
	Lasso	72.06	72.04	72.02	0.7203
Multitemplate-based methods	PC<con>	72.78	74.62	70.91	0.7245
	COMPARE<con>	73.35	75.76	70.83	0.7405
	t-test<con>	61.60	64.32	75.01	0.7163
	Lasso<con>	71.49	76.06	66.67	0.7136
	PC<ens>	73.92	73.38	72.32	0.7629
	COMPARE<ens>	75.56	75.75	73.48	0.7658
	t-test<ens>	63.36	60.60	71.74	0.6333
	Lasso<ens>	75.32	81.36	69.17	0.7602
	RIML	**79.25**	**87.92**	**75.54**	**0.8344**

From Table 9.7 and Fig. 9.13 (left), we can observe three main points. First, multitemplate-based methods generally achieve much better performance than single-template-based methods (ie, PC, COMPARE, *t*-test, and Lasso) in AD versus NC classification. For example, the best accuracy achieved by single-template-based methods is only 84.32% (achieved by Lasso), which is usually lower than those of multitemplate-based methods. This demonstrates that, compared with single-template-based methods, multitemplate-based methods can help promote the classification performance by taking advantage of richer representations for each subject. Second, using multiple templates, methods that adopt the proposed ensemble strategy (ie, PC<ens>, COMPARE<ens>, *t*-test<ens>, and Lasso<ens>) usually perform better than their counterparts that simply employ the feature concatenation strategy (ie, PC<con>, COMPARE<con>, *t*-test<con>, and Lasso<con>), in terms of four evaluation criteria. This implies that the feature concatenation strategy may not be a good choice to make use of multiple sets of features generated from multiple templates. Finally, RIML using the RIS FS algorithm achieves consistently better results than other methods in terms of classification accuracy, sensitivity, and AUC. To be specific, RIML achieves a classification accuracy of 93.06%, a sensitivity of 94.85%, and an AUC of 0.9579, while among all other methods the best accuracy is 87.27%, the best sensitivity is 85.44%, and the best AUC is 0.9279.

From Table 9.8 and Fig. 9.13 (right), one can observe again that multitemplate-based methods usually outperform single-template-based methods in pMCI versus sMCI classification. For example, the best accuracy of multitemplate-based methods (achieved by RIML) is 79.25%, which is much higher than the best accuracy of single-template-based methods, that is, 72.06% achieved by Lasso. In addition, among all nine multitemplate-based methods, RIML consistently achieves a better

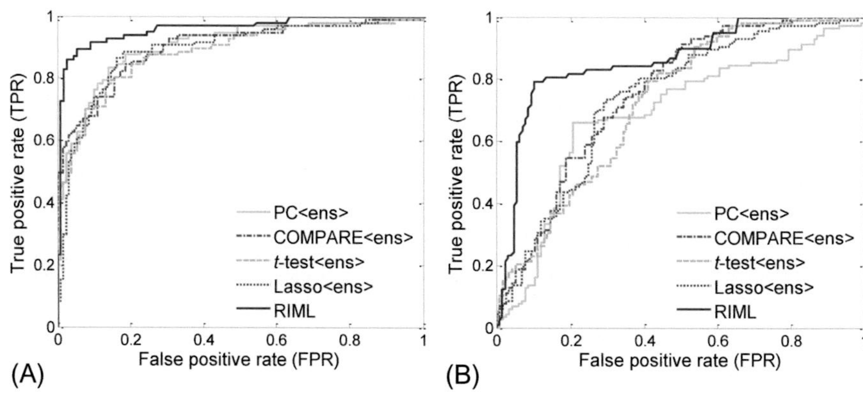

FIG. 9.13

ROC curves achieved by five ensemble-based methods using multiple templates in (A) AD versus NC classification and (B) pMCI versus sMCI classification.

performance than any other method, in terms of four evaluation criteria. In particular, RIML achieves an AUC of 0.8344, while the best AUC achieved by the comparison methods is only 0.7658 (achieved by COMPARE<ens>).

9.5 SUMMARY

In recent years, brain morphometric pattern analysis using MRI has been widely investigated for automatic diagnosis of AD and MCI. Existing MRI-based studies can be categorized into single-template-based and multitemplate-based approaches. It is widely accepted that different templates can convey complementary information, which is useful for AD and MCI diagnosis. In particular, by regarding each template as a specific view, some recent developments in multiview learning using multitemplate MRI data have been discussed in this chapter. This chapter first introduces a multiview feature representation method by using multiple templates (selected automatically from data). Then, four recent multiview learning approaches for AD/MCI diagnosis are presented. Specifically, a feature filter-based method provides a direct way to make use of multiview feature representations, with experimental results demonstrating significant improvements in AD and MCI diagnosis. For learning an optimal feature representation based on features generated from multitemplates, an MMRL method is proposed for improving the discriminative power of the original multiview features. To take advantage of the guidance information provided by different templates, a view-centralized learning method is developed to treat each template as a main view, while the other templates are used as side information source. Then, a relationship-induced multiview learning method is presented to model both the relationships among templates and those among subjects, to further boost the performance of the AD/MCI classification model. The increased accuracy, sensitivity, and specificity achieved by these approaches indicate that multiview learning methods are a viable alternative to clinical diagnosis of brain alterations associated with cognitive impairment.

REFERENCES

Argyriou, A., Micchelli, C.A., Pontil, M., Ying, Y., 2008. A spectral regularization framework for multi-task structure learning. In: Advances in Neural Information Processing Systems 20. MIT Press, MA, USA, pp. 25–32.

Ashburner, J., Friston, K.J., 2000. Voxel-based morphometry—the methods. NeuroImage 11, 805–821.

Ashburner, J., Hutton, C., Frackowiak, R., Johnsrude, I., Price, C., Friston, K., 1998. Identifying global anatomical differences: deformation-based morphometry. Hum. Brain Map. 6, 348–357.

Alzheimer's Association, 2013. Alzheimer's disease facts and figures. Alzheimer's Dement. 9, 208–245.

Basha, T., Moses, Y., Kiryati, N., 2013. Multi-view scene flow estimation: a view centered variational approach. Int. J. Comput. Vision 101, 6–21.

Baxter, J., 1997. A Bayesian/information theoretic model of learning to learn via multiple task sampling. Mach. Learn. 28, 7–39.

Bengio, Y., 2009. Learning deep architectures for AI. In: Foundations and Trends® in Machine Learning, vol. 2. Now Publishers, Boston, USA, pp. 1–127.

Bozzali, M., Filippi, M., Magnani, G., Cercignani, M., Franceschi, M., Schiatti, E., et al., 2006. The contribution of voxel-based morphometry in staging patients with mild cognitive impairment. Neurology 67, 453–460.

Burges, C.J., 1998. A tutorial on support vector machines for pattern recognition. Data Min. Knowl. Disc. 2, 121–167.

Caruana, R., 1997. Multitask learning. Mach. Learn. 28, 41–75.

Chan, D., Janssen, J.C., Whitwell, J.L., Watt, H.C., Jenkins, R., Frost, C., et al., 2003. Change in rates of cerebral atrophy over time in early-onset Alzheimer's disease: longitudinal MRI study. Lancet 362, 1121–1122.

Chang, C.C., Lin, C.J., 2011. Libsvm: a library for support vector machines. ACM Trans. Intell. Syst. Technol. 2, 27.

Chung, M., Worsley, K., Paus, T., Cherif, C., Collins, D., Giedd, J., et al., 2001. A unified statistical approach to deformation-based morphometry. NeuroImage 14, 595–606.

Cortes, C., Vapnik, V., 1995. Support-vector networks. Mach. Learn. 20, 273–297.

Cuingnet, R., Gerardin, E., Tessieras, J., Auzias, G., Lehéricy, S., Habert, M.O., et al., 2011. Automatic classification of patients with Alzheimer's disease from structural MRI: a comparison of ten methods using the ADNI database. NeuroImage 56, 766–781.

Davatzikos, C., 1998. Mapping image data to stereotaxic spaces: applications to brain mapping. Hum. Brain Map. 6, 334–338.

Davatzikos, C., Genc, A., Xu, D., Resnick, S.M., 2001. Voxel-based morphometry using the RAVENS maps: methods and validation using simulated longitudinal atrophy. NeuroImage 14, 1361–1369.

Davatzikos, C., Fan, Y., Wu, X, Shen, D., Resnick, S.M., 2008. Detection of prodromal Alzheimer's disease via pattern classification of magnetic resonance imaging. Neurobiol. Aging 29, 514–523.

Dickerson, B.C., Goncharova, I., Sullivan, M., Forchetti, C., Wilson, R., Bennett, D., et al., 2001. MRI-derived entorhinal and hippocampal atrophy in incipient and very mild Alzheimer's disease. Neurobiol. Aging 22, 747–754.

Fan, Y., Shen, D., Gur, R.C., Gur, R.E., Davatzikos, C., 2007. COMPARE: classification of morphological patterns using adaptive regional elements. IEEE Trans. Med. Imaging 26, 93–105.

Fan, Y., Resnick, S.M., Wu, X, Davatzikos, C., 2008. Structural and functional biomarkers of prodromal Alzheimer's disease: a high-dimensional pattern classification study. NeuroImage 41, 277–285.

Fox, N., Warrington, E., Freeborough, P., Hartikainen, P., Kennedy, A., Stevens, J., et al., 1996. Presymptomatic hippocampal atrophy in Alzheimer's disease: a longitudinal MRI study. Brain 119, 2001–2007.

Frey, B.J., Dueck, D., 2007. Clustering by passing messages between data points. Science 315, 972–976.

Frisoni, G., Testa, C., Zorzan, A., Sabattoli, F., Beltramello, A., Soininen, H., et al., 2002. Detection of grey matter loss in mild Alzheimer's disease with voxel based morphometry. J. Neurol. Neurosurg. Psychiat. 73, 657–664.

Gaser, C., Nenadic, I., Buchsbaum, B.R., Hazlett, E.A., Buchsbaum, M.S., 2001. Deformation-based morphometry and its relation to conventional volumetry of brain lateral ventricles in MRI. NeuroImage 13, 1140–1145.

Goldszal, A.F., Davatzikos, C., Pham, D.L., Yan, M.X., Bryan, R.N., Resnick, S.M., 1998. An image-processing system for qualitative and quantitative volumetric analysis of brain images. J. Comput. Assist. Tomograp. 22, 827–837.

Gong, Y., Ke, Q., Isard, M., Lazebnik, S., 2014. A multi-view embedding space for modeling internet images, tags, and their semantics. Int. J. Comput. Vision 106, 210–233.

Grau, V., Mewes, A., Alcaniz, M., Kikinis, R., Warfield, S.K., 2004. Improved watershed transform for medical image segmentation using prior information. IEEE Trans. Med. Imaging 23, 447–458.

Guyon, I., Weston, J., Barnhill, S., Vapnik, V., 2002. Gene selection for cancer classification using support vector machines. Mach. Learn. 46, 389–422.

He, X., Cai, D., Niyogi, P., 2005. Laplacian score for feature selection. In: Advances in Neural Information Processing Systems. MIT Press, MA, USA, pp. 507–514.

Hinrichs, C., Singh, V., Mukherjee, L., Xu, G., Chung, M.K., Johnson, S.C., 2009. Spatially augmented LPboosting for AD classification with evaluations on the ADNI dataset. NeuroImage 48, 138–149.

Ho, T.K., 1998. The random subspace method for constructing decision forests. IEEE Trans. Pattern Anal. Mach. Intell. 20, 832–844.

Hua, X., Leow, A.D., Lee, S, Klunder, A.D., Toga, A.W., Lepore, N., et al., 2008. 3D characterization of brain atrophy in Alzheimer's disease and mild cognitive impairment using tensor-based morphometry. NeuroImage 41, 19–34.

Hua, X., Hibar, D.P., Ching, C.R., Boyle, C.P., Rajagopalan, P., Gutman, B.A., et al., 2013. Unbiased tensor-based morphometry: improved robustness and sample size estimates for Alzheimer's disease clinical trials. NeuroImage 66, 648–661.

Jack, C.R., Bernstein, M.A., Fox, N.C., Thompson, P., Alexander, G., Harvey, D., et al., 2008. The Alzheimer's disease neuroimaging initiative (ADNI): MRI methods. J. Mag. Reson. Imaging 27, 685–691.

Jenkinson, M., Smith, S., 2001. A global optimisation method for robust affine registration of brain images. Med. Image Anal. 5, 143–156.

Jenkinson, M., Bannister, P., Brady, M., Smith, S., 2002. Improved optimization for the robust and accurate linear registration and motion correction of brain images. NeuroImage 17, 825–841.

Jolliffe, I., 2002. Principal Component Analysis. Wiley Online Library.

Joseph, J., Warton, C., Jacobson, S.W., Jacobson, J.L., Molteno, C.D., Eicher, A., et al., 2014. Three-dimensional surface deformation-based shape analysis of hippocampus and caudate nucleus in children with fetal alcohol spectrum disorders. Hum. Brain Map. 35, 659–672.

Kipps, C., Duggins, A., Mahant, N., Gomes, L., Ashburner, J., McCusker, E., 2005. Progression of structural neuropathology in preclinical Huntington's disease: a tensor based morphometry study. J. Neurol. Neurosurg. Psychiat. 76, 650–655.

Koikkalainen, J., Lötjönen, J., Thurfjell, L., Rueckert, D., Waldemar, G., Soininen, H., 2011. Multi-template tensor-based morphometry: application to analysis of Alzheimer's disease. NeuroImage 56, 1134–1144.

Kruskal, J.B., 1964. Multidimensional scaling by optimizing goodness of fit to a nonmetric hypothesis. Psychometrika 29, 1–27.

Leow, A.D., Klunder, A.D., Jack, C.R., Toga, A.W., Dale, A.M., Bernstein, M.A., et al., 2006. Longitudinal stability of MRI for mapping brain change using tensor-based morphometry. NeuroImage 31, 627–640.

Leporé, N., Brun, C., Chou, Y.Y., Lee, A., Barysheva, M., Zubicaray, G.I.D., et al., 2008. Multi-atlas tensor-based morphometry and its application to a genetic study of 92 twins. In: Medical Image Computing and Computer-Assisted Intervention Workshop on Mathematical Foundations of Computational Anatomy, New York, USA, pp. 48–55.

Li, S.Z., Zhu, L., Zhang, Z, Blake, A., Zhang, H., Shum, H., 2002. Statistical learning of multi-view face detection. In: European Conference on Computer Vision, pp. 67–81.

Liu, M., Zhang, D., Shen, D., 2012. Ensemble sparse classification of Alzheimer's disease. NeuroImage 60, 1106–1116.

Liu, M., Zhang, D., Shen, D., 2014. Hierarchical fusion of features and classifier decisions for Alzheimer's disease diagnosis. Hum. Brain Map. 35, 1305–1319.

Liu, M., Zhang, D., Shen, D., 2015. View-centralized multi-atlas classification for Alzheimer's disease diagnosis. Hum. Brain Map. 36, 1847–1865.

Liu, M., Zhang, D., Shen, D., 2016. Relationship induced multi-template learning for diagnosis of Alzheimer's disease and mild cognitive impairment. IEEE Trans. Med. Imaging, http://dx.doi.org/10.1109/TMI.2016.2515021.

Magnin, B., Mesrob, L., Kinkingnéhun, S., Pélégrini-Issac, M., Colliot, O., Sarazin, M., et al., 2009. Support vector machine-based classification of Alzheimer's disease from whole-brain anatomical MRI. Neuroradiology 51, 73–83.

Min, R., Wu, G., Cheng, J., Wang, Q., Shen, D., 2014a. Multi-atlas based representations for Alzheimer's disease diagnosis. Hum. Brain Map. 35, 5052–5070.

Min, R., Wu, G., Shen, D., 2014b, Maximum-margin based representation learning from multiple atlases for Alzheimer's disease classication. In: Presented at the Medical Image Computing and Computer-Assisted Intervention, New York, USA, pp. 635–642.

Mueller, S.G., Weiner, M.W., Thal, L.J., Petersen, R.C., Jack, C.R., Jagust, W., et al., 2005. Ways toward an early diagnosis in Alzheimer's disease: the Alzheimer's disease neuroimaging initiative (ADNI). Alzheimer's Dement. 1, 55–66.

Peng, H., Long, F., Ding, C., 2005. Feature selection based on mutual information criteria of max-dependency, max-relevance, and min-redundancy. IEEE Trans. Pattern Anal. Mach. Intell. 27, 1226–1238.

Pereira, F., Mitchell, T., Botvinick, M., 2009. Machine learning classifiers and fMRI: a tutorial overview. NeuroImage 45, S199–S209.

Shen, D., Davatzikos, C., 2002. HAMMER: hierarchical attribute matching mechanism for elastic registration. IEEE Trans. Med. Imaging 21, 1421–1439.

Shen, D., Davatzikos, C., 2003. Very high-resolution morphometry using mass-preserving deformations and HAMMER elastic registration. NeuroImage 18, 28–41.

Sled, J.G., Zijdenbos, A.P., Evans, A.C., 1998. A nonparametric method for automatic correction of intensity nonuniformity in MRI data. IEEE Trans. Med. Imaging 17, 87–97.

Sotiras, A., Davatzikos, C., Paragios, N., 2013. Deformable medical image registration: a survey. IEEE Trans. Med. Imaging 32, 1153–1190.

Tang, S., Fan, Y., Wu, G., Kim, M., Shen, D., 2009. RABBIT: rapid alignment of brains by building intermediate templates. NeuroImage 47, 1277–1287.

Teipel, S.J., Born, C., Ewers, M., Bokde, A.L., Reiser, M.F., Möller, H.J., et al., 2007. Multivariate deformation-based analysis of brain atrophy to predict Alzheimer's disease in mild cognitive impairment. NeuroImage 38, 13–24.

Thomas, A., Ferrar, V., Leibe, B., Tuytelaars, T., Schiel, B., Van Gool, L., 2006. Towards multi-view object class detection. In: IEEE Conference on Computer Vision and Pattern Recognition, New York, USA, pp. 1589–1596.

Thompson, P.M., Mega, M.S., Woods, R.P., Zoumalan, C.I., Lindshield, C.J., Blanton, R.E., et al., 2001. Cortical change in Alzheimer's disease detected with a disease-specific population-based brain atlas. Cerebral Cortex 11, 1–16.

Tibshirani, R., 1996. Regression shrinkage and selection via the lasso. J. R. Stat. Soc. B 267-288.

Vincent, L., Soille, P., 1991. Watersheds in digital spaces: an efficient algorithm based on immersion simulations. IEEE Trans. Pattern Anal. Mach. Intell. 13, 583–598.

Wang, Y., Nie, J., Yap, P.T., Shi, F., Guo, L., Shen, D., 2011. Robust deformable-surface-based skull-stripping for large-scale studies. In: Medical Image Computing and Computer-Assisted Intervention, Toronto, Canada, pp. 635–642.

Wang, Y., Nie, J., Yap, P.T., Li, G., Shi, F., Geng, X., et al., 2014. Knowledge-guided robust MRI brain extraction for diverse large-scale neuroimaging studies on humans and non-human primates. PLoS ONE 9 (1), e77810.

Whitford, T.J., Grieve, S.M., Farrow, T.F., Gomes, L., Brennan, J., Harris, A.W., et al., 2006. Progressive grey matter atrophy over the first 2-3 years of illness in first-episode schizophrenia: a tensor-based morphometry study. NeuroImage 32, 511–519.

Xu, C., Tao, D., Xu, C., 2014. Large-margin multi-view information bottleneck. IEEE Trans. Pattern Anal. Mach. Intell. 36, 1159–1572.

Yap, P.T., Wu, G., Zhu, H., Lin, W., Shen, D., 2009. Timer: tensor image morphing for elastic registration. NeuroImage 47, 549–563.

Zhang, D., Shen, D., 2012. Multi-modal multi-task learning for joint prediction of multiple regression and classification variables in Alzheimer's disease. NeuroImage 59, 895–907.

Zhang, Y., Brady, M., Smith, S, 2001. Segmentation of brain MR images through a hidden Markov random field model and the expectation-maximization algorithm. IEEE Trans. Med. Imaging 20, 45–57.

Zhang, D., Wang, Y., Zhou, L, Yuan, H., Shen, D., 2011. Multimodal classification of Alzheimer's disease and mild cognitive impairment. NeuroImage 55, 856–867.

Zou, H., Hastie, T., 2005. Regularization and variable selection via the elastic net. J. R. Stat. Soc. B 67, 301–320.

CHAPTER

Machine learning as a means toward precision diagnostics and prognostics

10

A. Sotiras, B. Gaonkar, H. Eavani, N. Honnorat, E. Varol, A. Dong, C. Davatzikos

University of Pennsylvania, Philadelphia, PA, United States

CHAPTER OUTLINE

10.1 INTRODUCTION

The advent of new imaging modalities providing high-resolution depictions of the anatomy (Hsieh, 2009; Atlas, 2009; Liang and Lauterbur, 2000) and function (Detre et al., 1992; Baxton, 2009; Phelps, 2000) of the brain in disease and health has resulted in medical imaging becoming increasingly indispensable for patients' healthcare. The way medical images are analyzed has been greatly shaped by machine learning, which has found application in numerous fields, including image segmentation (Pham et al., 2000; Heimann and Meinzer, 2009), image registration

Machine Learning and Medical Imaging. http://dx.doi.org/10.1016/B978-0-12-804076-8.00010-4

(Maintz and Viergever, 1998; Sotiras et al., 2013), image fusion, and computer-aided diagnosis (Sajda, 2006).

Among the reasons behind the success of machine learning in medical imaging are increased automation, high sensitivity and specificity. Machine learning has fueled automated approaches that provide measurements by circumventing the error-prone and labor-intensive manual procedures that are typically involved in traditional interest-based analyses. Moreover, contrary to conventional automated approaches, such as mass univariate analyses, high-dimensional multivariate pattern analysis (MPVA) (Norman et al., 2006; McIntosh and Mišić, 2013) driven methods fully harness the potential of high-dimensional data by examining statistical relationships between elements that span the whole image domain.

Integrating information from the whole domain while taking advantage of prior knowledge allows MVPA techniques to identify and measure subtle and spatially complex structural and functional changes in the brain that are induced by disease or pharmacological interventions, despite important normal variability. As a consequence, sophisticated pattern analysis techniques have been employed to identify disease-specific signatures and elucidate the selective vulnerability of different brain networks to different pathologies (Mour ao-Miranda et al., 2005; Zhu et al., 2008; Sun et al., 2009; Zeng et al., 2012; Davatzikos et al., 2008a; Klöppel et al., 2008; Vemuri et al., 2008; Duchesne et al., 2008). This has led to the construction of sensitive biomarkers that are able to quantify the risk of developing a disease, track the disease progression or the effect of pharmacological interventions in clinical trials, and deliver patient-specific diagnosis before measurable clinical effects occur.

Neurodegenerative diseases such as Alzheimer's disease (AD) have been in the epicenter of the development of computerized biomarkers. Machine learning diagnostic and prognostic tools have been developed to identify patients with neurodegenerative diseases such as dementia (Davatzikos et al., 2008a; Klöppel et al., 2008; Vemuri et al., 2008; Duchesne et al., 2008; Fan et al., 2008b; McEvoy et al., 2009; Hinrichs et al., 2009; Gerardin et al., 2009; Hinrichs et al., 2011; Zhang et al., 2011; Cuingnet et al., 2011), to differentially distinguish between AD and frontotemporal dementia (FTD) (Davatzikos et al., 2008b), or to predict clinical progression of patients (Fan et al., 2008a; Davatzikos et al., 2011). Studies into mental disorders have also benefited from the application of computer-assisted imaging techniques. Fully automated classification algorithms have been successfully applied to diagnose a wide range of neurological and psychiatric diseases, including schizophrenia (Davatzikos et al., 2005; Koutsouleris et al., 2015), psychosis (Koutsouleris et al., 2012), and depression (Mour ao-Miranda et al., 2011).

However, despite important advances and successes, there remain significant challenges to be addressed. Three of the most important challenges comprise (i) dimensionality reduction; (ii) interpreting the learned model; and (iii) elucidating disease heterogeneity.

The first challenge tackles a fundamental problem one encounters when training machine learning models to identify imaging signatures towards automated diagnosis and prognosis, namely the sheer dimensionality of imaging data along with the relatively small sample size that is typically available. This problem is further exacerbated by the increasing resolution of the imaging data, as well as the increasing availability of multiparametric imaging, which further increase the dimensionality and complexity of the available data. The main challenge here is summarizing the imaging information through a reduced number of features that is compatible with the sample size of a typical imaging study, while retaining the necessary information that will allow the learning system to recognize relevant imaging patterns.

The second challenge relates to the interpretability of the learned model. Machine learning models are generally treated as "black-boxes" that provide us with an index of the presence of a disease. While this index may be used to perform diagnosis, it does not inform us about how each brain region contributes to the construction of the discriminative multivariate pattern. This information is of significant importance since it provides key insight regarding the selective vulnerability of different brain systems to different pathologies, thus elucidating disease mechanisms, paving the road for more effective treatments.

The third challenge addresses the problem of elucidating disease heterogeneity. Most existing methodologies assume a single, unifying pathophysiological process and aim to reveal it by identifying a unique imaging pattern that can distinguish between healthy and diseased populations, or between two subgroups of patients. However, this assumption effectively disregards ample evidence for the heterogeneous nature of brain diseases. Neurodegenerative, neuropsychiatric and neurodevelopmental disorders are characterized by high clinical heterogeneity, which is likely due to the underlying neuroanatomical heterogeneity of various pathologies. Elucidating disease heterogeneity is crucial for deepening our understanding of the involved pathological mechanisms, and may lead to more precise diagnosis, prognosis, and specialized treatment.

In this chapter, we are going to present solutions for tackling the aforementioned challenges. In Section 10.2, we present a clustering and statistical-based approached for dimensionality reduction of both structural and functional data. In Section 10.3, we detail an efficient technique for deriving statistical significance maps in classification tasks using support vector machines (SVMs), while in Section 10.4 we present a palette of techniques to tackle disease heterogeneity under different methodological assumptions. In Section 10.5, we provide evidence of the usefulness of machine learning techniques at the clinical and research level, while Section 10.6 concludes the chapter.

10.2 DIMENSIONALITY REDUCTION

During the past decades, the advent of high-resolution imaging techniques has given rise to high-dimensional, complex clinical datasets consisting of hundreds of patient scans that comprise millions of voxels (Van Essen et al., 2013; Satterthwaite et al., 2014). The high dimensionality of the data, along with the relatively small sample size that is typically available, poses an important challenge when aiming to holistically analyze imaging patterns in association with brain diseases. This challenge is further exacerbated by the increasing availability of multiparametric imaging data, which results in an additional increase in both the dimensionality and complexity of the data. Moreover, the emergence of sophisticated imaging techniques, such as diffusion tensor imaging and functional magnetic resonance imaging that derive complex representations of the axonal anatomy and brain activity, not only emphasizes the aforementioned challenge but also calls for tailored analysis tools.

To address this challenge, dimensionality reduction is typically performed. The aim is to extract, in an optimal way, a few imaging features, thus reducing the dimensionality of the data to a level that is compatible with the sample size of a typical imaging study. Additionally, these features should retain the important image information that will allow for the identification of imaging patterns that offer good predictive value.

Numerous approaches have been proposed to reduce the dimensionality of imaging data. Dimensionality reduction methods can be typically categorized into two groups: (i) spatial grouping and (ii) statistically driven reduction, depending on the driving assumption behind its method. In the first case, one aims to group together elements that are spatially close and similar in terms of imaging measurements. In the second case, emphasis is put on considering together image elements that vary in consistent ways across the population. This taxonomy may be further refined by taking into account the nature of the imaging data the method handles.

10.2.1 DIMENSIONALITY REDUCTION THROUGH SPATIAL GROUPING

Methods of this class typically formulate the problem as clustering, and dimensionality reduction is achieved by summarizing the data through a restricted set of features that correspond to the estimated clusters. Features are typically extracted by computing a single average measure per estimated cluster, while clusters are obtained by segmenting the brain into contiguous regions that encompass elements with imaging measurements that are similar to each other. Defining an appropriate similarity measure is of significant importance for the success of these methods and should take into account the nature of the imaging signal, leading to data-specific algorithms. In the following, we summarize two such algorithms for structural magnetic resonance imaging (MRI) scans and resting-state functional MRI (rs-fMRI), respectively.

10.2.2 SPATIAL GROUPING OF STRUCTURAL MRI

Structural imaging based on magnetic resonance provides information regarding the integrity of gray and white matter structures in the brain, making it an integral part of the clinical assessment of patients with dementia, such as AD and FTD. Automated classification approaches applied on structural MRI data have shown promise for the diagnosis of AD and the identification of whole-brain patterns of disease-specific atrophy. In this scenario, when dimensionality reduction is performed prior to a supervised machine learning task, such as patient classification, it is appealing to adopt a *supervised* clustering approach. The goal is to exploit prior information (ie, disease diagnosis) in order to generate regions of interest that are adapted not only to the data, but also to the machine learning task, with the aim to improve its performance.

This supervised approach was adopted by the COMPARE method (Fan et al., 2007) that aims to perform classification of morphological patterns using adaptive regional elements. COMPARE extracts spatially smooth clusters that can be used to train a classifier to predict patient diagnosis by combining information stemming from both the imaging signal and subjects' diagnosis. The two types of information are integrated at each image location, p, in a multiplicative fashion though the score:

$$s(p) = P(p)C(p), \tag{10.1}$$

where $C(p)$ measures the spatial consistency of the imaging signal, while $P(p)$ measures discriminative power. More precisely, P is calculated as the following leave-one-out absolute Pearson correlation:

$$P(p) = \mathtt{argmin}_{i=1..n}|\rho(p,i)|, \tag{10.2}$$

where $\rho(p,i)$ denotes the Pearson correlation measured between the imaging signal at p and the classification labels when excluding the ith subject/sample. The consistency $C(p)$ is the intra-class coefficient measuring the proportion of neighboring feature variance that is explained by the inter-subject variability (McGraw and Wong, 1996; Fan et al., 2007). It takes values between 0 and 1, with higher values indicating that the variance of the measurements across neighboring brain location is small with respect to the inter-subject variability of the imaging signal. As a result, the score $s(p)$ is bounded between 0 and 1, with values close to 1 indicating that the imaging signal around p is simultaneously highly reliable and discriminative (ie, highly correlated or anticorrelated with patient diagnosis).

This score map is subsequently smoothed, and its gradient is used in conjunction with a watershed segmentation algorithm (Vincent and Soille, 1991) to partition the brain into different regions (Fig. 10.1 presents brain regions generated by watershed from white matter tissue density maps of demented and normally aging subjects (Shen and Davatzikos, 2002; Fan et al., 2007)). These regions are then refined by considering only locations that optimize the classification power of the extracted features. This is performed in a region growing fashion where, initially, only the node of the region with the highest discriminative score is selected, and adjacent locations are incrementally aggregated as long as the discriminative power does not decrease. The previous steps are summarized in Algorithm 10.1. This approach extracts a single connected component per watershed region. Each component comprises highly discriminative elements whose average imaging signal may be used as a feature for training a classifier, such as an SVM (Vapnik, 2000).

ALGORITHM 10.1 PSEUDO-CODE FOR COMPARE.

```
Data: Structural MRI images and group labels
Result: Regional elements optimized for classification
/* Initialization                                              */
for each image location p do
    compute discriminative power P(p) using Eq. 10.2;
    compute spatial consistency C(p);
    compute score s(p) using Eq. 10.1;
end
/* Region segmentation                                         */
Gaussian smoothing of s(p);
gradient computation of score map;
parcellate score map using watershed;
/* Region refinement/feature extraction                        */
for each parcel do
    select voxel with highest score s(p);
    /* Region growing                                          */
    while regional discriminative power increases do
        find voxel with highest discriminative
        power at the border of growing region;
        add voxel to growing region;
    end
end
```

FIG. 10.1

Coronal and sagittal cross-sectional views of a watershed segmentation generated by COMPARE.

The efficiency of this supervised dimensionality reduction scheme was demonstrated in classifying patients with clinical dementia versus normal individuals, as well as distinguishing between schizophrenic patients and normal controls (Fan et al., 2007). COMPARE is generic and can be readily extended to incorporate different forms of prior information, such as the ones provided in regression and multiclass classification settings.

10.2.2.1 Spatial grouping of rs-fMRI

Functional MRI is an imaging technique that reflects neural activity in the whole brain by detecting changes in oxygen consumption. Resting-state fMRI reveals brain networks (Biswal, 2012) by evaluating regional interactions that occur when the subjects are relaxed and do not perform a particular mental task during the brain scan. The dynamic nature of this imaging modality results in extremely voluminous and complex datasets, underlining the need for efficient dimensionality reduction.

Clustering approaches have received considerable attention towards reducing the dimensionality of functional data. This is due to the fact that clustering is not only an efficient way to reduce the spatial dimension of rs-fMRI data, but also a biologically meaningful one. Clustering sheds light on the mid-scale functional structure of the brain that is considered to follow a *segregation and integration principle*. In other words, information is thought to be processed by compact groups of neurons in the brain, or *functional units*, that collaborate together towards addressing complex tasks (Tononi et al., 1994).

Clustering approaches typically aim to divide the brain into spatially smooth areas that are likely to correspond to the functional units that constitute the brain. This is usually performed by first representing the brain in the form of a graph, where nodes represent brain locations and edges connect nodes that correspond to spatially adjacent locations. The weight of the edges represents the strength of the connectivity between nodes and is estimated by computing the similarity between the rs-fMRI signals that are measured at each node. The similarity is commonly measured by the Pearson correlation or partial correlation (Smith et al., 2011). Once the node is constructed, adjacent brain locations that are strongly connected are grouped together in the same parcel.

Numerous methods have been proposed for this task. Among the most popular methods, one may cite hierarchical clustering (Cordes et al., 2002), normalized clustering (Shen et al., 2010; Craddock et al., 2012), *k*-means (Bellec et al., 2010), region growing (Blumensath et al., 2013; Heller et al., 2006), and Markov random fields (MRFs) (Ryali et al., 2013; Golland et al., 2007; Honnorat et al., 2015). Different methods exhibit distinct advantages and disadvantages. Generally, many of the above methods are either initialization dependent (eg, region growing (Blumensath et al., 2013; Heller et al., 2006) and *k*-means (Bellec et al., 2010)), or rely on complex models that involve a large number of parameters (Golland et al., 2007). As a result, they are sensitive to initialization and suffer from limitations related to the employed heuristics (eg, hierarchical clustering may lead to the creation of poorly fit parcels at coarser scales (Cordes et al., 2002; Honnorat et al., 2015)), and the large number of inferred parameters that may negatively impact the quality of the locally optimal solution that is obtained (Golland et al., 2007). Moreover, not all methods produce contiguous parcels.

In order to address the aforementioned concerns, a discrete MRF approach, termed GraSP (Graph based segmentation with Shape Priors), was recently introduced in (Honnorat et al., 2015). This approach adopts an *exemplar-based clustering*

FIG. 10.2

Functional parcellation of the left hemisphere of the brain, projected on an inflated brain surface.

approach that allows for the reduction of the number of parameters by representing the rs-fMRI time series of each parcel by the signal of one of the nodes that are assigned to it. Thus, the clustering framework is simplified through the encoding of the parcels with their functional center. Only one parameter needs to be chosen by the user, the label cost K. This corresponds to the cost of introducing a new parcel into the clustering result, which indirectly determines the size of the produced parcels (Delong et al., 2012). Contrary to other MRF clustering methods (Ryali et al., 2013), these parcels are connected (Fig. 10.2 presents a functional parcellation that was produced for reducing the dimension of rs-fMRI scans from a neurodevelopmental study (Satterthwaite et al., 2014)). Parcel connectedness is promoted without any spatial smoothing by the inclusion of a shape prior term into the MRF energy formulation (Veksler, 2008; Gulshan et al., 2010). Lastly, the energy is optimized in a single step, thus removing the need for initialization and specification of a stopping criterion.

The MRF energy is summarized in the following form:

$$\min \sum_{p} V_p(l_p) + L_p(\{l_p\}) + S_p(\{l_p\}),$$

where p denotes a node of the brain graph, l_p the parcel that should contain this node, $V_p(l_p)$ is a cost that decreases when the node p is assigned to a parcel l_p with highly correlated rs-fMRI signal, $L_p(\{l_p\})$ penalizes by a positive cost K the introduction of a parcel of functional center p, and the $S_p(\{l_p\})$ are the shape priors that enforce the connectedness of each parcel p. This energy is optimized by exploiting advanced solvers (Delong et al., 2012) that could provide a substantial advance over existing methods. Experimental results on large datasets demonstrated that this approach is capable of generating parcels that are all highly coherent, while the overall parcellation is slightly more reproducible than the results produced by hierarchical clustering and normalized cuts (Honnorat et al., 2015).

10.2.3 STATISTICALLY DRIVEN DIMENSIONALITY REDUCTION

The second family of dimensionality reduction methods is based on exploiting statistical procedures to project the data in a space of lower dimension. This is typically performed within a regularized matrix factorization framework where a tall matrix $\mathbf{X}$ comprising N samples/images of dimension D, each one arrayed per column $\left(\mathbf{X} = [\mathbf{x}_1, \ldots, \mathbf{x}_N], \mathbf{x}_i \in \mathbb{R}^D\right)$, is approximated by a product of matrices ($\mathbf{X} \approx \mathbf{BC}$). $\mathbf{B}$ is a matrix of the basis vectors that span the estimated subspace, and $\mathbf{C}$ contains the loading coefficients that provide the low-dimensional description of the data. Depending on the implemented modeling assumptions, $\mathbf{B}$ and $\mathbf{C}$ exhibit different properties.

Among the most widely used methods of this class, one may cite principal component analysis (PCA) (Friston et al., 1993; Strother et al., 1995; Hansen et al., 1999) and independent component analysis (ICA). PCA maps the data to a lower dimensional space through an orthogonal linear transformation, while preserving the variance of the data. The transformation is performed in such a way that basis vectors (or principal components) are ordered in descending order according to the amount of the variance they explain. ICA (McKeown et al., 1998; Calhoun et al., 2001; Beckmann and Smith, 2004), on the other hand, maps the data into a set of components that are as statistically independent from each other as possible.

Despite their widespread use in neuroimaging, conventional factorization methods that are used for dimensionality reduction suffer from limitations related to the interpretability and the reproducibility of the derived representation. For example, both PCA and ICA estimate components and coefficients of mixed sign, thus approximating the data through complex mutual cancelation between component regions of opposite sign. This complex modeling of the data, along with the fact that the estimated components highly overlap due to their often global spatial support, results in representations that lack specificity. In other words, while it is possible to interpret individual components, it is difficult to associate a specific brain region with a specific effect. Lastly, conventional factorization methods, and especially PCA, aim to approximate the data as faithfully as possible, thus capturing both relevant and irrelevant sources of variation, resulting in poor generalization in unseen datasets.

Next, we summarize our group's work to derive efficient, interpretable, and reproducible statistically driven dimensionality reduction techniques for structural and functional MRI data. The key idea behind the developed frameworks is to derive highly parsimonious representations. The reason behind this choice is twofold: (i) sparse methods achieve a higher degree of specificity than conventional multivariate analysis methods (Lee and Seung, 1999); and (ii) they show improved generalizability (Avants et al., 2010). The above underline the importance of sparsity in brain modeling and analysis (Daubechies et al., 2009). Sparsity is introduced in a tailored way, taking into account the specific properties of different imaging modalities.

10.2.3.1 Statistically driven dimensionality reduction of structural MRI

Structural MRI scans typically encode the physical properties of the image tissue through the use of non-negative values. This fact allows us to derive parsimonious representations through the use of non-negative matrix factorization (NNMF) (Lee and Seung, 1999; Sotiras et al., 2015). NNMF was proposed as an analytical and interpretive tool in structural neuroimaging in Sotiras et al. (2015).

NNMF produces a factorization that constrains the elements of both the components and the loading coefficients matrix to be non-negative. This is achieved by minimizing the following energy:

$$\underset{\mathbf{B},\mathbf{C}}{\text{minimize}}\ ||\boldsymbol{X} - \mathbf{BC}||_F^2 \text{ subject to } \mathbf{B} \geq 0,\ \mathbf{C} \geq 0,$$

where $\mathbf{B} = [\mathbf{b}_1, \ldots, \mathbf{b}_K]$, $\mathbf{b}_i \in \mathbb{R}^D$, $\mathbf{C} = [\mathbf{c}_1, \ldots, \mathbf{c}_N], \mathbf{c}_i \in \mathbb{R}^K$, N is the number of images/samples, and K is the dimension of the estimated subspace. The non-negativity constraints lead to a sparse, parts-based representation (Lee and Seung, 1999). NNMF minimizes the reconstruction error by aggregating variance through positively weighting variables of the data matrix that tend to co-vary across the population. This provides a useful way of reducing the dimensionality of structural data. The structural data of each individual is approximated through an additive combination of the estimated components. In general, the estimated components identify regions that co-vary across individuals in a consistent way, thus forming patterns of structural co-variance that may potentially be parts of underlying networks or influenced by common mechanisms. The loading coefficients matrix $\mathbf{C}$ summarizes the integrity of each pattern of structural co-variance in each individual with a scalar value. These values provide an efficient and interpretable representation, and can be used for comparing the integrity of structural networks across individuals.

This method was applied in a cohort of normal aging adults and was compared against PCA and ICA in (Sotiras et al., 2015). It was shown to derive representations that are more parsimonious and coherent than the ones estimated by PCA and ICA. Moreover, the derived representation was quantitatively shown to be more relevant to age-related phenomena, while allowing for accurate age prediction as demonstrated through cross-validated age regression experiments. NNMF captured less of the variance in the data than PCA and ICA, resulting in higher reconstruction error. However, the high prediction accuracy suggests that the discarded information is not pertinent, leading to the conclusion that NNMF is able to retain important information while discarding irrelevant variations, which may potentially lead to increased generalizability. Indeed, split-sample experiments demonstrated that the non-negative components are more reproducible than the principal components.

Typical components estimated by NNMF are shown in Fig. 10.3. Note that the representation amounts to a soft clustering that segments the brain into structurally coherent units in a data-driven way by exploiting group statistics. The derived components are characterized by high spatial connectedness even though spatial smoothness was not explicitly enforced in the design of the method. Another important characteristic of the obtained representation is the symmetry of the

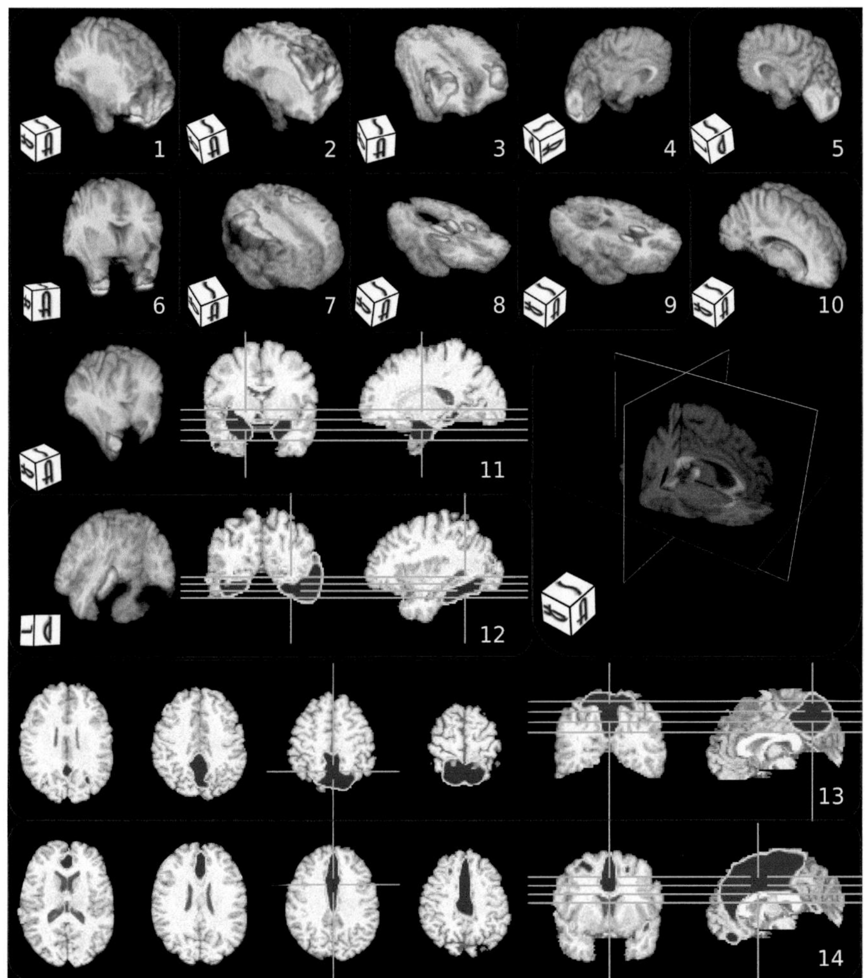

FIG. 10.3

Characteristic components estimated by NNMF. Different visualization strategies were used in order to enhance the visual perception of the components (note that the 2D images use radiographic convention). Warmer colors correspond to higher values. Note the alignment with anatomical regions: (1) prefrontal cortex; (2) superior frontal cortex; (3) superior lateral cortex; (4) left occipital lobe; (5) right occipital lobe; (6) inferior anterior temporal; (7) motor cortex; (8) thalamus and putamen; (9) head of caudate; (10) periventricular structures; (11) amygdala and hippocampus; (12) fusiform; (13) medial parietal including precuneus; (14) anterior and middle cingulate.

Source: Reprinted from Sotiras, A., Resnick, S.M., Davatzikos, C., 2015. Finding imaging patterns of structural covariance via Non-Negative Matrix Factorization. NeuroImage 108, 1–16.

estimated components. This symmetry is completely data-driven and it breaks when not supported by the group statistics. Lastly, and most importantly, the estimated components are not a solely statistical construct, but highly correspond to known structural and functional networks of the brain, or in some cases reflect underlying pathological processes.

10.2.3.2 Statistically driven dimensionality reduction of functional MRI

Resting-state functional MRI is typically used to analyze interactions between regions, aiming to reveal the brain's functional organization. Resting-state functional connectivity is used to reveal functional networks that can be found consistently in healthy populations by examining the connectivity between all pairs of regions in the brain. Pearson correlation is typically used to measure connectivity between different brain regions due to its simplicity and robustness (Smith et al., 2011; Lashkari et al., 2010). The resulting functional connectivity data is high dimensional and of mixed sign. The high dimensionality of the data makes subsequent group-wise analysis and interpretation of results difficult, underlining the need for an efficient and interpretable dimensionality reduction framework. However, the mixed sign nature of the data does not allow the application of the previously described non-negative framework. Instead, sparsity needs to be explicitly modeled through the inclusion of sparsity-inducing priors in the objective function of the matrix factorization framework.

A sparsity-based matrix factorization approach was proposed for functional connectivity data in Eavani et al. (2015). In this approach, each subject-specific correlation matrix $\mathbf{\Sigma}_n$ is approximated by a non-negative sum of sparse rank one matrices $\mathbf{b}_k\mathbf{b}_k^{\mathrm{T}}$. These sparse rank one matrices can be interpreted as functionally coherent subsets of brain regions, or sparse patterns of connectivity (SCPs), which occur in many of the subjects. A non-negative, subject-specific combination of SCPs, denoted by the set of coefficients $\mathbf{c}_n$, approximates the input correlation matrix $\mathbf{\Sigma}_n$:

$$
\begin{aligned}
&\underset{\mathbf{B},\mathbf{C}}{\text{minimize}} \sum_{n=1}^{N} \left\| \mathbf{\Sigma}_n - \mathbf{B}\,\mathrm{diag}(\mathbf{c}_n)\,\mathbf{B}^{\mathrm{T}} \right\|_F^2 \\
&\text{subject to} \\
&\qquad \|\mathbf{b}_k\|_1 \le \lambda, \quad k = 1, \dots, K, \\
&\qquad -1 \le \mathbf{b}_k(i) \le 1, \quad \max_i |\mathbf{b}_k(i)| = 1, \quad i = 1, \dots, P, \\
&\qquad \mathbf{c}_n \ge 0, \qquad n = 1, \dots, N,
\end{aligned}
$$

where $\mathbf{B} = [\mathbf{b}_1, \mathbf{b}_2, \dots, \mathbf{b}_K]$. Sparse connectivity patterns (SCPs) provide a useful manner of reducing the dimensionality of the connectivity data, while summarizing the connectivity within each SCP in each individual with a scalar SCP coefficient value. These values can be used for comparing functional connectivity across individuals.

Applied to a normative sample of young adults, the resulting SCPs were shown to be reproducible across datasets, while explaining more of the variance in the second-order connectivity data when compared to spatial and temporal ICA (Calhoun et al.,

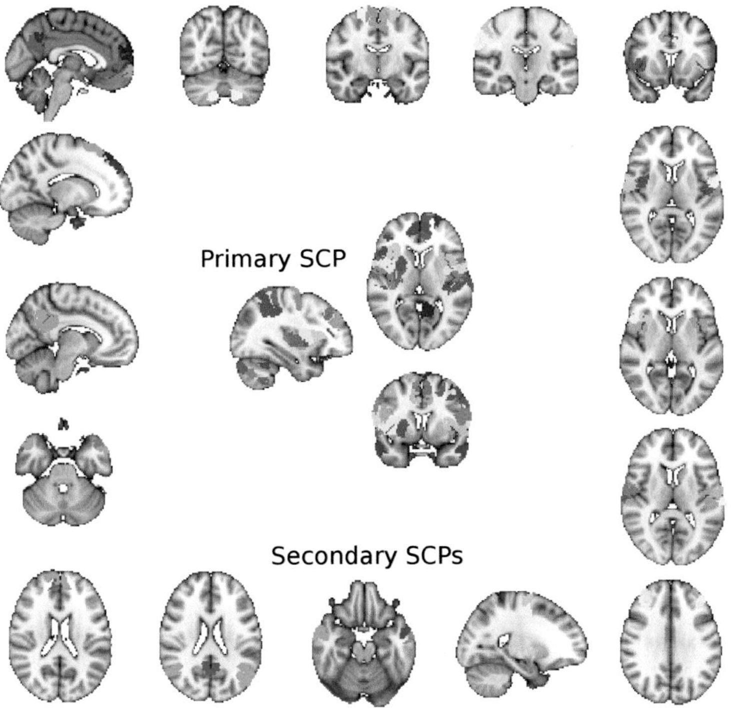

FIG. 10.4

Primary SCP (middle) showing the cingulum, operculum (red-yellow) and anticorrelated with the default mode (blue-light blue). Sixteen of its associated secondary SCPs are shown around it.

2003; Smith et al., 2012). This method can also be applied within a hierarchical framework, where each "primary" SCP with a large spatial extent can be split up into multiple smaller "secondary" SCPs, providing greater spatial specificity. Fig. 10.4 shows a large primary SCP with contributions from the operculum and anticorrelated with parts of the default mode. Its associated secondary SCPs, which represent a much smaller set of regions, are shown around it. Note the high specificity of the representation that is due to the sparsity of the derived networks.

10.3 MODEL INTERPRETATION: FROM CLASSIFICATION TO STATISTICAL SIGNIFICANCE MAPS

Once an appropriate set of features has been extracted, machine learning algorithms are employed to analyze neuroimaging data. This is typically performed by treating machine learning algorithms as "black-boxes" that are to able to integrate patterns of disease-induced morphological signals into subject-specific indices. Even though these indices carry significant prognostic and diagnostic value, this usage paradigm does not fully exploit the potential of machine learning methods. In order to fully harness this potential, it is important to be able interpret the learned model in terms of identifying brain regions that significantly contribute to the construction of the discriminative pattern. This could significantly improve our understanding of the disease mechanisms that selectively influence-specific brain systems, while at the same time making the automated system transparent to human expert-driven verification.

In this section, we present such a framework for SVMs (Burges, 1998; Vapnik, 2000). SVMs are significantly popular in neuroimaging (Fan et al., 2007; Cuingnet et al., 2011; Klöppel et al., 2008; Gaonkar and Davatzikos, 2013; Batmanghelich et al., 2012; Varol et al., 2013), mainly due to their simplicity and the fact that the resulting problem is convex, allowing for efficient and globally optimal solutions. The SVM operates by constructing a hyperplane in a high-dimensional space that separates samples from two classes (eg, disease group vs. healthy controls) by the largest possible margin (see Fig. 10.5 for an illustration of the principle). The hyperplane coefficients denoted by $\mathbf{w}^*$ and b^* are estimated by solving the following optimization problem:

$$\{\mathbf{w}^*, b^*\} = \min_{\mathbf{w},b} \frac{1}{2}||\mathbf{w}||^2 + C\sum_{i=1}^{m} \xi_i$$
$$\text{such that } y_i(\mathbf{w}^{\mathrm{T}}\mathbf{x}_i + b) \geq 1 - \xi_i \;\; \forall i = 1, \ldots, N$$
$$\xi_i \geq 0 \;\; \forall i = 1, \ldots, N,$$

where $\mathbf{x}_i \in \mathbb{R}^D$ denotes the vectorized image of the ith subject of the study, $y_i \in \{+1, -1\}$ denotes its respective binary label, ξ_i denotes the slack variable that accounts for the case where the classes are not separable, and C is a penalty parameter on the training error. The weight vector $\mathbf{w}^* \in \mathbb{R}^D$ describes the combination of all imaging elements that, along with the intercept b^*, best discriminates between the two classes.

It is tempting to use the weight image $\mathbf{w}^*$ to interpret the model by assigning more importance to elements that have higher weights. However, this is problematic (Haufe et al., 2014) and does not readily yield to a well-understood p-value-based statistical paradigm. One way to derive such a paradigm on the basis of SVM theory is to use permutation testing (see Fig. 10.6 for an illustration of the process). This is typically performed by generating a large number of shuffled instances of data labels by random permutations. Each shuffled instance is subsequently used for training one SVM, generating a new hyperplane parameterized by a vector $\mathbf{w}$. Thus for every element of $\mathbf{w}$, there is a set of possible values, each one corresponding to a specific shuffling of the labels. Collecting these values allows for the construction of the corresponding empirically obtained null distribution. Finally, comparing each component of $\mathbf{w}^*$ with the corresponding null distribution allows for the estimation of statistical significance. The number of permutations determines the minimal

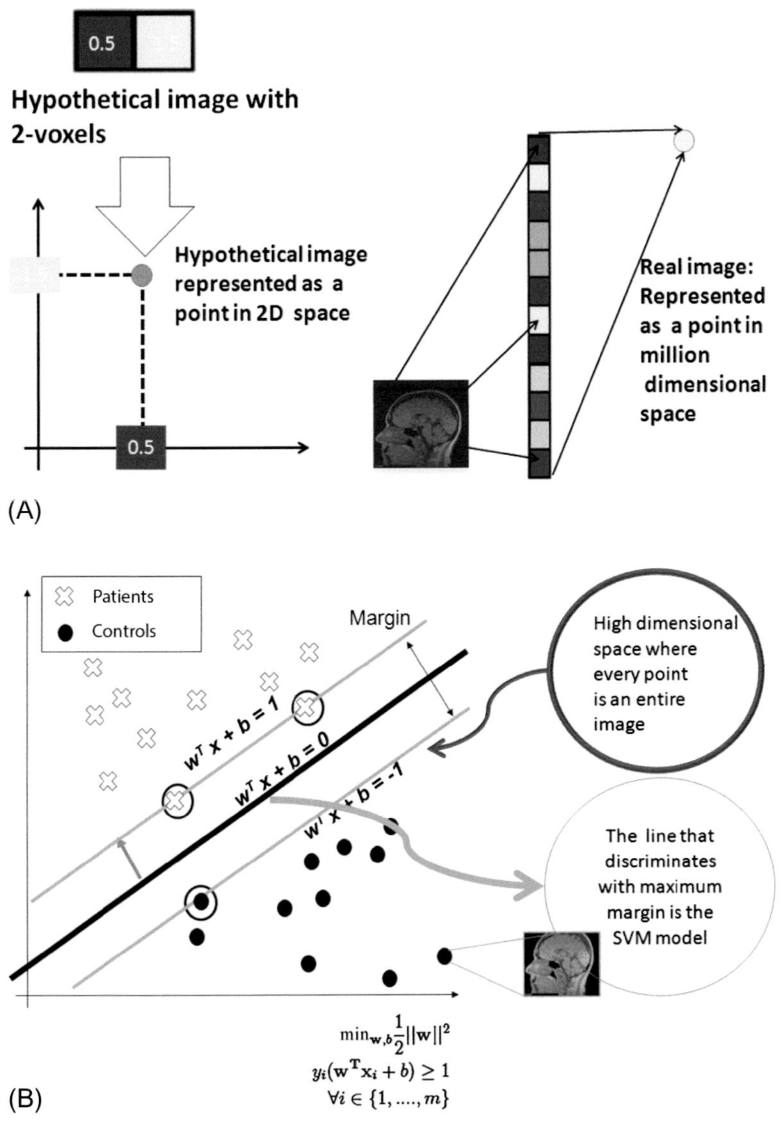

FIG. 10.5

The concept of imaging-based diagnosis using SVMs. (A) Images are treated as points located in a high-dimensional space. (B) The maximum margin principle of classification used in SVMs. Dots and crosses represent imaging scans taken from two groups. Even though the two groups cannot be separated on the basis of values along any single dimension, the combination of two dimensions gives perfect separation. This corresponds to the situation where a single anatomical region may not provide the necessary discriminative power between groups, whereas the multivariate SVM can still find the relevant hyperplane.

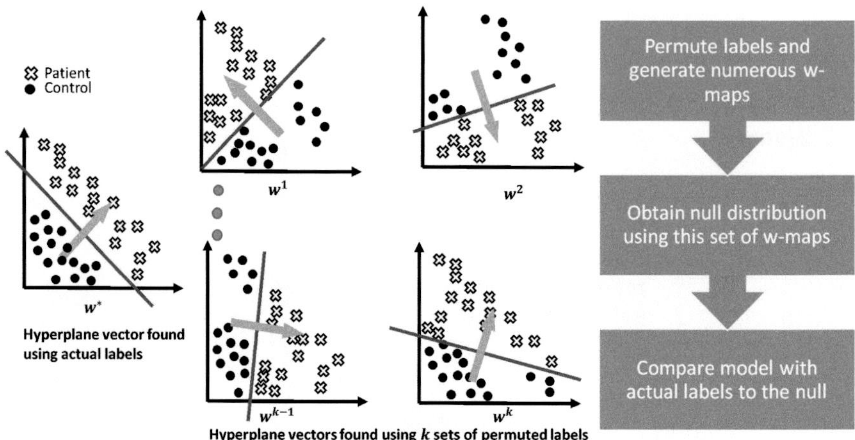

FIG. 10.6

Illustration of the permutation testing procedure.

Source: Reprinted from Gaonkar, B., Davatzikos, C., 2013. Analytic estimation of statistical significance maps for support vector machine based multi-variate image analysis and classification. NeuroImage 78, 270–283.

obtainable p-value as well as the resolution of the p-value. Increasing the number of permutations to a high number that will allow for the estimation of low p-values requires training a high number of support vector classifiers, which in turn requires a considerable amount of computational time and resources. Thus, a framework that would allow the analytic computation of the p-values in a computationally economic fashion would be of significant value.

Such a theoretical framework, that describes an analytic alternative to permutation testing, was introduced in Gaonkar and Davatzikos (2013) and Gaonkar et al. (2015). This analytical framework makes use of a certain set of simplifying assumptions that can be applied to the SVM formulations in high-dimensional spaces to derive an approximate null distribution, obviating the need for performing actual permutation testing. The first assumption regards the high-dimension, low-sample size setting that is typically encountered in medical imaging. In such a setting, it is always possible to find hyperplanes that can separate any possible labeling of points/samples. Thus when using linear SVMs, for any permutation of the labeling, one can always find a separating hyperplane that perfectly separates the training data. This allows us to use the hard margin SVM formulation. The second assumption regards the observation that, for most permutations, most data are support vectors. Taken together, these assumptions indicate that, for most permutations, it is possible to solve the following optimization problem:

$$\min_{\mathbf{w},b} \frac{1}{2}||\mathbf{w}||^2 \text{ such that } \mathbf{X}\mathbf{w} + \mathbf{J}b = \mathbf{y},$$

where **J** is a column matrix of ones, and **X** is a tall matrix with each row representing one image. Solving for **w** yields

$$\mathbf{w} = \underbrace{\mathbf{X}^{\mathrm{T}}\left[\left(\mathbf{X}\mathbf{X}^{\mathrm{T}}\right)^{-1} + \left(\mathbf{X}\mathbf{X}^{\mathrm{T}}\right)^{-1}\mathbf{J}\left(-\mathbf{J}^{\mathrm{T}}\left(\mathbf{X}\mathbf{X}^{\mathrm{T}}\right)^{-1}\mathbf{J}\right)^{-1}\mathbf{J}^{\mathrm{T}}\left(\mathbf{X}\mathbf{X}^{\mathrm{T}}\right)^{-1}\right]}_{=\mathbf{C}}\mathbf{y}.$$

Note that each element w_j of **w** is expressed as a linear combination of elements of **y**. Thus, it is possible to hypothesize about the probability distribution of the elements of **w** given the distributions of y_i. If y_i attains any of the labels with equal probability, then $E(y_i) = 0$ and $Var(y_i) = 1$, which in turns lead to $E(w_j) = 0$ and $Var(w_j) = \sum_{i=1}^{N} C_{ij}^2$. At this point, there is an analytical method to approximate the mean and the variance of the null distributions of components w_j of **w**. By taking advantage of the Lyapunov central limit theorem, it was demonstrated in Gaonkar and Davatzikos (2013) and Gaonkar et al. (2015) that the distribution of the individual components of **w** can be approximated using the normal distribution for a sufficiently large number of subjects. Thus, w_j^* computed by an SVM model using true labels can now simply be compared to the previous distribution and statistical inference can be made. The accuracy of this approximation is shown in Fig. 10.7. Note that the analytic and experimental p-maps are visually indistinguishable, while the scatter plot shows a good correspondence between the experimental and analytical p-values. Fig. 10.8 shows the regions that were identified by the method in Gaonkar et al. (2015) to be most statistically significant for classifying AD patients from controls. Note that the hippocampal complex, along with parahippocampal regions and amygdala, are clearly highlighted.

10.4 HETEROGENEITY

A common assumption behind automated group analysis methods applied in neuroimaging is that there is a single pattern that distinguishes the two contrasted groups. In other words, most approaches assume a single pathophysiological process that converts healthy controls to patients, and aim to reveal it through monistic analysis. However, this approach ignores ample evidence regarding the heterogeneous nature of diseases. For example, autism (Geschwind and Levitt, 2007; Jeste and Geschwind, 2014), schizophrenia (Buchanan and Carpenter, 1994; Koutsouleris et al., 2008; Zhang et al., 2015), Parkinson's disease (Graham and Sagar, 1999; Lewis et al., 2005), AD (Murray et al., 2011; Noh et al., 2014) or mild cognitive impairment (MCI) (Huang et al., 2003; Whitwell et al., 2007) are all characterized by clinical heterogeneity (see Fig. 10.9A for a graphical illustration of the problem).

Disentangling disease heterogeneity may greatly contribute to our understanding and lead to more accurate diagnosis, prognosis, and targeted treatment. We present here three recently proposed methods to tackle disease heterogeneity under different

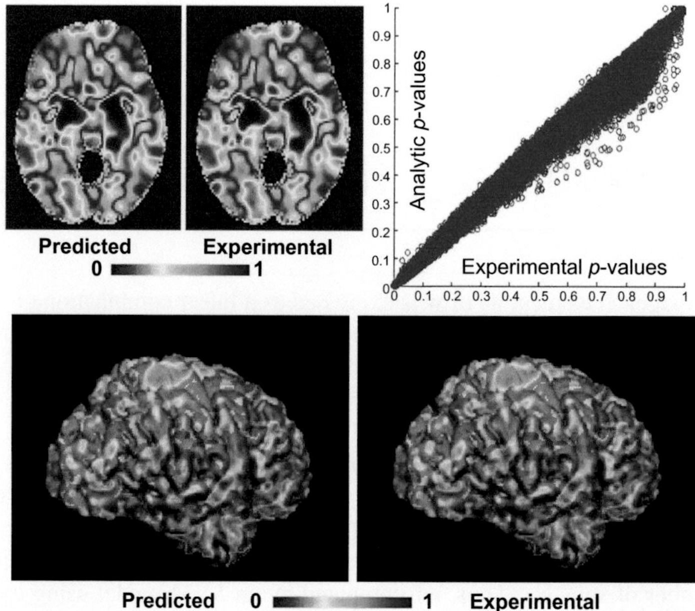

FIG. 10.7

(Top left) Analytic and experimental p-value maps thresholded at 0.01 are overlaid on the template brain. (Top right) A scatter plot of p-values comparing experimental and analytical p-values. (Bottom) A 3D rendering representing the predicted and experimental p-value maps.

Source: Reprinted from Gaonkar, B., Davatzikos, C., 2013. Analytic estimation of statistical significance maps for support vector machine based multi-variate image analysis and classification. NeuroImage 78, 270–283.

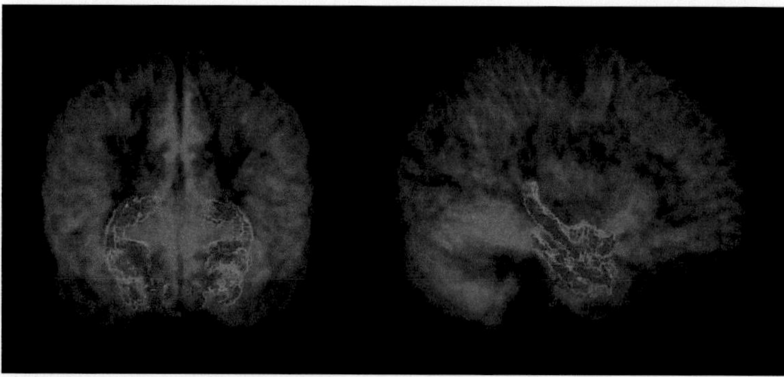

FIG. 10.8

3D views of the hippocampal and parahippocampal regions used by the SVM ($\alpha \leq 0.01$ FDR corrected).

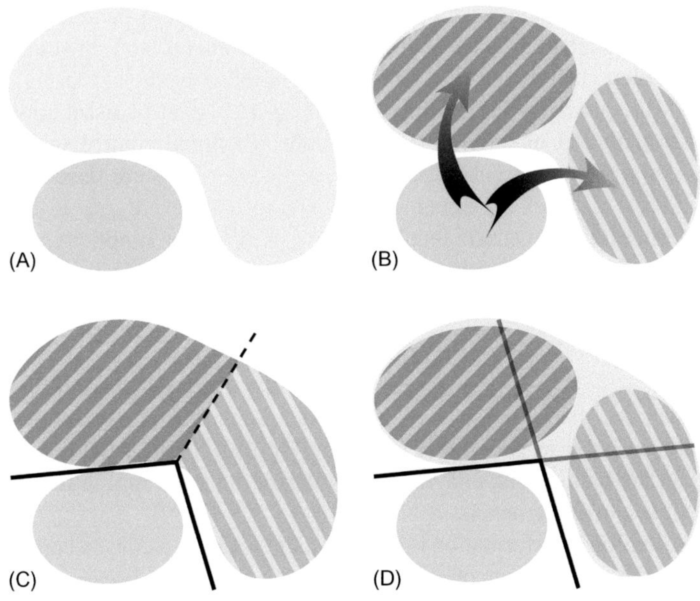

FIG. 10.9

Heterogeneity problem setting and different methods. (A) Problem setting. (B) CHIMERA. (C) HYDRA. (D) Mixture of Experts.

methodological assumptions. The first method is based on a generative clustering framework; the second adopts a purely discriminative approach, while the third combines discrimination and clustering.

10.4.1 GENERATIVE FRAMEWORK

The first method treats subjects as points in a high-dimensional feature space, where both the patient and the normal control group may be viewed as point distributions. In such a setting, the disease heterogeneity can be addressed by partitioning the patient distribution with a clustering method. However, directly clustering the patients would be driven by the distances between individuals, which would result in clustering the largest factor of data variability instead of the disease effect. In order to address this challenge, the generative approach proposed in Dong et al. (2016) considers the disease effect to be a transformation from the normal control distribution to the patient distribution (see Fig. 10.9b for a graphical illustration).

As a consequence, the patient distribution can be generated by transforming the normal control distribution with the assumption that if points of the patients had been spared from the disease, they would be covered by the normal control distribution. Heterogeneous disease effects are modeled by considering multiple distinct transformations. These transformations can be found by solving for a distribution matching of the true patient and generated patient distributions. The distribution matching takes into account both imaging and covariate features (known variables, such as age, sex, and height). In this way, the clustering of patient distribution is regularized by the structure of the normal control distribution.

More formally, let us assume that there are M normal control subjects, $\mathbf{X} = \{x_1, ..., x_M\}$, and N patient subjects, $\mathbf{Y} = \{y_1, ..., y_N\}$. They are described by two sets of features: a set of D_1-dimensional imaging features, $x_m^v, y_n^v \in \mathbb{R}^{D_1}$; and a set of D_2-dimensional covariate features, $x_m^c, y_n^c \in \mathbb{R}^{D_2}$. For simplicity, subjects are denoted in compact vector forms: $x_m = (x_m^v, x_m^c)$, $y_n = (y_n^v, y_n^c)$. The clustering model minimizes the following energy $\mathcal{E}$:

$$\mathcal{E}(\mathbf{X}, \mathbf{Y}, \Theta) = -\mathcal{L}(\mathbf{X}, \mathbf{Y}, \Theta) + \mathcal{R}(\Theta),$$

where Θ denotes the parameters of the model, such as transformations that are applied to $\mathbf{X}$ in order to generate $\mathbf{Y}$; $\mathcal{L}$ is the log-likelihood of the distributions $\mathbf{X}$ and $\mathbf{Y}$ given the parameters; and $\mathcal{R}$ is a regularization term aiming to improve the stability of the clustering results.

The distribution transformation is denoted as $\mathbf{T}$, which is a convex combination of K linear transformations, each one corresponding to a different disease effect. $\mathbf{T}$ maps the imaging feature x_m of a normal control sample to the patient distribution, while keeping its covariate feature unchanged: $\mathbf{T}(x_m) = (\sum_{k=1}^{K} \zeta_{km}(A_k x_m^v + b_k), x_m^c)$. The distribution matching is conducted as a variant of the coherent point drift algorithm Myronenko and Song (2010). Each transformed normal control point is considered as a centroid of a spherical Gaussian cluster, and patient points are treated as independent and identically distributed data generated by a Gaussian Mixture Model (GMM) with equal weights for each cluster. The data likelihood of this mixture model is optimized during the distribution matching, where covariate features are embedded in the distance between points with a multikernel setting. These model assumptions lead to the log-likelihood term $\mathcal{L}$ being:

$$\mathcal{L}(\mathbf{X}, \mathbf{Y}, \Theta) = \sum_{n=1}^{N} \log \sum_{m=1}^{M} \frac{1}{M} \frac{r^{D_2/2}}{(\sqrt{2\pi}\sigma)^{D_1+D_2}} \exp\left\{ \frac{\|y_n^v - \sum_{k=1}^{K} \zeta_{km}(A_k x_m^v + b_k)\|^2 + r\|y_n^c - x_m^c\|^2}{-2\sigma^2} \right\}.$$

The Frobenius norm of $A_k - \mathbf{I}$ and the ℓ_2 norm of b_k are to be regularized, where $\mathbf{I}$ is an identity matrix. This regularization is equivalent to posing Gaussian priors for the parameters:

$$\mathcal{R}(\Theta) = \frac{\lambda_1}{2\sigma^2} \sum_k \|b_k\|_2^2 + \frac{\lambda_2}{2\sigma^2} \sum_k \|A_k - \mathbf{I}\|_F^2$$

The energy objective $\mathcal{E}$ is optimized with an expectation-maximization Moon (1996) approach. The heterogeneous disease subgroups of patients are further clustered by the estimated transformations.

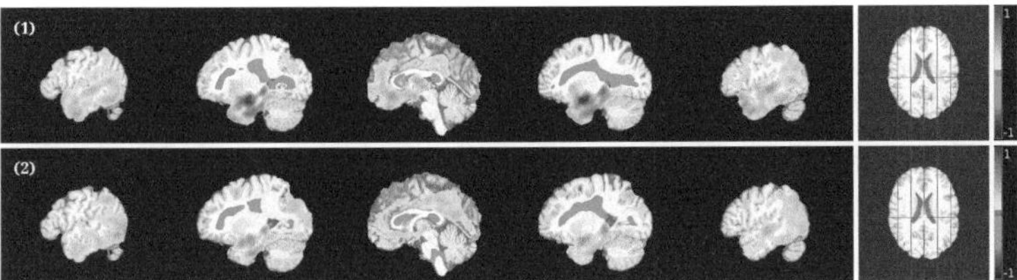

FIG. 10.10

VBM performed on gray matter RAVENS (Davatzikos, 1998) maps between (1) Subgroup 1 and Control group; (2) Subgroup 2 and Control group. Group comparison results are overlaid on the registration template image. Regions that are significant under a corrected threshold of FDR = 0.01 are shown. Color maps indicate the scale of the *t*-statistic. Warmer colors indicate volume loss, while colder colors indicate volume increase.

This method was applied to an AD dataset[1] comprising 390 T1 structural MRI scans with 177 AD patients and 213 normal controls. Multi-Atlas ROI volumes were generated and used as imaging features, while age and sex information were used as covariate features. With the cross-validated parameters, two subgroups were discovered. Voxel-based morphometry (VBM) Ashburner and Friston (2000) was employed to examine the differences between the estimated subgroups and the control population. The VBM results obtained from gray matter group comparisons are shown in Fig. 10.10. Subgroup 1 has more gray matter atrophy in limbic lobe and frontal insular regions, while it exhibits unique deep gray matter atrophy in basal ganglia. Subgroup 2 exhibits unique parietal and occipital gray matter atrophy on both lateral and medial structures.

10.4.2 DISCRIMINATIVE FRAMEWORK

The second method takes a purely discriminative approach. It is based upon the observation that, in high-dimensional spaces, the modeling capacity of linear SVMs is theoretically rich enough to discriminate between two homogeneous classes. However, while two classes may be linearly separable with high probability, the resulting margin could be small. This case arises, for example, when one class is generated by a multimodal distribution that models a heterogeneous process. This may be remedied by the use of nonlinear classifiers, allowing for larger margins and thus, better generalization. However, while kernel methods, such as Gaussian kernel SVM, provide nonlinearity, they lack interpretability when aiming to characterize heterogeneity.

[1] http://adni.loni.usc.edu/

In order to tackle the aforementioned limitations, a novel maximum margin nonlinear learning algorithm for simultaneous binary classification and subtype identification, termed HYDRA (HeterogeneitY through DiscRiminative Analysis) was introduced in Varol et al. (2015) and Varol et al. (2016). HYDRA aims to tackle disease subtype discovery in a principled machine learning framework. Neuroanatomical or genetic subtypes are effectively captured by multiple linear hyperplanes, which form a convex polytope that separates two groups (eg, healthy controls from pathologic samples); each face of this polytope effectively defines a disease subtype (see Fig. 10.9c for a graphical illustration).

More formally, let us assume an imaging (or genetic) dataset consisting of n binary labeled d-dimensional data points ($\mathcal{D} = (\mathbf{x}_i, y_i)_{i=1}^{n}, \mathbf{x}_i \in \mathbb{R}^d$ and $y_i \in \{-1, 1\}$). The maximum margin polytope that separates the assumed heterogenous patients from the controls can be solved by optimizing the following objective:

$$\min_{\substack{\{\mathbf{w}_j, b_j\}_{j=1}^{K} \\ \{s_{i,j}\}_{i,j}^{n^-,K}}} \sum_{j=1}^{K} \frac{\|\mathbf{w}_j\|_2^2}{2} + C \sum_{\substack{i|y_i=+1 \\ j}} \frac{1}{K} \max\{0, 1 - \mathbf{w}_j^{\mathrm{T}} \mathbf{x}_i - b_j\} + C \sum_{\substack{i|y_i=-1 \\ j}} s_{i,j} \max\{0, 1 + \mathbf{w}_j^{\mathrm{T}} \mathbf{x}_i + b_j\}.$$

The first term encourages maximum average margin across all K faces of the convex polytope classifier. The second term forces the control samples to be confined **inside** the polytope with slack. Lastly, the third term enforces the patient samples to lie **outside** the assigned face of the polytope with slack. The assignment of patient samples to the faces of the polytope is handled by the indicator variable $s_{i,j}$, which can be estimated by solving a linear program. The objective is optimized by following a two-step procedure that iterates between assigning samples to faces of the polytope, and solving for hyperplanes that maximize the overall margin. This is similar in spirit to unsupervised clustering methods, such as K-means, where centroids and assignments are iteratively solved.

This approach was applied to a genetic dataset comprising 53 AD patients and 68 cognitively normal (CN) older adults (see demographic information in Table 10.1), obtained from the ADNI study[2]. ADNI genotyping is performed using the Human610-Quad Bead-Chip (Illumina, Inc., San Diego, CA), which results in a set of 620,901 single nucleotide polymorphisms (SNPs) and copy number variation markers. Due to the weak, or spurious, signal in most of the genome, the features were pruned and only SNP loci that were found to be associated with AD in a recent large-scale genome-wide association study (Lambert et al., 2013) were kept. This resulted in a reduced set of 18 SNPs that were represented by using two binary variables that encode the presence of major-major or major-minor alleles, thus raising the total number of features to 36.

In order to estimate the optimal number of clusters, a reproducibility analysis was performed. The reproducibility of the clustering was evaluated at $K = 1, \ldots, 9$ by using the Adjusted Rand Index (Hubert and Arabie, 1985). This analysis suggested that

[2]http://adni.loni.usc.edu/data-samples/genetic-data/

Table 10.1 Demographic and Clinical Characteristics of Healthy Controls, AD Patients (Left) and the Estimated Genetic-Driven Subtypes of AD (Right)

Genetic Heterogeneity in Alzheimer's Disease						
	AD vs. CN (n = 121)			**AD Subgroups (n = 53)**		
	CN (n = 68)	AD (n = 53)	p-value[b]	Group 1 (n = 34)	Group 2 (n = 19)	p-value[c]
Age (years)	76.08 ± 4.672	76.08 ± 7.188	0.9944	75.27 ± 5.981	77.43 ± 8.872	0.3184
Sex (female), n (%)	33 (50)	25 (52.08)	0.828	15 (50)	10 (55.56)	0.7163
MMSE	28.44 ± 2.367	19.06 ± 5.05	1.228e-24	18.77 ± 5.71	19.56 ± 3.807	0.6057
Apoε-4 genotype[a], n (%)	20 (30.3)	31 (64.58)	0.0002108	29 (96.67)	2 (11.11)	1.901e-15

[a]Denotes subjects with at least one Apoε-4 allele present. [b]p-value estimated using two-tailed t-test to compare AD with CN. [c] p-value estimated using analysis of variance (ANOVA) to compare the two estimated AD subgroups.

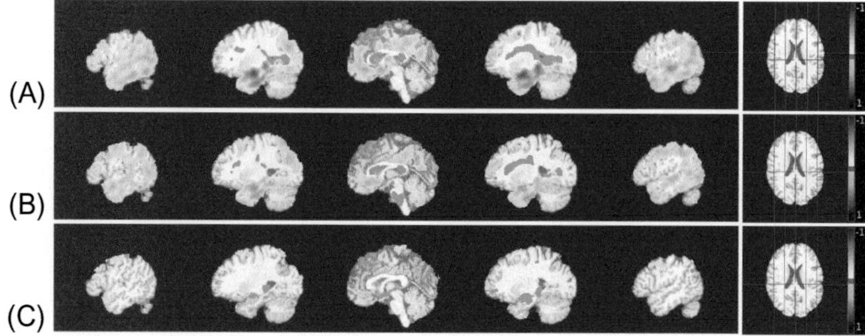

FIG. 10.11

The anatomic differences between the two genetic subtypes of AD: Axial views of gray matter group comparisons of (A) Controls vs. first AD subgroup; (B) Controls vs. second AD subgroup; and (C) first AD subgroup vs. second AD subgroup are visualized. For (A) and (B), colder colors indicate relative GM volume increases (CN < AD subgroups), while warmer colors correspond to relative GM volume decreases (CN > AD subgroups). Similarly for (C), colder colors indicate relative GM volume increases (first AD subgroup < second AD subgroup), while warmer colors correspond to relative GM volume decreases (first AD subgroup > second AD subgroup). Both groups exhibit atrophy in the temporal lobe and posterior medial cortex, while white matter lesions are present in the periventricular area. However, the first AD subgroup, which mainly comprises Apo-ε4 carriers, is characterized by significantly more hippocampal and entorhinal cortex atrophy.

two clusters were appropriate for capturing the intrinsic dimensionality of the genetic heterogeneity associated with AD. The optimal genotype clustering is visualized by contrasting the imaging phenotypes of the estimated subgroups against the healthy control population through morphometric analysis using RAVENS (see Fig. 10.11A and B). Correction for multiple comparisons was performed by controlling for False Discovery Rate (FDR). The results were thresholded at $q < 0.05$. It can be observed that at the $K = 2$ cluster level (see Fig. 10.11), the estimated subgroups were associated with distinct patterns of structural brain alterations. The first subgroup had increased temporal lobe atrophy (see Fig. 10.11A), including focal atrophy in the hippocampus and entorhinal cortex, as well as increased white matter lesion load. The second subgroup was characterized by diffuse temporal lobe atrophy (see Fig. 10.11B), including periventricular white matter lesions.

In summary, HYDRA seamlessly integrates clustering and discrimination in a coherent framework by solving a piecewise linear classifier that bears common geometric properties with convex polytopes. Discrimination is achieved by constraining one class in the interior of the polytope, while at the same time maximizing the margin between examples and class boundary. On the other hand, clustering is performed by associating disease samples with different faces of the polytope, and hence to different disease processes. Thus, each face of the polytope informs us about the distinct foci of disease effects that distinguish the patients from the healthy control subjects. This coupling between clustering and classification allows for segregating patients based on disease effects rather than global anatomy.

10.4.3 GENERATIVE DISCRIMINATIVE FRAMEWORK

The last approach that aims to identify heterogeneous subgroups in patient populations is based upon a mixture-of-experts (MOE) framework. The MOE framework was initially proposed for vowel discrimination within speech recognition (Jacobs et al., 1991) and, later on, as a fast and efficient alternative to "kernel" SVMs (Ladicky and Torr, 2011; Fu et al., 2010). While kernel SVMs can successfully model nonlinear separation boundaries between groups, they suffer from a major limitation in neuroimaging applications, namely the lack of interpretability of the results. In a kernel-based method, the data is projected into a higher dimensional space prior to being classified and the nonlinear separating boundary in the original feature space is not explicitly computed.

The presented joint generative-discriminative approach tackles this shortcoming by combining a generative clustering model with a discriminative classification/regression model (Eavani et al., 2016). Using this combination of unsupervised clustering (mixture) with supervised classification/regression (expert), it approximates the nonlinear boundary that separates the two classes with a piecewise linear separating boundary, providing us the identification of the subgroups as well as the multivariate patterns that discriminate each subgroup from the reference group (see Fig. 10.9d for a graphical illustration). The data is modeled using a mixture of distributions, such as fuzzy c-means, which assigns a soft subgroup membership to each subject in the affected group. The linear boundary between each affected subgroup and the reference group can be found using a linear classifier, such as a linear SVM.

This is a general framework that can be applied to any dataset, using any appropriate mixture model and expert classifier. Using a combination of fuzzy c-means and ℓ_2-loss linear SVMs, Eavani et al. (2016) found heterogeneity in the manner in which normal older individuals age in terms of functional connectivity. Of the two subgroups that were found within the older individuals (relative to a reference group of younger individuals), the authors found that one set of individuals had increased functional connectivity between the bilateral frontal and insula regions. Upon further investigation, the same set of individuals were found to have specific cognitive abilities (executive function and visual processing) comparable to that of the younger group, while the rest had worse cognition than the younger group, as expected due to aging. It is possible that the increased bilateral connectivity in the subset of older people acts as a compensatory mechanism, resulting in better cognitive performance for their age (Fig. 10.12).

These results produced using MOE have significant clinical implications in terms of identifying functional bio-markers of resilient aging, which is a very active topic of research in brain aging. These results provide important biological clues to the wide variation in cognitive performance that is normally seen in older individuals.

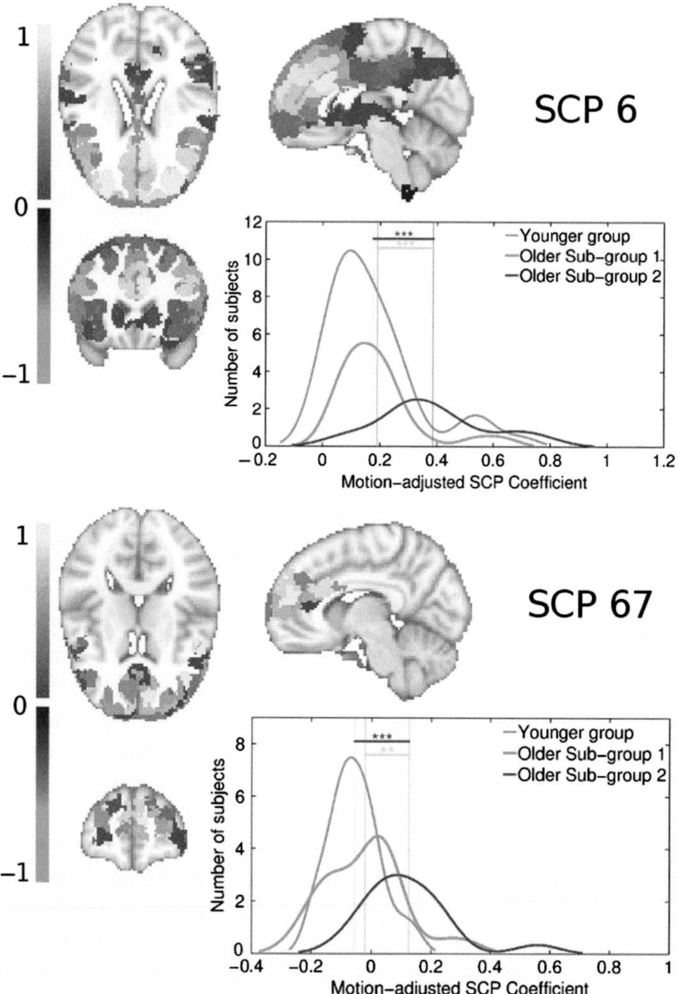

FIG. 10.12

Plot showing primary SCP 6, and its associated secondary SCP 67, whose average connectivity is increased in the second older subgroup, but not the first. SCP 6 highlights most of the prefrontal cortex. SCP 67 captures the bilateral paracingulate gyrus and inferior temporal gyrus. The distribution fit of the underlying SCP coefficient histograms are also shown, for each SCP and for each subgroup. Significance levels are indicated as follows: ***p-value < 0.001, **p-value < 0.01, and *p-value < 0.05.

Source: Reprinted from Eavani, H., Hsieh, M.K., An, Y., Erus, G., Beason-Held, L., Resnick, S., Davatzikos, C., 2016. Capturing heterogeneous group differences using mixture-of-experts: application to a study of aging. NeuroImage 125, 498–514.

10.5 APPLICATIONS

In this section, we present applications of machine learning tools towards tackling clinically relevant problems.

10.5.1 INDIVIDUALIZED DIAGNOSTIC INDICES USING MRI

The past 20 years have seen a wide acceptance of pattern analysis methods in neuroimaging as a means for capturing spatial patterns of morphological, functional, and pathologic signals. However, the vast majority of methods investigating disease effects on the brain have relied on voxel-based analysis (VBA) methods, which apply mass-univariate tests on a voxel-by-voxel basis in an attempt to elucidate the spatial patterns of imaging differences between patients and healthy controls. During the past decade, the use of machine learning to integrate and synthesize these patterns into indices of diagnostic and predictive value for each individual has gained a great deal of attention. This is due to its significance beyond understanding disease effects and into deriving individualized clinical indices of disease. Such machine learning-derived indices have been used in several diseases, including AD (Davatzikos et al., 2008a; Klöppel et al., 2008) and schizophrenia (Davatzikos et al., 2005). We now summarize our group's work on deriving the SPARE-AD index, an index that measures the presence of AD-like patterns of brain atrophy from brain MRI.

10.5.2 MRI-BASED DIAGNOSIS OF AD: THE SPARE-AD

In Fan et al. (2008a), the COMPARE algorithm was used on 122 MRI scans of cognitively normal (CN) older adults and AD patients, and the SPARE-AD index was derived: positive values reflect the presence of AD-like patterns of brain atrophy, and negative values indicate CN-like brain anatomy. The patterns used by the COMPARE algorithm to build the SPARE-AD score were fairly complex and distributed over several brain regions of gray matter (GM), white matter (WM), and cerebrospinal fluid (CSF). Fig. 10.13 indicates the regions with the most significant brain atrophy and ventricular expansion.

The histograms of the (cross-validated) SPARE-AD scores achieved in this classification are shown in Fig. 10.14, indicating excellent discrimination between CN individuals and AD patients. The SPARE-AD index is therefore an index that offers promise as a clinical score derived from sMRI and measuring the presence of AD patterns of brain atrophy.

10.5.3 INDIVIDUALIZED EARLY PREDICTIONS

As individualized diagnostic indices, like the SPARE-AD, are developed based on machine learning approaches, it is perhaps of greater interest to evaluate the predictive value of these indices at early disease stages or even preclinically. These are the stages where standard clinical evaluations might be less effective and hence

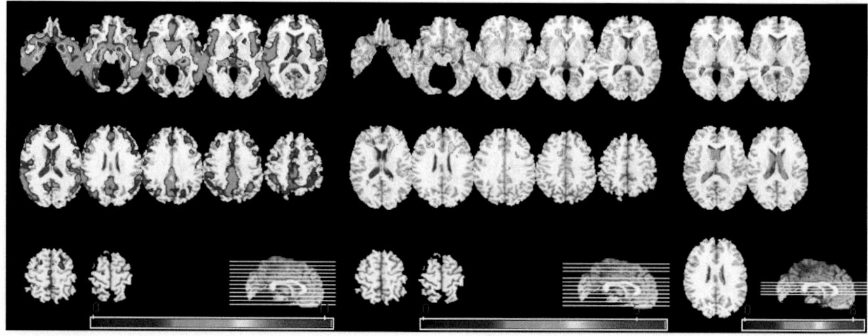

FIG. 10.13

From left to right, group comparison results on GM, WM, and CSF are shown. The color maps indicate the scale for the *t*-statistic. Images are displayed in radiological convention.

Source: Reprinted with permission from Fan, Y., Batmanghelich, N., Clark, C.M., Davatzikos, C., 2008a. Spatial patterns of brain atrophy in MCI patients, identified via high-dimensional pattern classification, predict subsequent cognitive decline. NeuroImage 39(4), 1731–1743.

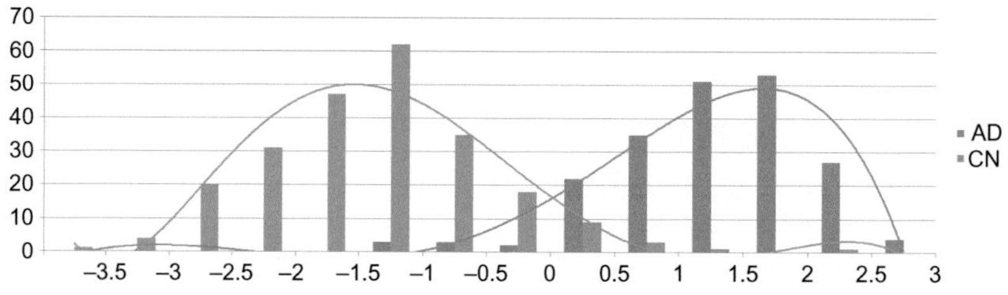

FIG. 10.14

Histograms of SPARE-AD scores obtained via cross-validation from the ADNI1 sample comprising CN and AD individuals.

likely to benefit from imaging-based biomarkers. In this vein, the SPARE-AD index was examined in individuals with mild cognitive impairment in Da et al. (2014) and Davatzikos et al. (2011), and it was found to predict, to a large extent, an individual's future progression to dementia. Fig. 10.15 shows survival curves obtained from baseline measures in 432 MCI patients of the ADNI1 study.

Looking at even earlier stages of the progression of patterns of brain atrophy evaluated via machine learning, the study in Davatzikos et al. (2009) investigated the predictive value of SPARE-AD in preclinical stages of cognitively normal aging. It was found that patterns of brain change at those stages are quite predictive of future cognitive decline. Fig. 10.16 shows the rates of SPARE-AD change for people who remained cognitively stable (left), and for people who progressed to MCI over an 8-year period; since conversion from MCI to AD also takes additional time

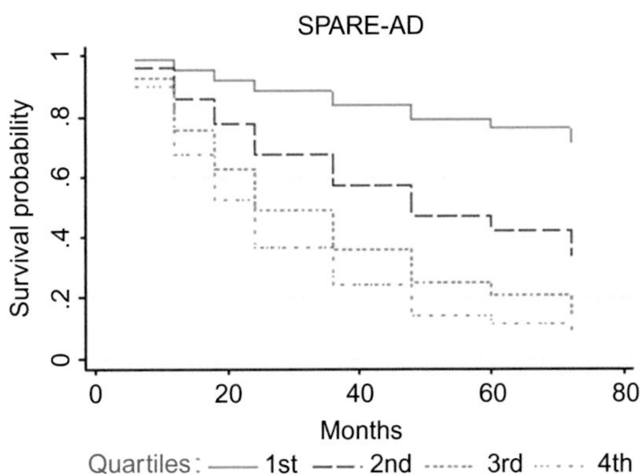

FIG. 10.15

Survival curves showing the predictive value of MRI-derived patterns of atrophy that were evaluated using machine learning (the SPARE-AD index).

Source: Reprinted with permission from Da, X., Toledo, J.B., Zee, J., Wolk, D.A., Xie, S.X., Ou, Y., Shacklett, A., Parmpi, P., Shaw, L., Trojanowski, J.Q., Davatzikos, C., 2014. Integration and relative value of biomarkers for prediction of MCI to AD progression: spatial patterns of brain atrophy, cognitive scores, APOE genotype and CSF biomarkers. NeuroImage: Clinical 4, 164–173.

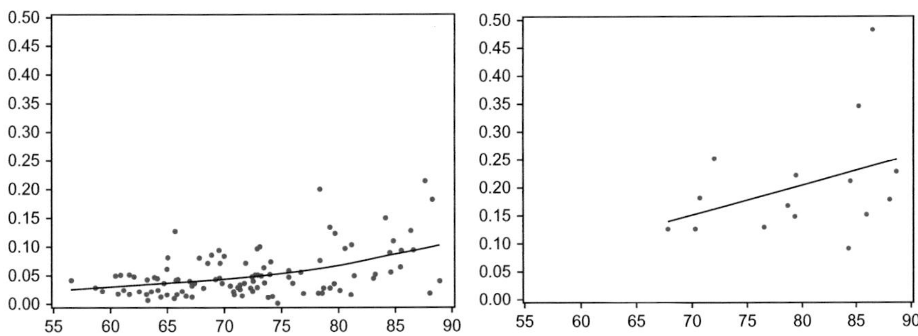

FIG. 10.16

Annual rates of SPARE-AD change at the Baltimore Longitudinal Study of Aging (BLSA). People who remained stable are shown on the left, and people who converted to MCI are shown on the right, displaying markedly higher rates of SPARE-AD change prior to cognitive decline.

Source: Adapted with permission from Davatzikos, C., Xu, F., An, Y., Fan, Y., Resnick, S.M., 2009. Longitudinal progression of Alzheimer's-like patterns of atrophy in normal older adults: the SPARE-AD index. Brain 132 (Pt 8), 2026–2035.

(conversion rate is about 15% annually), these studies indicate that patterns of brain atrophy captured by these machine learning approaches can evolve a decade or longer before dementia. The availability of such an early time window can prove critical for the success of future treatments.

10.6 CONCLUSION

In summary, machine learning approaches offer great promise in clinical research as a means for integrating complex imaging data into personalized indices of diagnostic and prognostic value. As imaging (and genomic) data becomes increasingly complex and multifaceted, such approaches promise to help reduce otherwise unmanageable data volumes down to relatively few clinically informed indices. One of the challenges faced ahead is the need to prove the generalization of these approaches in large samples of data obtained across different studies, scanners, or sites. This can be particularly challenging, in part due to the very ability of these methods to find subtle patterns. If these patterns become too specific to one type of data, then they might be less likely to generalize well across different clinics. Good imaging harmonization across clinics is essential, as is the need to regularize and cross-test machine learning methods sufficiently, to avoid data overfitting.

REFERENCES

Ashburner, J., Friston, K.J., 2000. Voxel-based morphometry—the methods. NeuroImage 11 (6), 805–821.

Atlas, S.W., 2009. Magnetic Resonance Imaging of the Brain and Spine. Lippincott Williams & Wilkins, Baltimore, 2256 .

Avants, B.B., Cook, P.A., Ungar, L., Gee, J.C., Grossman, M., 2010. Dementia induces correlated reductions in white matter integrity and cortical thickness: a multivariate neuroimaging study with sparse canonical correlation analysis. NeuroImage 50 (3), 1004–1016.

Batmanghelich, N.K., Taskar, B., Davatzikos, C., 2012. Generative-discriminative basis learning for medical imaging. IEEE Trans. Med. Imaging 31 (1), 51–69.

Baxton, R.B., 2009. Introduction to Functional Magnetic Resonance Imaging. Cambridge University Press, Cambridge, UK.

Beckmann, C.F., Smith, S.M., 2004. Probabilistic independent component analysis for functional magnetic resonance imaging. IEEE Trans. Med. Imaging 23 (2), 137–152.

Bellec, P., Rosa-Neto, P., Lyttelton, O.C., Benali, H., Evans, A.C., 2010. Multi-level bootstrap analysis of stable clusters in resting-state fMRI. NeuroImage 51, 1126–1139.

Biswal, B.B., 2012. Resting state fMRI: a personal history. NeuroImage 62 (2), 938–944.

Blumensath, T., Jbabdi, S., Glasser, M.F., Van Essen, D.C., Ugurbil, K., Behrens, T.E.J., Smith, S.M., 2013. Spatially constrained hierarchical parcellation of the brain with resting-state fMRI. NeuroImage 76, 313–324.

Buchanan, R.W., Carpenter, W.T., 1994. Domains of psychopathology: an approach to the reduction of heterogeneity in schizophrenia. J. Nerv. Ment. Dis. 182 (4), 193–204.

Burges, C.J., 1998. A tutorial on support vector machines for pattern recognition. Data Min. Knowl. Disc. 2 (2), 121–167.

Calhoun, V.D., Adali, and Pearlson, G.D., Pekar, J.J., 2001. A method for making group inferences from functional MRI data using independent component analysis. Hum. Brain Map. 14, 140–151.

Calhoun, V.D., Adali, T., Hansen, L.K., Larsen, J., Pekar, J.J., 2003. ICA of functional MRI data: an overview. In: Proceedings of the International Workshop on Independent Component Analysis and Blind Signal Separation, pp. 281–288.

Cordes, D., Haughton, V., Carew, J.D., Arfanakis, K., Maravilla, K., 2002. Hierarchical clustering to measure connectivity in fMRI resting-state data. Mag. Reson. Imaging 20, 305–317.

Craddock, R.C., James, G.A., Holtzheimer, P.E.I., Hu, X.P., Mayberg, H.S., 2012. A whole brain fMRI atlas generated via spatially constrained spectral clustering. Hum. Brain Map. 33 (8), 1914–1928.

Cuingnet, R., Gerardin, E., Tessieras, J., Auzias, G., Lehéricy, S., Habert, M.O., Chupin, M., Benali, H., Colliot, O., 2011. Automatic classification of patients with Alzheimer's disease from structural MRI: a comparison of ten methods using the ADNI database. NeuroImage 56 (2), 766–781.

Da, X., Toledo, J.B., Zee, J., Wolk, D.A., Xie, S.X., Ou, Y., Shacklett, A., Parmpi, P., Shaw, L., Trojanowski, J.Q., Davatzikos, C., 2014. Integration and relative value of biomarkers for prediction of MCI to AD progression: spatial patterns of brain atrophy, cognitive scores, APOE genotype and CSF biomarkers. NeuroImage: Clinical 4, 164–173.

Daubechies, I., Roussos, E., Takerkart, S., Benharrosh, M., Golden, C., Ardenne, K.D., Richter, W., Cohen, J.D., Haxby, J., 2009. Independent component analysis for brain fMRI does not select for independence. Proc. Natl. Acad. Sci. USA 106 (26), 10415–10422.

Davatzikos, C., 1998. Mapping image data to stereotaxic spaces: applications to brain mapping. Hum. Brain Map. 6 (5-6), 334–338.

Davatzikos, C., Shen, D., Gur, R.C., Wu, X., Liu, D., Fan, Y., Hughett, P., Turetsky, B.I., Gur, R.E., 2005. Whole-brain morphometric study of schizophrenia revealing a spatially complex set of focal abnormalities. Arch. Gen. Psychiat. 62 (11), 1218–1227.

Davatzikos, C., Fan, Y., Wu, X., Shen, D., Resnick, S.M., 2008a. Detection of prodromal Alzheimer's disease via pattern classification of magnetic resonance imaging. Neurobiol. Aging 29 (4), 514–523.

Davatzikos, C., Resnick, S.M., Wu, X., Parmpi, P., Clark, C.M., 2008b. Individual patient diagnosis of AD and FTD via high-dimensional pattern classification of MRI. NeuroImage 41 (4), 1220–1227.

Davatzikos, C., Xu, F., An, Y., Fan, Y., Resnick, S.M., 2009. Longitudinal progression of Alzheimer's-like patterns of atrophy in normal older adults: the SPARE-AD index. Brain 132 (Pt 8), 2026–2035.

Davatzikos, C., Bhatt, P., Shaw, L.M., Batmanghelich, K.N., Trojanowski, J.Q., 2011. Prediction of MCI to AD conversion, via MRI, CSF biomarkers, and pattern classification. Neurobiol. Aging 32 (12), 2322.e19–2322.e27.

Delong, A., Osokin, A., Isack, H.N., Boykov, Y., 2012. Fast approximate energy minimization with label costs. Int. J. Comput. Vision 96, 1–27.

Detre, J.A., Leigh, J.S., Williams, D.S., Koretsky, A.P., 1992. Perfusion imaging. Magn. Reson. Med. 23 (1), 37–45.

Dong, A., Honnorat, N., Gaonkar, B., Davatzikos, C., 2016. CHIMERA: clustering of heterogeneous disease effects via distribution matching of imaging patterns. IEEE Trans. Med. Imaging 35 (2), 612–621.

Duchesne, S., Caroli, a., Geroldi, C., Barillot, C., Frisoni, G.B., Collins, D.L., 2008. MRI-based automated computer classification of probable AD versus normal controls. IEEE Trans. Med. Imaging 27 (4), 509–520.

Eavani, H., Satterthwaite, T.D., Filipovych, R., Gur, R.E., Gur, R.C., Davatzikos, C., 2015. Identifying sparse connectivity patterns in the brain using resting-state fMRI. NeuroImage 105, 286–299.

Eavani, H., Hsieh, M.K., An, Y., Erus, G., Beason-Held, L., Resnick, S., Davatzikos, C., 2016. Capturing heterogeneous group differences using mixture-of-experts: application to a study of aging. NeuroImage 125, 498–514.

Fan, Y., Shen, D., Gur, R.C., Gur, R.E., Davatzikos, C., 2007. COMPARE: classification of morphological patterns using adaptive regional elements. IEEE Trans. Med. Imaging 26 (1), 93–105.

Fan, Y., Batmanghelich, N., Clark, C.M., Davatzikos, C., 2008a. Spatial patterns of brain atrophy in MCI patients, identified via high-dimensional pattern classification, predict subsequent cognitive decline. NeuroImage 39 (4), 1731–1743.

Fan, Y., Resnick, S.M., Wu, X., Davatzikos, C., 2008b, Structural and functional biomarkers of prodromal Alzheimer's disease: a high-dimensional pattern classification study. NeuroImage 41 (2), 277–285.

Friston, K.J., Frith, C.D., Liddle, P.F., Frackowiak, R.S., 1993. Functional connectivity: the principal-component analysis of large (PET) data sets. J. Cerebral Blood Flow Metabol. 13 (1), 5–14.

Fu, Z., Robles-Kelly, A., Zhou, J., 2010. Mixing linear SVMS for nonlinear classification. IEEE Trans. Neural Netw. 21 (12), 1963–1975.

Gaonkar, B., Davatzikos, C., 2013. Analytic estimation of statistical significance maps for support vector machine based multi-variate image analysis and classification. NeuroImage 78, 270–283.

Gaonkar, B., Shinohara, R.T., Davatzikos, C., Initiative, A.D.N., et al., 2015. Interpreting support vector machine models for multivariate group wise analysis in neuroimaging. Med. Image Anal. 24 (1), 190–204.

Gerardin, E., Chételat, G., Chupin, M., Cuingnet, R., Desgranges, B., Kim, H.S., Niethammer, M., Dubois, B., Lehéricy, S., Garnero, L., Eustache, F., Colliot, O., 2009. Multidimensional classification of hippocampal shape features discriminates Alzheimer's disease and mild cognitive impairment from normal aging. NeuroImage 47 (4), 1476–1486.

Geschwind, D.H., Levitt, P., 2007. Autism spectrum disorders: developmental disconnection syndromes. Curr. Opin. Neurobiol. 17 (1), 103–111.

Golland, P., Golland, Y., Malach, R., 2007. Detection of spatial activation patterns as unsupervised segmentation of fMRI data. In: Medical Image Computing and Computer-Assisted Intervention MICCAI 2007, vol. 4791. Springer, Berlin, pp. 110–118.

Graham, J.M., Sagar, H.J., 1999. A data-driven approach to the study of heterogeneity in idiopathic Parkinson's disease: identification of three distinct subtypes. Move. Disord. 14 (1), 10–20.

Gulshan, V., Rother, C., Criminisi, A., Blake, A., Zisserman, A., 2010. Geodesic star convexity for interactive image segmentation. In: IEEE Conference on Computer Vision and Pattern Recognition (CVPR), pp. 3129–3136.

Hansen, L.K., Larsen, J., Nielsen, F.A., Strother, S.C., Rostrup, E., Savoy, R., Lange, N., Sidtis, J., Svarer, C., Paulson, O.B., 1999. Generalizable patterns in neuroimaging: how many principal components? NeuroImage 9 (5), 534–544.

Haufe, S., Meinecke, F., Görgen, K., Dähne, S., Haynes, J.D., Blankertz, B., Bießmann, F., 2014. On the interpretation of weight vectors of linear models in multivariate neuroimaging. NeuroImage 87, 96–110.

Heimann, T., Meinzer, H.P., 2009. Statistical shape models for 3d medical image segmentation: a review. Med. Image Anal. 13 (4), 543–563.

Heller, R., Stanley, D., Yekutieli, D., Rubin, N., Benjamini, Y., 2006. Cluster-based analysis of fMRI data. NeuroImage 33 (2), 599–608.

Hinrichs, C., Singh, V., Mukherjee, L., Xu, G., Chung, M.K., Johnson, S.C., 2009. Spatially augmented LPboosting for AD classification with evaluations on the ADNI dataset. NeuroImage 48 (1), 138–149.

Hinrichs, C., Singh, V., Xu, G., Johnson, S.C., 2011. Predictive markers for AD in a multi-modality framework: an analysis of MCI progression in the ADNI population. NeuroImage 55 (2), 574–589.

Honnorat, N., Eavani, H., Satterthwaite, T.D., Gur, R.E., Gur, R.C., Davatzikos.C., 2015. GraSP: geodesic graph-based segmentation with shape priors for the functional parcellation of the cortex. NeuroImage 106, 207–211.

Hsieh, J., 2009. Computed tomography: principles, design, artifacts, and recent advances. SPIE, Bellingham, WA.

Huang, C., Wahlund, L.O., Almkvist, O., Elehu, D., Svensson, L., Jonsson, T., Winblad, B., Julin, P., 2003. Voxel- and VOI-based analysis of SPECT CBF in relation to clinical and psychological heterogeneity of mild cognitive impairment. NeuroImage 19 (3), 1137–1144.

Hubert, L., Arabie, P., 1985. Comparing partitions. J. Class. 2 (1), 193–218.

Jacobs, R.A., Jordan, M.I., Nowlan, S.J., Hinton, G.E., 1991. Adaptive mixtures of local experts. Neural Comput. 3 (1), 79–87.

Jeste, S.S., Geschwind, D.H., 2014. Disentangling the heterogeneity of autism spectrum disorder through genetic findings. Nat. Rev. Neurol. 10 (2), 74–81.

Klöppel, S., Stonnington, C.M., Chu, C., Draganski, B., Scahill, R.I., Rohrer, J.D., Fox, N.C., Jack, C.R., Ashburner, J., Frackowiak, R.S.J., 2008. Automatic classification of MR scans in Alzheimer's disease. Brain 131 (3), 681–689.

Koutsouleris, N., Gaser, C., Jäger, M., Bottlender, R., Frodl, T., Holzinger, S., Schmitt, G.J.E., Zetzsche, T., Burgermeister, B., Scheuerecker, J., Born, C., Reiser, M., Möller, H.J., Meisenzahl, E.M., 2008. Structural correlates of psychopathological symptom dimensions in schizophrenia: a voxel-based morphometric study. NeuroImage 39 (4), 1600–1612.

Koutsouleris, N., Davatzikos, C., Bottlender, R., Patschurek-Kliche, K., Scheuerecker, J., Decker, P., Gaser, C., Moller, H.J., Meisenzahl, E.M., 2012. Early Recognition and Disease Prediction in the At-Risk Mental States for Psychosis Using Neurocognitive Pattern Classification. Schizophrenia Bull. 38 (6), 1200–1215.

Koutsouleris, N., Meisenzahl, E.M., Borgwardt, S., Riecher-Rossler, A., Frodl, T., Kambeitz, J., Kohler, Y., Falkai, P., Moller, H.J., Reiser, M., Davatzikos, C., 2015. Individualized differential diagnosis of schizophrenia and mood disorders using neuroanatomical biomarkers. Brain 138 (7), 2059–2073.

Ladicky, L., Torr, P., 2011. Locally linear support vector machines. In: Proceedings of the 28th International Conference on Machine Learning (ICML-11), pp. 985–992.

Lambert, J.C., Ibrahim-Verbaas, C.A., Harold, D., Naj, A.C., Sims, R., Bellenguez, C., Jun, G., DeStefano, A.L., Bis, J.C., Beecham, G.W., et al., 2013. Meta-analysis of 74,046 individuals identifies 11 new susceptibility loci for Alzheimer's disease. Nat. Genet. 45 (12), 1452–1458.

Lashkari, D., Vul, E., Kanwisher, N., Golland, P., 2010. Discovering structure in the space of fMRI selectivity profiles. NeuroImage 50 (3), 1085–1098.

Lee, D.D., Seung, H.S., 1999. Learning the parts of objects by non-negative matrix factorization. Nature 401 (6755), 788–791.

Lewis, S.J.G., Foltynie, T., Blackwell, A.D., Robbins, T.W., Owen, A.M., Barker, R.A., 2005. Heterogeneity of Parkinson's disease in the early clinical stages using a data driven approach. J. Neurol. Neurosurg. Psychiat. 76 (3), 343–348.

Liang, Z.P., Lauterbur, P.C., 2000. Principles of Magnetic Resonance Imaging. SPIE Optical Engineering Press, Bellingham, WA.

Maintz, J.B.A., Viergever, M.A., 1998. A survey of medical image registration. Med. Image Anal. 2 (1), 1–36.

McEvoy, L.K., Fennema-Notestine, C., Roddey, J.C., Hagler, D.J., Holland, D., Karow, D.S., Pung, C.J., Brewer, J.B., Dale, A.M., 2009. Alzheimer disease: quantitative structural neuroimaging for detection and prediction of clinical and structural changes in mild cognitive impairment. Radiology 251 (1), 195–205.

McGraw, K.O., Wong, S.P., 1996. Forming inferences about some intraclass correlation coefficients. Psychol. Meth. 1, 30–46.

McIntosh, A.R., Mišić, B., 2013. Multivariate statistical analyses for neuroimaging data. Ann. Rev. Psychol. 64, 499–525.

McKeown, M.J., Makeig, S., Brown, G.G., Jung, T.P., Kindermann, S.S., Bell, A.J., Sejnowski, T.J., 1998. Analysis of fMRI data by blind separation into independent spatial components. Hum. Brain Map. 6 (3), 160–188.

Moon, T.K., 1996. The expectation-maximization algorithm. IEEE Signal Process. Mag. 13 (6), 47–60.

Mour ao-Miranda, J., Bokde, A.L.W., Born, C., Hampel, H., Stetter, M., 2005. Classifying brain states and determining the discriminating activation patterns: support Vector Machine on functional MRI data. NeuroImage 28 (4), 980–995.

Mour ao-Miranda, J., Hardoon, D.R., Hahn, T., Marquand, A.F., Williams, S.C.R., Shawe-Taylor, J., Brammer, M., 2011. Patient classification as an outlier detection problem: an application of the One-Class Support Vector Machine. NeuroImage 58 (3), 793–804.

Murray, M.E., Graff-Radford, N.R., Ross, O.A., Petersen, R.C., Duara, R., Dickson, D.W., 38 (6), 1200–1215 2011. Neuropathologically defined subtypes of Alzheimer's disease with distinct clinical characteristics: a retrospective study. Lancet Neurol. 10 (9), 785–796.

Myronenko, A., Song, X., 2010. Point set registration: coherent point drift. IEEE Trans. Pattern Anal. Mach. Intell. 32, 2262–2275.

Noh, Y., Jeon, S., Lee, J.M., Seo, S.W., Kim, G.H., Cho, H., Ye, B.S., Yoon, C.W., Kim, H.J., Chin, J., et al., 2014. Anatomical heterogeneity of Alzheimer disease based on cortical thickness on MRIs. Neurology 83 (21), 1936–1944.

Norman, K.A., Polyn, S.M., Detre, G.J., Haxby, J.V., 2006. Beyond mind-reading: multi-voxel pattern analysis of fMRI data. Trends Cogn. Sci. 10 (9), 424–430.

Pham, D.L., Xu, C., Prince, J.L., 2000. Current Methods in Medical Image Segmentation. Ann. Rev. Biomed. Eng. 2 (1), 315–337.

Phelps, M.E., 2000. Positron emission tomography provides molecular imaging of biological processes. Proc. Natl. Acad. Sci. USA 97 (16), 9226–9233.

Ryali, S., Chen, T., Supekar, K., Menon, V., 2013. A parcellation scheme based on von Mises-Fisher distributions and Markov random fields for segmenting brain regions using resting-state fMRI. NeuroImage 65 (0), 83–96.

Sajda, P., 2006. Machine Learning for Detection and Diagnosis of Disease. Ann. Rev. Biomed. Eng. 8 (1), 537–565.

Satterthwaite, T., Elliott, M.A., Ruparel, K., Loughead, J., Prabhakaran, K., Calkins, M.E., Hopson, R., Jackson, C., Keefe, J., Riley, M., Mentch, F.D., Sleiman, P., Verma, R., Davatzikos, C., Hakonarson, H., Gur, R.C., Gur, R.E., 2014. Neuroimaging of the Philadelphia neurodevelopmental cohort. NeuroImage 86, 544–553.

Shen, D., Davatzikos, C., 2002. HAMMER: hierarchical attribute matching mechanism for elastic registration. IEEE Trans. Med. Imaging 21 (11), 1421–1439.

Shen, X., Papademetris, X., Constable, R.T., 2010. Graph-theory based parcellation of functional subunits in the brain from resting-state fMRI data. NeuroImage 50, 1027–1035.

Smith, S.M., Miller, K.L., Salimi-Khorshidi, G., Webster, M., Beckmann, C.F., Nichols, T.E., Ramsey, J.D., Woolrich, M.W., 2011. Network modelling methods for fMRI. NeuroImage 54, 875–891.

Smith, S.M., Miller, K.L., Moeller, S., Xu, J., Auerbach, E.J., Woolrich, M.W., Beckmann, C.F., Jenkinson, M., Andersson, J., Glasser, M.F., Van Essen, D.C., Feinberg, D.A., Yacoub, E.S., Ugurbil, K., 2012. Temporally-independent functional modes of spontaneous brain activity. Proc. Natl. Acad. Sci. USA 109 (8), 3131–3136.

Sotiras, A., Davatzikos, C., Paragios, N., 2013. Deformable medical image registration: a survey. IEEE Trans. Med. Imaging 32 (7), 1153–1190.

Sotiras, A., Resnick, S.M., Davatzikos, C., 2015. Finding imaging patterns of structural covariance via Non-Negative Matrix Factorization. NeuroImage 108, 1–16.

Strother, S.C., Anderson, J.R., Schaper, K.A., Sidtis, J.J., Liow, J.S., Woods, R.P., Rottenberg, D.A., 1995. Principal component analysis and the scaled subprofile model compared to intersubject averaging and statistical parametric mapping: I. "Functional connectivity" of the human motor system studied with [15O]water PET. J. Cerebral Blood Flow Metabol. 15 (5), 738–753.

Sun, D., van Erp, T.G., Thompson, P.M., Bearden, C.E., Daley, M., Kushan, L., Hardt, M.E., Nuechterlein, K.H., Toga, A.W., Cannon, T.D., 2009. Elucidating a magnetic resonance imaging-based neuroanatomic biomarker for psychosis: classification analysis using probabilistic brain atlas and machine learning algorithms. Biol. Psychiat. 66 (11), 1055–1060.

Tononi, G., Sporns, O., Edelman, G.M., 1994. A measure for brain complexity: relating functional segregation and integration in the nervous system. Proc. Natl. Acad. Sci. USA 91, 5033–5037.

Van Essen, D.C., Smith, S.M., Barch, D.M., Behrens, T.E.J., Yacoub, E., Ugurbil for the WU-Minn HCP Consortium., K., 2013. The WU-Minn human connectome project: an overview. NeuroImage 80, 62–79.

Vapnik, V.N., 2000. The Nature of Statistical Learning Theory. Springer New York. 315.

Varol, E., Gaonkar, B., Davatzikos, C., 2013. Classifying medical images using morphological appearance manifolds. In: IEEE 10th International Symposium on Biomedical Imaging (ISBI), pp. 744–747.

Varol, E., Sotiras, A., Davatzikos, C., 2015. Disentangling disease heterogeneity with max-margin multiple hyperplane classifier. In: Medical Image Computing and Computer-Assisted Intervention—MICCAI 2015. Springer, pp. 702–709.

Varol, E., Sotiras, A., Davatzikos, C., 2016. HYDRA: revealing heterogeneity of imaging and genetic patterns through a multiple max-margin discriminative analysis framework. NeuroImage. ISSN: 1053–8119, doi: http://dx.doi.org/10.1016/j.neuroimage.2016.02.041, http://www.sciencedirect.com/science/article/pii/S1053811916001506.

Veksler, O., 2008. Star shape prior for graph-cut image segmentation. In: IEEE European Conference on Computer Vision (ECCV), pp. 454–467.

Vemuri, P., Gunter, J.L., Senjem, M.L., Whitwell, J.L., Kantarci, K., Knopman, D.S., Boeve, B.F., Petersen, R.C., Jack, C.R., 2008. Alzheimer's disease diagnosis in individual subjects using structural MR images: validation studies. NeuroImage 39 (3), 1186–1197.

Vincent, L., Soille, P., 1991. Watersheds in digital spaces: an efficient algorithm based on immersion simulations. IEEE Trans. Pattern Anal. Mach. Intell. 13 (6), 583–589.

Whitwell, J.L., Petersen, R.C., Negash, S., Weigand, S.D., Kantarci, K., Ivnik, R.J., Knopman, D.S., Boeve, B.F., Smith, G.E., Jack, C.R., 2007. Patterns of atrophy differ among specific subtypes of mild cognitive impairment. Arch. Neurol. 64 (8), 1130–1138.

Zeng, L.L., Shen, H., Liu, L., Wang, L., Li, B., Fang, P., Zhou, Z., Li, Y., Hu, D., 2012. Identifying major depression using whole-brain functional connectivity: a multivariate pattern analysis. Brain 135 (5), 1498–1507.

Zhang, D., Wang, Y., Zhou, L., Yuan, H., Shen, D., 2011. Multimodal classification of Alzheimer's disease and mild cognitive impairment. NeuroImage 55 (3), 856–867.

Zhang, T., Koutsouleris, N., Meisenzahl, E., Davatzikos, C., 2015. Heterogeneity of Structural Brain Changes in Subtypes of Schizophrenia Revealed Using Magnetic Resonance Imaging Pattern Analysis. Schizophrenia Bull. 41 (1), 74–84.

Zhu, C.Z., Zang, Y.F., Cao, Q.J., Yan, C.G., He, Y., Jiang, T.Z., Sui, M.Q., Wang, Y.F., 2008. Fisher discriminative analysis of resting-state brain function for attention-deficit/hyperactivity disorder. NeuroImage 40 (1), 110–120.

CHAPTER

11 Learning and predicting respiratory motion from 4D CT lung images

T. He, Z. Xue

Houston Methodist Research Institute, Houston, TX, United States

CHAPTER OUTLINE

11.1 INTRODUCTION

In radiology, interventional radiology, radiation therapy and surgery, medical imaging techniques have been widely used for capturing structural and functional images of human body/organs for diagnosis, treatment planning, and guidance of treatments. Medical imaging provides images of internal organs that are not directly accessible (Smith and Kim, 2011; Jacob et al., 2000; Klein et al., 1995) and, from these

Machine Learning and Medical Imaging. http://dx.doi.org/10.1016/B978-0-12-804076-8.00011-6

images, physicians can gather anatomical and functional information and interpret the pathological conditions of patients. Using computer-assisted diagnosis (CAD) tools, quantitative measures of pathological conditions such as lesion, tumor, blood flow, calcification, organ/tissue size and thickness, cardiac structural and functional dysfunctions, as well as brain connectivity can be collected to help perform diagnosis. Meanwhile, images of these conditions can help plan procedures such as intervention, radiotherapy, and surgery. Follow-up studies can also be performed to assess the treatment response, for example, radiotherapy and chemotherapy outcomes, recurrence of cancer, neuroradiological assessment of developmental and degenerative diseases.

Advances in imaging equipment have facilitated the development of new effective and efficient treatment options, converting complex open surgeries to possible minimally invasive procedures, shortening patients' recovery time, improving patient comfort, and eliminating the risk of complications. Minimally invasive intervention is now replacing more costly open surgery in oncology, neurology, pulmonology, and cardiology (Sadeghi Naini et al., 2010; Seinstra et al., 2010; Ukimura, 2010; de Gregorio et al., 2008; Yeung et al., 2006; DeLucia et al., 2006; Shamir et al., 2005; Westendorff et al., 2004; Cleary et al., 2002; Broaddus et al., 2001; Hall et al., 2001; Seibel, 1997; Yaniv et al., 2010; Cleary and Peters, 2010; Enquobahrie et al., 2008, 2007; Sun et al., 2006). During minimally invasive procedures clinicians rely on imaging patients and tracking of interventional devices and combining the device location with their knowledge of patients' anatomy and pathological conditions to build up a mental picture to direct their actions. Software fusion and visualization of images and devices together provide an effective tool to help physicians perform the procedure. In a typical interventional procedure, first a diagnostic or preprocedural image, such as computed tomography (CT) or magnetic resonance imaging (MRI), is captured for planning. Image segmentation and visualization tools can be used at this stage to highlight the details of pathological conditions and to plan the procedure (such as defining the entrance point and trajectory path for biopsy). In radiotherapy planning, tumor(s) are segmented precisely to define the gross tumor volume (GTV), the clinical target volume (CTV) and the planning target volume (PTV), and radiation beams are optimized to maximizing desired radiation doses to the tumor, while minimizing the damage to surrounding normal tissues. Then during radiotherapy, the planning data are automatically registered onto the intraprocedural images to guide the delivery of radiation beams. Typically, in image-guided procedures it is desirable that a real-time imaging technique can provide patients' anatomy information and track device location on-site. Ultrasound is preferred because it is real-time and without radiation exposure. However, ultrasound may have some limitations such as limited tissue contrast. Intraprocedural CT or CT fluoroscopy (CTF) are also often used but they may not be sufficient for dynamic organs, such as lung and cardiac imaging. Similar problems are also encountered in radiation therapy, and we are lacking real-time or radiation-free imaging devices to track the patients' motion.

To date, ultrasound is the most widely used real-time imaging modality for visualizing patients' anatomy in soft-tissue organs such as the breasts, liver, kidney, and prostate (Sadeghi Naini et al., 2010; Bouchet et al., 2001; Karnik et al., 2010;

Park et al., 2010). What is more, during intervention the device (needle) is also visible by ultrasound imaging. This makes it convenient for use by interventional radiologists, and in fact ultrasound is one of the most popular imaging techniques for guidance. But the disadvantage of ultrasound is that the echo signals are not as clear as CT and MRI for tissue structures. Many real-time image fusion methods can be used to register a preprocedure CT or MRI onto the ultrasound image, so that by superimposing the registered image onto the ultrasound images, the structures can be clearly visualized. Additionally, ultrasound also has limitations in assessing the lungs and deep organs because of the reflection on the tissue-air interface, the effect of rib bones, and lack of image contrast inside the lungs (Park et al., 2010).

Other real-time imaging modalities such as endoscopy, X-rays (Lin et al., 2009), and real-time MR (Cervino et al., 2011) can also be used for guidance. For example, in bronchoscopy or colonoscopy endoscope cameras are used for real-time visualization; in vascular intervention real-time X-ray or fluoroscopy coupled with contrast agents are used to highlight the vessels; in operation open MR is used to get real-time anatomy images (of the brain). While endoscopy gives real-time videos and fluoroscopy obtains real-time 2D projection images, they largely rely on clinicians' skills for navigating the device to the target. Concerns for fluoroscopy are that they provide only local 2D projection images and expose both clinicians and patients to radiation. To incorporate more global information, preprocedural 3D CT can be registered with 2D fluoroscopy images so not only local or 2D real-time feedback but also global 3D visualization about where the devices are located can be provided. On the other hand, although it is possible to perform image-guided intervention within open MR scanners, it is not commonly used in clinics due to the cost of the machine time and the complicated clinical setup that requires all devices to be MRI compatible (Seimenis et al., 2012; Yakar et al., 2011; Lang et al., 2011; Tokuda et al., Jan 2010; Pandya et al., 2009; Patriciu et al., 2007; DiMaio et al., 2007).

Besides the above-mentioned real-time imaging techniques, tracking of interventional devices is another task in order to visualize them in the context of anatomy. The most commonly used tracking techniques of interventional devices include electromagnetic (EM) tracking and optical tracking (Shah et al., 2011; Cala et al., 1996; Li et al., 2011). For EM tracking coil sensors are installed within the interventional devices so that their positions can be tracked in real-time within the human body covered by a magnetic field. For optical tracking, markers can be attached to the device allowing its position to be tracked with a stereo vision system by assuming the device is rigid. Other tracking techniques are being developed so that the interventional devices can be tracked within the imaging equipment. For example, coil-based sensors can be installed on the needle tip and radio-frequency signals can be detected using an MR machine while the intervention procedure is performed.

The key for image-guided procedures is to visualize the patient's anatomy and operative devices simultaneously in real-time. However, as mentioned above, real-time imaging may not be available together with device tracking in clinics. For example, in percutaneous lung intervention electromagnetic or optical tracking can be used for detecting the interventional probes, but there is a lack of real-time imaging technique for navigation. A similar situation also exists in lung cancer radiotherapy. The major problem herein is that using static preprocedural CT is not able to solve the discrepancies between the CT and the patient's respiratory stage. Thus for lung cancer treatment, monitoring respiratory phases has been the major focus in dealing with breathing or poor reproducibility of breath-holding.

It is in this context that motion compensation becomes a major task for accurate guidance: it is highly desirable that a precise lung motion model can be used as the roadmap for guiding the intervention during each breath-holding cycle. Recent advances in motion estimation show that by using regression models or machine learning methods it is possible to estimate the lung motion from partially measurable signals such as chest motion signals. In the literature, there are many works dealing with lung motion, and they can be classified into three categories. The first is lung motion deformation modeling with registration (Sundaram et al., 2004; Vandemeulebroucke et al., 2011; Handels et al., 2007), but they only construct the respiratory patterns and do not achieve motion modeling/estimation. The second uses an individual patient's dataset (Wu et al., 2011), for example, first extracting lung motion in preprocedural 4D scans and then applying this model during treatment (radiotherapy or intervention), where 4D scanning is required. Finally, a statistical model can be applied to incorporate both group and individual information for motion modeling. Respiratory patterns trained from a large number of subjects can be used to guide the estimation of dynamic images of individuals, even when the 4D CT images are not available for them (Klinder et al., 2010; Ehrhardt et al., 2011). For example, one can use the motion of a patient's external features in combination with a motion model to compensate for internal respiratory motion (Lu et al., 2006; Santelli et al., 2011; He et al., 2010). The challenge is that the limited dimension of respiratory sensor signals may not reflect the high-dimensional lung motion accurately, and although they monitor respiratory phases well, they may not estimate accurate lung tumor location and shape dynamics. With new machine learning and estimation technologies, the noncontact vision-based motion monitoring devices can be used for detecting the high-dimensional chest surface motion (Tan et al., 2010), and it would be promising to estimate dynamic lung motion by statistical model-based prediction between high-dimensional chest surface and lung motion vectors.

In this chapter, we introduce a method to estimate dynamic lung images from the 3D CT and high-dimensional chest surface signals. This estimation framework consists of two stages: the training and the estimating stages. In the training stage, after performing longitudinal registration of the 4D CT data of each training subject, the respiratory motion fields are calculated. Then all the images, motion fields, and chest surface motions of each subject are aligned onto a template space for training a statistical model. The relationship between the chest surface motion and lung respiratory motion is then established in the template space by using prediction or regression algorithms. During the motion estimation stage, the chest surface motion signals and the 3D CT of a patient will be captured and transformed to the template space through image registration, and we can apply the motion prediction model trained above to estimate patient-specific lung field motion from the chest surface signals. Finally, the estimated dynamic lung images are transformed back to the patient image space.

The chapter is organized according to the key steps of this motion estimation framework. First, we will discuss 3D and 4D CT lung image processing including segmentation and registration, which act as the key steps used to extract lung motion, in Section 11.2. Then we introduce the motion estimation method, which aims to model the relationship between the lung motion fields with the chest surface motion signals, in Section 11.3. Finally, in Section 11.4 we give an example of how the motion estimation can be used in a CTF-guided lung intervention system to register the patient's diagnostic CT image onto the real-time captured CTFs during an interventional procedure. Section 11.5 is the conclusion of this chapter.

11.2 3D/4D CT LUNG IMAGE PROCESSING

In this section two important steps for CT lung image processing, segmentation and registration, are introduced. After presenting lung field and vessel segmentation, we describe a joint image segmentation and registration algorithm for 4D images. This processing is essential for extracting longitudinal respiratory motion vectors and shape information for motion estimation.

11.2.1 LUNG FIELD AND VESSEL SEGMENTATION

11.2.1.1 Lung field extraction

Lung field extraction is a critical preprocessing stage in lung segmentation to separate the cavity, lung field, and free space from the input CT images. Many subsequent processes such as vessel segmentation and tumor segmentation utilize this step as initial preprocessing. For example, the segmented lung field provides a bounding area for vessel segmentation, and subsequently lung field surface motion can be used to characterize respiratory motion of the lung.

The lung field extraction method consists of the following two steps. First, the rough lung area is extracted using 3D region growing. In CT, since the inner lung field and airways have much lower density than pulmonary vessels and surrounding thoracic cavity, density-based segmentation methods like gray-level thresholding and region growing are generally used (Sluimer et al., 2006). The inverse operation of 3D region growing is an effective method of CT lung image segmentation (Yim and Hong, 2008). Therefore we start segmenting the thoracic cavity using region growing as follows: getting several seeds and extracting the chest wall by region growing; separating the air background using a similar region growing method. The seeds are selected automatically from the image boundary with intensities close to zero (points in the air). Then binary morphology and connected components analysis (Hu et al., Jun 2001) are applied to address the boundary indentations and bulges and to fill the cavities inside the lung area. Finally, the original CT image is masked by the binary lung field.

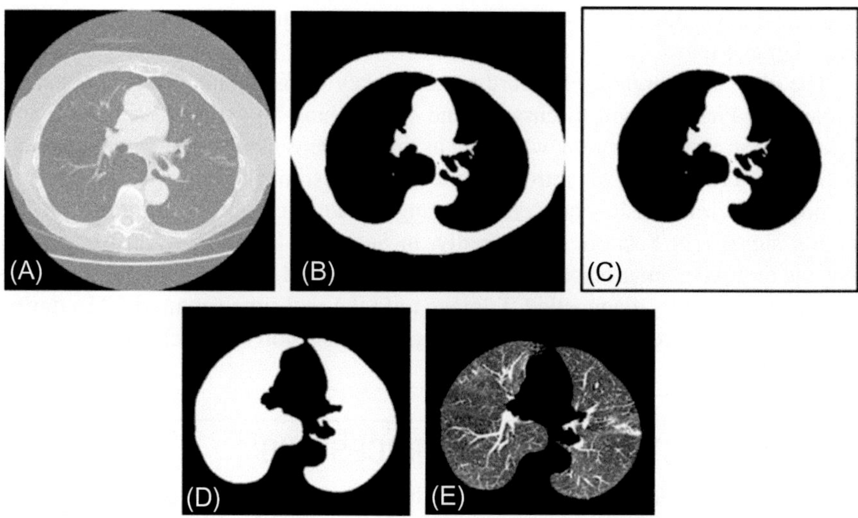

FIG. 11.1

Segmentation of lung field. (A) Input CT image. (B) 3D region growing on chest wall. (C) 3D region growing on air background. (D) Inverting region growing result and performing morphology and connected components analysis. (E) Masking the lung image to extract the lung field.

Fig. 11.1 shows an example of lung field segmentation. We can see that by first segmenting the chest wall and the empty space surrounding the patient, the lung field area can be filtered. Further morphological operations can help clean up and fine-tune the binary mask of the lung field.

11.2.1.2 Vessel segmentation using geometric active contour models

Precise segmentation of pulmonary vessels from CT lung images provides vital visualization for interventional guidance to avoid major vessel damage. It is also important to align the vessel structures when performing intra- and intersubject image registration, particularly in motion estimation of the lung. While simple thresholding and window/level setting can briefly segment different tissues, their results are not accurate because of the intensity variation. Recent studies showed that level-set methods have been successfully used in image segmentation. Let us look at the traditional level-set method (Osher and Sethian, 1988). Briefly, the evolving surface $c(t)$ that separates the object and background is represented by a propagating front embedded as the zero level-set of function $\psi(\mathbf{x}, t)$, such that $c(t) = \{\mathbf{x} | \psi(\mathbf{x}, t) = 0\}$. $\mathbf{x}$ represents an image voxel. The evolution rule for $\psi(\mathbf{x}, t)$ is

$$\frac{\partial \psi}{\partial t} + F |\nabla \psi| = 0. \tag{11.1}$$

The velocity function F is dependent on the image data and the current level-set function ψ. ψ is usually defined as the signed distance function (SDF), and it deforms iteratively according to F. Malladi et al. (1995) proposed a simple formulation for F:

$$F = g \cdot (v - \theta k), \tag{11.2}$$

where θ indicates an external propagation force leading to the surface contraction or expansion uniformly based on its sign. k is the local curvature of the evolving front. θ controls the influence of the curvature and acts as a regularization term to smooth out the high curvature part of the surface. g is the data consistency term. In image segmentation, if one needs to halt the evolution of the surface at object boundaries, g can be created as an edge stopping function as

$$g = \frac{1}{1 + |\nabla G_\sigma * I_{\mathbf{x}}|}. \tag{11.3}$$

In implementation, the front can propagate by updating the level-set function in its neighborhood, so we can only update ψ in a narrow band around the front surface.

11.2.1.3 Vascularity-oriented level-set (VOLES)

When applying the geometric active contour method in vessel segmentation, we encountered a major obstacle in propagating the zero level-set along the vessel direction. This is because the high curvature will receive a high penalty on the evolving front in order to yield a smooth surface of the objects, which makes the propagation slower along tubular structures than in blob structures. To solve this problem, Lorigo et al. (2001) extended level-set segmentation to higher codimension: an underlying vector field is developed for driving the curve evolution in 3D. On the other hand, vessel enhancement filters also attracted more attention. The vesselness measure based on the eigensystem of the Hessian matrix of image intensity has been one of the most popular methods, and it can be combined with various image matching approaches to detect and identify vascular structures (Masutani et al., 2001; Descoteaux et al., 2008). Combining the ideas of using different curvature penalty and the vesselness enhancement, we proposed a new strategy called VOLES that applies vesselness-based compensation on the curvature penalty of the traditional level-set. The advantage is that the evolving front can move faster along the vessel directions, while still maintaining smoothness on the vessel walls.

Notice that θ in Eq. (11.2) controls the influence of the curvature, but when using it to propagate the front inside vascular structures, the front grows much more slowly along the vessels due to the larger curvature along the vessel directions. Fig. 11.2 shows such cases. In VOLES, we adopt the vesselness measure (Frangi et al., 1998), which is small along the vessel boundary and becomes larger near the centerline. Hence θ is now weighted according to this vascularity-oriented information:

$$\frac{\partial \psi}{\partial t} + g \cdot (v - \theta \cdot k)\,|\nabla \psi| = 0. \tag{11.4}$$

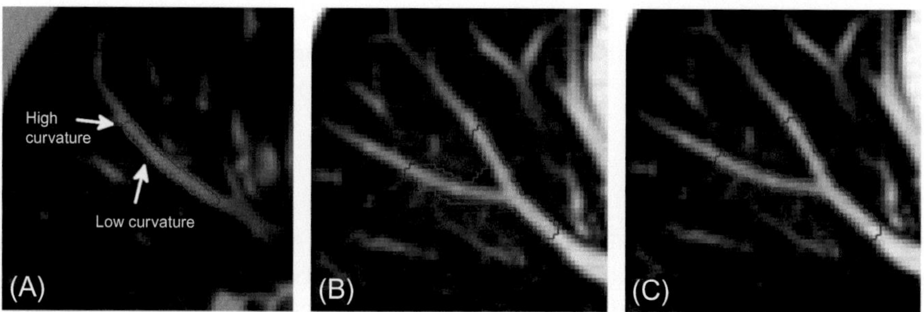

FIG. 11.2

Illustration of vessel propagation. (A) Different curvatures at different locations. (B) The same criterion results in slow grow along vessels and overflow in other regions. (C) Application of different constraints adaptively.

The curvature influence parameter θ is designed to be weighted by the vesselness measure V as

$$\theta = \alpha \cdot \mathrm{e}^{(-\beta V)^3}. \tag{11.5}$$

Vesselness V is large along the vessel direction where local curvature k is large, and in this case θ will be smaller to compensate the penalty of the large curvature. As shown in Fig. 11.2C, the propagation of the level-set will be faster along the vessel direction. Similarly, the vesselness measure can also be applied to the edge stopping function g,

$$g = \frac{1}{1 + |\nabla G_\sigma * V_{\mathbf{x}}|}. \tag{11.6}$$

This allows more curvature penalty when the front is near the boundary. Finally, the external propagation force that determines the local propagation direction at each point of the front is defined as

$$v = \mathrm{sign}\left\{\Lambda_{\mathrm{o}} p_{\mathrm{o}}(I, V) - \Lambda_{\mathrm{i}} p_{\mathrm{i}}(I, V)\right\}. \tag{11.7}$$

Instead of using a single method, either expanding or contracting the surface, we classify each point of the current surface $c(t)$ and let it expand if it locally belongs to (inside) the vessel or contract if it does not (Hall et al., 2001). $p\,(I, V)$ stands for a 2D distribution of image intensity and vesselness for the local regions either inside or outside $c(t)$.

11.2.2 SERIAL IMAGE SEGMENTATION AND REGISTRATION

Simultaneous serial image segmentation and registration methods have been studied for several image modalities (Wang et al., 2006; Xiaohua et al., 2005; Xue et al., 2006). The advantage is that segmentation can improve longitudinal image registration and longitudinal deformation fields can improve segment consistency of the serial images. This section introduces joint serial image registration and segmentation, wherein serial images are segmented based on the current temporal deformations so that the temporally corresponding tissues tend to be segmented into the same tissue type, and at the same time temporal deformations among the serial images are iteratively refined based on the updated segmentation results.

11.2.2.1 4D registration

Given a series of images $I_{t=1,...T}$, where I_1 is the baseline, the goal for serial image registration is to obtain the longitudinal deformation. We assume that the serial images have been globally aligned onto the baseline by applying rigid registration. To estimate the deformations from the baseline onto each image, that is, $\mathbf{f}_{1\to t}$ or simplified as $\mathbf{f}_t$, registration needs to be performed. Since no longitudinal information is used in traditional pairwise registration or group-wise registration, temporal stability of the resultant serial deformations cannot be preserved. For example, in group-wise registration if one registers all the subsequent images onto the first image, no temporal information such as motion has been used and there are no temporal constraints between $\mathbf{f}_t$ and $\mathbf{f}_{t+1}$, for example. Therefore to better use the temporal motion information we formulate the serial image registration in such a way that the registration of the current time-point image is related not only to the previous but also the following images (if available). No longitudinal smoothness constraints are applied to the serial deformations so that our algorithm can tolerate temporal anatomical and tissue property changes.

For the current image at time-point t, I_t, if the deformation of its previous image I_{t-1}, $\mathbf{f}_{t-1}$, and that of the next image I_{t+1}, $\mathbf{f}_{t+1}$, are known, the serial registration can be achieved by minimizing the following energy function:

$$\begin{aligned} E_{s,t}() = \sum\nolimits_{\mathbf{x}\in\Omega} \Big\{ & |\mathrm{e}(I_t\,(\mathrm{x}+\mathrm{f}_t(\mathrm{x}))) - \mathrm{e}(I_1(\mathrm{x}))|^2 \\ & + \sum\nolimits_{i=-1,1} \left|\mathrm{e}(I_{t+i}\left(\mathrm{X}+\mathrm{f}_{t+i}(\mathrm{X})\right)) - \mathrm{e}(I_t\,(\mathrm{x}+\mathrm{f}_t(\mathrm{x})))\right|^2 \Big\} + E_r(\mathrm{f}_t(\mathrm{x})), \end{aligned} \tag{11.8}$$

where $\mathbf{e}()$ is the operator for calculating image feature vectors and Ω is the image domain. The feature vector for each voxel consists of the intensity, gradient magnitude, and segmented tissue types, that is, $\mathbf{e}\,(\mathbf{x}) = [I\,(\mathbf{x})\,, \nabla I\,(\mathbf{x})\,, \mu_1\,(\mathbf{x})\,, \ldots, \mu_C(\mathbf{x})]$. E_r is the regularization energy of the deformation field, and it can be derived from the prior distribution of the deformation. If no prior distribution is available, the regularization term can be some continuity and smoothness constraints. When cubic B-spline is used to model the deformation field, the continuity and smoothness are guaranteed, thus the regularization term E_r is omitted.

The serial image registration algorithm then iteratively calculates the deformation field $\mathbf{f}_t$ at each time-point by minimizing the energy function in Eq. (11.8) until convergence. Notice that in the first iteration, since the registration results for neighboring images are not available, a pairwise registration can be used for

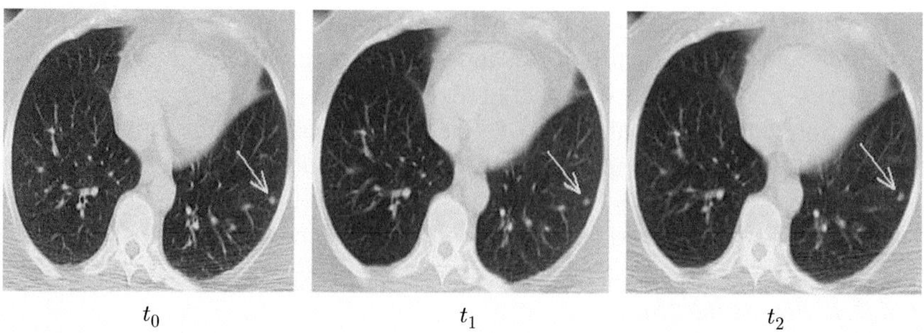

FIG. 11.3

Registered images of an image series.

initialization. Fig. 11.3 shows three consequent images after registration. It can be seen that the nodule has been aligned at the same position.

11.2.2.2 4D segmentation

Notice that in Eq. (11.8) $\mathbf{e}()$ stands for the image features extracting operator. It has been shown that by incorporating segmentation results such as tissue type, registration can be made more robust. With the estimated longitudinal deformation, 4D segmentation can be more accurate by applying the constraints that longitudinally corresponding voxels are likely to be segmented into the same tissue type; on the other hand, the same tissue types are likely to correspond longitudinally in order to refine the longitudinal deformation fields. Thus the purpose of the 4D segmentation is to calculate the segmented images by considering not only the spatial but also the temporal neighborhoods. A 4D clustering algorithm is used to classify each voxel of the serial image into C different tissue types by minimizing the objective function,

$$E(\mu, c) = \sum_{t=1}^{T} \sum_{\mathbf{x}\in\Omega} \left\{ \sum_{k=1}^{C} \left[\mu^q_{(t,\mathbf{x}),k} (I_{t,\mathbf{x}} - c_{t,k})^2 \right] + \frac{\alpha}{2} \rho^{(s)}_{(t,\mathbf{x})} \sum_{k=1}^{C} \left[\mu^q_{(t,\mathbf{x}),k} \overline{\mu}^{(s)}_{(t,\mathbf{x}),k} \right] + \frac{\beta}{2} \rho^{(t)}_{(t,\mathbf{x})} \sum_{k=1}^{C} \left[\mu^q_{(t,\mathbf{x}),k} \overline{\mu}^{(t)}_{(t,\mathbf{x}),k} \right] \right\}, \tag{11.9}$$

where voxel $\mathbf{x}$ in image I_1 corresponds to voxel $\mathbf{x} + \mathbf{f}_t(\mathbf{x})$ in image I_t, referred to as voxel $(t, \mathbf{x}) = \mathbf{x} + \mathbf{f}_t(\mathbf{x})$, and μ, c, q, C follow the fuzzy c-mean (FCM) clustering formulation. We choose $C = 4$ so the tissue types include bone, background, and low/high-intensity soft tissue. $\overline{\mu}^{(s)}_{(t,\mathbf{x}),k}$ and $\overline{\mu}^{(t)}_{(t,\mathbf{x}),k}$ are the spatial and temporal neighborhood average membership functions and are defined as

$$\overline{\mu}^{(s)}_{(t,\mathbf{x}),k} = \frac{1}{N_1}\sum\nolimits_{(t,\mathbf{u})\in N^s_{(t,\mathbf{x})}}\sum\nolimits_{m\in M_k}\mu^q_{(t,\mathrm{u}),m'} \text{and } \overline{\mu}^{(t)}_{(t,\mathrm{x}),k}$$
$$= \frac{1}{N_2}\sum\nolimits_{(\tau,\mathrm{w})\in N^{(t)}_{(t,\mathrm{x})}}\sum\nolimits_{m\in M_k}\mu^q_{(t,\mathrm{w})}, m, \tag{11.10}$$

where $N^{(s)}_{(t,\mathbf{x})}$ and $N^{(t)}_{(t,\mathbf{x})}$ are the spatial and temporal neighborhoods of voxel $(t,\mathbf{x})$and $M_k = \{m = 1,\ldots,C; m \neq k\}$. The fuzzy membership functions $\mu_{(t,\mathbf{x}),k}$ are subject to $\sum_{k=1}^{C}\mu_{(t,\mathbf{x}),k} = 1$, for all t and $\mathbf{x}$. The second term of Eq. (11.9) reflects the spatial constraints of the fuzzy membership functions. The difference is that an additional weight $\rho^{(s)}_{(t,\mathbf{x})}$ is used as an image-adaptive weighting coefficient, thus stronger smoothness constraints are applied in the image regions that have more uniform intensities, and vice versa. $\rho^{(s)}_{(t,\mathbf{x})}$ is defined as $\rho^{(s)}_{(t,\mathbf{x})} = \exp\left\{-\sum_r\left[(D_r * I_t)^2_{(t,\mathbf{x})}/2\sigma_s^2\right]\right\}$, where $(D_r * I_t)_{(t,\mathbf{x})}$ refers to first calculating the spatial convolution and then taking its value at location $(t,\mathbf{x})$, and D_r is a spatial differential operator along axis r. Similarly, the third term of Eq. (11.9) reflects the temporal consistency constraints, and $\rho^{(t)}_{(t,\mathbf{x})}$ is calculated as $\rho^{(t)}_{(t,\mathbf{x})} = \exp\left\{-(D_t * I_{(t,\mathbf{x})})^2_{(t)}/2\sigma_t^2\right\}$. It is worth noting that the temporal smoothness constraint herein does not mean that the serial deformations have to be smooth across different time-points.

Using Lagrange multipliers to enforce the constraint of fuzzy membership function in the objective function, we get the following two equations to iteratively update the fuzzy membership functions and calculate the clustering centroids:

$$\mu_{(t,\mathbf{x}),k} = \frac{\left[(I_{(t,\mathbf{x})} - c_{t,k})^2 + \alpha\rho^{(s)}_{(t,\mathbf{x})}\overline{\mu}^{(s)}_{(t,\mathbf{x}),k} + \beta\rho^{(t)}_{(t,\mathbf{x})}\overline{\mu}^{(t)}_{(t,\mathbf{x}),k}\right]^{\frac{-1}{q-1}}}{\sum_{m=1}^{C}\left[(I_{(t,\mathbf{x})} - c_{t,m})^2 + \alpha\rho^{(s)}_{(t,\mathbf{x})}\overline{\mu}^{(s)}_{(t,\mathbf{x}),m} + \beta\rho^{(t)}_{(t,\mathbf{x})}\overline{\mu}^{(t)}_{(t,\mathbf{x}),m}\right]^{\frac{-1}{q-1}}} \tag{11.11}$$

and

$$c_{t,k} = \frac{\sum_{\mathbf{x}\in\Omega}\mu^q_{(t,\mathbf{x}),k}I_{(t,\mathbf{x})}}{\sum_{\mathbf{x}\in\Omega}\mu^q_{(t,\mathbf{x}),k}}. \tag{11.12}$$

Fig. 11.4 shows an example of the 4D segmentation results for serial CT lung images. To iteratively perform the joint registration and segmentation algorithm, the segmented images are used together with image intensities and image intensity gradients as the feature vector **e**() to refine the longitudinal deformations. The advantage of the joint algorithm is that the temporal consistency can be improved to recover the longitudinal deformations. Moreover, no temporal smoothness about the deformation fields is enforced in order to tolerate larger or discontinuous temporal changes that often appear during image-guided diagnosis and treatment. In the experiments no anatomical information such as blood vessels are used for the 4D registration. Future works include combining sophisticated segmentation to better represent the lung information for more robust registration.

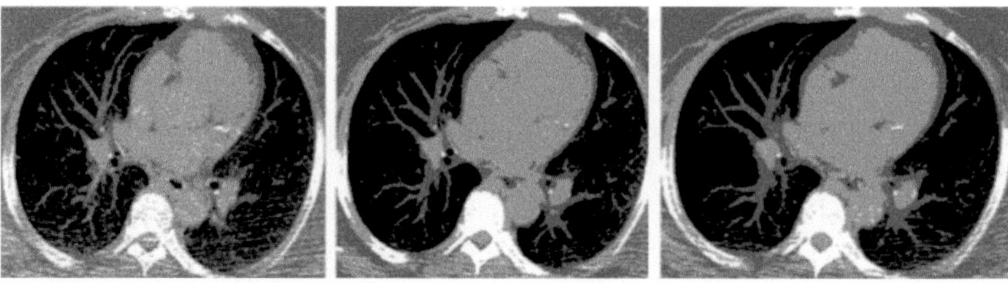

FIG. 11.4

Sample 4D segmentation results.

11.3 EXTRACTING AND ESTIMATING MOTION PATTERNS FROM 4D CT

11.3.1 A LUNG MOTION ESTIMATION FRAMEWORK

In this section, we introduce the framework of dynamic lung image estimation. The algorithm is based on the idea that the relationship between lung motion and chest signals can be learned from a number of 4D CT samples of different subjects. Once the prediction model is established, for a new patient whose 3D CT and chest signals are available, we can use this respiratory motion prediction model to estimate the patient-specific respiratory motion.

Fig. 11.5 shows the framework for lung motion estimation. The algorithm consists of training and prediction stages. In the training stage, 4D CT images from a number subjects are segmented, and their 4D motion fields are extracted by performing intrasubject registration. Then these images are registered onto a template subject. After registration the relationship between the chest motion signals extracted and the entire lung motion fields is trained using a machine learning algorithm. For example, we can train a joint distribution between the 4D deformation fields and the chest motion signals, and their relationship can be estimated by either using a regression model or the Bayesian estimation algorithm. In the motion prediction stage, the 3D CT and chest motion vectors of a new subject are obtained and registered onto the same template space. Then the trained motion estimation model is applied to generate a series of lung deformations based on the chest surface motion vectors. These estimated lung deformations are transferred onto the patient space to generate a series of 3D images from the patient's 3D CT.

For intrasubject registration the joint 4D segmentation and registration algorithm described in Section 11.2 is used to improve temporal consistency of motion vectors. Lung fields and vessels are also segmented as additional anatomical information to be used for the registration. For subject s, after registration the longitudinal deformations $\mathbf{f}^s_{1\to 2}, \mathbf{f}^s_{1\to 3}, \ldots, \mathbf{f}^s_{1\to T}$ for serial images $I^s_1, I^s_2, \ldots, I^s_T$ with T respiratory phases are obtained. To register the 4D respiratory motion fields of each subject onto the template, the baseline (first) image of the subject s and that of the template M, I^s_1 and

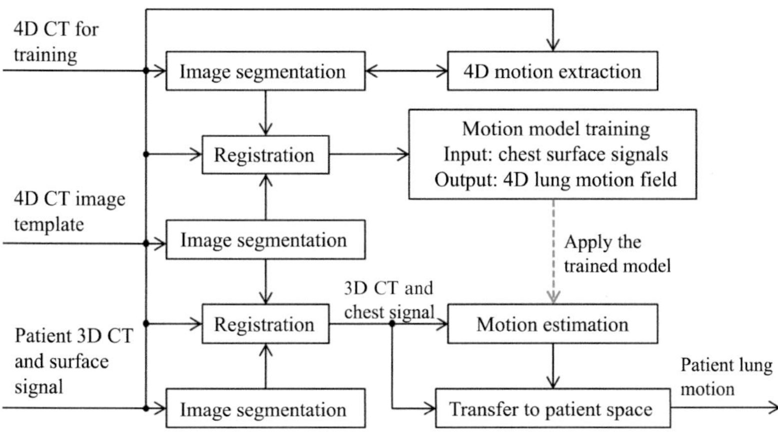

FIG. 11.5

The framework of the proposed 4D CT estimation algorithm.

I_1^M, are registered (Alexander et al., 2001), resulting in a global affine transformation $G_{M\to s}$ and a deformation field $\mathbf{f}_{M\to s}$. Therefore the baseline of the subject can be first transformed using $G_{M\to s}^{-1}$ (inverse of $G_{M\to s}$) and then deformed onto the template space based on the field $\mathbf{f}_{M\to s}$. To transform the respiratory motion fields of subject s onto the template space, we need to maintain the original motion patterns. Thus the global and local reorientation transformations between the subject and the template are applied to $\mathbf{f}_{1\to 2}^s, \mathbf{f}_{1\to 3}^s, \ldots, \mathbf{f}_{1\to T}^s$ to reflect the respiratory motion of the same subject on the template space. Denoting $g_{M\to s}$ as the global reorientation matrix of $G_{M\to s}$ without translation, whose inverse is $g_{M\to s}^{-1}$, the respiratory motion of subject s can be calculated first by $\mathbf{f}_{1\to t}^{s'} = g_{M\to s}^{-1}\mathbf{f}_{1\to t}^s, t = 2, \ldots T$, and then by performing a local reorientation:

$$\mathbf{u}_{1\to t}^s(\mathbf{x}) = R_{M\to S}^{-1}(\mathbf{x})\,\mathbf{f}_{1\to t}^{s'}(\mathbf{x} + \mathbf{f}_{M\to s}(\mathbf{x})), \tag{11.13}$$

where $R_{M\to S}(\mathbf{x})$ is the local rotation matrix at a template voxel $\mathbf{x}$, calculated from the Jacobian matrix $J_{M\to S}(\mathbf{x})$ of $\mathbf{f}_{M\to S}(\mathbf{x})$ as $D_{M\to S}(\mathbf{x}) = ID + J_{M\to S}(\mathbf{x})$ (ID refers to the identity matrix), ie

$$R_{M\to S}(\mathbf{x}) = (D_{M\to S}(\mathbf{x})\,D_{M\to S}^{\mathrm{T}}(\mathbf{x}))^{-1/2}D_{M\to S}(\mathbf{x})\,. \tag{11.14}$$

In this way, all the longitudinal deformation fields calculated from the 4D CT training samples are transformed onto the template space. Similar operations are applied to the chest surface motion vectors, denoted as $\mathbf{v}_{1\to t}^s, t = 2, \ldots T$. The goal of the motion estimation is therefore to estimate the relationship between $\mathbf{v}_{1\to t}^s$ and $\mathbf{u}_{1\to t}^s$ for all the respiratory phases $t = 1, \ldots, T$.

For a new subject/patient P, we need to first determine which respiratory phase the input 3D CT belongs to. This can be determined by using its associated chest motion signals. By performing smoothness filtering and detecting the exhale and inhale phases from the chest motion signals and dividing such signals into M phases, the phase of the synchronized 3D CT is determined. Without loss of generalization, let us assume the 3D CT corresponds to the first phase of the template image. Registering the patient data with the template image results in the global affine transformation $G_{M\to P}$ (reverse is $G_{P\to M}$) and deformation field $\mathbf{f}_{M\to P}$ (reverse is $\mathbf{f}_{P\to M}(\mathbf{x})$). In a similar way, the chest motion vectors of the patient can be transferred onto the template space, which can be denoted as $\mathbf{v}^P_{1\to t}, t = 2, \ldots, T$. Then the lung motion vector of the patient can be estimated using a motion estimation model:

$$\mathbf{u}^P_{1\to t} = \Theta\left(\mathbf{v}^P_{1\to t}, \mathbf{u}^s_{1\to t}, \mathbf{v}^s_{1\to t}\right), \tag{11.15}$$

where $\Theta()$ represents the motion estimation operator. $\mathbf{u}^P_{1\to t}$ can be transformed back to the patient space as

$$\mathbf{f}^P_{1\to t}(\mathbf{x}) = g_{M\to P} \circ R_{M\to P}(\mathbf{x})\, \mathbf{u}^P_{1\to t}\left(G_{P\to M}\left(\mathbf{x} + \mathbf{f}_{P\to \mathrm{M}}(\mathbf{x})\right)\right), \tag{11.16}$$

which can be applied to the input 3D image to generate a series of 3D images of the patient that match his/her chest motion signals. In the following, we introduce both linear and nonlinear estimation to determine the estimation model used in Eq. (11.15).

11.3.2 MOTION ESTIMATION MODELS

In this section, we introduce a kernel-principal component analysis (K-PCA)-based nonlinear statistical modeling method for estimating lung motion from chest surface motion. We first introduce the principal component analysis (PCA) method and then present the K-PCA-based estimation.

11.3.2.1 PCA model

The PCA algorithm can be used. According to PCA, a new vector can be calculated by:

$$\begin{bmatrix} \mathbf{u} \\ \mathbf{v} \end{bmatrix} = \begin{bmatrix} \bar{\mathbf{u}} \\ \bar{\mathbf{v}} \end{bmatrix} + \mathcal{M} \begin{bmatrix} \mathbf{b}_u \\ \mathbf{b}_v \end{bmatrix} \tag{11.17}$$

where $\mathcal{M}$ is the matrix formed by the eigenvectors of the covariance matrix corresponding to the M largest eigenvalues. Thus given a new feature vector, $\mathbf{b} = [\mathbf{b}_u^{\mathrm{T}}, \mathbf{b}_v^{\mathrm{T}}]^{\mathrm{T}}$, a lung respiratory motion field and the surface motion vector can be generated using Eq. (11.17). The underlying assumption of PCA is that a multidimensional Gaussian distribution is used to model these feature vectors:

$$p(\mathbf{b}) = \frac{1}{\sigma} \exp^{\left\{-\sum_{m=1}^{M} b_m^2/2\lambda_m\right\}}. \tag{11.18}$$

In order to estimate the lung respiratory motion field $\mathbf{u}^P$ from the chest surface motion $\mathbf{v}^P$ we need to solve the best the feature vector $\mathbf{b}$, so that $\mathbf{v}$ calculated from $\mathbf{b}$ using

Eq. (11.17) matches the patient's chest surface motion $\mathbf{v}^P$, and at the same time the prior distribution of Eq. (11.18) is maximized. The energy function is defined as

$$E(\mathbf{b}) = ||\mathbf{v}^P - \overline{\mathbf{v}} - \mathcal{M}_v \mathbf{b}||^2 + \xi \sum_{m=1}^{M} b_m^2 / 2\lambda_m, \tag{11.19}$$

where $\mathcal{M}_v$ is the matrix formed by the last rows of $\mathcal{M}$ corresponding to $\overline{\mathbf{v}}$. After $\mathbf{b}$ is estimated, the corresponding lung motion vector $\mathbf{u}^P$ can be calculated by

$$\mathbf{u}^P = \Theta(\ldots) = \overline{\mathbf{u}} + \mathcal{M}_u \mathbf{b}_P, \tag{11.20}$$

with $\mathcal{M}_u$ as the top part of $\mathcal{M}$ corresponding to the lung respiratory motion field $\mathbf{u}$.

11.3.2.2 Kernel-PCA model

K-PCA is a nonlinear statistical modeling method and can capture the variations of shapes more accurately than PCA. The basic idea of K-PCA is to compute PCA in a high-dimensional implicit mapping function of the motion vectors $\mathbf{u}$ and $\mathbf{v}$. Let K denote the kernel matrix of N samples, $k_{i,j} = k(\mathbf{u}_i^s, \mathbf{u}_j^s)$. K-PCA can be computed in a closed form by finding the first M eigenvalues and eigenvectors of K. Therefore given a motion vector $\mathbf{u}$, it can be projected onto the K-PCA space by $\lambda = A^{\mathrm{T}}(\mathbf{k}(\mathbf{u}, \mathbf{u}_{i=1,\ldots,N}^s) - \overline{\mathbf{k}})$, where $\overline{\mathbf{k}}$ is the mean of the kernel vectors, and $\mathbf{k}(\mathbf{u}, \mathbf{u}_{i=1,\ldots,N}^s)$ is the vector obtained by calculating the kernel function between $\mathbf{u}$ and $\mathbf{u}_i^s$, $i = 1, \ldots, N$. Because in K-PCA the feature space is induced implicitly, reconstruction of a new vector $\mathbf{u}$ from the feature vector λ is not trivial. Herein, Kwok, and Tsang's algorithm can be used for reconstruction (Kwok and Tsang, 2004).

11.3.2.3 Motion prediction using LS-SVM

Now that the lung motion vectors $\mathbf{u}^s$ are projected onto the K-PCA space as $\boldsymbol{\lambda}^s$, and the goal of motion estimation is to establish the relationship between the lung motion $\boldsymbol{\lambda}^s$ and chest surface motion $\mathbf{v}_{1\to t}^s$. Because $\mathbf{v}_{1\to \mathrm{t}}^s$ has relatively low dimension, we did not apply K-PCA to it. Given N training sample-pairs, we can employ the least squares support vector machine (LS-SVM) model to estimate this relationship. In this case the estimator can be written as:

$$\boldsymbol{\lambda}_t^s = \Theta\left(\mathbf{v}_{1\to t}^P\right) + \epsilon. \tag{11.21}$$

Because the elements of $\boldsymbol{\lambda}_t^s$ are independent of each other in the K-PCA space, we can estimate each element separately using SVM:

$$\lambda = \mathbf{w}^{\mathrm{T}} \varphi(\mathbf{v}) + \mathbf{b}, \tag{11.22}$$

where $\varphi(\mathbf{v})$ denotes a potential mapping function. $\mathbf{w}$ is the weighting vector and $\mathbf{b}$ is the shifting vector. The regularized cost function of the LS-SVM is given by An et al. (2007)

$$\min \xi(\mathbf{w}, \boldsymbol{\epsilon}) = \frac{1}{2}\mathbf{w}^{\mathrm{T}}\mathbf{w} + \frac{\gamma}{2}\sum\nolimits_{i=1}^{N} ||\epsilon_i||^2 \tag{11.23}$$

s.t.

$$\lambda = \mathbf{w}^{\mathrm{T}}\varphi(\mathbf{v}_i) + \mathbf{b} + \epsilon_i, \quad i = 1, \ldots, N, \tag{11.24}$$

where γ is referred to as the regularization constant. The Lagrangian method is utilized to solve the constrained optimization problem. Using typical radial basis function (RBF) kernel Π we get the parameters α and $\mathbf{b}$ to estimate $\boldsymbol{\lambda}$:

$$\lambda = \sum \alpha \, \Pi(\mathbf{v}, \mathbf{v}_i) + \mathbf{b}. \tag{11.25}$$

Notice that because different elements of the lung motion feature vector $\boldsymbol{\lambda}$ are independent, all of the elements of $\boldsymbol{\lambda}$ at different time-points are calculated by this model separately, similar to modeling the motion according to different lung capacity.

11.3.3 EXPERIMENTS

We introduced a method to estimate lung motion from a patient's 3D CT and chest signal after training the respiratory motion prediction models using 4D CT from a number of training samples. During intervention, 3D CT can be scanned on-site, and the key is to track the chest motion signals. Chest fiducial points and entire chest surface motion are commonly tracked. The difference is that for chest fiducial points four or more fiducial signals were used to track the chest motion, while for surface tracking much higher dimensional respiratory motion signals are used for estimation.

In the experiments, we used 4D CT datasets from 40 lung cancer patients undergoing radiotherapy planning. The images for each subject consist of 10 respiratory phases with the first and the last images being the exhale images and the fifth and the sixth the inhale images. The in-plane resolution of each image is 0.98 mm × 0.98 mm and the slice thickness is 1.5 mm. One subject was randomly selected as the template, 29 were used for training and 10 were used for testing. The algorithm was implemented on a Dell workstation with Intel Core 2 Quad Processor Q9300 and 8 GB RAM.

The quantitative performance regarding how well the method recovers the longitudinal deformations for each testing image series is evaluated. For each testing subject, we register the 4D CT images and obtain the longitudinal deformations $\mathbf{f}^*_{1\to t}(\mathbf{x})$. Then only the first frame and the chest surface motion signals extracted from the 4D CT images are used for estimation. After estimation, the estimated longitudinal deformations are denoted as $\hat{\mathbf{f}}_{1\to t}(\mathbf{x}), t = 2, \ldots, T$. The average lung respiratory motion estimation error for each subject can be calculated as

$$Err = \frac{1}{|\Omega|}\sum\nolimits_{t=2,\ldots,T}\sum\nolimits_{\mathbf{x}\in\Omega}\left|\mathbf{f}^*_{1\to t}(\mathbf{x}) - \hat{\mathbf{f}}_{1\to t}(\mathbf{x})\right|, \tag{11.26}$$

where Ω represents the domain of the voxels of the first time-point image of each subject. $|\Omega|$ is the number of voxels. The results showed that the estimation errors using surface-based estimation method are between 0.92 and 1.63 mm, with an average of 1.17 mm, while the errors using the fiducial-based estimation method are between 1.22 and 2.16 mm, with an average of 1.61 mm.

Generally, the proposed surface-based estimation has higher performance than the estimation based on EM-tracked chest fiducials (He et al., 2010). The idea is that high-dimensional deformation is involved in the prediction problem, and compared to fiducial signals with four sensors high-dimensional chest surface motion would provide more accurate estimation. For the fiducial-based estimation, because of the limited number of fiducials, it may be feasible to predict the respiratory phases and to monitor the respiratory motion during intervention, but it may not have sufficient degrees-of-freedom to predict the precise shape and location of the internal anatomical structures, such as lung field, bone, diaphragm, tumor, etc.

In another evaluation, we calculated how well the shape of lung tumor is estimated. Fig. 11.6 shows some sample results. Fig. 11.6A is the inhale CT for a subject and Fig. 11.6B is the exhale CT. Their overlay is shown in Fig. 11.6C. To measure the tumor estimation accuracy, we estimated the inhale images from exhale images and compared quantitatively whether the estimated tumor shape in the estimated inhale phase matches that in the original inhale image. Fig. 11.6D–F shows the images of three other subjects with red curves (dark gray in print versions) as the manually marked tumor from the original inhale images, and the green curves

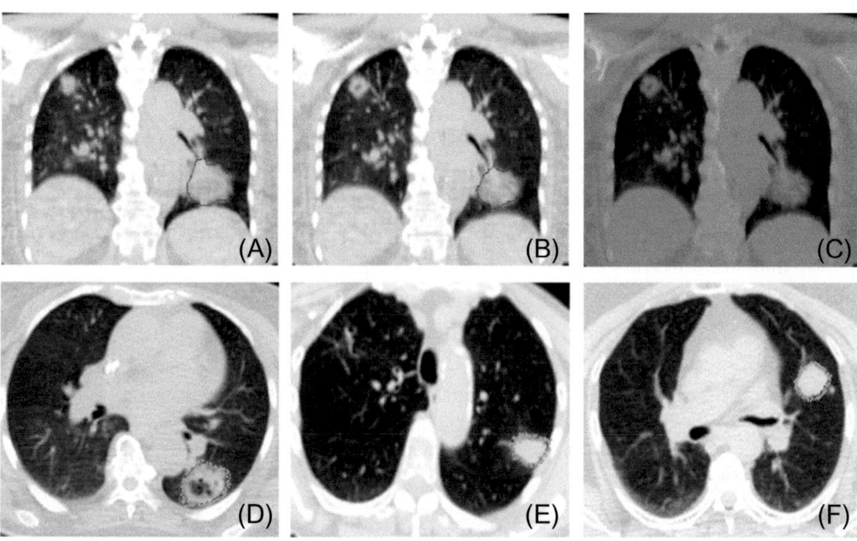

FIG. 11.6

Lung tumor estimation with dynamic modeling. (A) Exhale CT. (B) Inhale CT. (C) Overlaying (A) and (B). (D–F) Comparisons of tumor estimation between actual and estimated CT images.

show the estimated tumor shapes. We can see that the tumor positions and shapes under different respiratory phases can be estimated well. The DICE coefficient was used for quantitatively measuring estimation results of tumor positions. In our experiments, the average value of DICE coefficients over the 10 testing subjects is 88.1±2.8% using the proposed algorithm, while the result is 82.6±5.7% for fiducial-based estimation. Overall, these results demonstrate promising dynamic lung motion modeling.

11.4 AN EXAMPLE FOR IMAGE-GUIDED INTERVENTION

A CT-fluoroscopy (CTF)-guided intervention system is introduced as an example of how the previously mentioned motion estimation can be applied in clinical settings. The objective is to register a preprocedural or diagnostic 3D CT image onto the real-time captured CTF image for 3D lung intervention guidance. Although there are many papers addressing the registration between 2D-2D, 3D-2D, or 3D-3D images in the literature (Yaniv et al., 2010; Cleary and Peters, 2010; Enquobahrie et al., 2008, 2007; Sun et al., 2006; Bouchet et al., 2001; Karnik et al., 2010; Park et al., 2010), they do not fit such a novel task: there is a limited number of slices for CTF, and the preprocedural 3D CT covers the entire or most of the lung region. To solve this problem, we customized the cubic B-spline model so that the deformation in the transverse plane is modeled using 2D B-spline, and the deformation in the z-direction is regularized using the general smoothness criterion. To improve registration accuracy, a respiratory motion compensation framework is incorporated into the registration procedure. Finally, parallel implementation allows for on-site application.

11.4.1 CTF GUIDANCE WITH MOTION COMPENSATION

Because the preprocedural inhale and exhale 3D CT images cover a larger lung region than the intraprocedural CTF that only has 4–10 slices, the current 3D-3D registration or 3D-2D registration algorithms do not fit this application. For example, the 3D cubic B-spline model is not applicable for the deformation in the z-direction given the limited number of slices for CTF. To solve this problem, we use the idea that the deformation model in the transverse plane should be handled differently from that in the z-direction. In addition, each CTF image is subject to a deformation from the inhale 3D CT, consisting of both respiratory motion and subtle anatomical deformation. Therefore the framework of motion compensation (MC) is utilized to better estimate the deformation between the current CTF image and the inhale 3D CT.

Fig. 11.7 illustrates the workflow of the CT-CTF registration algorithm with MC. First, the preprocedural or diagnostic CT scans are captured, and together with the intraprocedural CTF, they are used for estimating a series of images based on the previously introduced motion estimation methods. Then the best phase that matches the CTF image is picked and a fast registration is performed between the CTF and

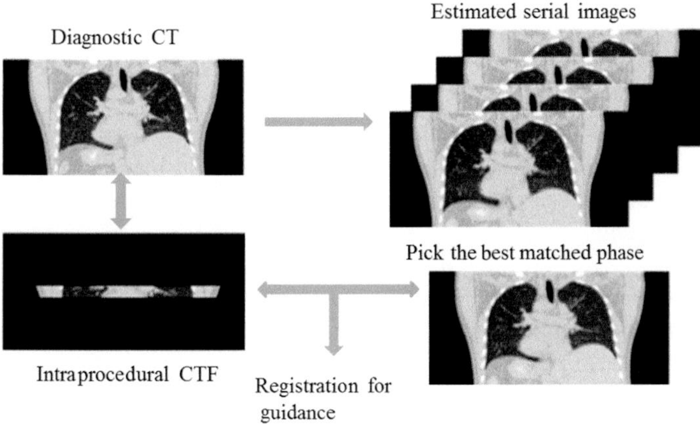

FIG. 11.7

The workflow of the CT-CTF registration algorithm with motion compensation.

this selected image. The overlay of the registered 3D image on the CTF provides 3D guidance for the intervention. In this framework, the deformation between the phase-matched intermediate CT and the inhale CT is known, and our goal is to estimate the elastic deformation between this intermediate CT and the CTF. Compared to registering the diagnostic CT directly onto the CTF, the deformation between the intermediate CT and the CTF is much smaller because of the motion compensation, so the registration can be more accurate.

11.4.2 THE CT-CTF REGISTRATION ALGORITHM

The formulation of CT-CTF registration is in line with that of the common deformable image registration framework (Bai and Brady, 2009; Ferrant et al., 2000; Huang et al., 2006; Jacobson and Murphy, 2011; Johnson and Christensen, 2002; Mattes et al., 2003; Noblet et al., 2005; Rueckert et al., 2006, 2003, 1999; Shackleford et al., 2010; Sorzano et al., 2005; Warfield et al., 2002; Xue et al., 2010), with the objective of solving a deformation field $\mathbf{f}(\mathbf{x})$ to align the source 3D image $I_T(\mathbf{x})$ to the CTF image $I_{\text{CTF}}(\mathbf{x})$. The basic registration framework can be formulated by

$$\hat{\mathbf{f}} = \text{argmin}_{\mathbf{f}}(E_s(I_{\text{CTF}}, I_T, \mathbf{f}) + \lambda \Re(\mathbf{f})), \quad (11.27)$$

where $E_s(I_{\text{CTF}}, I_T, \mathbf{f})$ represents the image difference measure between the CTF image and the 3D CT under the current deformation field $\mathbf{f}$, and $\Re(\mathbf{f})$ is the regularization term for the deformation field. $\lambda > 0$ is the weighting coefficient of the regularization term. We use the modified cubic B-spline to model the deformations:

$$\mathbf{f}(\mathbf{x}) = \sum_{l=0}^{3}\sum_{m=0}^{3} B_l(u)\, B_m(v)\, c_{i+l,j+m,z}, \tag{11.28}$$

where B_l represents the lth basis function of the cubic B-spline. $\mathbf{x} = (x_1, x_2, x_3)$ is a voxel in the CTF image space Ω and c represents the deformation values of control points. Here, the control points are uniformly distributed on the voxel grids in the image domain, and $i = \lceil x_1/n_{x_1} \rceil - 1, j = \lceil x_2/n_{x_2} \rceil - 1, u = x_1/n_{x_1} - \lceil x_1/n_{x_1} \rceil$, and $v = x_2/n_{x_2} - \lceil x_2/n_{x_2} \rceil$. n_{x_1} and n_{x_2} are the pixel spacing between the control points in the x_1 and x_2 directions, respectively. Smaller control point spacing will result in a finer deformation but the computation time will increase slightly as the number of control points increases.

Notice that in Eq. (11.28), $\mathbf{f}$ and c are defined in 3D space to model volumetric deformation, but the B-spline model is only used in the transverse plane. Therefore smoothness in the x–y plane is ensured due to the property of the B-spline. In the z-direction, an additional smoothness constraint is required, thus $\Re(\mathbf{f})$ in Eq. (11.27) can be defined as

$$\Re(\mathbf{f}) = \frac{1}{2}\sum_{\mathbf{x}\in\Omega} \|\partial\mathbf{f}(\mathbf{x})/\partial x_3\|^2. \tag{11.29}$$

For image difference measurement, the sum of squared differences (SSD) of both image intensities and gradients are used. It has been shown that realistic deformation should not involve folding of the field and the neighboring voxels should not cross each other, and hence the local Jacobian determinant of the field should be positive. Therefore after registration the topology of the deformation field is checked (Karacali and Davatzikos, 2004).

In CT-CTF registration, the deformation field is defined on the CTF image space with a limited number of slices, hence the deformation on other slices of the 3D CT images can only be estimated. One way to estimate this deformation is to expand the deformation field outside the boundary of the CTF slices in the z-direction. However, if there is a large respiratory motion discrepancy between the preprocedural CT images and the CTF, a gradual transition or gap might be noticed. In order to estimate this deformation and accomplish more accurate registration, we propose to incorporate motion compensation into our CT-CTF registration.

Fig. 11.8 shows the framework for CT-CTF registration with MC, which consists of two steps. The first step is to generate a series of estimated intermediate respiratory phase CT images and to determine the intermediate CT that best matches the CTF. To do this, an inhale CT I_T is acquired in advance. Then CTF and I_T can be used to estimate an intermediate CT $\mathrm{I_t}$ using the motion estimation techniques described in Section 11.3. The second step is to register the inhale CT I_T with the CTF I_{CTF} by considering the respiratory differences. Denoting the deformation from the CTF to the phase-matched intermediate CT as $\mathbf{f}_{\mathrm{CTF}\rightarrow t}$, the deformation of the whole procedure can be formulated as

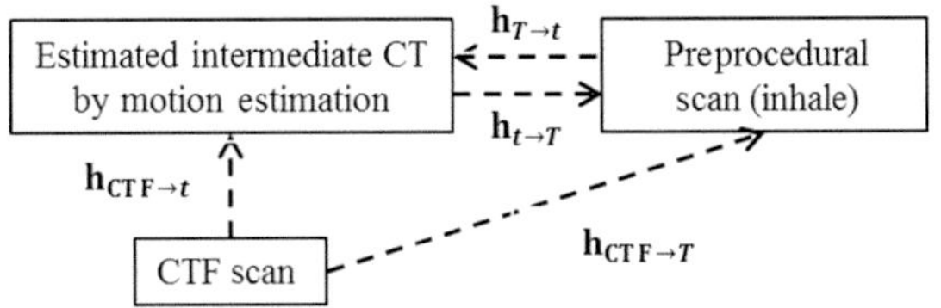

FIG. 11.8

CT-CTF registration with motion compensation.

$$\mathbf{f} = \mathbf{f}_{t\to T} {}^{\circ} \mathbf{f}_{\mathrm{CTF}\to t}. \tag{11.30}$$

The deformation field from t to T, $\mathbf{f}_{t\to T}$, can be estimated through inverse operation. The intermediate CT image I_t can be generated by warping I_T using $\mathbf{f}_{t\to T}$. We can precalculate T intermediate CT images, and once a CTF is captured during the procedure, the phase-matched intermediate CT can be determined as the one that has the smallest least squares error (LSE) with the CTF after global rigid-body registration.

It can be seen that rather than registering the inhale CT onto the CTF image directly, we first deform the inhale CT I_T onto the respiratory phase-matched intermediate CT I_t and then further register it onto the CTF I_{CTF}. Because the deformation for respiratory motion $\mathbf{f}_{t\to T}$ is known, our goal is to estimate the deformation $\mathbf{f}_{\mathrm{CTF}\to t}$. Compared to solving the deformation $\mathbf{f}$ directly, the new strategy generates more accurate results because the unknown deformation $\mathbf{f}_{\mathrm{CTF}\to t}$ is relatively small.

In fact, the phase-matched intermediate image I_t may not necessarily be used explicitly in the registration to prevent accumulation of image resampling errors. This is because $\mathbf{f}_{t\to T}$ is known, and we can embed it into the energy function in Eq. (11.27) as

$$\hat{\mathbf{f}} = \mathrm{argmin}_{\mathbf{f}}(E_s\left(I_{\mathrm{CTF}}, I_T, \mathbf{f}_{t\to T} {}^{\circ} \mathbf{f}_{\mathrm{CTF}\to t}\right) + \lambda \Re\left(\mathbf{f}_{\mathrm{CTF}\to t}\right)). \tag{11.31}$$

Alternatively, a linear interpolation can be used for motion estimation if both exhale I_1 and inhale I_T images are available. We can linearly scale the deformation $\mathbf{f}_{T\to 0}$ to estimate intermediate CTs I_t corresponding to respiratory phases between inhale and exhale. Let $T \in \mathbb{Z}^+$ be the number of estimated respiratory phases (ie, $T = 10$), and $t \in \mathbb{Z}^+$, $1 \leq t \leq T$, is an intermediate respiratory phase. The larger T is, the more intermediate CTs we can estimate. So the deformation from T to t, $\mathbf{f}_{T\to t}$can be expressed as

$$\mathbf{f}_{T\to t} = \frac{T-t}{T}\mathbf{f}_{T\to 1}. \tag{11.32}$$

The deformation field from t to T, $\mathbf{f}_{t\to T}$, can be estimated through inverse operation.

11.4.3 EXPERIMENTS

A parallel and multiresolution strategy for the CT-CTF registration was implemented. Three resolutions were used and at each resolution images were partitioned into sub-blocks in the x–y plane, and then the registration of each block was performed in parallel. The finite differential method was used for optimization, and the change of the deformation value of each control point was calculated locally according to the B-spline. The advantage of using the finite differential method comes from being able to hierarchically adjust the number of control points involved in the registration. For example, in each iteration we only update the deformation of control points whose local image differences are ranked in a percentage over all the control points. We started from 50% and then gradually increased to 90% of control points in this hierarchical procedure. Using C++ language, we showed that the registration procedure finished within 2 s for the CT images of size 512×512×70 and CTF image with size of 512×512×4 on a quad-core Intel Xeon 2.8 GHz machine.

Fig. 11.9 shows an example of the CT-CTF registration algorithm with linear motion compensation. Fig. 11.9A is the CTF captured during breath holding,

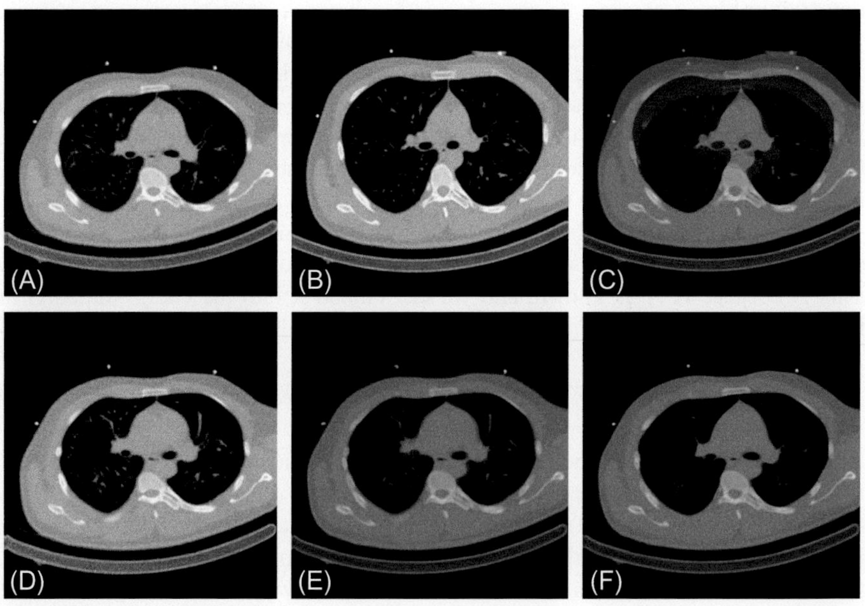

FIG. 11.9

An example of the CT-CTF registration algorithm with motion estimation. (A) CTF. (B) Inhale CT. (C) Overlay CTF (red) with inhale CT (gray scale) before registration. (D) The estimated intermediate image which matches the respiratory phase of CTF. (E) Overlay CTF (red) with (D) (gray scale). (F) The final registration result by overlaying the registered CT (gray scale) with CTF (red).

Fig. 11.9B is the inhale CT data, and Fig. 11.9C shows the overlay between these two images after global rigid-body registration. It is clearly seen that the discrepancy caused by the difference between breath-holding CTF image and inhale CT image is considerable. Using Eq. (11.32) a series of images between inhale and exhale are generated and the phase-matched intermediate CT image is chosen based on the least squares difference of the image intensity, shown in Fig. 11.9D. Fig. 11.9E shows the overlay of (A) and (D). It is shown that after applying motion compensation, the CT image matches CTF and a lot of respiratory motion-related deformation is removed. Fig. 11.9F gives the final registration results showing that the inhale CT image is nicely aligned with the CTF. Compared to registering (A) and (B) directly (the result is shown in Fig. 11.10), the deformation to be solved in our proposed framework is much smaller, rendering more accurate registration because the chance of being stuck in local minima during optimization is low.

We also applied the CT-CTF registration with MC to register the preprocedural inhale CT images onto CTFs and verified the registration by visual inspection for patient data during intervention. Fig. 11.11 shows examples of the overlay of the deformed inhale CT images (gray scale) with the intraprocedure CTFs (red; dark gray in print versions). From Fig. 11.11B it can be seen that the tumor and other parts (liver and lung field boundaries) are aligned well, while Fig. 11.11D and H show that the bones, airway, and lung field boundaries are also aligned well. We had gone through all the registered images by superimposing corresponding CTFs on them. The visual inspection by our radiologist showed that the aligned results on real-patient data are similar to those on the 4D CT data. For percutaneous lung

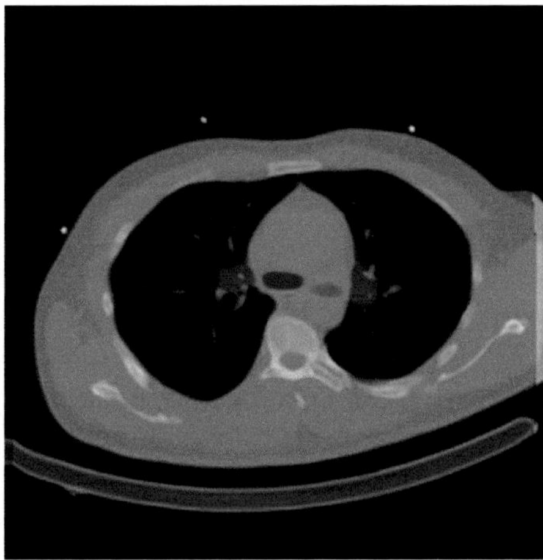

FIG. 11.10

The result by directly registering the CTF image in Fig. 11.9A and the CT image in Fig. 11.9B.

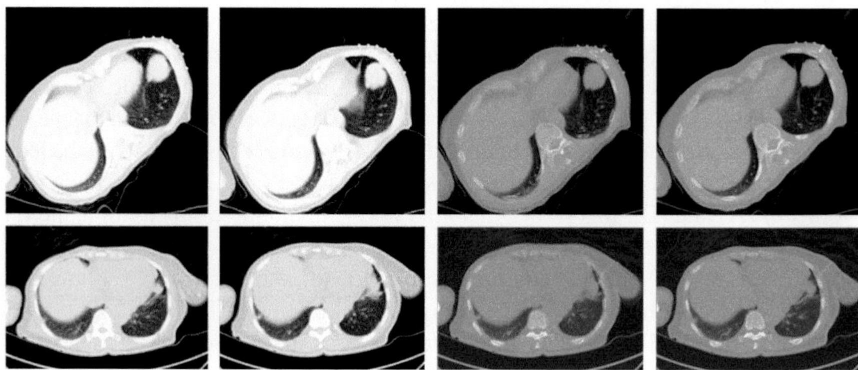

FIG. 11.11

Results of registered CT images by overlaying onto the CTF images of patient data collected from lung biopsy. From left to right: exhale CT image; inhale CT image; fused images before registration; fused images after registration.

intervention, it was reported that the average navigation error (3.3 mm) based on the registration between planning and confirmation images can successfully accomplish the intervention task (Yaniv et al., 2010). In our results, after motion compensation, the average registration error was 1.36 mm. We believe that the accuracy of the CT-CTF registration with MC is accurate enough for CT-guided percutaneous lung intervention. In future work, we plan to test the performance of the method during lung intervention (He et al., 2012).

11.5 CONCLUDING REMARKS

Using CT lung image motion compensation as an example, we showed that it is possible to estimate the dynamics of the lung by using available real-time chest motion signals. In addition to the necessary medical image processing techniques needed for motion compensation, we addressed the general question, ie, given the variability of the whole shape, how to estimate the shape from partially known information. In fact many engineering problems fall into this estimation category, and various prediction and regression models can be used. In this chapter, we showed that a general PCA model or K-PCA model can be used to project the high-dimensional signals and fields onto low-dimensional spaces. Then, Bayesian estimation or a support vector machine are used for estimation. Finally, using a CTF-guided intervention as an example, we illustrate how motion estimation can be applied in a clinical application. Future works include evaluating how the estimation model can be more robust and reliable to handle different sizes and breathing patterns of the patients, and how to apply them in real-time estimation of location and shape of tumor for intervention and radiotherapy.

ACKNOWLEDGMENT

This work was supported by NIH grant 1R03EB018977 (ZX).

REFERENCES

Alexander, D.C., Pierpaoli, C., Basser, P.J., Gee, J.C., 2001. Spatial transformations of diffusion tensor magnetic resonance images. IEEE Trans. Med. Imaging 20, 1131–1139.

An, S.J., Liu, W.Q., Venkatesh, S., 2007. Fast cross-validation algorithms for least squares support vector machine and kernel ridge regression. Pattern Recogn. 40, 2154 2162.

Bai, W., Brady, M., 2009. Regularized b-spline deformable registration for respiratory motion correction in pet images. Phys. Med. Biol. 54, 2719–2736.

Bouchet, L.G., Meeks, S.L., Goodchild, G., Bova, F.J., Buatti, J.M., Friedman, W.A., 2001. Calibration of three-dimensional ultrasound images for image-guided radiation therapy. Phys. Med. Biol. 46, 559–577.

Broaddus, W.C., Gillies, G.T., Kucharczyk, J., 2001. Minimally invasive procedures. Advances in image-guided delivery of drug and cell therapies into the central nervous system. Neuroimaging Clin. N. Am. 11, 727–735.

Cala, S.J., Kenyon, C.M., Ferrigno, G., Carnevali, P., Aliverti, A., Pedotti, A., Macklem, P.T., Rochester, D.F., 1996. Chest wall and lung volume estimation by optical reflectance motion analysis. J. Appl. Physiol. 81,2680–2689.

Cervino, L.I., Du, J., Jiang, S.B., 2011. MRI-guided tumor tracking in lung cancer radiotherapy. Phys. Med. Biol. 56,3773–3785.

Cleary, K., Peters, T.M., 2010. Image-guided interventions: technology review and clinical applications. Ann. Rev. Biomed. Eng. 12, 119–142.

Cleary, K., Clifford, M., Stoianovici, D., Freedman, M., Mun, S.K., Watson, V., 2002. Technology improvements for image-guided and minimally invasive spine procedures. IEEE Trans. Inform. Tech. Biomed. 6, 249–261.

de Gregorio, M.A., Laborda, A., Ortas, R., Higuera, T., Gomez-Arrue, J., Medrano, J., Mainar, A., 2008. Image-guided minimally invasive treatment of pulmonary arterial hypertension due to embolic disease. Arch. Bronconeumol. 44, 312–317.

DeLucia, P.R., Mather, R.D., Griswold, J.A., Mitra, S., 2006. Toward the improvement of image-guided interventions for minimally invasive surgery: three factors that affect performance. Hum. Fact. 48, 23–38.

Descoteaux, M., Collins, D.L., Siddiqi, K., 2008. A geometric flow for segmenting vasculature in proton-density weighted MRI. Med. Image Anal. 12, 497–513.

DiMaio, S.P., Pieper, S., Chinzei, K., Hata, N., Haker, S.J., Kacher, D.F., Fichtinger, G., Tempany, C.M., Kikinis, R., 2007. Robot-assisted needle placement in open MRI: system architecture, integration and validation. Comput. Aided Surg. 12, 15–24.

Ehrhardt, J., Werner, R., Schmidt-Richberg, A., Handels, H., 2011. Statistical modeling of 4d respiratory lung motion using diffeomorphic image registration. IEEE Trans. Med. Imaging 30, 251–265.

Enquobahrie, A., Cheng, P., Gary, K., Ibanez, L., Gobbi, D., Lindseth, F., Yaniv, Z., Aylward, S., Jomier, J., Cleary, K., 2007. The image-guided surgery toolkit IGSTK: an open source C++ software toolkit. J. Digital Imaging 20 (1), 21–33.

Enquobahrie, A., Gobbi, D., Turek, M., Cheng, P., Yaniv, Z., Lindseth, F., Cleary, K., 2008. Designing tracking software for image-guided surgery applications: IGSTK experience. Int. J. Comput. Assist. Radiol. Surg. 3, 395–403.

Ferrant, M., Warfield, S., Nabavi, A., Jolesz, F., Kikinis, R., 2000. Registration of 3D intraoperative MR images of the brain using a finite element biomechanical model. In: Medical Image Computing and Computer-Assisted Intervention—MICCAI 2000, pp. 249–258.

Frangi, A.F., Niessen, W.J., Vincken, K.L., Viergever, M.A., 1998. Multiscale vessel enhancement filtering. In: Medical Image Computing and Computer-Assisted Intervention—MICCAI'98 1496, 130–137.

Hall, W.A., Liu, H., Martin, A.J., Truwit, C.L., 2001. Minimally invasive procedures: interventional MR image-guided neurobiopsy. Neuroimaging Clin. N. Am. 11, 705–713.

Handels, H., Werner, R., Schmidt, R., Frenzel, T., Lu, W., Low, D., Ehrhardt, J., 2007. 4d medical image computing and visualization of lung tumor mobility in spatio-temporal CT image data. Int. J. Med. Inform. 76 (3), S433–S439.

He, T., Xue, Z., Xie, W., Wong, S.T., 2010. Online 4-d CT estimation for patient-specific respiratory motion based on real-time breathing signals. Med. Image Comput. Comput. Assist. Interv. 13, 392–399.

He, T., Xue, Z., Lu, K., Valdivia, Y.A.M., Wong, K.K., Xie, W., Wong, S.T., 2012. A minimally invasive multimodality image-guided (MIMIG) system for peripheral lung cancer intervention and diagnosis. Comput. Med. Imaging Graph. 36, 345–355.

Hu, S., Hoffman, E.A., Reinhardt, J.M., Jun 2001. Automatic lung segmentation for accurate quantitation of volumetric X-ray CT images. IEEE Trans. Med. Imaging 20, 490–498.

Huang, X., Paragios, N., Metaxas, D.N., 2006. Shape registration in implicit spaces using information theory and free form deformations. IEEE Trans. Pattern Anal. Mach. Intell. 28, 1303–1318.

Jacob, A.L., Messmer, P., Kaim, A., Suhm, N., Regazzoni, P., Baumann, B., 2000. A whole-body registration-free navigation system for image-guided surgery and interventional radiology. Investigat. Radiol. 35, 279–288.

Jacobson, T.J., Murphy, M.J., 2011. Optimized knot placement for b-splines in deformable image registration. Med. Phys. 38, 4579–4582.

Johnson, H.J., Christensen, G.E., 2002. Consistent landmark and intensity-based image registration. IEEE Trans. Med. Imaging 21, 450–461.

Karacali, B., Davatzikos, C., 2004. Estimating topology preserving and smooth displacement fields. IEEE Trans. Med. Imaging 23, 868–880.

Karnik, V.V., Fenster, A., Bax, J., Cool, D.W., Gardi, L., Gyacskov, I., Romagnoli, C., Ward, A.D., 2010. Assessment of image registration accuracy in three-dimensional transrectal ultrasound guided prostate biopsy. Med. Phys. 37, 802–813.

Klein, J.S., Schultz, S., Heffner, J.E., 1995. Interventional radiology of the chest: image-guided percutaneous drainage of pleural effusions, lung abscess, and pneumothorax. Am. J. Roentgenol. 164, 581–588.

Klinder, T., Lorenz, C., Ostermann, J., 2010. Prediction framework for statistical respiratory motion modeling. Med. Image Comput. Comput. Assist. Interv. 13, 327–334.

Kwok, J.T.Y., Tsang, I.W.H., 2004. The pre-image problem in kernel methods. IEEE Trans. Neural Netw. 15, 1517–1525.

Lang, M.J., Greer, A.D., Sutherland, G.R., 2011. Intra-operative robotics: Neuroarm. Acta Neurochirurg. Suppl. 109, 231–236.

Li, G., Ballangrud, A., Kuo, L.C., Kang, H., Kirov, A., Lovelock, M., Yamada, Y., Mechalakos, J., Amols, H., 2011. Motion monitoring for cranial frameless stereotactic radiosurgery using video-based three-dimensional optical surface imaging. Med. Phys. 38, 3981–3994.

Lin, T., Cervino, L.I., Tang, X., Vasconcelos, N., Jiang, S.B., 2009. Fluoroscopic tumor tracking for image-guided lung cancer radiotherapy. Phys. Med. Biol. 54, 981–992.

Lorigo, L.M., Faugeras, O.D., Grimson, W.E., Keriven, R., Kikinis, R., Nabavi, A., Westin, C.F., 2001. Curves: curve evolution for vessel segmentation. Med. Image Anal. 5, 195–206.

Lu, W., Song, J.H., Christensen, G.E., Parikh, P.J., Zhao, T., Hubenschmidt, J.P., Bradley, J.D., Low, D.A., 2006. Evaluating lung motion variations in repeated 4d CT studies using inverse consistent image registration. Int. J. Radiat. Oncol. Biol. Phys. 66, S606–S607.

Malladi, R., Sethian, J.A., Vemuri, B.C., 1995. Shape modeling with front propagation - a level set approach. IEEE Trans. Pattern Anal. Mach. Intell. 17, 158–175.

Masutani, Y., MacMahon, H., Doi, K., 2001. Automated segmentation and visualization of the pulmonary vascular tree in spiral CT angiography: an anatomy-oriented approach based on three-dimensional image analysis. J. Comput. Assist. Tomog. 25, 587–597.

Mattes, D., Haynor, D.R., Vesselle, H., Lewellen, T.K., Eubank, W., 2003. PET-CT image registration in the chest using free-form deformations. IEEE Trans. Med. Imaging 22, 120–128.

Noblet, V., Heinrich, C., Heitz, F., Armspach, J.P., 2005. 3-d deformable image registration: a topology preservation scheme based on hierarchical deformation models and interval analysis optimization. IEEE Trans. Image Process. 14, 553–566.

Osher, S., Sethian, J.A., 1988. Fronts propagating with curvature-dependent speed—algorithms based on Hamilton-Jacobi formulations. J. Comput. Phys. 79, 12–49.

Pandya, S., Motkoski, J.W., Serrano-Almeida, C., Greer, A.D., Latour, I., Sutherland, G.R., 2009. Advancing neurosurgery with image-guided robotics. J. Neurosurg. 111, 1141–1149.

Park, B.K., Kim, C.K., Choi, H.Y., Lee, H.M., Jeon, S.S., Seo, S.I., Han, D.H., 2010. Limitation for performing ultrasound-guided radiofrequency ablation of small renal masses. Eur. J. Radiol. 75, 248–252.

Patriciu, A., Petrisor, D., Muntener, M., Mazilu, D., Schar, M., Stoianovici, D., 2007. Automatic brachytherapy seed placement under MRI guidance. IEEE Trans. Biomed. Eng. 54, 1499–1506.

Rueckert, D., Sonoda, L.I., Hayes, C., Hill, D.L., Leach, M.O., Hawkes, D.J., 1999. Nonrigid registration using free-form deformations: application to breast MR images. IEEE Trans. Med. Imaging 18, 712–721.

Rueckert, D., Frangi, A.F., Schnabel, J.A., 2003. Automatic construction of 3-d statistical deformation models of the brain using nonrigid registration. IEEE Trans. Med. Imaging 22, 1014–1025.

Rueckert, D., Aljabar, P., Heckemann, R.A., Hajnal, J.V., Hammers, A., 2006. Diffeomorphic registration using b-splines. Med. Image Comput. Comput. Assist. Interv. 9, 702–709.

Sadeghi Naini, A., Patel, R.V., Samani, A., 2010. Ct-enhanced ultrasound image of a totally deflated lung for image-guided minimally invasive tumor ablative procedures. IEEE Trans. Biomed. Eng. 57, 2627–2630.

Santelli, C., Nezafat, R., Goddu, B., Manning, W.J., Smink, J., Kozerke, S., Peters, D.C., 2011. Respiratory bellows revisited for motion compensation: preliminary experience for cardiovascular MR. Magn. Reson. Med. 65, 1098–1103.

Seibel, R.M., 1997. Image-guided minimally invasive therapy. Surg. Endosc. 11, 154–162.

Seimenis, I., Tsekos, N.V., Keroglou, C., Eracleous, E., Pitris, C., Christoforou, E.G., 2012. An approach for preoperative planning and performance of mr-guided interventions demonstrated with a manual manipulator in a 1.5t MRI scanner. CardioVasc. Interven. Radiol. 35, 359–367.

Seinstra, B.A., van Delden, O.M., van Erpecum, K.J., van Hillegersberg, R., Mali, W.P., van den Bosch, M.A., 2010. Minimally invasive image-guided therapy for inoperable hepatocellular carcinoma: what is the evidence today? Insights Imaging 1, 167–181.

Shackleford, J.A., Kandasamy, N., Sharp, G.C., 2010. On developing b-spline registration algorithms for multi-core processors. Phys. Med. Biol. 55, 6329–6351.

Shah, A.P., Kupelian, P.A., Willoughby, T.R., Meeks, S.L., 2011. Expanding the use of real-time electromagnetic tracking in radiation oncology. J. Appl. Clin. Med. Phys. 12, 34–49.

Shamir, R., Freiman, M., Joskowicz, L., Shoham, M., Zehavi, E., Shoshan, Y., 2005. Robot-assisted image-guided targeting for minimally invasive neurosurgery: planning, registration, and in-vitro experiment. Med. Image Comput. Comput. Assist. Interv. 8, 131–138.

Sluimer, I., Schilham, A., Prokop, M., van Ginneken, B., 2006. Computer analysis of computed tomography scans of the lung: a survey. IEEE Trans. Med. Imaging 25, 385–405.

Smith, K.A., Kim, H.S., 2011. Interventional radiology and image-guided medicine: interventional oncology. Semin. Oncol. 38, 151–162.

Sorzano, C.O.S., Thévenaz, P., Unser, M., 2005. Elastic registration of biological images using vector-spline regularization. IEEE Trans. Biomed. Eng. 52, 652–663.

Sun, D., Willingham, C., Durrani, A., King, P., Cleary, K., Wood, B., 2006. A novel end-effector design for robotics in image-guided needle procedures. Int. J. Med. Robot. Comput. Assist. Surg. 2, 91–97.

Sundaram, T.A., Avants, B.B., Gee, J.C., 2004. A dynamic model of average lung deformation using capacity-based reparameterization and shape averaging of lung MR images. presented at the MICCAI 2004.

Tan, K.S., Saatchi, R., Elphick, H., Burke, D., 2010. Real-time vision based respiration monitoring system. In: 7th International Symposium on Communication Systems Networks and Digital Signal Processing (CSNDSP), 2010, pp. 770–774.

Tokuda, J., Fischer, G.S., DiMaio, S.P., Gobbi, D.G., Csoma, C., Mewes, P.W., Fichtinger, G., Tempany, C.M., Hata, N., Jan 2010. Integrated navigation and control software system for MRI-guided robotic prostate interventions. Comput. Med. Imaging Grap. 34, 3–8.

Ukimura, O., 2010. Image-guided surgery in minimally invasive urology. Curr. Opin. Urol. 20, 136–140.

Vandemeulebroucke, J., Rit, S., Kybic, J., Clarysse, P., Sarrut, D., 2011. Spatiotemporal motion estimation for respiratory-correlated imaging of the lungs. Med. Phys. 38, 166–178.

Wang, F., Vemuri, B.C., Eisenschenk, S.J., 2006. Joint registration and segmentation of neuroanatomic structures from brain MRI. Acad. Radiol. 13, 1104–1111.

Warfield, S.K., Talos, F., Tei, A., Bharatha, A., Nabavi, A., Ferrant, M., Black, P.M., Jolesz, F.A., Kikinis, R., 2002. Real-time registration of volumetric brain MRI by biomechanical simulation of deformation during image guided neurosurgery. Comput. Vis. Sci. 5, 3–11.

Westendorff, C., Hoffmann, J., Troitzsch, D., Dammann, F., Reinert, S., 2004. Ossifying fibroma of the skull: interactive image-guided minimally invasive localization and resection. J. Craniofac. Surg. 15, 854–858.

Wu, G., Wang, Q., Lian, J., Shen, D., 2011. Estimating the 4d respiratory lung motion by spatiotemporal registration and building super-resolution image. Med. Image Comput. Comput. Assist. Interv. 14, 532–539.

Xiaohua, C., Brady, M., Lo, J.L., Moore, N., 2005. Simultaneous segmentation and registration of contrast-enhanced breast MRI. Inform. Process. Med. Imaging 19, 126–137.

Xue, Z., Shen, D., Davatzikos, C., 2006. Classic: consistent longitudinal alignment and segmentation for serial image computing. NeuroImage 30, 388–99.

Xue, Z., Wong, K., Wong, S.T., 2010. Joint registration and segmentation of serial lung CT images for image-guided lung cancer diagnosis and therapy. Comput. Med. Imaging Grap. 34, 55–60.

Yakar, D., Schouten, M.G., Bosboom, D.G., Barentsz, J.O., Scheenen, T.W., Futterer, J.J., 2011. Feasibility of a pneumatically actuated mr-compatible robot for transrectal prostate biopsy guidance. Radiology 260, 241–247.

Yaniv, Z., Cheng, P., Wilson, E., Popa, T., Lindisch, D., Campos-Nanez, E., Abeledo, H., Watson, V., Cleary, K., Banovac, F., 2010. Needle-based interventions with the image-guided surgery toolkit (IGSTK): from phantoms to clinical trials. IEEE Trans. Biomed. Eng. 57, 922–933.

Yeung, R.W., Xia, J.J., Samman, N., 2006. Image-guided minimally invasive surgical access to the temporomandibular joint: a preliminary report. J. Oral Maxillofac. Surg. 64, 1546–1552.

Yim, Y., Hong, H., 2008. Correction of segmented lung boundary for inclusion of pleural nodules and pulmonary vessels in chest CT images. Comput. Biol. Med. 38, 845–857.

CHAPTER

12

Learning pathological deviations from a normal pattern of myocardial motion: Added value for CRT studies?

N. Duchateau[1], G. Piella[2], A. Frangi[3], M. De Craene[4]

Inria Sophia Antipolis, Sophia Antipolis, France[1] Universitat Pompeu Fabra, Barcelona, Spain[2] University of Sheffield, Sheffield, United Kingdom[3] Philips Medisys, Suresnes, France[4]

CHAPTER OUTLINE

Machine Learning and Medical Imaging. http://dx.doi.org/10.1016/B978-0-12-804076-8.00012-8

12.1 INTRODUCTION

12.1.1 CARDIAC RESYNCHRONIZATION THERAPY

In the last decade, cardiac resynchronization therapy (CRT) has become the recommended procedure to treat heart failure in patients with asynchronous contraction of the cardiac chambers (Yu et al., 2006). A biventricular pacing device optimizes the left/right ventricular delay, and eventually the atrioventricular delay. When deficiencies in the conduction system are compensated, synchronous contraction is expected to improve the heart efficiency and patient survival. Established international guidelines recommend pacing patients with symptomatic heart failure, electrical abnormalities, and decreased left ventricular function (Brignole et al., 2013). Nonetheless, these selection criteria are suboptimal. The therapy fails to "improve enough" patient condition in approximately 30% of the cases (clinical response), and reverse remodeling in 50% of the cases (volume response) (Bleeker et al., 2006) This issue is of primary concern given the prevalence of the symptoms (between 25% and 50% of heart failure patients, at least 15 million people in Europe (van Veldhuisen et al., 2009)), and the associated costs (better implantation of 1% of the devices in the United States between 2003 and 2007 (Laskey et al., 2012) already means an impact of 105M€ to 15K€ per device, for 7000 devices).

Currently, no consensus exists on better selection criteria. Publications abound on new predictive indexes. Paradoxically, they focus too much on the blind assessment of mechanical dyssychrony (mainly through peak or time-to-event measurements), instead of better understanding the mechanisms of therapy response. A change towards a more comprehensive strategy was claimed more recently (Fornwalt, 2011) Some studies highlighted the relevance of distinguishing between specific types of dyssynchrony, each one associated with a pattern correctable by CRT (Parsai et al., 2009; Doltra et al., 2014) They pave the way towards interpreting those mechanical patterns in the light of precise electrical and structural abnormalities (Vernooy et al., 2014).

According to these findings, improving patient selection should therefore increase the recognition of these different types of dyssynchrony in new candidates.

12.1.2 PATTERNS OF MOTION/DEFORMATION

Myocardial motion and deformation can be extracted from cardiac sequences by combining segmentation and temporal registration/tracking. Local speckles attached to the myocardium allow tracking along 2D or 3D echocardiographic sequences and computing these parameters locally (Duchateau et al., 2013a) However, once temporal and anatomical variability are removed, the comparison of spatiotemporal patterns of motion and deformation is challenging.

Quantifying the distance from a patient to a normal motion pattern is an intuitive way of encoding pathological patterns. Clinicians actually do so when diagnosing a new patient: given the representation of a normal contraction (previously learnt), where and how much does the observed patient differ from normality? Then, once categorized as abnormal, can a similar process be extended to quantify the distance to a known (ab)normal pattern?

This philosophy guided our work on dyssynchrony patterns amenable to CRT response. In a first instance, an atlas of normal motion was built from healthy subjects (Duchateau et al., 2011). The mean and covariance of myocardial velocities were encoded at each point of the myocardium. The motion of CRT candidates was locally compared to the atlas by a statistical distance to normality (p-value associated with the Mahalanobis distance). Motion abnormalities were coherent across subgroups of dyssynchrony patterns (Duchateau et al., 2012b) Notably, a specific pattern of intraventricular dyssynchrony (also called *septal flash*) predominated among the responders to the therapy, and mostly disappeared at mid-term follow-up. Qualitative observations previously reported that this pattern is associated with high response rates, if corrected (Parsai et al., 2009; Doltra et al., 2014).

In a second instance, we therefore intended to represent this specific pattern as a progressive deviation from normality (Duchateau et al., 2012a). We hypothesized that pattern variations were encoded in a nonlinear space of lower dimensionality, which could be modeled as a manifold. Then we proposed to quantify the distance between a new patient and this specific pattern, as a complement to its distance to normality. We expected that such distances could (before CRT) reflect the probability of response, and (after CRT) link the reduction in abnormality with the actual response (Duchateau et al., 2013c)

12.1.3 SUMMARY OF THE CHALLENGES

Due to the choice of a machine learning methodology for such a clinical application, we believe that three challenges should be highlighted:

- First, determining the *amount of preprocessing required* to study motion and deformation patterns. Features quality strongly impacts the performance of machine learning algorithms, although exaggerated efforts are often dedicated to their extraction. The literature abounds with methods to estimate myocardial motion via cardiac segmentation and tracking. Specific concrete physiological knowledge could refine the feature extraction in our application: local anatomical coordinates (radial, circumferential, and longitudinal), spatiotemporal a priori (phases of the cycle and anatomy should match), quasi-incompressibility of the cardiac motion (De Craene et al., 2012), etc. In addition to these specificities, we decided to precompute distances to a reference for normality to imitate the process of learning of clinicians, and enhance specific features of the motion patterns (Duchateau et al., 2012a).
- Then, knowing how to *"learn" a representation of a given pathology*. We believe that different grades of the same disease, for a patient or within a population, can be modeled as progressive impairments of a normal condition. We hypothesized that motion patterns belong or lie close to a nonlinear manifold that can be learnt from data by means of nonlinear dimensionality reduction.

Specific issues consist of choosing relevant learning techniques, depending on which feature characteristics to highlight, and comparing new subjects to this learnt representation.

- Finally, evaluating whether this representation is *useful* for the clinical application As indicated in Parsai et al. (2009) and Doltra et al. (2014), relating pattern changes to patient response is valuable for therapy planning. However, their assessment is currently qualitative or based on simple quantitative indexes. We therefore focused on demonstrating the value of our method to go beyond these limitations and quantify pattern-specific changes, to investigate if the pattern was fully or partially corrected after treatment.

12.2 FEATURES EXTRACTION: STATISTICAL DISTANCE FROM NORMAL MOTION

12.2.1 CONSTRUCTION OF ABNORMALITY MAPS

We chose to use myocardial velocity as a local descriptor for cardiac motion. This corresponds to small transformations that do not require elaborate computational anatomy techniques (Duchateau et al., 2011). Furthermore, the signature of the septal flash pattern shows greater contrast for velocities than displacements or strain patterns (Duchateau et al., 2014) These velocities are extracted from echocardiographic sequences by spatiotemporal registration (De Craene et al., 2012) or commercial speckle-tracking. These velocities are normalized both spatially and temporally to take into account differences in anatomy and in the timing of physiological events (Duchateau et al., 2011).

We decided to postprocess these data by highlighting values significantly different from normality (Fig. 12.1). This consists of three steps. At each location of the myocardium, the mean and covariance of velocities over a set of healthy volunteers encode a representation of local normal motion (the atlas in Duchateau et al. (2011)). Then, at the same locations, the velocity of each individual is compared to the distribution of velocity vectors for this reference population, using the Mahalanobis distance. Finally, the p-value associated with this distribution (assumed as Gaussian) encodes abnormality at each location: a low p-value indicates high abnormality and a p-value close to 1 stands for normal motion.

In our application, contrast in the abnormality maps is enhanced by using the logarithm of this p-value On top of this, the maps are multiplied by the sign of the radial velocity so that the characteristic pattern of septal flash is highlighted (the inward and outward motions of the septum during early diastole, Fig. 12.1).

12.2.2 WHICH STATISTICS FOR MOTION PATTERNS?

Voxel-based statistics are often used to analyze group-wise and inter-group differences after alignment to a common reference. Each voxel is considered independently

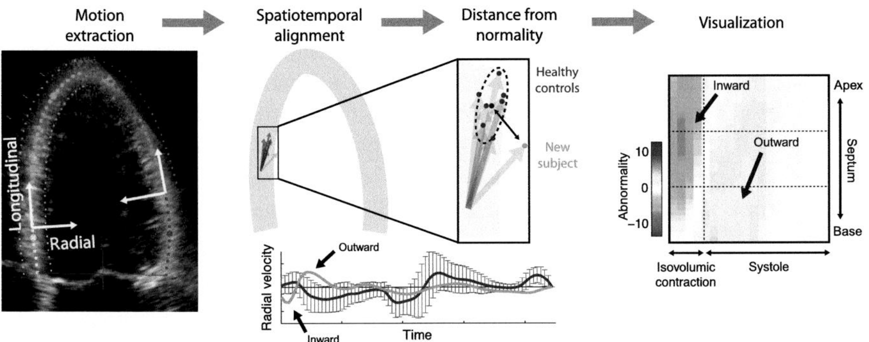

FIG. 12.1

Overview of the extraction of motion features, leading to the computation of spatiotemporal abnormality maps.

from the others, which may affect the statistical power of the results and bias conclusions. In our application, the link between neighboring regions cannot be discarded due to the nature of the data: mechanical tissue properties, regional noise patterns such as speckles, smoothness and temporal consistency of the extracted motion, etc.

Global statistics are also often retained. Methods search for an optimal space to compare subjects, according to specific criteria (eg, explaining the data variance, or discriminating groups of subjects). However, they often neglect the constraints of imaging data, in the sense that generalizations of the model should remain physiologically coherent.

When the structure of the data space is known (eg, the manifold of diffusion tensors or diffeomorphic transformations), mathematical operators exist to define statistics compliant with this space. In contrast, in our application, the structure of such a manifold should be learnt from data. We chose a specific nonlinear dimensionality reduction algorithm that is suitable when the data distribution is not clustered (Isomap; Tenenbaum et al., 2000). However, our framework is flexible and can easily be adapted to other embedding algorithms, which share similar principles (Yan et al., 2007).

12.3 MANIFOLD LEARNING: CHARACTERIZING PATHOLOGICAL DEVIATIONS FROM NORMALITY

Our dataset consists of 109 maps of abnormality like the one in Fig. 12.1, where rows and columns respectively correspond to the position along the septum and the time along the cycle. They were extracted from 2D echocardiographic sequences in a four-chamber view. Focus was kept on the isovolumic contraction (where septal flash appears) and the systole. One map, completely synthetic, was filled with 0 values and served to encode true normality. The *learning set* was composed of this synthetic map and 50 real maps corresponding to a septal flash (Section 12.1). The *testing set* was made up of the remaining 58 real maps, corresponding to six cases with septal flash, 31 CRT candidates without septal flash, and 21 healthy volunteers.

The abnormality maps were considered as high-dimensional objects (20 dimensions for time × 31 dimensions for space), handled as column vectors by our method.

12.3.1 LEARNING PART: PATHOLOGICAL DEVIATIONS FROM NORMALITY (MANIFOLD LEARNING)

12.3.1.1 From high-dimensional motion patterns to low-dimensional coordinates (training set)

The Isomap algorithm (Tenenbaum et al., 2000) is used to reduce the dimensionality of the data while preserving a specific data arrangement. It maps each (high-dimensional) input sample to a space of (low-dimensional) coordinates where the Euclidean distance approximates the geodesic distance between two samples. One hypothesis is required: there is a low-dimensional manifold that can explain the main variations in the data.

The algorithm consists of three steps. First, input samples are connected via a nearest-neighbors search, using the Euclidean distance as metric. Then, the shortest path between each pair of samples is taken as surrogate for the geodesic distance, and stored in a matrix. Finally, this matrix is centered and diagonalized. Due to the matrix diagonalization and our initial hypotheses, dimensions of lower significance can be removed. This leads to a space of low-dimensional coordinates into which the geodesic distance is approximated by the Euclidean distance (Fig. 12.2). The top

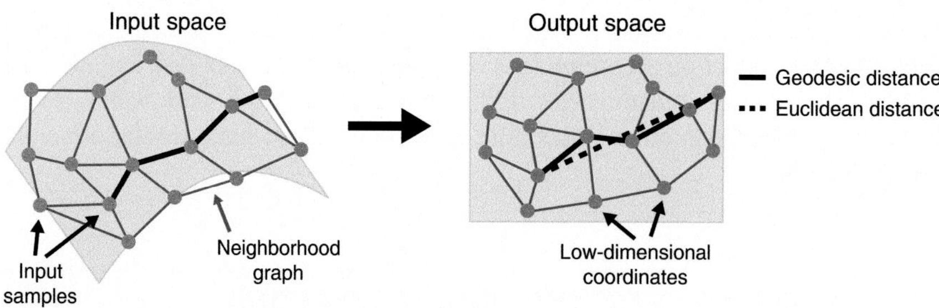

FIG. 12.2

Nonlinear dimensionality reduction via the Isomap algorithm.

Source: Adapted from Duchateau, N., De Craene, M., Piella, G., Frangi, A.F., 2012a. Constrained manifold learning for the characterization of pathological deviations from normality. Med. Image Anal. 16 (8), 1532–1549.

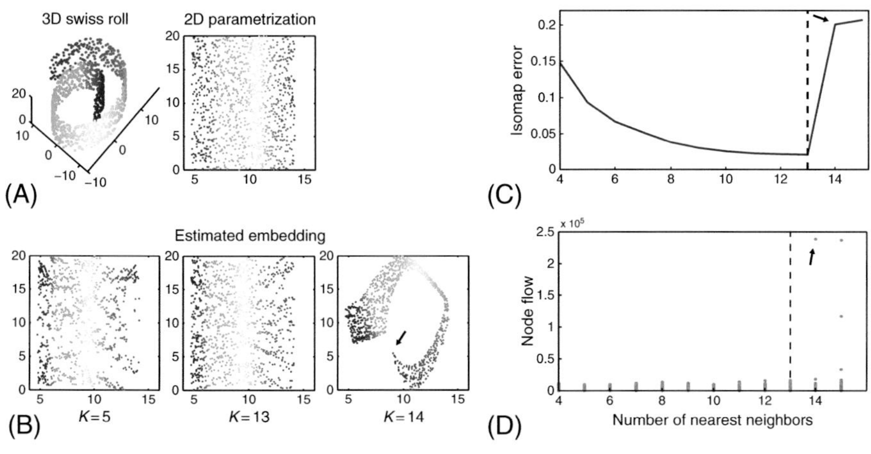

FIG. 12.3

(A) 3D Swiss roll and its associated 2D parameterization. (B)–(D) Influence of the number of nearest neighbors K (B) against the Isomap error (C) and the node flow (D), highlighting the apparition of a short-circuit (*black arrows*).

Source: Adapted from Duchateau, N., De Craene, M., Piella, G., Frangi, A.F., 2012a. Constrained manifold learning for the characterization of pathological deviations from normality. Med. Image Anal. 16 (8), 1532–1549.

N eigenvectors from the matrix diagonalization define the coordinates in this new N-dimensional space.

12.3.1.2 Tuning parameters

The algorithm starts building a graph of the training samples, by connecting nearest neighbors according to a given metric (in our case, the Euclidean distance). In this procedure, the number of neighbors needs to be determined. Synthetic experiments on the Swiss roll (Fig. 12.3) can show that the accuracy of the manifold estimation increases with the number of neighbors until a short-circuit occurs, which breaks the data arrangement. The ratio between the geodesic and the Euclidean distances (the *Isomap error*), computed in the output space, highlights the apparition of such a short-circuit (Fig. 12.3B). However, this measure may not be optimal on real data, in particular when the number of training samples is low.

We complemented this measure by the *node flow on a graph* (Choi and Choi, 2007), which reflects the number of shortest paths passing through each node of the graph. This measurement is directly performed on the graph, before any dimensionality reduction, and is therefore independent of the choice of dimensionality. A uniform arrangement of the training samples is reflected by the uniformity of the node flow distribution. In contrast, much higher node flow is observed at the nodes where a short-circuit occurs, as most of the shortest paths pass by this point (Fig. 12.3C).

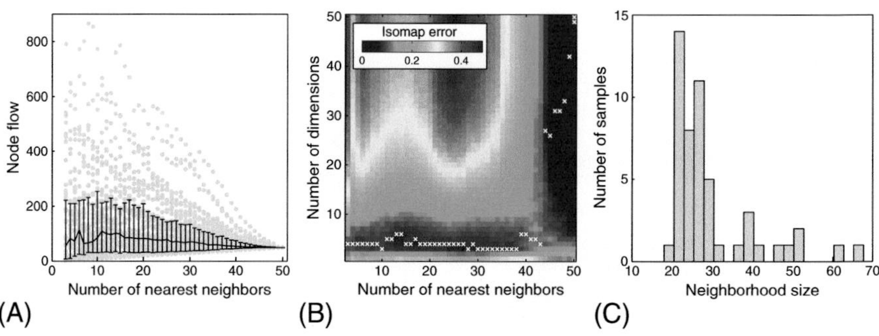

FIG. 12.4

(A) Node flow distribution against the number of nearest neighbors. Error bars indicate the median and first/third quartiles over the training set. (B) Isomap error against the dimensionality of the output space and the number of nearest neighbors. White crosses indicate the minimum value of each column. (C) Training set distribution of the neighborhood sizes (average distance between the five nearest neighbors).

Source: Adapted from Duchateau, N., De Craene, M., Piella, G., Frangi, A.F., 2012a. Constrained manifold learning for the characterization of pathological deviations from normality. Med. Image Anal. 16 (8), 1532–1549.

However, in this synthetic example, the dimensionality of the output space is known—the Swiss roll is a 2D plane embedded in a 3D space This is not the case for real data, and the number of dimensions of the output space should be estimated together with the number of nearest neighbors. Thus parameter estimation should start by checking the presence of a short-circuit by measuring the node flow, and then jointly evaluating the Isomap error against the dimensionality of the output space and the number of nearest neighbors. For our data, an increase in the number of neighbors did not point out the apparition of a short-circuit (Fig. 12.4A). According to the Isomap error (Fig. 12.4B), the dimensionality was set to 4. Few error changes were observed when using less than 30 neighbors, and this number was set to 5 to minimize the computational time.

The Isomap algorithm may be greatly affected by a heterogeneous density in the sample distribution, and in particular in the presence of "holes" in the distribution. This is partially the case for our data, as seen in Fig. 12.4C. For this aspect of learning, our method could be improved by algorithms that specifically target this robustness, such as diffusion maps (Coifman and Lafon, 2006). Note that similar concerns arise for the mapping part of our method, addressed in the following section.

12.3.1.3 Visualization: data spread and main directions

The first necessary checking consists of visualizing the low-dimensional embedding of the data. Two-dimensional representations such that in Fig. 12.5A are easily readable. The depicted patterns correspond to the abnormality maps of the training set.

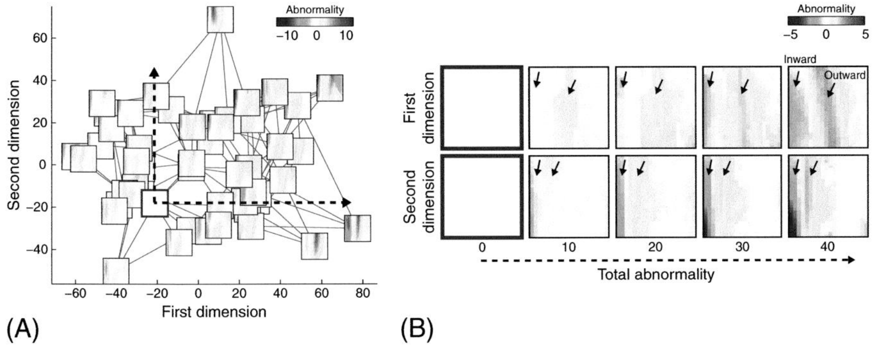

FIG. 12.5

(A) 2D visualization of the estimated embedding. The bold-framed map corresponds to the perfectly normal pattern used to constrain the manifold. (B) Progressive deviations from normality along the first two principal directions. The abnormal inward/outward motion pattern is preserved.

Source: Adapted from Duchateau, N., De Craene, M., Piella, G., Frangi, A.F., 2012a. Constrained manifold learning for the characterization of pathological deviations from normality. Med. Image Anal. 16 (8), 1532–1549.

We can qualitatively check that patterns are arranged in a coherent way, with more abnormal patterns on the border zone of the graph.

This information is complemented by Fig. 12.5B, where synthetic patterns were generated from coordinates evolving from normality along the first two dimensions (the dashed lines in Fig. 12.5A). Notably, the characteristic inward/outward events of the septal flash are preserved, while this is not the case with linear dimensionality reduction (Duchateau et al., 2012a) Note that this figure requires being able to reconstruct motion patterns from the low-dimensional coordinates, using the techniques described in Section 12.3.2.

12.3.2 TESTING PART: DISTANCES TO THE MODELED PATHOLOGY AND TO NORMALITY

12.3.2.1 From high-dimensional motion patterns to low-dimensional coordinates (testing set)

The low-dimensional coordinates associated with the motion pattern of a new subject can be estimated by means of nonlinear regression (Fig. 12.6A). We formulated it as an inexact matching problem using kernels—also referred to as ridge regression. The optimization looks for a "smooth enough" interpolating function (regularization term) that gets "close enough" to the training samples (similarity term). The kernel formulation both constrains the search into a given space of smooth functions (fixed by the kernel bandwidth) and provides an analytical solution to the

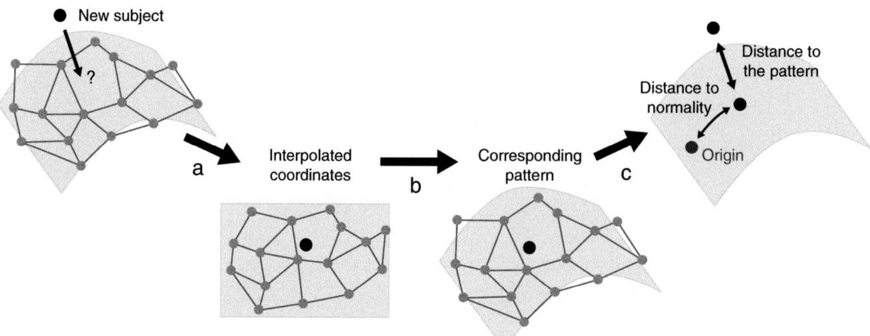

FIG. 12.6

Steps for comparing a new subject to the modeled pattern and to normality.

Source: Adapted from Duchateau, N., De Craene, M., Piella, G., Frangi, A.F., 2012a. Constrained manifold learning for the characterization of pathological deviations from normality. Med. Image Anal. 16 (8), 1532–1549.

interpolation problem. A scalar weight balances the contribution of the regularization and similarity terms.

To improve the quality of the mapping of new subjects, we adapted the methods in two ways.

First, our dataset includes a synthetic zero-valued abnormality map that encodes true normality (the bold map in Fig. 12.5). The interpolation problem is therefore adapted to an exact matching for this specific sample and an inexact matching for the remaining ones (Fig. 12.7A). This is simply addressed by modifying one matrix constraint in the analytical solution of the kernel regression.

Then, similarly to the learning algorithm of Section 12.3.1, this subpart of the algorithm is also sensitive to the nonuniform distribution of samples. Artifacts may arise from the use of a single-scale kernel, generally set to the average density of the samples Two methods were therefore tested:

- A *locally adjustable kernel* (Duchateau et al., 2012a), whose bandwidth is adapted to the local density of the samples, defined as the local neighborhood size. Although this approach provides robustness to local density variations, it deviates from the original formulation where a given kernel determines the space of functions where the problem is solved.
- A *multiscale strategy* (Duchateau et al., 2013b), where the interpolation process is iterated across scales, by dividing the kernel bandwidth by two at each iteration and interpolating the residual (Fig. 12.7B). This approach also has the advantage of removing one parameter from the method, as the kernel bandwidth is automatically determined, starting from the data spread to the average density of the samples.

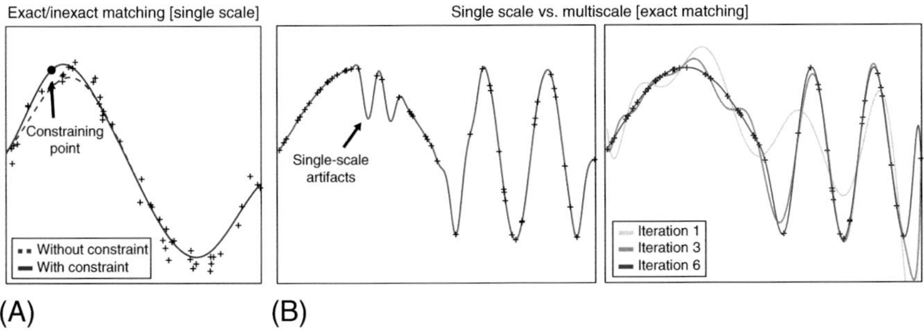

FIG. 12.7

(A) Single-scale interpolation without/with forcing the curve to pass by a given point. (B) Robustness to local density variations: single- vs. multi-scale interpolation.

Source: Adapted from Duchateau, N., De Craene, M., Sitges, M., Caselles, V., 2013b. Adaptation of multiscale function extension to inexact matching. application to the mapping of individuals to a learnt manifold. In: SEE International Conference on Geometric Science of Information, LNCS. Springer, New York, with permission of Springer.

12.3.2.2 From low-dimensional motion coordinates to high-dimensional motion patterns (testing set)

The second part of the algorithm solves the reverse problem: it estimates the motion pattern that corresponds to the low-dimensional coordinates determined by the first step (Fig. 12.6B). The problem is formulated similarly, by switching the role of the motion patterns and the coordinates in the previous formulation.

12.3.2.3 Projection to the manifold and distances computation

As coordinates belong to the low-dimensional space associated with the manifold of motion patterns, all the patterns generated from these coordinates belong to the manifold. In the case of our application, the manifold is learnt from a population with a specific abnormal pattern, and all reconstructed data are expected to show this pattern (Fig. 12.5B). This means that any motion abnormality pattern can be "projected" to the manifold by the combination of these two interpolations. In other words, the method estimates the element of the manifold that shares coordinates with a tested individual, possibly out of this manifold.

Two distances are defined (Fig. 12.6C):

- A distance between any abnormality map and the manifold (*distance to the manifold*), which quantifies how far a given subject is from the modeled abnormal pattern. This corresponds to the total error between a given abnormality map and its reconstruction.
- A *distance to normality*. This corresponds to the Euclidean distance between the coordinates of the new subject and the origin—the synthetic map defining true normality.

Table 12.1 Determination of the Main Parameters for Our Method

Parameter	Estimation Method	Concerned Set
Number of nearest neighbors	Node flow + Isomap error	Training
Dimensionality of the output space	Node flow + Isomap error	Training
Kernel bandwidth	Distribution of neighborhood sizes + adaptive kernel or multiscale formulation	Testing
Interpolation weight	Generalization ability	Testing
Sample size	Convergence of the results	Training + testing

12.3.2.4 Tuning parameters

As for the learning part of the algorithm, the method parameters need to be estimated. Each interpolation depends on a kernel bandwidth and a regularization weight. Kernel bandwidths are automatically determined (Section 12.3.2). The regularization weights for the two consecutive interpolations are jointly estimated by heuristic tests. Their optimal values are determined based on the generalization ability over the training set—the reconstruction error when using a leave-one-out approach (Davies et al., 2010) In our application, they corresponded to 10^1 and $10^{0.5}$ respectively (Duchateau et al., 2012a).

Naturally, the performance of the methods depends on the size of the training population. The two estimated distances (to the manifold and to normality) were computed from randomly generated datasets of smaller size. Convergence was assessed by the number of subjects above which the results stabilize to their final value ±5%: above 45 subjects for the distance to the manifold and above 41 subjects for the distance to normality (Duchateau et al., 2012a). Both indicate that results on our population of 50 real subjects can be trusted.

A summary of the estimation of the method parameters is given in Table 12.1.

12.4 BACK TO THE CLINICAL APPLICATION: UNDERSTANDING CRT-INDUCED CHANGES

The motion patterns of our population were analyzed with respect to the two proposed distances: the distance to the manifold, which models the abnormal pattern of septal flash, and the distance to normality along the manifold. These distances define a 2D space where subgroup differences can easily be visualized (Fig. 12.8), as discussed below.

12.4.1 ANALYSIS PER POPULATION

We first checked that the two distances provided a meaningful organization of all the data at baseline.

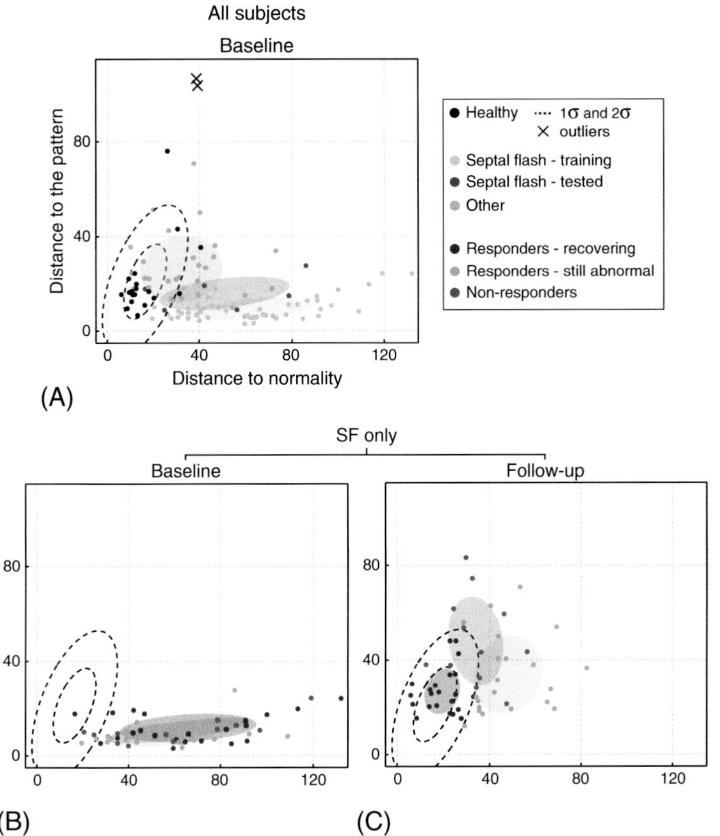

FIG. 12.8

Subjects distribution according to the distances to the modeled pattern and to normality. Dashed lines indicate the normality range defined by the healthy population.

Source: Adapted from Duchateau, N., Piella, G., Doltra, A., Mont, L., Bijnens, B.H., Sitges, M., De Craene, M., 2013c, Manifold learning characterization of abnormal myocardial motion patterns: application to crt-induced changes. In: Ourselin, S., Rueckert, D., Smith, N. (Eds.), Proceedings of Functional Imaging and Modeling of the Heart, LNCS, vol. 7945. Springer, New York, pp. 450–457. With permission of Springer.

The training set, made of septal flash cases, has a low distance to the modeled pattern. This actually reflects the reconstruction error of the interpolations, formulated as an inexact matching. Such cases spread all along the horizontal axis from low to high abnormal patterns, which also indicates that the whole support of the manifold defines the distance to normality. Few of them are close to the origin for normality, which may arise from the accuracy of the septal flash diagnosis and the accuracy of the abnormality maps to detect low-grade abnormalities.

Among the testing set, patients with septal flash are mapped close to the training cases, according to their respective abnormality. Volunteers lie close from the origin for normality. Two subjects are outliers (black crosses out of the dashed ellipse). They correspond to cases with high abnormalities due to high velocities despite a normal pattern, lying out of the range of the other healthy subjects. Their pattern of abnormality cannot be reconstructed by the training set of septal flash cases, which leads to a large distance to the modeled pattern. Similarly, patients with an abnormal pattern different from septal flash are mapped far from the manifold, and out of the normality range.

12.4.2 LINK WITH CRT RESPONSE

The manifold is exclusively learnt from septal flash patterns. When looking at individual evolutions with CRT, our method only evaluates if a studied pattern (not necessarily septal flash) got closer to septal flash or normality. If the pattern is different from septal flash, we cannot tell if this pattern is still present or not. Thus we focused the outcome analysis on the subjects diagnosed with septal flash at baseline (Fig. 12.8B and C).

Patients were separated according to the volume response to CRT, measured by a clinical expert as a reduction $\geq$15% in left-ventricular end-systolic volume, without heart transplantation Three subgroups were identified: responders who recovered normal motion at follow-up (who were within the normality range defined by the healthy subjects), responders for which abnormal motion is still present, and nonresponders.

These subgroups cannot be distinguished at baseline: the presence of septal flash at baseline is not a sufficient condition for CRT response. This coincides with the philosophy of Parsai et al. (2009) and Doltra et al. (2014), which interpreted response at follow-up in light of the correction of the identified abnormal mechanisms. In our population, significant changes are visible at follow-up. Patients within the range of normal motion are mostly responders. The other patients have higher distance to the manifold, suggesting that septal flash may have disappeared. However, their abnormalities are still high, which may explain their lack of response. Beyond these qualitative interpretations, the number of changes can be quantified with our method and can further be related with quantitative indexes of therapy outcome.

Further results and discussion can be found in Duchateau et al. (2013c). They notably include the evolution of the whole population, and the outcome immediately after pacing Complementary interpretations are also given in Duchateau et al. (2012a), which points out the limitations of a linear analysis (principal compound analysis, PCA) and the meaning of the proposed distances against the total abnormality in each map.

12.5 DISCUSSION/FUTURE WORK

We presented a method to model pathological myocardial motion patterns as a smooth deviation from normality, along a nonlinear manifold structure, learnt from data. This representation gives a comparison of new samples to the modeled pattern and to normality, and provides quantitative insights into the mechanisms of CRT response.

12.5.1 PATTERN-BASED COMPARISONS

Regarding technical aspects, our main contribution resides in the methodology to compare myocardial motion patterns. Previous work used atlas tools to align data in time and space, and to enhance the contrast in the features of interest by a statistical comparison to normality at each voxel (Duchateau et al., 2011). These tools already facilitated the quantitative analysis of CRT response, by highlighting the prevalence and the correction of specific abnormal patterns in groups of responders and nonresponders. In this chapter, we illustrated the potential of the pattern-wise comparisons proposed in Duchateau et al. (2012a). Pattern variations now model the spatiotemporal interdependences in the data. Furthermore, the proposed distances provide physiologically coherent support to the interpretation of the patterns' evolution with therapy.

The methodology is based on standard manifold learning and nonlinear regression. Improvements could be made on several parts of the pipeline, depending on the application targeted: getting more power out of the metrics, adding robustness to the density of the samples during learning, combining features of different types (Sanchez-Martinez et al., 2015), modeling several abnormal patterns at once, etc. The pipeline is easily transposable to other applications, imaging modalities, and features of interest (not necessarily in cardiac imaging). However, the main decision should be taken as early as possible, to decide whether using a similar but complex framework (statistical atlas and learning a manifold) can add value to the clinical application. In our case, there was a clear claim for the quantitative analysis of dyssynchrony patterns amenable to CRT response (Fornwalt, 2011), and a convergence of clinical studies towards mechanistic approaches (Parsai et al., 2009), which was subsequently confirmed (Doltra et al., 2014). 2D echocardiographic data were retained due to the wide implantation of the modality in clinical routine and CRT monitoring. The high temporal resolution and the easy checking of 2D echocardiographic data outputs also facilitated this decision, when 3D data were not straightforward to handle.

12.5.2 GOING BEYOND PARSAI'S PAPER?

The results in Parsai et al. (2009) had already led to a great improvement in CRT, as the decision algorithm improved patient selection by removing ambiguity on three out of five groups, which always or never respond. Simple criteria defined the inclusion to a given group and facilitated the reproducibility of the results. Our methodology is more difficult to implement in clinical routines, and simpler but comprehensive indexes that characterize dyssynchrony patterns may be more relevant in this context. However, our approach allows quantification of pattern evolution with therapy, which would not be straightforward with the indexes in (Parsai et al., 2009), in particular if the pattern is highly modified.

The results in Parsai et al. (2009) also offered a broader view on the way to treat dyssynchrony. Intraventricular dyssynchrony is purely electrical, meaning that resynchronization should lead to a response in subjects with septal flash, therefore defining a hyperresponders population. In contrast, structural disease on top of the electrical abnormalities worsens the outcome: although electrical dyssynchrony is corrected, mechanical abnormalities (myocardial infarct) still exist and the probability of response is highly decreased Going beyond these interpretations, confirmed by our results, should consider additional variety in these mechanisms. External factors could affect the clinical condition, such as the presence of atrial fibrillation, the lack of contractile reserve, or the lead position, and should be considered in future studies. Nonetheless, from a learning perspective, the user should always keep in mind the amount of effort required to extract features of interest for these data.

Finally, one of the strongest limitations to CRT resides in the definition of response. Paradoxically, no agreement exists (Fornwalt et al., 2010) We hope that our approach, based on the quantification of abnormality, can open the way towards less binary views of the success of the therapy. Further studies should evaluate the added value of a continuous spectrum of responses, which may include measures of abnormality evolution with the therapy.

ACKNOWLEDGMENTS

This work was done between 2008 and 2013 with the partial support of the Spanish Industrial and Technological Development Center (CDTeam and cvREMOD CEN-20091044), the Spanish Ministry of Science and Innovation, Plan E and ERDF (STIMATH TIN2009-14536-C02-01), and the European Commission's 7th Framework Program (euHeart FP7-ICT-224495). The authors acknowledge the contribution of their co-authors on related publications, in particular those from Bart Bijnens (ICREA, Universitat Pompeu Fabra, Barcelona, Spain) and Marta Sitges (Hospital Clínic, Barcelona, Spain) on clinical applications, and Vicent Caselles (Universitat Pompeu Fabra, Barcelona, Spain) on aspects of density invariance.

REFERENCES

Bleeker, G.B., Bax, J.J., Fung, J.W., van der Wall, E.E., Zhang, Q., Schalij, M.J., Chan, J.Y., Yu, C.M., 2006. Clinical versus echocardiographic parameters to assess response to cardiac resynchronization therapy. Am. J. Cardiol. 97 (2), 260–263.

Brignole, M., Auricchio, A., Baron-Esquivias, G., Bordachar, P., Boriani, G., Breithardt, O.A., Cleland, J., Deharo, J.C., Delgado, V., Elliott, P.M., Gorenek, B., Israel, C.W., Leclercq, C., Linde, C., Mont, L., Padeletti, L., Sutton, R., Vardas, P.E., ESC Committee for Practice Guidelines (CPG), Zamorano J.L., Achenbach, S., Baumgartner, H., Bax, J.J., Bueno, H., Dean, V., Deaton, C., Erol, C., Fagard, R., Ferrari, R., Hasdai, D., Hoes, A.W., Kirchhof, P., Knuuti, J., Kolh, P., Lancellotti, P., Linhart, A., Nihoyannopoulos, P., Piepoli, M.F., Ponikowski, P., Sirnes, P.A., Tamargo, J.L., Tendera, M., Torbicki, A.,

Wijns, W., Windecker, S., Document Reviewers: Kirchhof, P., Blomstrom-Lundqvist, C., Badano, L.P., Aliyev, F., Bansch, D., Baumgartner, H., Bsata, W., Buser, P., Charron, P., Daubert, J.C., Dobreanu, D., Faerestrand, S., Hasdai, D., Hoes, A.W., Le Heuzey, J.Y., Mavrakis, H., McDonagh, T., Merino, J.L., Nawar, M.M., Nielsen, J.C., Pieske, B., Poposka, L., Ruschitzka, F., Tendera, M., Van Gelder, I.C., Wilson, C.M., 2013. 2013 ESC guidelines on cardiac pacing and cardiac resynchronization therapy. Eur. Heart J. 34 (29), 2281–2329.

Choi, H., Choi, S., 2007. Robust kernel isomap. Pattern Recogn. 40 (3), 853–862.

Coifman, R., Lafon, S., 2006. Diffusion maps. Appl. Comput. Harmon. Anal. 21 (1), 5–30.

Davies, R.H., Twining, C.J., Cootes, T.F., Taylor, C.J., 2010. Building 3-d statistical shape models by direct optimization. IEEE Trans. Med. Imaging 29 (4), 961–981.

De Craene, M., Piella, G., Camara, O., Duchateau, N., Silva, E., Doltra, A., D'hooge, J., Brugada, J., Sitges, M., Frangi, A.F., 2012. Temporal diffeomorphic free-form deformation: application to motion and strain estimation from 3d echocardiography. Med. Image Anal. 16 (2), 427–450.

Doltra, A., Bijnens, B., Tolosana, J.M., Borràs, R., Khatib, M., Penela, D., De Caralt, T.M., Castel, M.Á., Berruezo, A., Brugada, J., Mont, L., Sitges, M., 2014. Mechanical abnormalities detected with conventional echocardiography are associated with response and midterm survival in CRT. JACC Cardiovasc. Imaging 7 (10), 969–979.

Duchateau, N., De Craene, M., Piella, G., Silva, E., Doltra, A., Sitges, M., Bijnens, B.H., Frangi, A.F., 2011. A spatiotemporal statistical atlas of motion for the quantification of abnormalities in myocardial tissue velocities. Med. Image Anal. 15 (3), 316–328.

Duchateau, N., De Craene, M., Piella, G., Frangi, A.F., 2012a. Constrained manifold learning for the characterization of pathological deviations from normality. Med. Image Anal. 16 (8), 1532–1549.

Duchateau, N., Doltra, A., Silva, E., De Craene, M., Piella, G., Castel, M.Á., Mont, L., Brugada, J., Frangi, A.F., Sitges, M., 2012b. Atlas-based quantification of myocardial motion abnormalities: added-value for understanding the effect of cardiac resynchronization therapy. Ultrasound Med. Biol. 38 (12), 2186–2197.

Duchateau, N., Bijnens, B.H., D'hooge, J., Sitges, M., 2013a. Three-dimensional assessment of cardiac motion and deformation, second ed. CRC Press, Boca Raton, FL. pp. 201–213.

Duchateau, N., De Craene, M., Sitges, M., Caselles, V., 2013b, Adaptation of multiscale function extension to inexact matching. Application to the mapping of individuals to a learnt manifold, LNCS. In: SEE International Conference on Geometric Science of Information. Springer, New York.

Duchateau, N., Piella, G., Doltra, A., Mont, L., Bijnens, B.H., Sitges, M., De Craene, M., 2013c, Manifold learning characterization of abnormal myocardial motion patterns: application to CRT-induced changes. In: Ourselin, S., Rueckert, D., Smith, N. (Eds.), Proceedings of Functional Imaging and Modeling of the Heart, LNCS, vol. 7945. Springer, New York, pp. 450–457.

Duchateau, N., Sitges, M., Doltra, A., Fernández-Armenta, J., Solanes, N., Rigol, M., Gabrielli, L., Silva, E., Barceló, A., Berruezo, A., Mont, L., Brugada, J., Bijnens, B., 2014. Myocardial motion and deformation patterns in an experimental swine model of acute LBBB/CRT and chronic infarct. Int. J. Cardiovasc. Imaging 30 (5), 875–887.

Fornwalt, B.K., 2011. The dyssynchrony in predicting response to cardiac resynchronization therapy: a call for change. J. Am. Soc. Echocardiogr. 24 (2), 180–184.

Fornwalt, B.K., Sprague, W.W., BeDell, P., Suever, J.D., Gerritse, B., Merlino, J.D., Fyfe, D.A., León AR, O.J., 2010. Agreement is poor among current criteria used to define response to cardiac resynchronization therapy. Circulation 121 (18), 1985–1991.

Laskey, W., Awad, K., Lum, J., Skodacek, K., Zimmerman, B., Selzman, K., Zuckerman, B., 2012. An analysis of implantable cardiac device reliability. the case for improved postmarketing risk assessment and surveillance. Am. J. Therap. 19 (4), 248–254.

Parsai, C., Bijnens, B., Sutherland, G.R., Baltabaeva, A., Claus, P., Marciniak, M., Paul, V., Scheffer, M., Donal, E., Derumeaux, G., Anderson, L., 2009. Toward understanding response to cardiac resynchronization therapy: left ventricular dyssynchrony is only one of multiple mechanisms. Eur. Heart J. 30 (8), 940–949.

Sanchez-Martinez, S., Duchateau, N., Bijnens, B., Erdei, T.and Fraser, A., Piella, G., 2015. Characterization of myocardial motion by multiple kernel learning: application to heart failure with preserved ejection fraction. In: van Assen, H., Bovendeerd P., Delhaas, T. (Eds.), Proceedings of Functional Imaging and Modeling of the Heart, LNCS vol. 9126. springer, New York, pp. 65–73.

Tenenbaum, J.B., de Silva, V., Langford, J.C., 2000. A global geometric framework for nonlinear dimensionality reduction. Science 290 (5500), 2319–2323.

van Veldhuisen, D.J., Maass, A.H., Priori, S.G., Stolt, P., van Gelder, I.C., Dickstein, K., Swedberg, K., 2009. Implementation of device therapy (cardiac resynchronization therapy and implantable cardioverter defibrillator) for patients with heart failure in Europe: changes from 2004 to 2008. Eur. J. Heart Fail. 11 (12), 1143–1151.

Vernooy, K., van Deursen, C.J., Strik, M., Prinzen, F.W., 2014. Strategies to improve cardiac resynchronization therapy. Nat. Rev. Cardiol. 11 (8), 481–493.

Yan, S., Xu, D., Zhang, B., Zhang, H.J., Yang, Q., Lin, S., 2007. Graph embedding and extensions: a general framework for dimensionality reduction. IEEE Trans. Pattern Anal. Mach. Intell. 29 (1), 40–51.

Yu, C.M., Hayes, D.L., Auricchio, A., 2006. Cardiac Resynchronization Therapy. Wiley-Blackwell, New York.

CHAPTER

From point to surface: Hierarchical parsing of human anatomy in medical images using machine learning technologies 13

Y. Zhan[1], M. Dewan[2], S. Zhang[3], Z. Peng[1], B. Jian[4], X.S. Zhou[1]

Siemens Healthcare, Malvern, PA, United States[1] Flipkart, Palo Alto, CA, United States[2] University of North Carolina at Charlotte, Charlotte, NC, United States[3] Google, Mountain View, CA, United States[4]

CHAPTER OUTLINE

Machine Learning and Medical Imaging. http://dx.doi.org/10.1016/B978-0-12-804076-8.00013-X

13.1 INTRODUCTION

In various medical imaging applications, from body-part identification to lung nodule detection, one critical task is to localize and interpret specific anatomical structures in medical images. For an experienced professional, this task is often accomplished by leveraging the "anatomy signatures" in medical images—the unique shape or appearance characteristics of the anatomies of interest. In fact, during clinical training and practice, a medical professional learns many of these "anatomy signatures" from seeing a large number of medical images. While computer-aided algorithms have become more and more popular in the medical imaging domain, it is not surprising that the research community has become interested in developing algorithms to automatically localize and interpret different anatomies in medical images. The success of these algorithms is essentially dependent on if/how many anatomy signatures can be well extracted by the algorithms. While earlier studies often crafted specific image filters to extract anatomy signatures, a more recent research trend shows the prevalence of learning-based approaches for two reasons: (1) Machine learning technologies have matured to solve real-world problems. (2) More and more medical image datasets are available to facilitate the "exploration" of complex medical imaging data.

In this chapter, we will present a learning-based framework that is able to "learn" anatomy signatures and localize anatomical structures in medical images. As shown in Fig. 13.1, this framework has a hierarchical structure and parses human anatomies in a coarse-to-fine fashion. Specifically, we start from the smallest anatomical

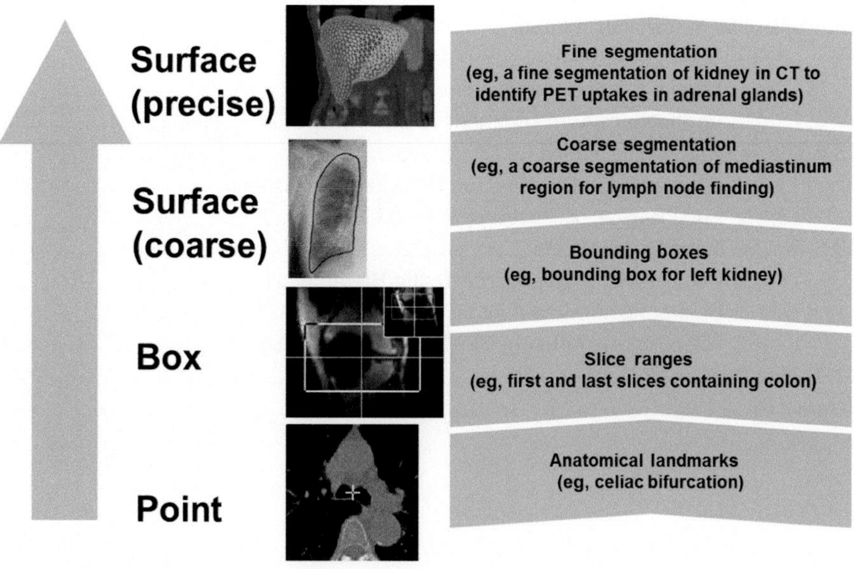

FIG. 13.1

Infrastructure of our anatomy interpretation system.

entity—anatomical landmarks. Our system is then extended to the detection of anatomy boxes (region of interests), coarse organ segmentation, and finally reaches the precise delineation of organ boundaries. On one hand, algorithms developed on each layer may benefit different image analysis applications, for example, bounding box detection of knee meniscus helps to automate magnetic resonance (MR) imaging planning. On the other hand, algorithms developed on lower layers also provides intermediate results for the higher layers. For example, landmark detection provide basic image appearance cues for coarse organ segmentation. It is worth noting that since algorithms in this framework aim to extract anatomy signatures through learning, they are not restricted to specific organ characteristics. Thus the entire framework is scalable to different imaging modalities and organ systems.

The rest of this chapter is organized as follows. We start in Section 13.2 by reviewing some related works. In Sections 13.3–13.6, we present algorithms from landmark detection to precise organ segmentation. In each of these sections, we start by introducing the methods followed by some relevant medical imaging use cases. Conclusions are drawn in Section 13.7.

13.2 LITERATURE REVIEW

Since our framework includes anatomy parsing at different levels, from landmarks to organ surfaces, the literature review will cover all these different aspects.

Since anatomical landmarks are used in various clinical use cases, automatic landmark detection has gained significant interests in the medical imaging research community. While traditional methods aim to design filters or templates for specific landmarks (Betke et al., 2003), more recent studies have used machine learning technologies for a solution scalable to different landmarks in different imaging modalities. Representative works include Liu and Zhou (2012) and Criminisi et al. (2011), which employ probabilistic boosting tree and random forests to learn the appearance characteristics of landmarks.

Automatic detection of anatomical boxes has also been extensively investigated. A key problem is how to address the appearance variability resulting from different object orientations. In Schneiderman and Kanade (2000), object detectors for different poses are exclusively performed and the pose corresponding to the detector with the strongest response is regarded as the aligned pose. Lv et al. (2000) employs local orientation analysis to estimate object poses. Recently, marginal space learning (MSL) (Zheng et al., 2008) has shown promise in estimating rough transformations of anatomical structures in medical images. The basic idea is to decompose the transformation parameters into location, orientation, scale, etc., and train detectors to estimate each of them sequentially. Since this method treats the target anatomy (eg, a heart chamber or the knee menisci) as a whole, the burden of learning is high, especially for the first round of learning (location detectors).

The success of coarse organ segmentation relies on the modeling of shape priors. Following the pioneering use of the active shape model (Cootes et al., 1995), various methods have been proposed to improve the effectiveness of shape prior modeling in different circumstances. For example, Cootes and Taylor (1997) and Etyngier et al. (2007) aims to model the nonlinear shape variations through a mixture of Gaussians and manifold learning. Yan et al. (2010) tried to use partial active shape model (ASM) to make the shape model robust to missing boundaries. Sjostrand and et.al. (2007) employed a sparse principal component analysis (PCA) to obtain sparser modes and produce near-orthogonal components. Davatzikos et al. (2003) divides the shape model into several independently modeled parts for hierarchical shape representation. Both of these works aim to capture small shape variations and preserve shape details.

For precise organ segmentation, deformable models have been extensively studied, particularly in the area of medical image segmentation. The widely recognized potency of deformable models comes from their ability to segment anatomic structures by exploiting constraints derived from the image data (bottom-up) together with prior knowledge about these structures (top-down). The deformation process is usually formulated as an optimization problem whose objective function consists of an external (image) term and an internal (shape) term. While internal energy is designed to preserve the geometric characteristics of the organ under study, the external energy is defined to move the model toward organ boundaries. Traditionally, the external energy term usually comes from edge information (Xu and Prince, 1998), for example, image gradient. In recent years, more effort has been invested on the integration of other image features, for example, local regional information (Vese and Chan, 2002; Ronfard, 1994) and texture models (Huang et al., 2004). By combining different image features as the external energy, deformable models have achieved tremendous success in various clinical practices. Machine learning technologies have opened the door to a more generic external energy design. By using learning-based methods, boundary characteristics can be learned from training data (Zhan and Shen, 2006; Zheng et al., 2008). In this way, the "design" of external energy becomes data driven and can be extended to different imaging modalities.

13.3 ANATOMY LANDMARK DETECTION

13.3.1 LEARNING-BASED LANDMARK DETECTION

Anatomical landmarks are biologically meaningful points existing in various organ systems. They are defined to describe the morphological characteristics of anatomical structures and facilitate communication between scientists in the fields of biology and medicine, etc. In the medical imaging community, while most anatomical landmarks become visible in vivo, they play important roles in the interpretation of medical images. In the same way that geographic landmarks guide travelers in exploring the earth, anatomical landmarks provide guidance to navigate the medical images. For example, anatomical content within a medical image can be determined by the locations of anatomical landmarks. In addition, some anatomical landmarks also provides critical clues to diagnose diseases. For example, the posterior junction point of a lumbar vertebra and the spinal cord can be used to diagnose spondylolisthesis and evaluate the disease stage. Hence algorithms that can automatically detect anatomical landmarks can directly benefit various clinical use cases. Besides the direct impact on clinical use cases, automatic detection of anatomical landmarks also paves the way for other medical image analysis tasks. For example, automatically detected landmarks can be used to initialize deformable models for organ segmentation (Zhan et al., 2009; Zhang et al., 2012). They can also provide an initial transformation for image registration (Zhan et al., 2011).

Due to the complex appearance of different anatomical landmarks, we use a learning-based approach for individual landmark detection. Because of its data-driven nature, learning-based approaches also make our method highly scalable to different imaging modalities, for example, computed tomography (CT), MR, positron emission tomography (PET), ultrasound, etc. We formulate landmark detection as a voxel-wise classification problem. Specifically, voxels close to the landmark are considered as positive samples and voxels away from the landmark are regarded as negative ones. To learn a specific landmark detector, we first annotate the landmark in a set of training images. The positive and negative samples/voxels are determined based on their distances to the annotated landmark. For each training sample (voxel), a set of elementary features is extracted in its neighborhood. Our elementary features are generated by a set of mother functions, $\{H_l(\mathbf{x})\}$, extended from the Haar wavelet basis. As shown in Eq. (13.1) and Fig. 13.2, each mother function consists of one or more 3D rectangle functions with different polarities:

$$H(\mathbf{x}) = \sum_{i=1}^{N} p_i R(\mathbf{x} - \mathbf{a_i}), \tag{13.1}$$

where polarities $p_i = \{-1, 1\}$,

$$R(\mathbf{x}) = \begin{cases} 1, & \|\mathbf{x}\|_\infty \leq 1, \\ 0, & \|\mathbf{x}\|_\infty > 1 \end{cases}$$

denotes rectangle functions, and $\mathbf{a_i}$ is the translation.

By scaling the mother functions and convoluting them with the original image, a set of spatial-frequency spaces are constructed:

$$F_l(\mathbf{x}, s) = H_l(s\mathbf{x}) * I(\mathbf{x}), \tag{13.2}$$

where s and l denote the scaling factor and index of mother functions respectively.

Finally, for any voxel $\mathbf{x_0} \in \Re^3$, its feature vector $\mathfrak{F}(\mathbf{x_0})$ is obtained by sampling these spatial-frequency spaces in the neighborhood of $\mathbf{x_0}$ (Eq. 13.3). It provides cross-scale appearance descriptions of voxel $\mathbf{x_0}$.

$$\mathfrak{F}(\mathbf{x_0}) = \bigcup_{l=1...L} \{F_l(\mathbf{x_i}, s_j) | \mathbf{x_i} \in \mathbb{N}(\mathbf{x_0}), s_{\min} < s_j < s_{\max}\}. \tag{13.3}$$

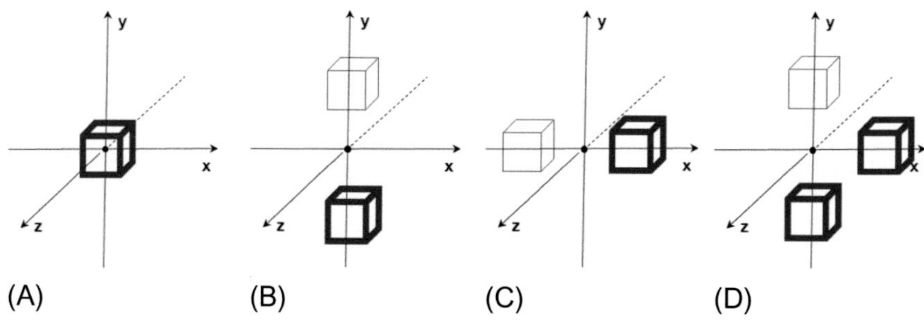

FIG. 13.2

Some examples of Haar-based mother functions.

Compared to the standard Haar wavelet, the mother functions we employed are not orthogonal. However, they provide more comprehensive image features to characterize different anatomy primitives. For example, as shown in Fig. 13.2, mother function (A) potentially works as a smoothing filter, which is able to extract regional features. Mother functions (B) and (C) can generate horizontal or vertical "edgeness" responses, which are robust to local noises. More complicated mother functions like (D) are able to detect "L-shaped" patterns, which might be useful to distinguish some anatomy primitives. In addition, our features can be quickly calculated through integral images (Crow, 1984). This paves the way for an efficient landmark detection system.

All elementary features are then fed into a cascade classification framework (Viola and Jones, 2004), as shown in Fig 13.3. The cascade framework is designed to address the highly unbalanced positive and negative samples. In fact, since only voxels around landmarks under study are positives and all other voxels are negatives, the ratio of positives to negatives is often less than $1{:}10^5$. In the training stage, all positives and a small proportion of negatives are used at every cascade level. The training algorithm is "biased" to positives, such that each positive has to be correctly classified but the negatives are allowed to be misclassified. These misclassified negatives, that is, false positives, will be further used to train the following cascades. At run-time, while positives are expected to go through all cascades, most negatives can be rejected in the first several cascades and do not need further evaluation.

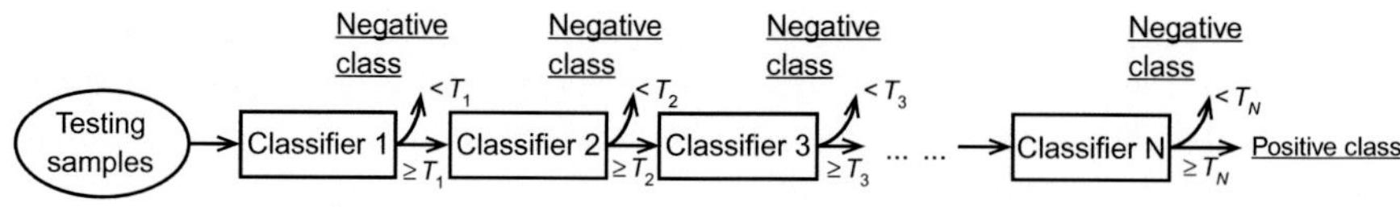

FIG. 13.3

Schematic explanation of cascade Adaboost classifers.

In this way, the run-time speed can be dramatically increased. In our study, we use *Adaboost* (Freund and Schapire, 1997) as the basic classifier in the cascade framework. The output of the learned classifier $\mathcal{A}(\mathfrak{F}(\mathbf{x}))$ indicates the likelihood of the specific landmark appearing at $\mathbf{x}$.

At run-time, a sliding window approach is employed to apply the learned classifiers on each voxel in the image. The voxel that has the highest response of $\mathcal{A}(\mathfrak{F}(\mathbf{x}))$ can be considered as the detected landmark.

13.3.2 APPLICATION: AUTOMATIC LOCALIZE AND LABEL OF VERTEBRAE

The spine is one of the major organs in the human body. It includes the vertebral column and spinal cord. A human vertebral column typically consists of 33 vertebrae. Twenty-four of them are articulating (7 cervical, 12 thoracic, and 5 lumbar vertebrae) and nine of them are fused vertebrae in the sacrum and the coccyx. As the spine is strongly correlated to both neural and skeletal systems, various neurological, orthopedic, and oncological studies involve investigations of spinal anatomies. In addition, due to the strong spatial correlations between specific vertebrae and their surrounding organs, the spine may also be used as a vertical reference framework to describe the locations of other organs in the trunk, for example, the transpyloric plane. In spine image analysis, localization and labeling of vertebrae are often the first steps, and are tedious and time consuming for manual operators. Accordingly, we employ the landmark detection algorithm to automatically localize and label vertebrae. (Besides the algorithms introduced in this section, we also treat different vertebrae as "anchor" and "bundle" and train vertebrae detectors in a hierarchical way. In addition, an articulated model is used to model the spine geometry. Please refer to Zhan et al. (2012) for more details.)

Our method is evaluated on both CT and MR datasets. The CT dataset includes 189 randomly selected CT cases with partial or whole spine coverage. A wide range of imaging parameters are used for this dataset, including different reconstruction kernels and slice thicknesses. Our MR dataset includes 300 T1-weighted 3D MR scout scans. These scout scans have relatively low but isotropic resolution of 1.7 mm and large fields of view. These datasets come from different clinical sites and were generated by different types of Siemens MR Scanners (Avanto 1.5T, Verio 3T, Skyra 3T, etc.). Both MR and CT datasets include partial or whole spines.

The automatic vertebrae labeling results were shown to experienced radiologists and rated as "perfect" (no manual editing required), "acceptable" (minor manual editing required), and "rejected" (major manual editing required).

As shown in Table 13.1, our method achieves "perfect" detection in 95+% of cases. It is worth noting that our testing set includes CT/MR scans with severe diseases or imaging artifacts. However, our method can still detect vertebrae and discs robustly. Figs. 13.4 and 13.5 show the results of our method on challenging CT and

Table 13.1 Evaluation of Spine Detection in CT and MR Scans

	Number of Cases	Perfect	Acceptable	Reject
CT	189	180 (95.2%)	6 (3.2%)	3 (1.6%)
MR	300	293 (97.7%)	4 (1.3%)	3 (1.0%)

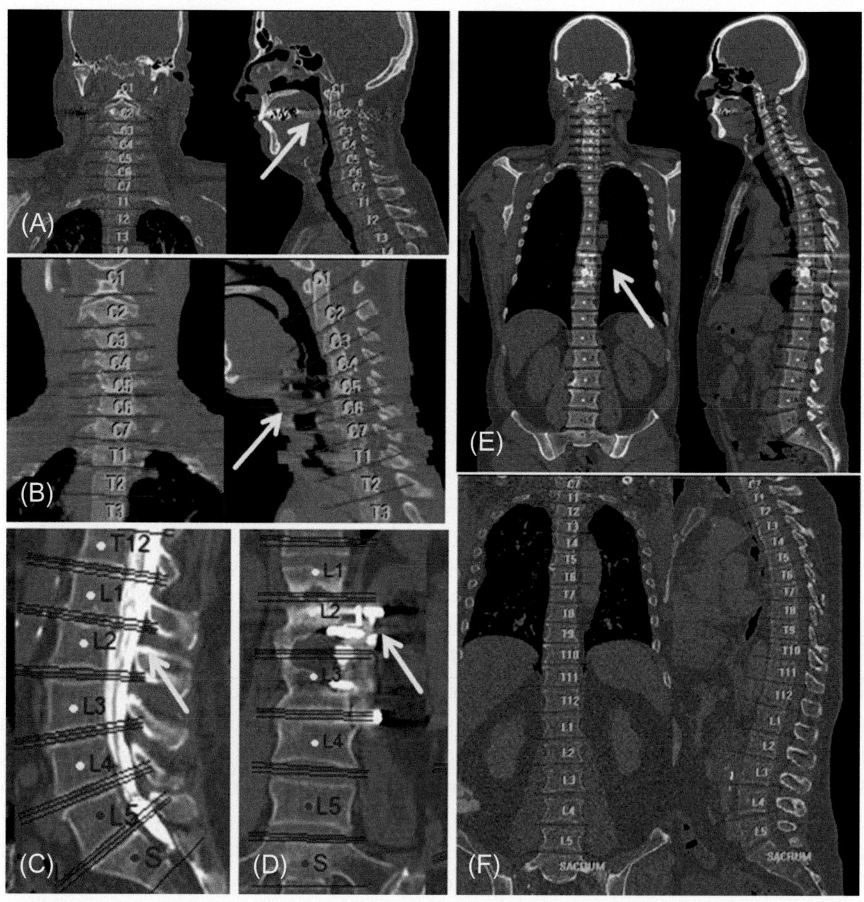

FIG. 13.4

Examples of vertebrae labeling results in challenging CT scans. (A) C-spine scan with metal artifacts. (B) C-spine scan with motion artifacts. (C) L-spine scan with spinal cord disease. (D) L-spine scan with metal implant. (E) Whole-spine scan with metal artifacts. (F) Whole-spine scan with large imaging noises.

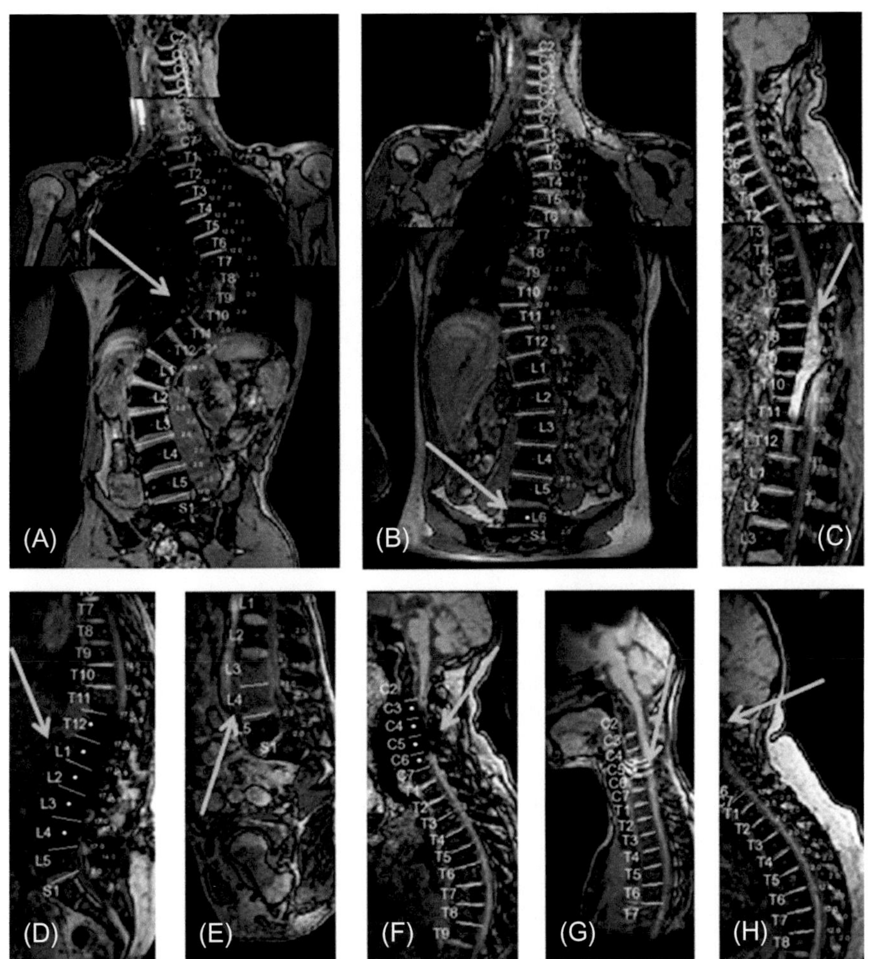

FIG. 13.5

Examples of vertebrae labeling results in challenging MR scans. (A) Whole-spine scan with strong scoliosis. (B) Whole-spine scan with congenital abnormality (six lumbar vertebrae). (C) Whole-spine scan with folding artifact. (D) L-spine scan with metal implant. (E) L-spine scan with vertebra pathology. (F) C-spine scan with metal artifact. (G) C-spine scan with ring artifact. (H) C-spine scan where anchor vertebra is out of field of view.

MR scans. As shown in these cases, our method is robust to different kinds of imaging artifacts (Figs. 13.4A, B and 13.5C, G), large imaging noises (Figs. 13.4F), metal implants (Figs. 13.4D, E and 13.5D, F), severe scoliosis (Fig. 13.5A), pathologies (Figs. 13.4C and 13.5E) congenital abnormality (Fig. 13.5B) and scans that have anchor vertebrae out of field of view (Fig. 13.5H).

13.4 DETECTION OF ANATOMICAL BOXES

13.4.1 ROBUST ANATOMICAL BOX DETECTION USING AN ENSEMBLE OF LOCAL SPATIAL CONFIGURATIONS

An anatomical box is one level beyond anatomical landmarks. It is usually defined by a rectangle (2D) or a cuboid (3D) that contains anatomy of interest and is oriented according to specific anatomical structures, for example, knee meniscus plane, femoral neck axis, intervertebral discs, etc. Autodetection of anatomical boxes is necessary to improve radiological workflow. For example, MR imaging processes often consist of scout scanning and high-resolution scanning stages. In between these two stages, technicians need to manually position high-resolution slice group boxes in scout scans. Appropriate positioning of the slice group is very critical to the quality of the high-resolution MR images (more details will be presented in Section 13.4.2). If an algorithm can automatically detect these high-resolution slice group boxes, which are essentially anatomical boxes, the speed and quality of MR imaging process can be significantly improved.

Since an anatomical box is often defined by anatomical landmarks, it is natural to derive it based on the landmarking technology introduced in Section 13.3. However, a straightforward extension of landmarking may not be enough to derive the anatomical box robustly. For example, in theory, a cuboid can be derived by four noncoplanar points. If we invoke the four landmark detectors independently to derive the box, assuming the detection rate of each landmark is as high as 95% (which is quite high given the variability of diseases and imaging artifacts in medical images), the detection rate of the box falls to 81%. To achieve high robustness, we go beyond the "necessary" number of landmarks and employ "more than enough" landmarks to derive anatomical boxes. The key principle here is to detect "redundant" landmarks and prune the erroneous detections using a spatial model of multiple landmarks.

In contrast to the active shape model that learns global shape statistics, we propose to learn distributed spatial configuration models, that is, for each landmark p_i, we aim to learn its spatial relations with other landmarks in a group-wise fashion. Assume p_i is the landmark under study, then $U(\mathbf{p}\backslash p_i)$ is a subset of landmarks which does not contain p_i, that is, $U(\mathbf{p}\backslash p_i) \subset \{\mathbf{p}\backslash p_i\}$. The group-wise spatial configuration between p_i and $U(\mathbf{p}\backslash p_i)$ is modeled as a conditional probability following a multivariant Gaussian distribution:

$$s(p_i|U(\mathbf{p}\backslash p_i)) = \frac{1}{(2\pi)^{(3/2)}|\Sigma|^{1/2}}\exp\left(-\frac{1}{2}(p_i-\mu)^T\Sigma^{-1}(p_i-\mu)\right), \tag{13.4}$$

where μ and Σ are two statistical coefficients that are learned as follows.

We employ a linear model to capture the spatial correlation between p_i and $U(\mathbf{p}\backslash p_i)$:

$$p_i = \mathfrak{C}\cdot\mathbb{U}, \tag{13.5}$$

where $\mathbb{U}$ is a vector concatenated by $\{p_j | p_j \in U(\mathbf{p} \backslash p_i)\}$ and $\mathfrak{C}$ denotes the linear correlation matrix. Given a set of training samples, $\mathfrak{C}$ can be learned by solving a least squares problem. Furthermore, μ and Σ are calculated as

$$\mu = E[\mathfrak{C} \cdot \mathbb{U}]$$
$$\Sigma = E[(\mathfrak{C} \cdot \mathbb{U} - \mu)(\mathfrak{C} \cdot \mathbb{U} - \mu)^{\mathrm{T}}]. \tag{13.6}$$

At run-time, we will first detect a "redundant" set of landmarks. Each landmark p_i will then receive a set of "votes" from other landmark groups $U(\mathbf{p} \backslash p_i)$ (Eq. 13.4) based on their relative spatial locations. It is worth noting that a correctly detected p_i may still receive a low conditional probability from a subset $U(\mathbf{p} \backslash p_i)$ that includes erroneous detections. In contrast, the conditional probabilities received by an erroneously detected p_i will *all* be low. (The only exception is that the erroneous detections happen to construct a correct spatial configuration, which very rarely happens in reality.) Therefore the maximum value of all votes received by p_i is used to measure the "eligibility" of the spatial location of the p_i. A landmark p_i whose maximum "vote" value is less than a threshold will be considered as an erroneous detection and excluded from the detected landmark set. This process is iteratively conducted until erroneous detections are gradually "peeled" away.

The advantage of our local spatial correlation modeling is twofold. First, instead of learning global spatial statistics, we learn spatial correlations within small groups of landmarks. At run-time, the decisions from these distributed models are assembled in a "democratic" way. This makes our method robust to missing or gross detection failures. Even when only a minority of landmarks are correctly detected, they will form a consistent clique to robustly remove erroneous detections. Second, by constraining the cardinality of U_j and using a linear model, our spatial model will not overfit erroneous detections. The run-time efficiency is also guaranteed in this way.

13.4.2 APPLICATION: AUTOALIGNMENT FOR MR KNEE SCAN PLANNING

Magnetic resonance imaging (MRI) has been successfully used to diagnose various knee diseases ranging from acute knee injuries to chronic knee dysfunction, for example, ligament tears, patella dislocation, meniscal pathology, and arthropathies (Ostlere, 2007). The inherent imaging physics and speed limitations of MR typically constrain the diagnostic MR images to have isotropic high resolution in 3D. Instead, the diagnostic MR image is a 2.5*D* modality with high in-slice resolution and low through-slice resolution (Fig. 13.6C). Hence their diagnostic quality is highly dependent upon the positioning accuracy of the slice groups (imaging planes). Good centering and orientation ensures that the anatomy of interest is optimally captured within the high-resolution (high-res) imaging plane.

Traditionally, this achieved by acquiring 2D scout images (called "localizers"): technicians can then plan out high-res slice groups based on relevant anatomies visible in these scouts. Recently, the 3D knee scout scan has been introduced to

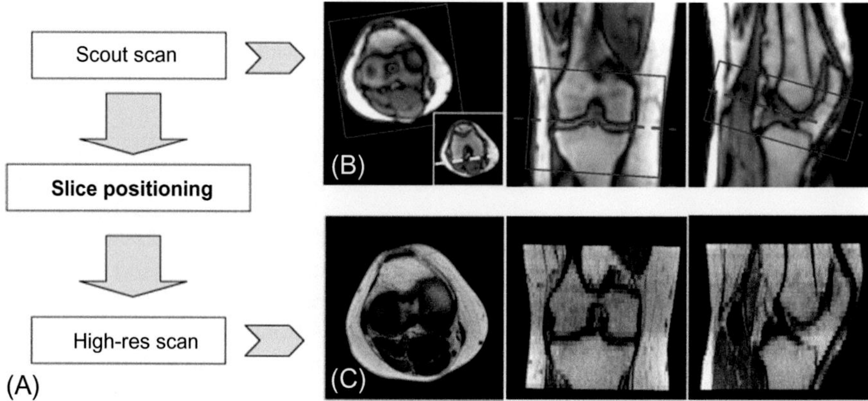

FIG. 13.6

MR slice positioning using scout scans. (A) Workflow of knee MR scans. (B) An MR knee scout scan. Rectangular boxes: coverage of MR slice group. Circles: imaging centers. Yellow dashed line (in the small thumbnail) defines the in-plane rotation. (C) A typical high-resolution diagnostic MR knee transversal scan.

improve the quality of the workflow. A 3D knee scout scan has low but isotropic resolution. Although it might not be of diagnostic quality, it provides complete 3D context, which enables a human operator, or a computer, to plan out all required high-res slice groups without any additional scout scans. Fig. 13.6 shows an example of how to plan a high-res slice group using the 3D scout in a reproducible way: to image the menisci, the transversal slice group should be parallel to the meniscus plane, and the in-plane rotation is determined by aligning with the line connecting the two lower edges of the femoral condyle (the yellow dashed line).

Technically, the slice positioning problem is equivalent to detecting a 3D double oblique anatomical box based on several anatomical landmarks. Therefore we employed the aforementioned anatomical box detection framework for the auto-slice positioning. (Besides the aforementioned algorithms, hierarchial learning is also used in the system. Please refer to Zhan et al. (2011) for more details.)

We evaluate our method on 744 knee scout scans. The alignment results were visually checked and evaluated by two experienced professionals as "accurate (AC)": accurate positioning that is acceptable by technicians; "reasonable (RE)": reasonable good positioning but still has the space to improve; and "gross failure (GF)". Our method can achieve "AC" in 736(98.9%) cases, and has only 1(0.1%) "GF". As shown in Fig. 13.7, our system is able to achieve robust knee alignment even for challenging cases, for example, severe bone diseases, metal implants, etc.

We also quantitatively evaluate the accuracy of the alignment using the detected landmarks on 50 cases with manual alignment. As shown in Table 13.2, the errors are very limited and our system can satisfy clinical requirements.

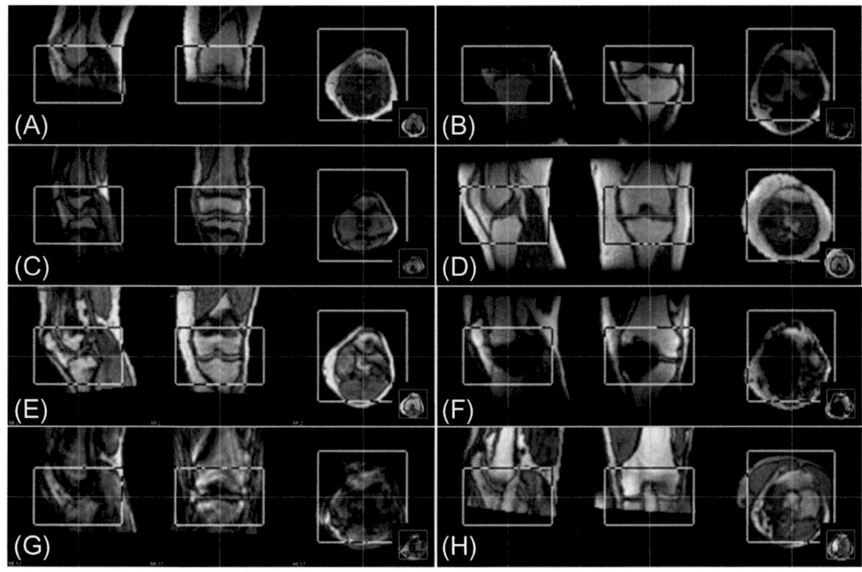

FIG. 13.7

Robust slice positioning of our algorithm on "stress testing" cases. Each part shows the three orthogonal multiplanar reconstructions (MPRs) of the aligned volume at the center of the knee meniscus slice box: (A) tibia out of field of view, (B) femur out of field of view, (C) a pediatric patient, (D) an old patient, (E) a patient with bone disease, (F) a patient with metal implant, (G) motion artifacts, and (H) folding artifacts.

Table 13.2 Quantitative Errors (Average and Standard Deviation) of Our Alignment Method Compared to Manual Alignment

	Knee Meniscus	Femur Cartilage	Patella Cartilage
Trans. (mm)	0.93±0.27	1.53±0.42	1.53±0.38
Rot. tra>>cor (°)	0.83±0.31	1.99±0.60	2.01±0.77
Rot. sag>>tra (°)	0.54±0.28	0.72±0.25	3.14±1.16
Rot. sag>>cor (°)	0.63±0.27	0.82±0.35	1.35±0.52

Rot. tra>>cor, sag>>tra and sag>>cor denote the rotation angles from transversal to coronal, from sagittal to transversal, and from sagittal to coronal respectively.

13.5 COARSE ORGAN SEGMENTATION

13.5.1 SPARSE SHAPE COMPOSITION FOR ORGAN LOCALIZATION

While the detection of anatomical boxes benefits radiological workflow in many ways, some use cases require more precise interpretation of anatomical characteristics—organ shapes. For example, the localization of lung regions in chest X-ray images (not necessary to be very precise) not only directly assists clinicians, for example, auto-measurement of cardiac-thoracic ratio, but also helps following intelligent algorithms, for example, reduction of false positives of lung CAD (computer aided detection).

Organ localization in medical images (in the remainder of this chapter, "coarse organ segmentation" and "organ localization" refer to the same thing by default) requires both appearance and shape information. In our framework, the appearance cues come from the learning-based landmark detectors introduced in Section 13.3. These pretrained detectors essentially capture local appearance cues and automatically identify anatomical landmarks at run-time. Based on these detected landmarks, an organ shape model can be transformed to the subject image under study and considered as a coarse organ segmentation. The remaining question is how to effectively model shape priors and make the organ localization robust to missing/misleading local appearance cues (see Fig. 13.8).

Effective modeling of shape priors faces three challenges: (1) complex shape variation cannot always be modeled by a parametric probability distribution; (2) local image appearance cues (input shape) may have gross errors, for example, detected lung landmarks may be wrong; and (3) local details of the input shape are difficult to preserve if they are not statistically significant in the training data.

We propose a sparse shape composition model (SSC) to address the aforementioned challenges in a unified framework. Given that a shape repository consists of a number of annotated shape instances, instead of explicitly learning shape priors from these shapes offline, we propose to adaptively approximate the input shape on-the-fly, by a sparse linear combination of a subset of shapes in the shape repository. Hence the shape prior constraint is implicitly applied. Our method is inspired by recently proposed sparsity theories in the compressive sensing community, that is, the

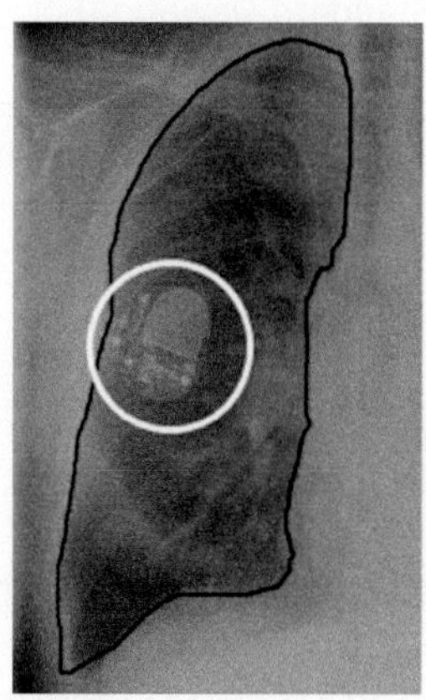

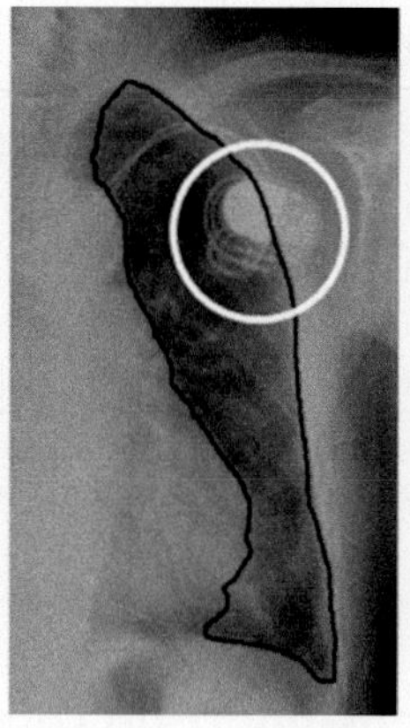

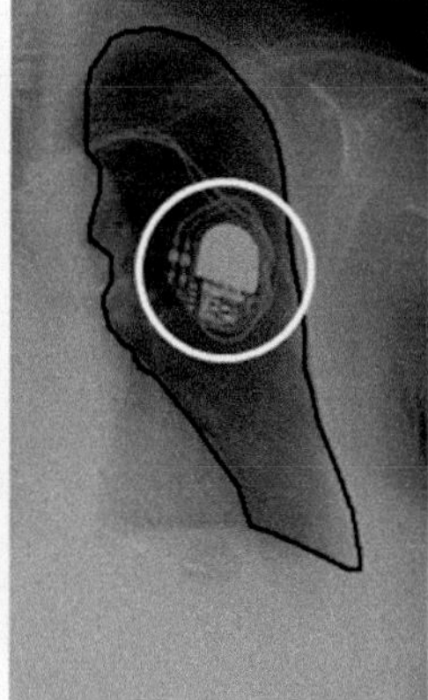

FIG. 13.8

Chest X-ray images with annotated boundaries. The appearance cue is misleading because of the instruments in the marked regions.

problem of computing sparse linear representations with respect to an overcomplete dictionary of base elements (Candes and Tao, 2006; Donoho, 2004). It has been successfully applied in many computer vision applications, such as, but not limited to, face recognition (Wright et al., 2009), image restoration (Mairal et al., 2009), and background subtraction (Huang et al., 2009).

Our method is designed based on two "sparsity" observations. First, given a large enough training dataset (repository), an instance can be approximately represented by a sparse linear combination of instances in the shape repository. Similarly, in our application each given shape is approximated by a sparse linear combination of annotated shapes. Without any assumption of a parametric distribution model (eg, a unimodal distribution assumption in ASM), it becomes general to objects whose shape statistics can be very complex. Moreover, such a setting is able to recover detailed information even if the detail of the input shape is only present in a small proportion of the shape repository and is not statistically significant. Second, the given shape information may contain gross errors resulting from the wrong detections, but such errors are often very sparse, for example, there is an object occluded in the image or a point missing in the input shape. Combining these two, we formulate the shape prior task as a sparse learning problem, and efficiently solve it by an expectation-maximization (EM) type of framework.

In the SSC model, we employ explicit parametric shape representation to model a shape instance, that is, a curve (2D) or a triangular mesh (3D) consisting of a set of vertices. To describe the ith shape in the training data, the coordinates of all its vertices are concatenated into a vector $d_i \in \mathbb{R}^n$, where n is the product of the number of vertices in each shape by the dimension. Thus the training repository can be represented as a matrix $D = [d_1, d_2, \ldots, d_k] \in \mathbb{R}^{n \times k}$, where k is the number of shapes. Here, all d_i, $i = 1, 2, 3, \ldots, k$, are prealigned using generalized Procrustes analysis (Goodall, 1991). $y \in \mathbf{R}^n$ is the vector of a newly input shape which needs to be constrained or refined. Our basic framework assumes that after proper alignment, any input shape y can be approximately represented as a weighted linear combination of existing data d_i, $i = 1, 2, 3, \ldots, k$, and the parts which cannot be approximated are noises. We denote $x = [x_1, x_2, \ldots, x_k]^{\mathrm{T}} \in \mathbb{R}^k$ as the combination coefficients or weights.

To incorporate the two "sparsity" observations, we derive x by minimizing the loss function as

$$\begin{aligned} &\arg\min_{x,e,\beta} \|T(y, \beta) - Dx - e\|_2^2, \\ &\text{s.t. } \|x\|_0 \le k_1, \|e\|_0 \le k_2. \end{aligned} \tag{13.7}$$

Here, $T(y, \beta)$ is a global transformation operator with parameter β. It aligns the input shape y to the mean shape of existing data D. $\|x\|_0 \le k_1$ defines the first sparsity constraint, that is, an instance can be approximately represented by a sparse linear combination of instances in the shape repository. $\|e\|_0 \le k_2$ corresponds to the second sparsity constraint; the given shape information may contain gross errors, but such errors are often very sparse.

The constraints in Eq. (13.7) are not directly tractable because of the nonconvexity of L^0 norm. Greedy algorithms can be applied to this NP-hard L^0 norm minimization problem, but there is no guarantee of capturing the global minima. In the general case, no known procedure can correctly find the sparsest solution more efficiently than exhausting all subsets of the entries for x and e. Furthermore, in practice the sparsity numbers k_1 and k_2 may change for different data in the same application. For example, some data have errors while others do not. Fortunately, recent developments in sparse representation provide a theorem to efficiently solve this kind of problem through L^1 norm relaxation (Starck et al., 2004). Thus Eq (13.7) is reformulated as

$$\arg\min_{x,e,\beta} \|T(y,\beta) - Dx - e\|_2^2 + \lambda_1\|x\|_1 + \lambda_2\|e\|_1, \tag{13.8}$$

where λ_1 and λ_2 respectively control how sparse x and e are. After relaxation, $\lambda_1\|x\|_1 + \lambda_2\|e\|_1$ is nonsmooth but continuous and convex. Eq. (13.8) is our objective function of our proposed SSC. The deviation from Eq. (13.7) to Eq. (13.8) relaxes the absolute sparseness constraints of the objective function (L^0 norm to L^1 norm). From the shape modeling perspective, we might use more shape instances for shape composition by optimizing Eq. (13.8). However, since this deviation converts an NP-hard problem to a continuous and convex optimization problem which can be solved efficiently using an expectation-maximization optimization framework (refer to (Zhang et al., 2012) for more details).

It is interesting to look into Eq. (13.8) by adjusting λ_1 and λ_2 into some extreme values:

- If λ_2 is extremely large, e will be all zeros. Thus SSC is similar to methods which do not model non-Gaussian errors.
- If both λ_1 and λ_2 are large enough, e will be all zeros and x may have only one nonzero element. Thus SSC becomes the nearest neighbor method.
- If λ_2 is extremely large and λ_1 is small, a dense linear combination of shapes is used, which is able to perfectly approximate the transformed input shape. Thus SSC degenerates to the Procrustes analysis.

The insight of Eq. (13.8) indeed reveals the connections of our SSC with some other popular methods. Those methods can be regarded as special cases of SSC. In other words, SSC provides a unified framework to deal with different challenges of shape prior modeling simultaneously. SSC can also provide flexibility to meet the requirements of different applications by adjusting the sparsity of x and e.

13.5.2 LUNG REGION LOCALIZATION IN CHEST X-RAY

Radiography (X-ray) is the most frequently used medical imaging modality due to its fast imaging speed and low cost. About one-third of radiograph exams are chest radiographs. It is used to reveal various pathologies including abnormal cardiac sizes, pneumonia shadow, and mass lesions. The automation of pathology detection often requires robust and accurate lung segmentation. The major challenges of lung segmentation in radiography come from large variations of lung shapes, lung disease, and pseudo-boundary close to the diaphragm. In chest X-ray, the position, size, and shape of lungs often provide important clinical information. Therefore in this experiment we try to locate the left or right lung using landmark detection and shape inference. Of 367 X-ray images (all images are from different patients), 200 are used as training data, and the other 167 are used for testing purposes. In this study, we select training samples to ensure a good coverage of different ages and genders (according to information from DICOM header). The number of training samples is determined empirically. The ground truths are binary masks of manual segmentation results. A 2D contour is extracted from each mask. To obtain the landmarks for training purposes, we manually select six specific points (eg, corner points) on the contour, and then evenly and automatically interpolate a fixed amount of points between two neighboring landmarks along the contour. Thus a rough one-to-one correspondence is obtained for both landmarks and shapes. Since the detected landmarks may not be accurate or complete, a shape prior is necessary to infer a shape from them. When applying this model, we constantly use the same parameter values for all X-ray images, that is, $\lambda_1 = 50$ and $\lambda_2 = 0.15$.

In this study, we compare the proposed sparsity-based shape prior modeling with other state-of-the-art methods listed below:

1. PA: Procrustes analysis is used to find a similarity transformation to fit a mean shape to detected landmarks.
2. SMS: This is the shape model search module in ASM, which employs the PCA method to refine the input shape. Note that we are not using the entire ASM framework including boundary detection and iterative fitting.
3. R-SMS: The shape model search step in the robust ASM (Rogers and Graham, 2002) method uses the RANSAC framework to remove the influence of erroneous detections.
4. SI-NN: This stands for shape inference using k nearest neighbors. It is similar to Georgescu et al. (2005), which uses nearest neighbors to find the closest prototypes in the expert's structure annotations. The distance metric we used is based on the $L2$ distance between corresponding points.
5. TPS: Thin-plate-spline (Bookstein, 1989) is used to deform the mean shape to fit detected landmarks.
6. SSC*: This is a variant of the proposed SSC algorithm without modeling e.

Some representative and challenging cases are shown in Figs. 13.9–13.11. In Fig. 13.9, there are some misdetections which are considered as gross errors. The Procrustes analysis, SMS method, SI-NN algorithm, and TPS cannot handle such cases. R-SMS is not sensitive to outliers and performs better. SSC* also fails to handle such non-Gaussian errors since e is not modeled. SSC can successfully capture such misdetections points in e and generate a reasonable shape. In Fig. 13.10,

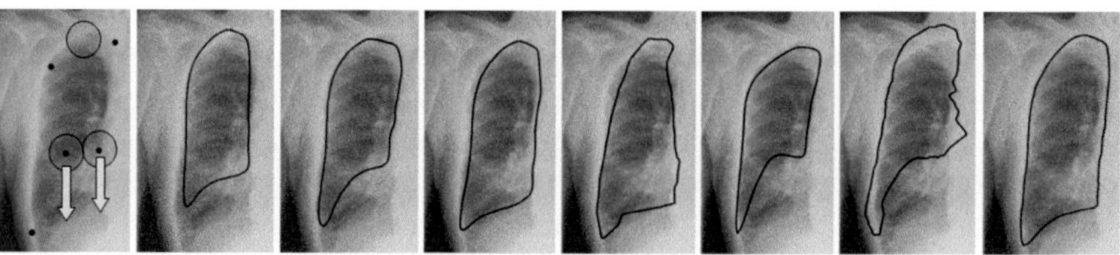

FIG. 13.9

Comparisons of right lung localization. (A) Detected landmarks are marked as black dots. There are two detection errors and one point missing (marked as circles, and the arrows point to the correct positions). (B) Similarity transformation from Procrustes analysis. (C) Shape model search module in ASM, using a PCA-based method. (D) Shape model search in robust ASM, using RANSAC to improve the robustness. (E) Shape inference method using nearest neighbors. (F) Thin-plate-spline. (G) Sparse representation without modeling *e*. (H) The proposed method by solving Eq. (13.8).

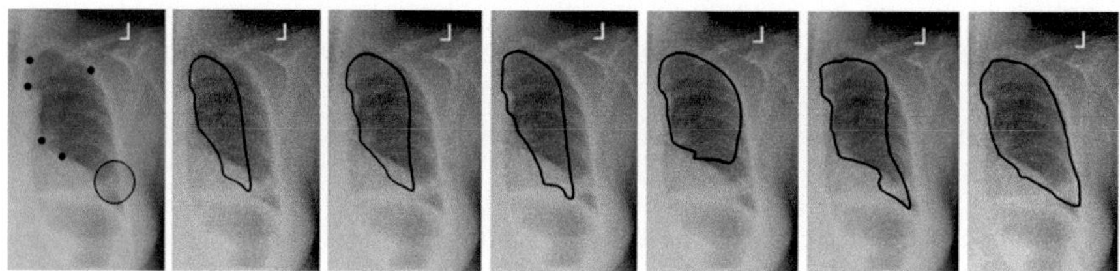

FIG. 13.10

Comparisons of left lung localization. There is one point missing (marked by a circle), and this lung has a very special shape, which is not captured by the mean shape or its variations. Compared methods are the same as in Fig. 13.9.

the underlying shape of the lung is special and different from most other lung shapes (see the mean shape in Fig. 13.10). Furthermore, there is a missing point. Neither a transformed mean shape nor its variations can represent such a shape. TPS is very flexible and able to generate special shapes. However, it fails to handle the missing point. SSC roughly captures the correct shape and generates a better result than the others. In Fig. 13.11, all six detections are correct. However, the shape's details are not preserved using the mean shape or its variations. Both SSC* and SSC discover more detailed information than the other methods. Thus a sparse linear combination is sufficient to recover such details even when the gross error e is not modeled. Fig. 13.12 shows some results from our proposed method for challenging cases with medical instruments. The shape prior contributes to the stability of the system. It still generates reasonable results with such misleading appearance cues (Table 13.3).

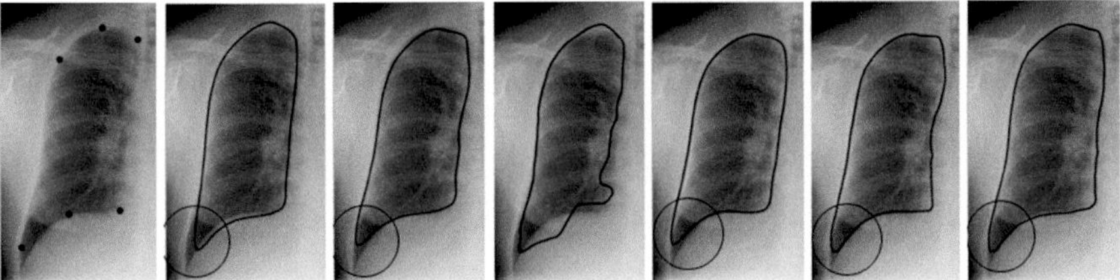

FIG. 13.11

Comparisons of right lung localization. All six detections are roughly accurate. Thus there is no gross error. The regions marked by circles show the difference of preserved details. Compared methods are the same as in Fig. 13.9.

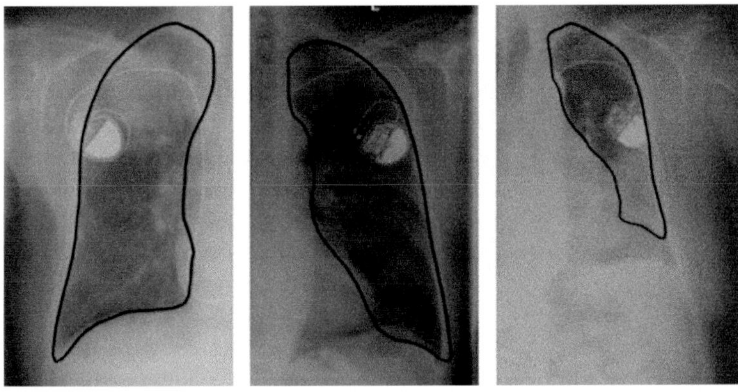

FIG. 13.12

Some localization results from our proposed method on challenging cases with medical instruments. Note that the localized shape may not be exactly on the boundary, since the shape module does not use image information. However, such results are good enough for the input of a CAD program or initialization of segmentation algorithms.

13.6 PRECISE ORGAN SEGMENTATION

13.6.1 DEFORMABLE SEGMENTATION USING HIERARCHICAL CLUSTERING AND LEARNING

Although rough organ boundaries/shapes from coarse segmentation approaches can benefit multiple use cases, they are still one step from precise delineation of organ boundaries, which are important for quantitative studies. Therefore, at the final level of our anatomical parsing system, we developed a precise organ segmentation algorithm based on a deformable model.

Table 13.3 Quantitative Comparisons of Seven Methods

	Fig. 13.9			Fig. 13.10			Fig. 13.11		
Method	**P**	**Q**	**DSC**	**P**	**Q**	**DSC**	**P**	**Q**	**DSC**
PA	62	99	76	50	99	64	93	99	94
SMS	66	99	78	61	99	72	93	99	95
R-SMS	81	99	88	61	99	72	93	99	95
SI-NN	81	99	87	63	98	73	87	99	90
TPS	59	99	74	75	99	79	97	98	94
SSC*	63	98	71	73	99	79	97	99	96
SSC	**87**	**99**	**91**	**92**	**99**	**91**	**98**	**99**	**96**

The sensitivity (P%), specificity (Q%), and Dice Similarity Coefficient (DSC%) are reported for the cases in Figs. 13.9–13.11.

Deformable modeling is a vigorously studied model-based approach in the area of medical image segmentation. The widely recognized potency of deformable models comes from their ability to segment anatomic structures by exploiting constraints derived from the image data (bottom-up) together with prior knowledge about these structures (top-down). The deformation process is usually formulated as an optimization problem whose objective function consists of an external (image) term and an internal (shape) term. While internal energy is designed to preserve the geometric characteristics of the organ under study, the external energy is defined to describe the appearance characteristics of the organ under study and drive the model toward organ boundaries. Recently, machine learning technologies have opened the door to a more generic external energy design. By using learning-based methods, boundary characteristics can be learned from training data (Zhan and Shen, 2006). In other words, the originally hand-crafted external energy is derived in a data-driven fashion and can be extended to different imaging modalities and organ systems. A potential problem is that the boundary characteristics of organs may show heterogeneous characteristics along organ boundaries. As shown in Fig. 13.13, the appearance characteristics of liver-lung boundaries is quite different from those of liver-heart boundaries. Hence, it is very challenging to learn the organ boundary characteristics using a single classifier. To address this problem, a "divide-and-conquer" strategy is needed. More specifically, the deformable model should be decomposed into a set of subsurfaces with relatively similar boundary characteristics (Fig. 13.14).

To this end, we propose a learning-based hierarchical deformable model. A key hallmark of our model is that its hierarchical structure is constructed through an iterative clustering and feature selection method. As shown in Fig. 13.15, every node of the hierarchical structure represents a subsurface of the deformable model. For each primitive subsurface, that is, leaf nodes in the hierarchical tree, a boundary

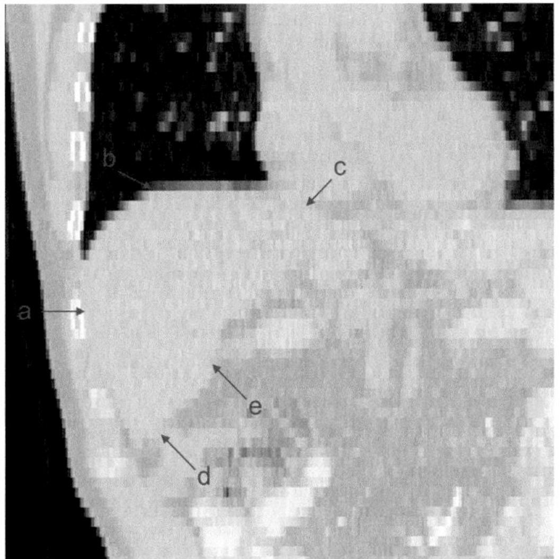

FIG. 13.13

An example of liver CT images. Arrows a–e point to boundaries between liver and rib, lung, heart, abdomen, and colon that show heterogeneous appearance.

detector is learned using a cascade boosting method. The ensemble of these learned boundary detectors actually captures the appearance characteristics of a specific organ in a specific imaging modality. Their responses guide the deformable model to the desired organ boundary.

Our deformable model is represented by a triangular mesh: $S \equiv (V, T)$, where $V = \{v_i | i = 1, \ldots, N\}$ denotes the vertices of the surface and $T = \{t_j | j = 1, \ldots, M\}$ denotes the triangles defined by vertices. Mathematically, the segmentation problem is formulated as the minimization of an energy function defined as

$$E(S) = E_{\text{ext}}(S) + E_{\text{int}}(S) = \sum_{i=1}^{N} E_{\text{ext}}(v_i) + E_{\text{int}}(S), \tag{13.9}$$

where E_{ext} and E_{int} are the image (external) energy term and shape (internal) energy term. A hierarchical deformation strategy is employed to solve this high-dimensional optimization problem (refer to Zhan and Shen (2006) for details).

The external energy is defined by the responses of a set (or ensemble) of boundary detectors built upon the hierarchical deformable model. The following steps are used to generate the hierarchial model and the boundary detectors.

13.6.1.1 Affinity propagation clustering

The construction of model hierarchy is equivalent to the clustering of vertices. In this study, we employ "affinity propagation," a generic clustering method proposed by

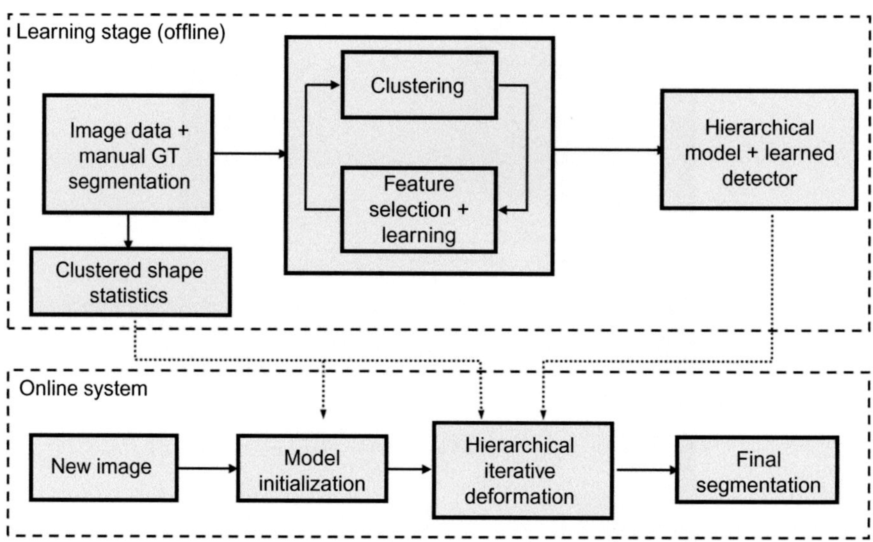

FIG. 13.14

Flowchart of the learning-based hierarchical model showing both the offline learning and the online testing system.

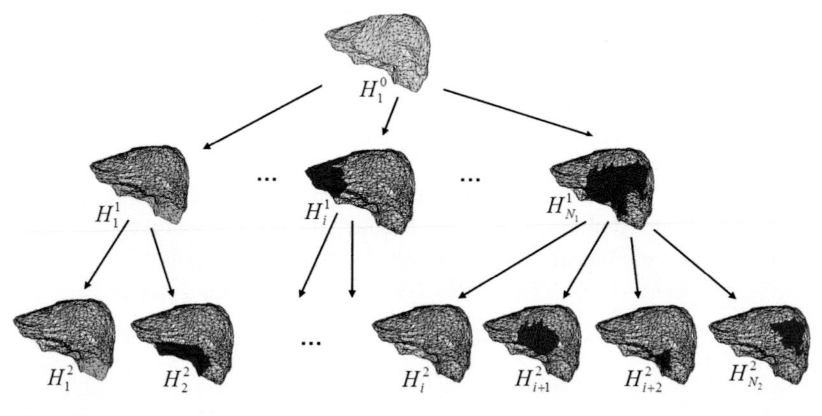

FIG. 13.15

Hierarchical structure of the deformable model. Color patches depict the subsurfaces (H_i^j) at the jth hierarchical level comprised of vertices in the ith cluster.

Frey and Dueck (2007). The affinity propagation method models each data point as a node in a network. During the clustering process, real-valued messages are recursively exchanged between data points until a high quality set of exemplars and corresponding clusters emerges.

In the "affinity propagation" framework, the quality of clustering is determined by the affinity between data points—the similarity between vertices in our study. We design the similarity based on two rationales. First, to facilitate the characterization of the heterogeneous boundary, vertices in the same cluster should have relatively similar image features. Second, the hierarchical deformable model requires the vertices within a cluster to be proximal to each other on the surface. In this way, the cluster center can be treated as a "driving vertex" and drive its neighborhood in the deformation process. Mathematically, the similarity between vertices is defined as follows:

$$s(v_i, v_j) = 1 - (1/K) \sum_{k=1}^{K} [\alpha G(v_i^k, v_j^k) + (1-\alpha) C(\mathcal{F}(v_i^k), \mathcal{F}(v_j^k))]. \tag{13.10}$$

Here, K is the number of training subjects and v_i^k denotes the ith vertex of the kth subject. $G(v_i^k, v_j^k)$ denotes the geodesic distance between v_i^k and v_j^k. $C(\mathcal{F}(v_i^k), \mathcal{F}(v_j^k))$ denotes the Euclidean distance between image feature vectors calculated at v_i^k and v_j^k.

13.6.1.2 Iterative feature selection/clustering

To construct the hierarchical structure of the deformable model, vertices are recursively clustered. Assume H_i^j is the ith cluster at the jth hierarchical level, then vertices belonging to H_i^j are further clustered to a set of subclusters $\{H_k^{j+1}, k = 1, \ldots, N_i\}$:

$$H_i^j = \bigcup_{k=i_1}^{i_{N_i}} H_k^{j+1} \text{ and } \bigcap_{k=i_1}^{i_{N_i}} H_k^{j+1} = \emptyset. \tag{13.11}$$

The remaining problem is the selection of appropriate $\mathcal{F}(.)$ in Eq. (13.10). This is actually an "egg-and-chicken" problem. On one hand, to achieve the desired clusters, we need to know the distinctive feature sets for boundary description. On the other hand, distinctive features for a local boundary can be obtained only after we have the vertex cluster. To address this problem, we propose an iterative clustering and feature selection method.

For the first level of cluster, we use the intensity profile along the normal of the vertices as $\mathcal{F}(.)$. After that, assume $H_i^j = \{v_l\}$, and we use the *Adaboost* method to select the features that are most powerful to distinguish $\{v_l\}$ from the points along their normal directions, both inside and outside of the surface. The selected feature set is used as $\mathcal{F}(.)$ in Eq. (13.10) to further cluster $\{v_l\}$ to a set of subclusters $\{H_k^{j+1}, k = i_1, \ldots, i_{N_i}\}$. Feature selection and clustering are iteratively executed until boundary characteristics within a cluster become learnable.

13.6.1.3 Learn boundary detectors

For each primitive cluster, that is, the leaf node of the hierarchical tree, a boundary detector is learned to characterize the local boundary. We use the similar

Haar-like features and a cascade *Adaboost* method as our landmarking method (see Section 13.3) to learn a boundary detector. Given an image I, $\mathfrak{F}(\mathbf{x}; I)$ denotes the redundant feature vector of $\mathbf{x}$. (In practice, we use 2D, 3D, or 4D Haar-like features depending on the dimensionality of different image modalities.) In the run-time system, each learned classifier generates a boundary probability map $P(\mathbf{x}|I)$. Hence, the external energy term in Eq. (13.9) is defined as

$$E_{\text{ext}}(v_i) = 1 - P(v_i|I) = 1 - C_{\hbar_{v_i}}(\mathfrak{F}(v_i; I)), \tag{13.12}$$

where $\hbar_{v_i}$ is the cluster index of v_i and $C_{\hbar_.}$ defines the corresponding classifier.

13.6.2 LIVER SEGMENTATION IN WHOLE-BODY PET-CT SCANS

PET-CT is a medical imaging modality that combines PET and an X-ray CT. As the acquired PET and CT images are inherently co-registered, PET-CT provides fused morphological and functional information, which potentially benefits various medical studies. After the first PET-CT prototype was introduced to clinical practice in 1998, PET-CT has triggered a revolution in image-based diagnosis for cancer patients. Many clinical studies reported that PET-CT has superior diagnostic value than mono-modalities, for example, CT (Shim et al., 2005), MR (Antoch et al., 2003) and PET (Bar-Shalom et al., 2003), and separate dual-modalities, for example, PET+morphological image (Pelosi et al., 2004).

In PET-CT-based cancer diagnosis, ^{18}F-fluoro-deoxyglucose (FDG) is the most widely used tracer. Since glucose utilization is known to be enhanced in many malignant tissues, uptake of FDG is usually considered as an indicator of malignance. In particular, the standardized uptake value (SUV) is employed to provide semiquantitative interpretation of malignance (Thie, 2004). However, glucose utilization is not entirely specific to malignant tissues. Normal tissues, such as the cerebral cortex, left ventricular myocardium and renal system, may also have high FDG uptake. A more troublesome fact is that FDG uptake shows large variations across different organs due to different glucose utilization. This is one of the major pitfalls of PET-CT interpretation. For example, while a spot in a lung with SUV = 3.0 is a highly suspicious lesion, a spot in the liver with the same level of SUV is completely normal.

An effective solution is to segment PET-CT and interpret it in an organ-specific fashion. Based on different physiological characteristics of organs, radiologists can use different thresholds/strategies to detect hot spots in different organs. The organ-specific PET-CT interpretation presumes the organs to be investigated have been segmented. Since the liver is one of the most important organs in oncology studies, we apply the learning-based hierarchial model to segment the liver in whole-body PET-CT scans. It is worth noting that the whole-body CT often has larger slice thickness and low dose (lower contrast), both of which make the segmentation task difficult. We trained the learning-based hierarchial model using 20 whole-body PET-CT scans with manually delineated liver surfaces. (To jointly exploit CT-PET information, 4D Haar-like filters are used for feature extractor (Zhan et al., 2008).) As shown in Fig. 13.16, the generated model has two hierarchical levels with 8 and

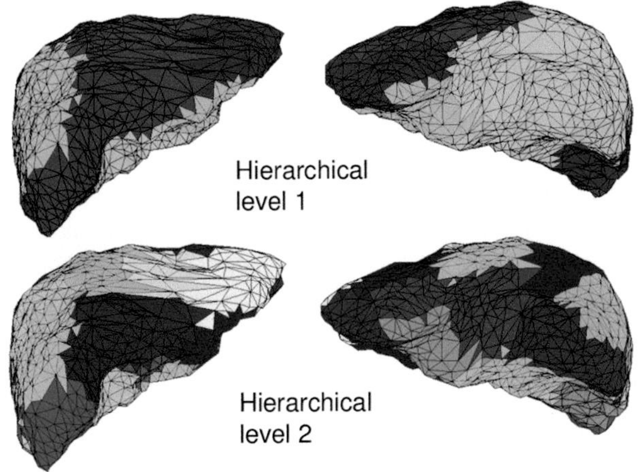

FIG. 13.16

3D rendering of the hierarchical structure of a liver model. Color patches denote the vertices belonging to the same cluster. Left: anterior view. Right: posterior view.

25 vertex clusters respectively. The automatic segmentation results on 30 testing datasets (PET: 5 mm × 5 mm × 5 mm; CT: 1.3 mm × 1.3 mm × 5 mm) are compared with manually delineated organ surfaces. Accuracy measurements include median distance between surfaces, average distance between surfaces, volume difference, and volume overlap difference. In Table 13.4, we compare our proposed method (Method 1) with Method 2, which is a learning-based deformable model with *heuristically* designed hierarchical structure. More specifically, in Method 2 the hierarchical structure is determined by clustering neighboring vertices, based only on geodesic information.[1] Therefore the spatially clustered vertices in Method 2 might have larger appearance variation, which is difficult to learn. Hence Method 2 shows inferior performance.

Table 13.4 Quantitative Comparison of Learning-Based Methods on PET-CT Liver Segmentation

	Med Surf. Dist. (voxel)	**Avg Surf. Dist. (voxel)**	**Vol. Diff. (%)**	**Overlap Diff. (%)**
Method 1	0.84	1.01	3.13	7.61
Method 2	1.27	1.61	5.16	12.1

Method 1: Our hierarchical deformable model. Method 2: A heuristically designed hierarchical deformable model.

[1]Note that Method 2 only differs from Method 1 in the way their hierarchical structures are built.

13.7 CONCLUSION

In this chapter, we introduced a framework to parse human anatomies in medical images. Our framework is designed in a coarse-to-fine fashion to localize and identify human anatomies from landmarks to precise organ boundaries. In each layer, different machine learning technologies, including Adaboost, sparse representation, iterative clustering, etc., are used to learn "anatomical signatures." In this way, our framework can be extended to diverse clinical applications in different imagine modalities and can be easily adapted to new applications. Although the different layers independently benefit different clinical use cases, they also help each other from a technical point of view. For example, the landmarking algorithm provides appearance cues for coarse organ segmentation and coarse organ segmentation paves the way for precise segmentation.

We have prototyped different applications based on this framework. Some representative examples are presented in this chapter. These applications are very diverse, covering different imaging modalities (CT, MR, PET, radiography, etc.) and clinical fields (oncology, orthopedics, neurology, etc.). The preliminary results show that our methods are able to achieve robust and accurate results and can potentially benefit clinical workflow in various ways.

REFERENCES

Antoch, G., Vogt, F.M., Freudenberg, L.S., Nazaradeh, F., Goehde, S.C., Barkhausen, J., Dahmen, G., Bockisch, A., Debatin, J.F., Ruehm, S.G., 2003. Whole-body dual-modality PET/CT and whole-body MRI for tumor staging in oncology. J. Am. Med. Assoc. 290, 3199–3206.

Bar-Shalom, R., Yefremov, N., Guralnik, L., Gaitini, D., A.Frenkel, Kuten, A., Altman, H., Keidar, Z., Israel, O., 2003. Clinical performance of PET/CT in evaluation of cancer: additional value for diagnostic imaging and patient management. J. Nucl. Med. 44 (8), 1200–1209.

Betke, M., Hong, H., Thomas, D., Prince, C., Ko, J.P., 2003. Landmark detection in the chest and registration of lung surfaces with an application to nodule registration. Med. Image Anal. 7 (3), 265–281.

Bookstein, F.L., 1989. Principal warps: thin-plate splines and the decomposition of deformations. IEEE Trans. Pattern Anal. Mach. Intell. 11 (6), 567–585.

Candes, E.J., Tao, T., 2006. Near-optimal signal recovery from random projections: universal encoding strategies? IEEE Trans. Inform. Theory 52 (12), 5406–5425.

Cootes, T.F., Taylor, C.J., 1997. A mixture model for representing shape variation. In: Image and Vision Computing, pp. 110–119.

Cootes, T.F., Taylor, C.J., Cooper, D.H., Graham, J., 1995. Active shape model—their training and application. Comput. Vision Image Understanding 61, 38–59.

Criminisi, A., Shotton, J., Robertson, D., Konukoglu, E., 2011. Regression forests for efficient anatomy detection and localization in CT studies. In: Medical Computer Vision. Recognition Techniques and Applications in Medical Imaging. Springer, New York, pp. 106–117.

Crow, F.C., 1984. Summed-area tables for texture mapping. In: ACM SIGGRAPH Computer Graphics 18.3, 207–212.

Davatzikos, C., Tao, X., Shen, D., 2003. Hierarchical active shape models, using the wavelet transform. IEEE Trans. Med. Imaging 22 (3), 414–423.

Donoho, D.L., 2004. For most large underdetermined systems of equations, the minimal l1-norm near-solution approximates the sparsest near-solution. Commun. Pure Appl. Math. 59, 797–829.

Etyngier, P., Segonne, F., Keriven, R., 2007. Shape priors using manifold learning techniques. In: International Conference on Computer Vision, pp. 1–8.

Freund, Y., Schapire, R.E., 1997. A decision-theoretic generalization of on-line learning and an application to boosting. J. Comput. System Sci. 55, 119–139.

Frey, B.J., Dueck, D., 2007. Clustering by passing messages between data points. Science 315, 972–976.

Georgescu, B., Zhou, X.S., Comaniciu, D., Gupta, A., 2005. Database-guided segmentation of anatomical structures with complex appearance. In: IEEE Conference on Computer Vision and Pattern Recognition, vol. 2, pp. 429–436.

Goodall, C., 1991. Procrustes methods in the statistical analysis of shape. J. R. Stat. Soc. 53, 285–339.

Huang, X., Metaxas, D., Chen, T., 2004. Metamorphs: deformable shape and texture models. In: Computer Vision and Pattern Recognition 2004, pp. 496–503.

Huang, J., Huang, X., Metaxas, D., 2009. Learning with dynamic group sparsity. In: International Conference on Computer Vision, pp. 64–71.

Liu, D., Zhou, S.K., 2012. Anatomical landmark detection using nearest neighbor matching and submodular optimization. In: Medical Image Computing and Computer-Assisted Intervention-MICCAI 2012. Springer, New York, pp. 393–401.

Lv, X., Zhou, J., Zhang, C.S., 2000. A novel algorithm for rotated human face detection. In: Computer Vision and Pattern Recognition 2000.

Mairal, J., Bach, F., Ponce, J., Sapiro, G., Zisserman, A., 2009. Non-local sparse models for image restoration. In: International Conference on Computer Vision, pp. 2272–2279.

Ostlere, S., 2007. Imaging the knee. Imaging 15, 217–241.

Pelosi, E., Messa, C., Sironi, S., Picchio, M., Landoni, C., Bettinardi, V., Gianolli, L., Maschio, A.D., Gilardi, M.C., Fazio, F., 2004. Value of integrated PET/CT for lesion localisation in cancer patients: a comparative study. Eur. J. Nucl. Med. Mol. Imaging 31 (7), 1619–7070.

Rogers, M., Graham, J., 2002. Robust active shape model search. In: European Conference on Computer Vision, pp. 517–530.

Ronfard, R., 1994. Region-based strategies for active contour models. Int. J. Comput. Vision 13 (2), 229–251.

Schneiderman, H., Kanade, T., 2000. A statistical method for 3d object detection applied to faces and cars. In: Computer Vision and Pattern Recognition novel algorithm for rotated human face detection 2000.

Shim, S.S., Lee, K.S., Kim, B.T., Chung, M.J., Lee, E.J., Han, J., Choi, J.Y., Kwon, O.J., Shim, Y.M., Kim, S., 2005. Nonsmall cell lung cancer: prospective comparison of integrated FDG PET/CT and CT alone for preoperative staging. Radiology 236, 1011–1019.

Sjostrand, K., et.al., 2007. Sparse decomposition and modeling of anatomical shape variation. IEEE Trans. Med. Imaging 26 (12), 1625–1635.

Starck, J.L., Elad, M., Donoho, D.L., 2004. Image decomposition via the combination of sparse representations and a variational approach. IEEE Trans. Image Process. 14, 1570–1582.

Thie, J., 2004. Understanding the standardized uptake value, its methods, and implications for usage. J. Nucl. Med. 45 (9), 1431–1434.

Vese, L.A., Chan, T.F., 2002. A multiphase level set framework for image segmentation using the Mumford and Shah model. Int. J. Comput. Vision 50 (3), 271–293.

Viola, P., Jones, M.J., 2004. Robust real-time face detection. Int. J. Comput. Vision 57, 137–154.

Wright, J., Yang, A.Y., Ganesh, A., Sastry, S.S., Ma, Y., 2009. Robust face recognition via sparse representation. IEEE Trans. Pattern Anal. Mach. Intell. 31 (2), 210–227.

Xu, C., Prince, J.L., 1998. Snakes, shapes, and gradient vector flow. IEEE Trans. Image Process. 7 (3), 359–369.

Yan, P., Xu, S., Turkbey, B., Kruecker, J., 2010. Discrete deformable model guided by partial active shape model for trus image segmentation. IEEE Trans. Biomed. Eng. 57 (5), 1158–1166.

Zhan, Y., Shen, D., 2006. Deformable segmentation of 3-d ultrasound prostate images using statistical texture matching method. IEEE Trans. Med. Imaging 25, 256–272.

Zhan, Y., Peng, Z., Zhou, X. 2008. Towards organ-specific PET-CT interpretation:generic organ segmentation using joint PET-CT information. In: MICCAI2008 Workshop on Analysis of Functional Medical Images.

Zhan, Y., Dewan, M., Zhou, X.S., 2009. Cross modality deformable segmentation using hierarchical clustering and learning. In: Medical Image Computing and Computer-Assisted Intervention, pp. 1033–1041.

Zhan, Y., Dewan, M., Harder, M., Krishnan, A., Zhou, X.S., 2011. Robust automatic knee MR slice positioning through redundant and hierarchical anatomy detection. IEEE Trans. Med. Imaging 30, 2087–2100.

Zhan, Y., Dewan, M., Harder, M., Zhou, X.S., 2012. Robust MR spine detection using hierarchical learning and local articulated model. In: Medical Image Computing and Computer-Assisted Intervention, pp. 141–148.

Zhang, S., Zhan, Y., Dewan, M., Huang, J., Metaxas, D.N., Zhou, X.S., 2012. Towards robust and effective shape modeling: sparse shape composition. Med. Image Anal. 16, 265–277.

Zheng, Y., Barbu, A., Georgescu, B., Scheuering, M., Comaniciu, D., 2008. Four-chamber heart modeling and automatic segmentation for 3D cardiac CT volumes using marginal space learning and steerable features. IEEE Trans. Med. Imaging 27.

CHAPTER

Machine learning in brain imaging genomics 14

J. Yan[a], L. Du, X. Yao, L. Shen

Indiana University School of Medicine, Indianapolis, IN, United States

CHAPTER OUTLINE

14.1 INTRODUCTION

Brain imaging genomics has attracted increasing attention in recent years. It is an emerging research field that has arisen with the advances in high-throughput genotyping and multimodal imaging techniques. Its major task is to examine the association between genetic markers such as single nucleotide polymorphisms (SNPs) and quantitative traits (QTs) extracted from multimodal neuroimaging data (eg, anatomical, functional, and molecular imaging scans). Given the well-known importance of imaging and genomics in the brain study, bridging these two factors and exploring their connections have the potential to provide a better mechanistic understanding of normal or disordered brain functions. Also, changes in imaging phenotypes usually precede those in disease status and cognitive outcomes, and are believed to be closer to the underlying genetic mechanisms. Therefore associating

[a]Equal contributions by J. Yan, L. Du, and X. Yao. J. Yan contributed to Sections 14.1, 14.2.1, 14.2.2, and 14.4. L. Du contributed to Section 14.2.3. X. Yao contributed to Section 14.3.

Machine Learning and Medical Imaging. http://dx.doi.org/10.1016/B978-0-12-804076-8.00014-1

genetic data with imaging phenotypes, rather than disease status, is highly promising for discovery of influential genetic architecture and has the potential to help reveal the earliest brain changes for prognosis.

Early attempts in imaging genomics were mostly pairwise univariate analyses for quantitative genome-wide association studies (GWASs), which were performed to correlate high-throughput SNP data to large-scale image QT data (Shen et al., 2010; Stein et al., 2010). A simple regression model, for example, that implemented in PLINK (Purcell et al., 2007), was typically used to examine the additive effect of each single SNP on each single imaging QT. These simple regression models were often coupled with hypothesis testing, in which the significances of regression coefficients were learned simultaneously. Pairwise univariate analysis was used in traditional association studies to quickly provide important association information between SNPs and QTs. However, both SNPs and QTs were treated as independent and isolated units, and therefore the underlying correlating structure between the units was ignored.

Multiple regression models were later introduced to study the multilocus effects on imaging phenotypes. Instead of discovering individual SNPs with significant effects, multivariate models have enabled the identification of SNPs that jointly affect phenotypic changes. GCTA, a popular heritability analysis tool, utilizes a linear mixed regression model and has successfully demonstrated that $\sim$45% of the phenotypic variance for human height can be explained by the joint effect of common SNPs (Yang et al., 2011, 2013). Also considering the intercorrelated nature and high-dimensional setting of imaging and genetic data, sparse regression models are particularly favored. These models can not only address the correlation problem, but also help identify a small number of biologically meaningful genetic markers for easy interpretation. In addition, many recent efforts have been devoted to prior knowledge guided regression models, and many studies have confirmed the beneficial role of prior data structure in capturing more accurate imaging genomic relationships (Wang et al., 2012a; Silver et al., 2012a,b).

Bi-multivariate association analysis has recently received increasing attention for exploration of complex multi-SNP-multi-QT relationship. Existing bi-multivariate models widely used in imaging genomic studies can generally be classified into two types: canonical correlation analysis (CCA) type and reduced rank regression (RRR) type. Based on the assumption that a real imaging genomic signal typically involves a small number of SNPs and QTs, both CCA and RRR have their sparse versions. These sparse models are designed to better fit the imaging genomic study as they yield sparse patterns for easy interpretation. Example studies applying these methods in imaging genomic applications include Chi et al. (2013), Wan et al. (2011), and Vounou et al. (2012). Prior knowledge has also been examined in bi-multivariate models recently, and its beneficial role has also been reported in many studies (Yan et al., 2014a; Du et al., 2014; Lin et al., 2014).

Enrichment analysis has been widely studied in gene expression data analysis, and has recently been modified to analyze GWAS data to extract biological insights based on functional annotation and pathway databases

(Ramanan et al., 2012; Younesi and Hofmann-Apitius, 2013). Recently, it has been extended to the imaging genomics domain, to discover high-level associations based on prior knowledge, including meaningful gene sets (GSs) and brain circuits (BCs), which typically contain multiple genes and multiple QTs, respectively (Yao et al., 2015). By jointly considering the complex relationships between the interlinked genetic markers and correlated imaging phenotypes, imaging genomic enrichment analysis (IGEA) provides additional power for extracting biological insights on neurogenomic associations at a systems biology level.

In this chapter, we will summarize the widely used traditional and state-of-the-art statistical and machine learning methods, ranging from univariate (Section 14.2.1), multilocus (Section 14.2.2), bi-multivariate (Section 14.2.3) models to recent enrichment models (Section 14.3), and discuss their applications in brain imaging genomics.

14.2 MINING IMAGING GENOMIC ASSOCIATIONS VIA REGRESSION OR CORRELATION ANALYSIS

Genotype-phenotype associations are typically explored in three ways based on different assumptions: (1) one genotype affects one or more phenotypes independently, (2) multiple genotypes jointly affect one phenotype, and (3) multiple genotypes jointly affect a BC with multiple phenotypes. The first assumption is the simplest and is the most common practice for mining imaging genomic associations. Generally, traditional correlation or linear regression models are sufficient for the analysis, and mostly coupled with hypothesis testing. Regression models and bi-multivariate association models are appropriate methods for Assumptions 2 and 3, respectively, and have already been widely applied. In this section, we will discuss these statistical and machine learning models in detail and summarize their applications in the field of brain imaging genomics.

14.2.1 SINGLE-LOCUS ANALYSIS

Early brain imaging genomic studies mostly focused on exploring paired relationships between a single SNP and a single imaging QT. A simple linear regression model (Eq. 14.1), for example, that implemented in PLINK (Purcell et al., 2007), is usually used to examine the additive effect of each single SNP on each imaging QT. Example studies include Potkin et al. (2009) and Shen et al. (2010). Let X_{cov} indicate the covariates, such as age and gender, X_{S} be the genotype of the SNP to be examined, and y be the imaging QT as the response to be associated with the SNP. SNPs are typically coded as 0, 1, or 2, the number of minor alleles at a particular genome position.

$$y = X_{\text{cov}} w_{\text{cov}} + X_{\text{S}} w_{\text{S}} + \epsilon. \tag{14.1}$$

This additive model is generally applied for allelic association tests that only examine the association between each SNP and each imaging QT. The dominant, recessive, or their combination can also be investigated in the same way. A common practice of GWAS is to examine the allelic effects only, as it has the reasonable power to detect both dominant and recessive effects. But there are also plenty of studies examining all of them individually, the result of which will be subject to a further multiple comparison correction. In order to capture more phenotypic variance, the interaction effects between SNPs and environmental factors (ie, $X_s X_{env}$) or just between SNPs (ie, $X_{s1} X_{s2}$) have also been investigated in some studies. Another similar approach is the generalized linear model (GLM), a procedure that uses ANOVA, which considers allelic, dominant, and recessive effect measures as a categorical predictor variable.

Typically, linear regression is performed together with hypothesis testing, in which the regression coefficients and corresponding statistical value will be obtained simultaneously. For example, assuming the normal distribution of the error term ϵ, a t-statistic of w_s can be calculated through:

$$t = \frac{w_s}{\mathrm{SE}}, \tag{14.2}$$

where SE is the standard deviation of w_s, ie, $\mathrm{SE} = \frac{\sqrt{\frac{1}{n-2}\Sigma(y_i - \bar{y})^2}}{\sqrt{\Sigma(w_{si} - \overline{w_s})^2}}$. The p-value generated from the statistic evaluates the probability of observing an equal or greater statistic by chance. The smaller the p-value is, the more significant the genotype-phenotype relationship is considered to be. Fig. 14.1 shows an example of single-locus analysis across the entire genome.

14.2.1.1 Multiple comparison correction

In large univariate tests of all the pairwise SNP-QT associations, the p-value obtained from each single test is generally further corrected using various strategies. Bonferroni correction is the simplest one, which works by multiplying the p-value by the test number (ie, the number of SNPs $\times$ the number of QTs). However, this method hypothesizes that all experiments are independent and thus is considered to be overly conservative. It is more common in practical applications to combine it with the estimated number of independent tests (Bush and Moore, 2012; Yan et al., 2015b). Also, the false positive rate (FDR) introduced by Benjamini and Hochberg (1995) is also quite favored due to its less stringent threshold. Permutation, though potentially time-consuming, is another approach to acquire the significance with dependent genotypes, in which imaging QTs are randomly reassigned to other subjects to break the underlying SNP-QT relationships. By repeating this permutation procedure many times, it helps simulate a null distribution of test statistics for w_s. An empirical p-value can then be generated as the proportion of those random statistics equal to or greater than the original one.

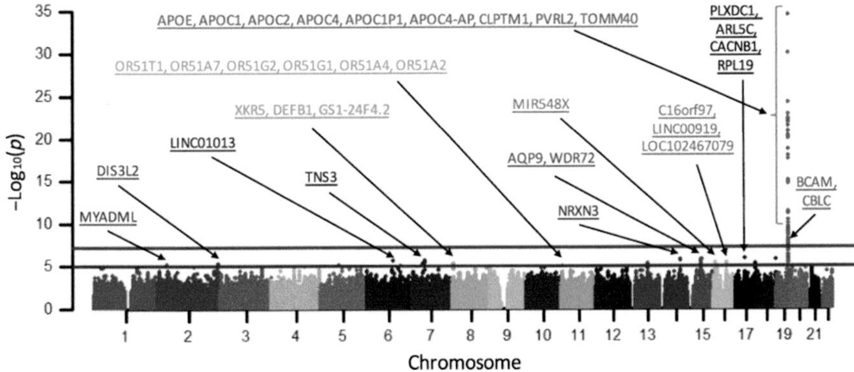

FIG. 14.1

Manhattan plot of a genome-wide association study (GWAS) of the average amyloid burden measure in right precuneus, using an imaging genomic data set from the Alzheimer's Disease Neuroimaging Initiative (ADNI); see Yao et al. (2015) for details of the analysis. The x-axis represents the chromosomes and the y-axis represents $-\log_{10}(p)$, where p is the SNP-based significance. Selected top hits are labeled with their corresponding gene names.

14.2.2 MULTILOCUS EFFECTS

Despite the great success of single-locus analysis, which has led to the identification of hundreds of candidate SNPs conferring the genetic contribution of complex disorders and diseases, the statistical model is relatively simple and not designed for identifying multilocus effects on imaging QTs. Due to the correlated structure of genetic data, a straightforward linear regression is rarely used for multilocus effect detection and thus is not included here. Note that most machine learning models are not capable of handling either genome-wide genotypes or brain-wide voxel phenotypes. Most existing multilocus imaging genomic studies using machine learning approaches are targeted analyses.

Principal component regression (PCR) is one of those early attempts to address the correlated structure. Unlike traditional regression techniques which operate directly on the original data, in PCR the principal components (PCs) are firstly extracted and linear regression is performed on top of them rather than the original features. Typically only a subset of all the PCs is selected for regression, which has higher variances represented by the large eigenvalues. With genome-wide SNP data, these top PCs extracted in PCR possibly represent the population structure helpful for population stratification (Liu et al., 2010a; Price et al., 2006). The primary advantage of PCR is to tackle the collinearity problem among predictors, since components with low variances are normally excluded in the final analysis. Also, by including only a subset of all the PCs, PCR can result in significant dimension reduction and therefore is very desirable in high-dimensional settings.

The linear mixed model (LMM) is an alternative method designed for purpose of correlated structure. The mixed model refers to the combination of both fixed and random effects. It has a similar form to linear regression:

$$Y = Xw + U\gamma + \epsilon = Xw + \epsilon^*, \tag{14.3}$$
$$\gamma \sim N(0, G), \epsilon \sim N(0, I\sigma^2),$$
$$\epsilon^* = XW + \epsilon \sim N(0, UGU^T + I\sigma^2),$$

where X are the fixed factors, eg, age and gender; w are coefficients to be estimated, which indicate the fixed effects; U are the random factors; and γ indicates the corresponding random effects. Instead of estimating γ as a fixed vector in linear regression, LMM assumes an underlying distribution for γ. Since the random effect is modeled as the deviation of a fixed effect, the mean value of γ is set to zero and is usually assumed to have the distribution $\beta \sim N(0, G)$. Here, G works as a prior that explicitly models the variance-covariance structure of random factors, and is the key of LMM for handling correlated genotype data. Note that in LMM only the coefficients of fixed factors are shared in all individuals, representing the population level significance, and those of random factors are different for each subject, indicating individual-level variations. GCTA (Yang et al., 2011, 2013) is one example tool implementing this model for QT heritability analysis. It has been successfully applied to various phenotypes in many disease studies, such as schizophrenia (Ripke et al., 2013), Alzheimer's disease (Ridge et al., 2013), and Parkinson's disease (Do et al., 2011).

Another group of regression models is known to excel in sparsity, which conforms to our assumption that only a few rather than a large portion of SNPs are responsible for the changes of a specific phenotype. The least absolute shrinkage and selection operator (LASSO or lasso) (Eq. 14.4) is a typical example, which achieves the sparsity goal by penalizing the sum of absolute values of all coefficients, also called ℓ_1-norm. This penalized regression problem can be easily solved using a soft-thresholding technique as proposed in Tibshirani (1996), or through an iterative procedure with smooth approximation of ℓ_1-norm (Lee et al., 2006; Schmidt et al., 2007)

$$Y = Xw + \gamma ||w||_1. \tag{14.4}$$

While lasso has shown excellent performances in many studies to identify a small set of SNPs associated with imaging QTs, the sparsity constraint, on the other hand, also leads to competition between correlated SNPs and ultimately yields possibly unstable sparse patterns (Bach, 2008). To address this random selection problem, the elastic net (Zou and Hastie, 2005) was proposed via introducing an extra ℓ_2 penalty term:

$$Y = Xw + \gamma_1 ||w||_1 + \gamma_2 ||w||_2. \tag{14.5}$$

With these two penalty terms, it is capable of seeking a balance point between sparsity and grouping correlated SNPs. Wan et al. (2011) used this model to explore

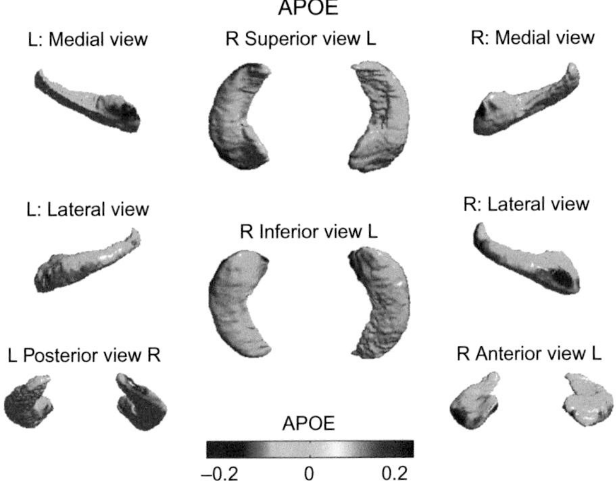

FIG. 14.2

Hippocampal surface map of genetic effects of the *APOE* SNP rs429358 estimated by an elastic net; see Wan et al. (2011) for details.

the genetic risk factors affecting the hippocampal surface and found that *APOE* and *TOMM40* were associated with hippocampal surface changes in anterior and middle regions (see Fig. 14.2 for the *APOE* result).

Other similar approaches include group lasso (Eq. 14.6), fused lasso (Eq. 14.7) and graphical fused lasso, all of which explicitly model the data structure as a prior in the penalty term so that grouped or correlated genetic factors can obtain similar weights in the training procedure:

$$Y = Xw + \gamma \sum_{i=1}^{g} \sqrt{\sum_{j \in G(i)} w_j^2}, \tag{14.6}$$

$$Y = Xw + \gamma \sum_{i<j} |w_i - w_j|. \tag{14.7}$$

Graphical fused lasso is simply an extension of fused lasso that makes constraints only on paired nodes that are connected in a prior graph. Silver et al. (2012b) discussed the group lasso with overlapping problem so that pathway information can be incorporated as a prior for SNP data. Dummy variables were proposed to be added into the weight matrix so that it reverts to a normal group lasso problem. Later, in another work (Silver et al., 2012a), they examined the same term in a multivariate model, which successfully identified several causal pathways associated with longitudinal structural change in the brains of patients with Alzheimer's disease.

Note that in these models estimating the multilocus effect on multiple imaging QTs is equivalent to performing separate analysis for each of them, and does not make use of the fact that multiple imaging QTs are usually highly correlated. This is possibly attributed to partial common underlying genetic architecture. Next, we will introduce several state-of-the-art methods utilizing advanced multivariate regression models to tackle this problem.

L1/L2, also referred to as $L_{2,1}$ (Eq. 14.8), is a typical multivariate model commonly used for identifying the multilocus effects shared across multiple related phenotypes:

$$Y = XW + \gamma||W||_{2,1} = XW + \gamma \sum_i \sqrt{\sum_j w_{ij}^2}. \tag{14.8}$$

Like group lasso, it also takes advantage of ℓ_1-norm and ℓ_2-norm but distributes the constraints differently. Instead of imposing the ℓ_2-norm constraints within prior groups, here ℓ_2-norm is firstly applied to each row of W, which is the coefficients of one SNP across all QTs, so that they can be pulled together for similarity. The ℓ_1-norm is then applied across rows of W on top of their ℓ_2-norm to guarantee the global sparse patterns of the ultimate result. In this way, it is capable of capturing the multilocus patterns shared across correlated imaging QTs. Inspired by group lasso and this L1/L2 model, Wang et al. (2012a) later proposed a more advanced model, group-sparse multi-task regression and feature selection (G-SMuRFS), in an attempt to find a tradeoff between global sparsity and group level sparsity. As shown in Eq. (14.9),

$$Y = XW + \gamma_1||W||_{G2,1} + \gamma_2||W||_{2,1}. \tag{14.9}$$

G-SMuRFS has an extra penalty term $||W||_{G2,1} = \sum_{k=1}^{g} \sqrt{\sum_{i \in G(k)} \sum_j w_{ij}^2}$, which can explicitly model the genetic linkage disequilibrium (LD) structure so that SNPs within an LD block tend to be extracted together. On the other hand, $||W||_{2,1}$ helps guarantee the global sparsity so that insignificant SNPs within a generally significant LD block can be removed. Their group also developed another joint classification and regression model with the same penalty terms (Wang et al., 2012b). Both models have been proved to be successful in an AD study, in which a compact set of correlated SNPs was identified with much less root mean square error.

14.2.3 MULTI-SNP-MULTI-QT ASSOCIATIONS

In the brain imaging genomics area, we have both the genetics data and the imaging data at hand. The importance is that they are collected from the same population. Therefore there are essential demands to identify the associations between two multidimensional data, that is, the imaging data and genetics data that come from the same population. Regression techniques usually address problems with only one or very few responses, indicating that they identify the associations between multiple predictors and one or few responses. Therefore they cannot fully mine the knowledge behind the complex brain imaging genomic data. Although we could regress multiple times for different responses using the same predictors, we still lose the relationship among those responses, especially for the brain structure which is of great interest and importance.

Bi-multivariate correlation models may be an appropriate fit to address this issue. We are aware of many models that have employed the bi-multivariate strategy to bioinformatics studies. Among those techniques, CCA (Hotelling, 1935) is a classical statistical technique that can find the associations between two sets of multidimensional variables. The CCA model is formally defined as follows:

$$\max_{u,v} u^{\mathrm{T}} X^{\mathrm{T}} Y v$$

subject to $u^{\mathrm{T}}X^{\mathrm{T}}Xu = 1$ and $v^{\mathrm{T}}Y^{\mathrm{T}}Yv = 1$, where u and v are canonical loadings (or canonical weights) for genetic markers and imaging QTs, respectively, in this chapter.

The standard CCA does not perform feature selection. However, feature selection is an important concern in the brain imaging genomics area, since the data is usually of quite high dimension. A model lacking the ability of feature selection may be inadequate. In order to make use of the advantage of CCA and overcome its weakness, Witten et al. (2009) introduce the sparse CCA (SCCA) using a regularization technique to induce sparsity into the model. They propose the penalized SCCA method using the ℓ_1-norm (lasso) to penalize two canonical loadings. For ease of description, here we only present the penalty function used by the SCCA model. The ℓ_1-norm penalty is defined as

$$\Omega_{\mathrm{Lasso}}(u) = \sum |u_i| \leq c_1,$$

where c_1 is a parameter which can control the sparsity of the corresponding canonical loading. The lasso penalty can result in a larger number of covariates being zero if given suitable parameters. From the point of view of interpretation, this sparsity is desirable because it only captures the most important covariates among a huge amount (Zou and Hastie, 2005).

The SNPs have been known to be correlated other than be perfectly independent from each other. The LD, that is, nonrandom association of alleles at different loci, are common within the gene. This is also the case for the imaging data. The covariates within an imaging modality are not independent, implying that they are correlated. L1-SCCA (lasso-based SCCA) mainly focuses on feature selection, which is insufficient for the association study in brain imaging genomics. In the situation where those covariates are dependent, the L1-SCCA tends to randomly select one covariate from them, and discard the rest. This incurs an unstable solution and makes the interpretation hard, especially on the smoothed data (Lu, 2010). Therefore the models that can capture the entire group of covariates rather than one of them are of great interest.

On the contrary, we can make use of the unstable selection of L1-SCCA. Yan et al. (2015a) propose the bootstrap L1-SCCA (BoSCCA) by repeatedly running the L1-SCCA on different data partitions. Since L1-SCCA randomly selects one of the correlated covariates, it has a high probability of selecting all the correlated covariates as long as the L1-SCCA is run for a sufficient number of times. The BoSCCA is validated on the SNPs of the top 22 AD risk genes and the amyloid imaging measures. The top three components with the highest probability are the frontal medial orbital gyrus, anterior cingulate gyrus, and posterior cingulate gyrus. For the genotypes, *APOE* gene and its neighbors *APOC1* and *TOMM40* are selected with high probability. All of the genotypes and imaging phenotypes are associated with AD.

In order to accommodate other types of structures in the data, several structured SCCA methods (Du et al., 2014, 2015a; Yan et al., 2014a; Witten et al., 2009) have been used. We classify these SCCA methods into two types according to their distinct regularization terms. One type used the group lasso penalty, and the other used the graphical fused lasso penalty to conduct feature selection and feature grouping. The first type of structured SCCA methods, that is, the group lasso-based SCCA, required prior knowledge to define the group structure. The group lasso penalty is defined as follows:

$$\Omega_{\mathrm{GL}}(u) = \sum_{i=1}^{g} \sqrt{\sum_{j \in G_i} u_j^2} \leq c_1$$

This penalty function has two aspects: (1) the intragroup ℓ_2-norm constraint, which enables covariates in the same group to have equal weights; (2) the intergroup ℓ_1-norm constraint, which assures sparsity in terms of groups. That is, for a group of covariates, all of them will be selected or discarded together; and for covariates of different groups, they are independent of each other.

Du et al. (2014) proposed structure-aware SCCA (S2CCA) using group lasso, and incorporated both the covariance matrix information and the prior knowledge information to recover group-level bi-multivariate associations. The authors confirmed that a strong association exists between the *APOE* gene and a VBM imaging marker (hippocampus).

The second kind of structured SCCA methods use graph/network-guided fused lasso penalties. The SCCA methods above will lose efficacy because they are prior knowledge dependent. These methods can perform well on any given prior knowledge. If the prior knowledge is not available, these methods can also work by using the sample correlation to define the graph/network constraint. Apart from the L1-SCCA, Witten et al. (2009) also introduced a smoothed penalized SCCA which utilized the fused lasso as penalty, that is,

$$\Omega_{\mathrm{FL}}(u) = \sum |u_i| + \lambda \sum_{i \neq j} |u_i - u_j| \leq c_1$$

Here u_i and u_j are two neighboring covariates after ordering the covariates. The fused lasso-based SCCA (FL-SCCA) can be viewed as a degenerate network-guided method since it imposes pairwise constraint to ensure smoothness. However, it requires these covariates to be ordered before running it, which limits its performance greatly. This is because it loses the relationship for those covariates that are not adjacent. Du et al. (2015a) have tested the FL-SCCA on the *APOE* gene and the amyloid burden measure of the AD database, and its results are acceptable.

Yan et al. (2014a) find that the brain structural constrait could be made up of a network instead of several nonoverlapping groups. However, the S2CCA can only identify the nonoverlapping group structure, indicating that it cannot deal with a more complex group or graph/network structure. Yan et al. introduce the knowledge-guided SCCA (KG-SCCA) where the network-guided penalty is defined as

$$\Omega_{\mathrm{NG}}(u) = \sum |u_i| + \lambda \sum_{(i,j)\in E, i<j} \tau(w_{ij})||u_i - \mathrm{sign}(w_{ij})u_j||_2 \leq c_1$$

where u_i and u_j are two connected nodes on a prior network; $\mathrm{sign}(w_{ij})$ indicates the sign of their correlation and $\tau(w_{ij})$ stands for the strength of their connection. This penalty is an extension to the fused lasso; however, it does not require the covariates to be ordered first. After using the AD-related genetic markers and imaging markers, they validate that the SNPs in the *APOE* gene are strongly associated with the amyloid burden measure.

Du et al. (2015b) proposed another network-guided SCCA which utilized the GraphNet regularization term of Grosenick et al. (2013). The GraphNet-based SCCA (GN-SCCA) employs the Laplacian matrix-based GraphNet, which can be defined as

$$\Omega_{\mathrm{GraphNet}}(u) = \sum |u_i| + \lambda \sum_{(i,j)\in E, i<j} (u_i - u_j)^2 \leq c_1$$

GraphNet is an extension to the elastic net penalty, which has wider applications. The GN-SCCA can identify structure associations even if we have little prior knowledge or we do not have it. Based on the correlation matrix of either modality, GN-SCCA can incorporate the correlation matrices and use them to guide the structure identification. They apply the SCCA on the data set with respect to the AD-related genetic markers and the human brain amyloid burden. The result confirms that the marker *APOE* gene has a strong relationship between the frontal measurements in the brain.

Recently, Du et al. (2015a) extend the network-guided SCCA to a more robust one. Their SCCA utilizes the graph OSCAR penalty (Yang et al., 2012):

$$\Omega_{\mathrm{GOSCAR}}(u) = \sum |u_i| + \lambda \sum_{(i,j)\in E, i\neq j} \max(|u_i|, |u_j|) \leq c_1$$

The GOSCAR penalty employs the pairwise ℓ_∞-norm to constrain every pair of covariates, and expects them to be equal if they are highly correlated or dissimilar if the correlation between them is quite low. The importance of this penalty is that it is sign independent, and thus it only focuses on whether $|u_i|$ equals $|u_j|$. Therefore the GOSC-SCCA will not suffer from the sign directionality issue and is more robust than those SCCAs above. They also assure an association between the *APOE* gene and the brain amyloid burden measurement.

Another way to handle the correlation between both multiple genetic markers and multiple imaging markers is to take advantage of the RRR technique. The RRR can be viewed as an extension to traditional regression techniques, which is capable of predicting multiple responses simultaneously. In traditional RRR, for the weight matrix $W \in \mathbb{R}^{p*q}$ satisfying rank $(W) \leq \min(p, q)$, it can be written as a product of two full rank r matrices $W = BA^{\mathrm{T}}$, where $B \in \mathbb{R}^{p*r}$ and $A \in \mathbb{R}^{q*r}$. The RRR can then be reformulated to solve the minimization problem:

$$\min_{A,B} \mathrm{Tr}\{(Y - XBA^{\mathrm{T}})\Gamma(Y - XBA^{\mathrm{T}})^{\mathrm{T}}\}$$

This factorization of W not only helps decrease the parameters to be estimated, but also enables the sparsity constraints on both SNPs and imaging QTs, respectively. Sparse reduced rank regression (sRRR) (Vounou et al., 2010) is a method in which the ℓ_1-norm is utilized to penalize the coefficients A and B, so that sparse variable selection can be achieved simultaneously on both sides. By assuming that $X^{\mathrm{T}}X = I$ and $\Gamma = I$, the minimization problem of sRRR can be formulated as

$$\min_{A,B} -2\mathrm{Tr}\{A^{\mathrm{T}}Y^{\mathrm{T}}XB\} + \mathrm{Tr}\{A^{\mathrm{T}}AB^{\mathrm{T}}B\} + \gamma_1||A||_1 + \gamma_2||B||_1$$

Like the SCCA, the sRRR is also a biconvex problem, and thus can be solved by an iterative procedure by alternately fixing one coefficient and updating another using soft-thresholding in each step (Tibshirani, 1996). The sRRR method uses multiple phenotypes as responses and multiple genotypes as predictors. They successfully detect the most important variables in both the genetic and imaging domains.

The generalized low rank regression (GLRR) model proposed recently by Zhu et al. (2014) is another approach making use of the matrix rank to induce sparsity, but under Bayesian infrastructure. Although both SCCA and sRRR can handle multiple genotypes and imaging phenotypes, they are different in terms of recovering the association. SCCA works symmetrically and there is no difference between the genotypes and the imaging phenotypes; however, the sRRR performs more like traditional regression and we cannot swap the genotypes and the imaging phenotypes to obtain the same findings.

14.3 MINING HIGHER LEVEL IMAGING GENOMIC ASSOCIATIONS VIA SET-BASED ANALYSIS

Higher level association analysis has been demonstrated that it can yield biologically meaningful findings by integrating prior knowledge (eg, pathway information) into a set of genetic findings. The prior knowledge could be from gene ontology (GO), functional annotation databases and pathway analysis systems. Various gene-based association tests have been proposed to construct genetic candidates based on phenotype-associated variants from GWAS. Recently, it has been applied to brain imaging genomic applications, by adopting prior knowledge from both genetic and imaging domains, to explore the Gene Set (GS) effects on specific imaging QT, or a predefined BC containing multiple imaging QTs.

In this section, we discuss two classes of higher level imaging genomic association tests: context-based test and context-free test, categorized by whether they use background information or not; and further discuss a two-dimensional enrichment analysis paradigm that jointly explores meaningful GS-BC modules.

14.3.1 CONTEXT-BASED TEST

Given a set of candidate genes, a context-based method tests if there is a trait-association difference between this GS and a random GS of the same size. By comparing the proportion of trait-associated signals from candidate and random GSs, a context-based test tells how important the set of candidate genes is compared to random GSs. Because of this, methods applying the context-based test require the significance results (eg, p-values) of not only the candidate genes but also all the other genes in the relevant context. The most common test is the pathway enrichment test, which can be classified into two types: over-representation analysis and rank-based analysis. Below we briefly discuss both types, including their applications in brain imaging genomics.

14.3.1.1 Over-representation analysis

In the over-representation test, a threshold is used to determine the set of candidate genes L (ie, genes with QT-associated p-values exceeding the threshold), such that all genes in L are significantly associated with specific imaging QT. We test if a predefined GS S (eg, a pathway) is over-represented in L. This can be formulated as an independence test problem. The most commonly used over-representation tests are based on hypergeometric (Fisher's exact test), binomial, and/or χ^2 distribution (Draghici et al., 2003; Goeman and Buhlmann, 2007).

Assume we have all N genes in the analysis. Of these, $n = |L|$ genes in the set L are significant ones (ie, p-value exceeds a threshold), $m = |T|$ genes are from a given GS T of interest (eg, a pathway), and k out of n significant genes are from T. Using a hypergeometric test (Fisher's exact test) for illustration, the over-represented p-value of having k or more genes from T in L can be calculated by summing the probabilities of a random set of n genes having $k, k+1, \ldots, n$ genes from T:

$$p = \Pr\left(|L \cap T| \geq k\right) = \sum_{i \geq k} \left(\binom{m}{i} \binom{N-m}{n-i} \Big/ \binom{N}{n} \right). \tag{14.10}$$

The hypergeometric distribution is rather difficult to calculate when the number of genes involved is large. However, it tends to be a binomial distribution when N is large.

There are a number of studies that have performed over-representation analyses and identified meaningful functional GSs with significant associations to the relevant phenotype or disease. For example, Perez-Palma et al. (2014) applied hypergeometric test to construct a network-based pathway enrichment using meta-analysis statistics of GWAS, and identified the over-representation of the glutamate signaling pathway in Alzheimer's disease.

There are several issues in the over-representation test. First, the enrichment statistic is based on an arbitrarily selected threshold (eg, $p \leq 0.05$) that is used for determining the significant candidate set. Interesting signals might be missed when there are many modest trait-associated genes that do not pass the threshold. Second, the over-representation tests consider only the number of significant genes but ignores their strength of associations. Third, the genes are not independent from the others. Over-representation tests treat each gene as an independent unit, and ignore their correlation structure, which may yield a biased enrichment estimation.

14.3.1.2 Rank-based enrichment analysis

To overcome the limitations of over-representation analysis, rank-based enrichment analysis has been developed to include all gene-level p-values in the analysis. One widely used method is Gene Set enrichment analysis (GSEA), which was originally devised for gene expression data analyses and then extended to GWAS analyses. The GSEA tests whether genes from a predefined set S (eg, a pathway) are distributed in the top (or bottom) of a ranked gene list L ordered by gene-level p-values, and thus is significantly associated with the GWAS trait. The implementation of the GSEA algorithm is briefly described below.

Given a predefined GS S and a sorted list of genes L with gene-level statistics (eg, gene-level p-values from the GWAS of a specific imaging trait), an enrichment score (ES) of S is calculated using a Kolmogorov-Smirnov (K-S) like statistic with weight 1. That is, by walking down the list L, a running-sum statistic is increased when encountering a gene in S, and is decreased when encountering a gene not in S. The ES is then provided by the maximum deviation from zero of the running sum. Statistical significance of ES (empirical p-value) is then estimated by performing phenotype-based permutation.

In brain imaging genomics, original GSEA and its modifications have been widely applied in pathway or network enrichment analysis. Ramanan et al. (2012) performed genome-wide pathway analysis of memory impairment in the Alzheimer's Disease Neuroimaging Initiative (ADNI) cohort using GSA-SNP, a GSEA-based GS analysis software. They identified 27 pathways with significant ESs against the composite memory score, of which most are involved in memory consolidation. One of the pathways displays colocalized expression in normal brain tissue along with known AD risk genes. Younesi and Hofmann-Apitius (2013), adopted GSEA to validate the functional association with a brain region-specific protein-protein interaction subnetwork extracted by text mining.

Rank-based methods overcome some limitations of several over-representation tests by using all gene-level statistics, without requiring a user-specified threshold. Furthermore, a phenotype-based permutation can keep the correlation structure among genes, and thus provide a more reasonable assessment of significance than permuting genes. However, rank-based methods also have several limitations. First, like over-representation methods, rank-based tests consider pathways independently, which often overlap with one another. Because of this, a pathway may be significantly enriched due to the common genes it shares with a real enriched pathway. Second, rank-based methods take into account the ranks of genes but ignore the strength of associations between genes and phenotypes. Some modifications have been proposed to improve this problem by adding weights to ranked genes based on their association strengths (Mooney and Wilmot, 2015).

14.3.2 CONTEXT-FREE TEST

In contrast to context-based tests, context-free tests use another strategy to formulate a null hypothesis. They have been applied in gene-based association analysis. Gene-based association analysis tests the association between a gene and a phenotype, based on the statistics of SNPs within this gene and without needing information from outside of the gene. Below we briefly discuss two groups of gene-based association analyses.

Genome-wide analysis has been employed in brain imaging genomics to identify individual susceptibility loci of neurodegenerative diseases, which however explain only a modest proportion of the total variance in liability to imaging trait or disease. Gene-based analysis jointly considers all variants within a gene to obtain a single statistic representing the association significance of the entire gene, and thus has three advantages. First, the gene is the functional unit of the genome and highly consistent across individuals. Second, compared to SNP-level association tests, gene-based analysis reduces the number of multiple corrections (from $\sim$0.5 million to 20,000–30,000). Third, the findings from gene-based analysis can be directly adopted by further analysis such as protein-protein interaction, pathway enrichment analysis, and so on.

A number of gene-based association tests have been proposed and can be categorized into two groups. One group incorporates the full set of SNPs within the gene, to test their association with a specified phenotype. The other tests the null hypothesis that no SNPs within the given gene show association with phenotype, whereas the alternative is that at least one SNP in the gene is associated with the phenotype. In other words, the first group simultaneously consider all SNPs and the second one focuses on the best SNP. Typical methods of the first group include regression, PLINK set-test (Purcell et al., 2007), Fisher's combination (Curtis et al., 2008), and VEGAS test (Liu et al., 2010b). The second group includes the VEGAS-Max test (Liu et al., 2010b), GATES (LI et al., 2011), and so on.

Linear regression (for QTs) and logistic regression (for binary traits) are straightforward methods to evaluate the overall association between a gene and a trait. In regression, all SNPs from the gene are treated as predictors simultaneously, with the phenotype as response. The statistical power of regression might be decreased when a lot of SNPs are involved, due to high degrees of freedom. Various strategies have been proposed to reduce the dimensionality by collapsing correlated SNPs, like by Fourier transformation (Wang and Elston, 2007), principal component analysis (PCA) (Gauderman et al., 2007; Wang and Abbott, 2008), and clustering analysis (Buil et al., 2009). Raw genotyping and phenotyping data are required for all regression methods.

Set-based tests implemented in PLINK use an LD structure within the SNPs to select a subset of representative ones. Association significance of the gene is then calculated as the mean statistic of the subset. The empirical p-value is obtained by repeating the permutation procedure many times. It should be noted that the LD structure among genotype variants should be kept in each permutation, such that the phenotype label can be changed. PLINK set-based tests are time consuming when applied to genome-wide results due to the permutations.

Another strategy is Fisher's combination method that combines all SNP-level p-values using the following formula:

$$X^2 \sim -2\sum_{i=1}^{n} \ln(p_i). \tag{14.11}$$

From Fisher's combination, the combination of n independent tests under the same null hypothesis will follow a χ^2 distribution with $2n$ degrees of freedom (n is the number of the SNP p-value). However, in gene-based analysis, the SNP association tests are not independent due to their LD. Permutation is still required to gain empirical significance, which takes computational intensity. Fisher's combination test only requires SNP-level p-values but not raw genotyping and phenotyping data.

Versatile gene-based association study (VEGAS), developed by Liu et al. (2010b), replaces permutation with simulation based on Fisher's combination, to reduce computation time. In VEGAS, all SNP p-values from a given gene are firstly transformed to upper-tail χ^2 statistics with one degree of freedom (*df*). The gene-based statistic is then calculated as the sum of all (or a predefined subset) of these χ^2 1 *df* statistics according to Fisher's combination. As mentioned before, the statistic will follow a χ^2 distribution with n *df* if all SNPs are independent. However, this is unlikely to be the case in real genetic data.

Instead of performing permutations many times to gain an empirical p-value, VEGAS adopts the Monte Carlo approach that makes use of simulations from the multivariate normal distribution to do this. The details are as follows. Given a gene with n SNPs, Σ is an $n \times n$ matrix of pairwise LD (r). An n-element multivariate normally distributed vector with mean 0 and variance Σ is simulated in the next two steps: (1) generate n independent, standard, normally distributed random variables, (2) multiply n variables from (1) by the Cholesky decomposition matrix of Σ (the lower triangular matrix C such that $CC^{\mathrm{T}} = \Sigma$). The new random vector $Z = (z_1, z_2, \ldots, z_n)$ will have a multivariate normal distribution $Z \sim N_n(0, \Sigma)$. It then transforms Z into a vector of correlated χ^2 1 *df* variables, $Q = (q_1, q_2, \ldots, q_n), q_i = z_i^2$. Simulated gene-based test statistics are then calculated as $\sum_{i=1}^{n} q_i$, which will have the same approximate distribution as the observed statistics. The empirical *p*-value can be calculated by repeating this simulation procedure many times.

GATES is one of the most common methods in the second group. Instead of considering all SNPs within the gene, GATES extends the Simes test based on the null hypothesis that no SNPs within the gene are associated with the disease. The gene-based *p*-value is calculated using

$$p_{\text{value}} = m_{\text{e}} * \text{Min}\left(\frac{p_{(j)}}{m_{\text{e}(j)}}\right), \tag{14.12}$$

where m_{e} is the effective number of independent *p*-values among the *m* SNPs and $m_{\text{e}(j)}$ is the effective number of independent *p*-values among the top *j* SNPs.

In VEGAS, one can also only use the best SNP (ie, the most significant SNP) within the gene to calculate gene-level statistics, called VEGAS-Max and belonging to the second gene-based analysis group.

In imaging genomics, GWAS have been widely performed on imaging QTs and identified a number of risk genetic variants (Saykin et al., 2010; Shen et al., 2010). While both groups of gene-based analysis have been widely used to measure the associations between gene and imaging, Perez-Palma et al. (2014) employed GATES to calculate gene-level brain imaging associations extracted from meta-analysis. Liang et al. (2015) used VEGAS to assign SNP *p*-values to respective genes in their network-based high-level imaging genomic association study, where 200 genes were identified to be associated with all 14 subcortical imaging measures, including left hippocampus and right amygdala, which were previously demonstrated to be related to Alzheimer's disease.

14.3.3 TWO-DIMENSIONAL IMAGING GENOMIC ENRICHMENT ANALYSIS

Brain imaging genomics is an emerging field that studies how genetic variation influences brain structure and function. In the genetics domain, GWAS have been performed to identify genetic markers such as SNPs that are associated with brain imaging QTs (Saykin et al., 2010; Shen et al., 2010). From Sections 14.3.1 and 14.3.2, using biological pathways and networks as prior knowledge, enrichment analysis has been performed to discover pathways or network modules enriched by GWAS findings to enhance statistical power and aid biological interpretation.

However, present analytic methods used in GWAS of imaging QTs typically ignore either the interrelated structure between genes or the correlation between imaging QTs, and are insufficient to provide insight into the mechanisms of complex diseases that could involve multiple genes and multiple QTs.

Recently, IGEA (Yao et al., 2015), a new enrichment analysis paradigm, has been proposed to jointly consider sets of interest (ie, GS and BC) in both genetic and imaging domains and examine whether any given GS-BC module is enriched in a list of gene-QT findings. IGEA aims to discover higher level associations between meaningful GS and BC, which typically include multiple genes and multiple QTs. Fig. 14.3 shows a brief overview of the IGEA framework.

In IGEA, there are many types of prior knowledge that can be used to determine meaningful GS and BC entities. In the genomic domain, the prior knowledge could be based on GO or functional annotation databases; in the imaging domain, the prior knowledge could be neuroanatomic ontology or brain databases. Gene-QT association can be identified from brain imaging GWAS results, or existing imaging genomic findings. Yao et al. (2015) in their study used brain-wide expression data from the Allen Human Brain Atlas (AHBA, Allen Institute for Brain Science, Seattle, WA; available from http://www.brain-map.org/) to extract GS and BC modules such that genes within a GS share similar expression profiles and so do ROIs

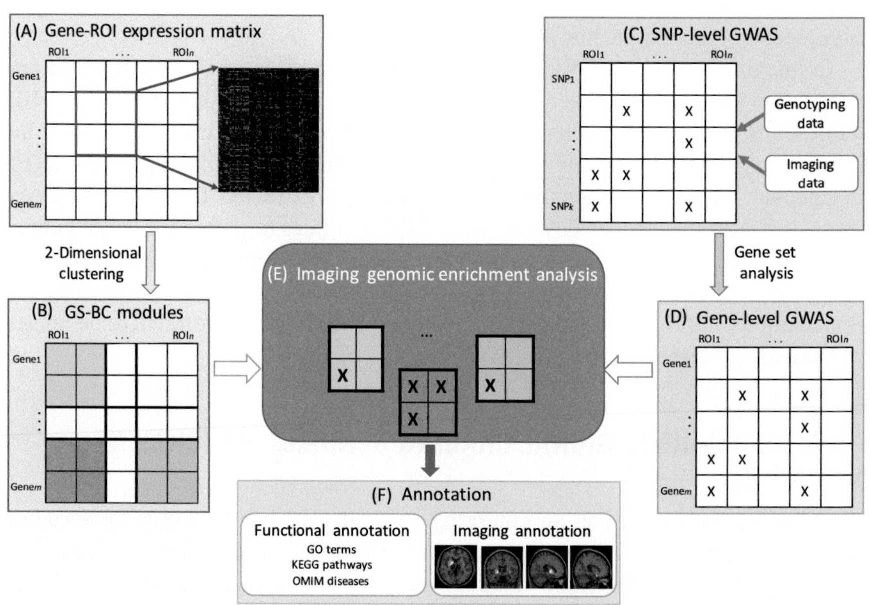

FIG. 14.3

Overview of the IGEA framework proposed in Yao et al. (2015). (A) Construct gene-ROI expression matrix from the Allen Human Brain Atlas (AHBA data). (B) Construct GS-BC modules by performing 2D hierarchical clustering, and then filter out nonsignificant 2D clusters. (C) Perform SNP-level GWAS of brain-wide imaging measures. (D) Map SNP-level GWAS findings to gene-based results. (E) Perform IGEA by mapping gene-based GWAS findings to the identified GS-BC modules. (F) For each enriched GS-BC module, examine the GS using GO terms, KEGG pathways, and OMIM disease databases, and map the BC to the brain.

within a BC, and extracted gene-QT associations from brain-wide genome-wide association analysis.

The enrichment test can be performed by adopting either over-representation or rank-based test described in Section 14.3.1. Yao et al. (2015) used the imaging genomic data from the ADNI as test beds, and identified 12 significant GS-BC modules. From their results, most identified GSs had significant functional enrichment, and several could be related to the neurodegenerative disease and its development; identified BCs also involved structures responsible for neurodegenerations, including motivated behaviors, sensory information processing, executive functions, major spots for amyloid accumulation in AD, and so on. By jointly considering the complex relationships between interlinked genetic markers and correlated brain imaging phenotypes, higher level imaging genomic association analysis can provide additional power for extracting biological insights on neurogenomic associations at a systems biology level.

14.4 DISCUSSION

14.4.1 PROMINENT FINDINGS

In the last decade, statistical and machine learning has been playing an essential role in imaging genomic studies and has successfully promoted the discoveries of biologically meaningful biomarkers as well as underlying association patterns. Potkin et al. used PLINK to examine the genetic basis of hippocampal atrophy in Alzheimer patients and successfully identified five risk genes involved in the regulation of protein degradation, apoptosis, neuronal loss, and neurodevelopment. Instead of structural changes, another group recently reported a GWAS result between brain connectivity and genetic variants, in which one risk gene *SPON1* was identified and further replicated in an independent cohort (Jahanshad et al., 2013). In Kohannim et al. (2012), multilocus effects were investigated against the temporal lobe volume using the elastic net method, where two SNPs in genes *RBFOX1* and *GRIN2B* were, respectively, found to be highly contributory genotypes. By applying a variant of PCR across all genes and a large database of voxel-wise imaging data, Hibar et al. (2011) identified 10 significant SNPs in *GRB-associated binding protein 2 gene* (*GAB2*), to significantly associate with all voxels. Similarly, Vounou et al. (2012) used sRRR, after validating its superior performance on a large-scale dataset, to examine the multilocus effects over the voxel-wise imaging data but using their longitudinal changes. Their findings confirmed the key role of *APOE* and *TOMM40* in AD and highlighted some other potential associations as well. In addition to these variant level findings, prior knowledge guided methods also help to reveal candidate pathways, whose perturbation may possibly lead to changes in imaging phenotypes. For example, in Silver et al. (2012a), an overlapping group lasso penalty added to RRR was used to model the pathway belonging to SNPs and several pathways were reported to be associated with longitudinal structure changes in brain, such as the insulin signaling pathway, Chemokine signaling pathway, and Alzheimer's disease pathway. Recently, with the help of S2CCA, Du et al. (2014, 2015a) examined the multiple-to-multiple relationship between *APOE* and brain-wide amyloid accumulations, and reported localized amyloid patterns affected by the joint effect of *APOE* SNPs. At the same time, in Yan et al. (2014a) the same experiment was performed using KG-SCCA, where a transcriptome co-expression network was applied as a prior, and similar amyloid accumulation patterns were identified.

14.4.2 FUTURE DIRECTIONS

The advent of brain imaging and genotyping techniques has brought unprecedented opportunities for discoveries of the underlying disease mechanisms. Whole genome sequencing, longitudinal imaging, and brain connectome data are now widely accessible, requiring more complicated models to capture the reality hidden behind the data. Existing methods, though successful, still work on a relatively small scale and will have limited power as the number of datasets becomes ever larger. Regression or association models capable of dealing with superdimensionality are highly desired in the near future.

Big data is another promising future direction. Plenty of efforts have been recently made to introduce big data science into the brain imaging genomics field, which is believed to hold great promise for overcoming the computation bottlenecks. Most attempts now are focusing on boosting the performance of traditional GWAS by taking advantage of supercomputing techniques. Example high-performance software tools include FaST-LMM (Lippert et al., 2013), EpiGPU (Hemani et al., 2011), and GBOOST (Yung et al., 2011). Yan et al. (2014b) made an initial attempt to accelerate the SCCA implementation by combining Intel Math Kernel Library (MKL) and the offload model for Intel Many Integrated Core (MIC), in which they observed consistent twofold speedup without any code modification. Another recent study (Wang et al., 2013) coupled the Map/Reduce framework with Random Forest for associating SNPs and imaging phenotypes, which achieved at most 10-fold improvement in running time. Despite little work for now, these initial efforts and their promising results show the great potential of big data techniques. More applications of machine learning in brain- and genome-wide studies are expected in future imaging genomics research.

REFERENCES

Bach, F.R., 2008. Consistency of the group lasso and multiple kernel learning. J. Mach. Learn. Res. 9, 1179–1225.

Benjamini, Y., Hochberg, Y., 1995. Controlling the false discovery rate—A practical and powerful approach to multiple testing. J. R. Stat. Soc. B 57, 289–300.

Buil, A., Martinez-Perez, A., Perera-Lluna, A., Rib, L., Caminal, P., Soria, J.M., 2009. A new gene-based association test for genome-wide association studies. BMC Proc. 3 (7), S130.

Bush, W.S., Moore, J.H., 2012. Chapter 11: Genome-wide association studies. PLOS Comput. Biol. 8, e1002822.

Chi, E.C., Allen, G.I., Zhou, H., Kohannim, O., Lange, K., Thompson, P.M., 2013. Imaging genetics via sparse canonical correlation analysis. IEEE Int. Symp. Biomed. Imaging 2013, 740–743.

Curtis, D., Vine, A.E., Knight, J., 2008. A simple method for assessing the strength of evidence for association at the level of the whole gene. Adv. Appl. Bioinform. Chem. 1, 115–120.

Do, C.B., Tung, J.Y., Dorfman, E., Kiefer, A.K., Drabant, E.M., Francke, U., et al., 2011. Web-based genome-wide association study identifies two novel loci and a substantial genetic component for Parkinson's disease. PLoS Genet. 7, e1002141.

Draghici, S., Khatri, P., Martins, R.P., Ostermeier, G.C., Krawetz, S.A., 2003. Global functional profiling of gene expression. Genomics 81, 98–104.

Du, L., Jingwen, Y., Kim, S., Risacher, S.L., Huang, H., Inlow, M., et al., 2014. A novel structure-aware sparse learning algorithm for brain imaging genetics. Med. Image Comput. Comput. Assis. Interven. 17, 329–336.

Du, L., Huang, H., Yan, J., Kim, S., Risacher, S.L., Inlow, M., et al., 2015a. Structured sparse CCA for brain imaging genetics via graph OSCAR. In: The International Conference on Intelligent Biology and Medicine, Indianapolis, USA.

Du, L., Yan, J., Kim, S., Risacher, S., Huang, H., Inlow, M., et al., 2015b. GN-SCCA: GraphNet Based Sparse Canonical Correlation Analysis for Brain Imaging Genetics. Springer, New York.

Gauderman, W.J., Murcray, C., Gilliland, F., Conti, D.V., 2007. Testing association between disease and multiple SNPs in a candidate gene. Genet. Epidemiol. 31, 450–450.

Goeman, J.J., Buhlmann, P., 2007. Analyzing gene expression data in terms of gene sets: methodological issues. Bioinformatics 23, 980–987.

Grosenick, L., Klingenberg, B., Katovich, K., Knutson, B., Taylor, J.E., 2013. Interpretable whole-brain prediction analysis with GraphNet. NeuroImage 72, 304–321.

Hemani, G., Theocharidis, A., Wei, W., Haley, C., 2011. EpiGPU: exhaustive pairwise epistasis scans parallelized on consumer level graphics cards. Bioinformatics 27, 1462–1465.

Hibar, D.P., Stein, J.L., Kohannim, O., Jahanshad, N., Jack, C.R., Weiner, M.W., et al., 2011. Principal components regression: multivariate, gene-based tests in imaging genomics. In: 2011 IEEE International Symposium on Biomedical Imaging: From Nano to Macro, Chicago, IL, pp. 289–293.

Hotelling, H., 1935. The most predictable criterion. J. Edu. Psychol. 26, 139.

Jahanshad, N., Rajagopalan, P., Hua, X., Hibar, D.P., Nir, T.M., Toga, A.W., et al., 2013. Genome-wide scan of healthy human connectome discovers SPON1 gene variant influencing dementia severity. Proc. Natl. Acad. Sci. USA 110, 4768–4773.

Kohannim, O., Hibar, D.P., Jahanshad, N., Stein, J.L., Hua, X., Toga, A.W., et al., 2012. Predicting temporal lobe volume on MRI from genotypes using L(1)-L(2) regularized regression. In: Proceedings of the 9th International Symposium on Biomedical Imaging: ISBI 2012, Barcelona, Spain, 1160–1163.

Lee, S.I., Lee, H., Abbeel, P., Ng, A.Y., 2006. Efficient L1 regularized logistic regression. In: Proceedings of the 21st National Conference on Artificial Intelligence (AAAI-06).

Li, M.X., Gui, H.S., Kwan, J.S., Sham, P.C., 2011. GATES: a rapid and powerful gene-based association test using extended Simes procedure. Am. J. Hum. Genet. 88, 283–293.

Liang, H., Meng, X., Chen, F., Zhang, Q., Yan, J., Yao, X., et al., 2015. A network-based framework for mining high-level imaging genetic associations. In: MICCAI Workshop on Imaging Genetics, October 9, 2015.

Lin, D., Calhoun, V.D., Wang, Y.P., 2014. Correspondence between fMRI and SNP data by group sparse canonical correlation analysis. Med. Image Anal. 18, 891–902.

Lippert, C., Listgarten, J., Davidson, R.I., Baxter, S., Poon, H., Kadie, C.M., et al., 2013. An exhaustive epistatic SNP association analysis on expanded Wellcome Trust data. Sci. Rep. 3, 1099.

Liu, J.Y., Hutchison, K., Perrone-Bizzozero, N., Morgan, M., Sui, J., Calhoun, V., 2010a. Identification of genetic and epigenetic marks involved in population structure. PLoS ONE 5 (10), e13209.

Liu, J.Z., Mcrae, A.F., Nyholt, D.R., Medland, S.E., Wray, N.R., Brown, K.M., et al., 2010b. A versatile gene-based test for genome-wide association studies. Am. J. Hum. Genet. 87, 139–145.

Lu, Z.Q.J, 2010. The elements of statistical learning: data mining, inference, and prediction, second ed. J. R. Stat. Soc. A 173, 693–694.

Mooney, M.A., Wilmot, B., 2015. Gene set analysis: a step-by-step guide. Am. J. Med. Genet. B 168, 517–527.

Perez-Palma, E., Bustos, B.I., Villaman, C.F., Alarcon, M.A., Avila, M.E., Ugarte, G.D., et al., 2014. Overrepresentation of glutamate signaling in Alzheimer's disease: network-based pathway enrichment using meta-analysis of genome-wide association studies. PLoS ONE 9, e95413.

Potkin, S.G., Guffanti, G., Lakatos, A., Turner, J.A., Kruggel, F., Fallon, J.H., et al., 2009. Hippocampal atrophy as a quantitative trait in a genome-wide association study identifying novel susceptibility genes for Alzheimer's disease. PLoS ONE 4, e6501.

Price, A.L., Patterson, N.J., Plenge, R.M., Weinblatt, M.E., Shadick, N.A., Reich, D., 2006. Principal components analysis corrects for stratification in genome-wide association studies. Nat. Genet. 38, 904–909.

Purcell, S., Neale, B., Todd-Brown, K., Thomas, L., Ferreira, M.A., Bender, D., et al., 2007. PLINK: a tool set for whole-genome association and population-based linkage analyses. Am. J. Hum. Genet. 81, 559–575.

Ramanan, V.K., Kim, S., Holohan, K., Shen, L., Nho, K., Risacher, S.L., et al., 2012. Genome-wide pathway analysis of memory impairment in the Alzheimer's Disease Neuroimaging Initiative (ADNI) cohort implicates gene candidates, canonical pathways, and networks. Brain Imaging Behav. 6, 634–648.

Ridge, P.G., Mukherjee, S., Crane, P.K., Kauwe, J.S., 2013. Alzheimer's disease: analyzing the missing heritability. PLoS ONE 8, e79771.

Ripke, S., O'dushlaine, C., Chambert, K., Moran, J.L., Kahler, A.K., Akterin, S., et al., 2013. Genome-wide association analysis identifies 13 new risk loci for schizophrenia. Nat. Genet. 45, 1150–1159.

Saykin, A.J., Shen, L., Foroud, T.M., Potkin, S.G., Swaminathan, S., Kim, S., et al., 2010. Alzheimer's Disease Neuroimaging Initiative biomarkers as quantitative phenotypes: genetics core aims, progress, and plans. Alzheimer's Dement. 6, 265–273.

Schmidt, M., Fung, G., Rosales, R., 2007. Fast optimization methods for l1 regularization: a comparative study and two new approaches. In: Proceedings of the 18th European Conference on Machine Learning: ECML 2007, Warsaw, Poland, vol. 4701, pp. 286–297.

Shen, L., Kim, S., Risacher, S.L., Nho, K., Swaminathan, S., West, J.D., et al., 2010. Whole genome association study of brain-wide imaging phenotypes for identifying quantitative trait loci in MCI and AD: a study of the ADNI cohort. NeuroImage 53, 1051–1063.

Silver, M., Janousova, E., Hua, X., Thompson, P.M., Montana, G., Alzheimer's Disease Neuroimaging Initiative, 2012a. Identification of gene pathways implicated in Alzheimer's disease using longitudinal imaging phenotypes with sparse regression. NeuroImage 63, 1681–1694.

Silver, M., Montana, G., Alzheimer's Disease Neuroimaging Initiative, 2012b. Fast identification of biological pathways associated with a quantitative trait using group lasso with overlaps. Stat. Appl. Genet. Mol. Biol. 11, 7.

Stein, J.L., Hua, X., Lee, S., Ho, A.J., Leow, A.D., Toga, A.W., et al., 2010. Voxelwise genome-wide association study (vGWAS). NeuroImage 53, 1160–1174.

Tibshirani, R., 1996. Regression shrinkage and selection via the Lasso. J. R. Stat. Soc. B 58, 267–288.

Vounou, M., Nichols, T.E., Montana, G., Alzheimer's Disease Neuroimaging Initiative, 2010. Discovering genetic associations with high-dimensional neuroimaging phenotypes: a sparse reduced-rank regression approach. NeuroImage 53, 1147–1159.

Vounou, M., Janousova, E., Wolz, R., Stein, J.L., Thompson, P.M., Rueckert, D., et al., 2012. Sparse reduced-rank regression detects genetic associations with voxel-wise longitudinal phenotypes in Alzheimer's disease. NeuroImage 60, 700–716.

Wan, J., Kim, S., Inlow, M., Nho, K., Swaminathan, S., Risacheri, S.L., et al., 2011. Hippocampal surface mapping of genetic risk factors in AD via sparse learning models. Med. Image Comput. Comput. Assis. Interven. 14, 376–383.

Wang, T., Elston, R.C., 2007. Improved power by use of a weighted score test for linkage disequilibrium mapping. Am. J. Hum. Genet. 80, 353–360.

Wang, K., Abbott, D., 2008. A principal components regression approach to multilocus genetic association studies. Genet. Epidemiol. 32, 108–118.

Wang, H., Nie, F., Huang, H., Kim, S., Nho, K., Risacher, S.L., et al., 2012a. Identifying quantitative trait loci via group-sparse multitask regression and feature selection: an imaging genetics study of the ADNI cohort. Bioinformatics 28, 229–237.

Wang, H., Nie, F., Huang, H., Risacher, S.L., Saykin, A.J., Shen, L., 2012b. Identifying disease sensitive and quantitative trait-relevant biomarkers from multidimensional heterogeneous imaging genetics data via sparse multimodal multitask learning. Bioinformatics 28, i127–i136.

Wang, Y., Goh, W., Wong, L., Montana, G., Alzheimer's Disease Neuroimaging Initiative, 2013. Random forests on Hadoop for genome-wide association studies of multivariate neuroimaging phenotypes. BMC Bioinformatics 14 (Suppl. 16), S6.

Witten, D.M., Tibshirani, R., Hastie, T., 2009. A penalized matrix decomposition, with applications to sparse principal components and canonical correlation analysis. Biostatistics 10, 515–534.

Yan, J., Du, L., Kim, S., Risacher, S.L., Huang, H., Moore, J.H., et al., 2014a. Transcriptome-guided amyloid imaging genetic analysis via a novel structured sparse learning algorithm. Bioinformatics 30, i564–i571.

Yan, J., Zhang, H., Du, L., Wernert, E., Saykin, A.J., Shen, L., 2014b. Accelerating sparse canonical correlation analysis for large brain imaging genetics data. In: Proceedings of the 2014 Annual Conference on Extreme Science and Engineering Discovery Environment, Atlanta, GA, USA, 2616515: ACM, pp. 1–7.

Yan, J., Du, L., Kim, S., Risacher, S.L., Huang, H., Moore, J.H., et al., 2015a. BoSCCA: mining stable imaging and genetic associations with implicit structure learning. In: MICGen 2015: MICCAI Workshop on Imaging Genetics.

Yan, J., Kim, S., Nho, K., Chen, R., Risacher, S.L., Moore, J.H., et al., 2015b. Hippocampal transcriptome-guided genetic analysis of correlated episodic memory phenotypes in Alzheimer's disease. Front. Genet. 6, 117.

Yang, J., Lee, S.H., Goddard, M.E., Visscher, P.M., 2011. GCTA: a tool for genome-wide complex trait analysis. Am. J. Hum. Genet. 88, 76–82.

Yang, S., Yuan, L., Lai, Y.C., Shen, X., Wonka, P., Ye, J., 2012. Feature grouping and selection over an undirected graph. In: Proceedings of the 18th ACM SIGKDD International Conference on Knowledge Discovery and Data Mining: KDD 2012, Beijing, China, pp. 922–930.

Yang, J., Lee, S.H., Goddard, M.E., Visscher, P.M., 2013. Genome-wide complex trait analysis (GCTA): methods, data analyses, and interpretations. Meth. Mol. Biol. 1019, 215–236.

Yao, X., Yan, J., Kim, S., Nho, K., Risacher, S.L., Inlow, M., et al., 2015. Two-dimensional enrichment analysis for mining high-level imaging genetic associations. In: Proceedings of the 8th International Conference on Brain Informatics & Health: BIH 2015, London, UK, Lecture Notes in Artificial Intelligence, 9250, 115–124.

Younesi, E., Hofmann-Apitius, M., 2013. Biomarker-guided translation of brain imaging into disease pathway models. Sci. Rep. 3, 3375.

Yung, L.S., Yang, C., Wan, X., Yu, W., 2011. GBOOST: a GPU-based tool for detecting gene-gene interactions in genome-wide case control studies. Bioinformatics 27, 1309–1310.

Zhu, H., Khondker, Z., Lu, Z., Ibrahim, J.G., Alzheimer's Disease Neuroimaging Initiative, 2014. Bayesian generalized low rank regression models for neuroimaging phenotypes and genetic markers. J. Am. Stat. Assoc. 109, 997–990.

Zou, H., Hastie, T., 2005. Regularization and variable selection via the elastic net. J. R. Stat. Soc. B 67, 301–320.

CHAPTER

Holistic atlases of functional networks and interactions (HAFNI) 15

X. Jiang[1], D. Zhu[2], T. Liu[1]

The University of Georgia, Athens, GA, United States[1] University of Southern California, Los Angeles, CA, United States[2]

CHAPTER OUTLINE

15.1 INTRODUCTION

An unrelenting human quest in neuroscience is to understand the organizational architecture of brain function, which to a great extent defines what we are and who we are. After decades of active research, functional magnetic resonance imaging (fMRI) (Biswal et al., 1995; Ogawa et al., 1990a,b) became a popular neuroimaging technique due to its noninvasive and in vivo nature. For example, blood oxygenation level-dependent (BOLD) contrast (Ogawa et al., 1990a,b), a commonly used form of fMRI, can detect local changes in deoxyhemoglobin concentration and is believed to represent a vascular coupling from neuronal activities. In general, fMRI aims to explore the brain's functional activities by leveraging the relations between brain neural activity and hemodynamics (Friston, 2009; Logothetis, 2008). Since its inception in 1990 (Ogawa et al., 1990a,b), modern fMRI techniques have been applied widely and have become the most important way to study brain functions (Friston, 2009; Logothetis, 2008) in the past decade. In particular, task fMRI (tfMRI) has been commonly adopted as a benchmark approach in mapping and localizing

Machine Learning and Medical Imaging. http://dx.doi.org/10.1016/B978-0-12-804076-8.00015-3

functionally specialized brain areas under specific task stimulus (Friston, 2009; Logothetis, 2008). Meanwhile, resting state fMRI (rsfMRI) has been increasingly applied to explore and map intrinsic connectivity networks (ICNs) based on the fact that correlated brain activity patterns have been reported to have similar low-frequency oscillations within rsfMRI time series (Cohen et al., 2008; Fox and Raichle, 2007; Van Den Heuvel et al., 2008).

In the computational neuroscience community, a variety of computational and statistical methods targeting fMRI BOLD signals analysis have been developed for activation detection and latent functional network modeling, such as correlation analysis (Bandettini et al., 1993), principal component analysis (PCA) (Anderson et al., 1999), general linear model (GLM) (Friston et al., 1994; Worsley, 1997), Markov random field (MRF) models (Descombes et al., 1998), independent component analysis (ICA) (McKeown et al., 1998, 2003; Calhoun and Adali, 2006), mixture models (Hartvig and Jensen, 2000), autoregressive spatial models (Woolrich et al., 2001), wavelet algorithms (Bullmore et al., 2003; Shimizu et al., 2004), empirical mean curve decomposition (Deng et al., 2012), and Bayesian approaches (DuBois Bowman et al., 2008). Of all these methods, GLM and ICA are the most widely used methods for tfMRI and rsfMRI, due to their effectiveness, robustness, simplicity, and wide availability. GLM considers the relationship of observed data and given factors: As a linear model, the observations or responses are formed as a linear combination of a series of known factors. For tfMRI, the external stimulus can be used to construct the regressors (design matrix) by convolution with hemodynamic response function (hrf) (Wagenmakers, 2015). The outcome of GLM analysis is a statistic (eg, z score) for each voxel, which indicates if the response of that voxel is task-related. For rsfMRI, model-free approaches (like ICA) are widely accepted since, given that there is no external stimulus, it is difficult to infer or design a model which can effectively interpret resting-state BOLD fluctuations. In essence, ICA is based on matrix factorization and its objective function is to maximize the statistical independence (eg, minimize the mutual information) between any pair of components. ICA-based methods aim to explore a mixture of underlying sources that can explain the brain activity patterns in the resting state (van den Heuvel and Pol, 2010).

Despite the significant neuroscientific insights and remarkable successes of the above-mentioned methods in the past decades, we are still facing challenges as our understanding becomes deeper regarding the organization of brain architecture. For example, it has been pointed out that spatially overlapping networks subserving different brain functions may be unnoticed by the traditional blocked subtraction paradigms and the related analysis methods, such as GLM (Krekelberg et al., 2006; Logothetis, 2008). It also has been widely argued that a variety of cortical areas and brain networks exhibit remarkable functional diversity and heterogeneity (Anderson et al., 2013; Duncan, 2010; Fedorenko et al., 2013; Kanwisher, 2010; Pessoa, 2012). That is, the human brain is widely considered as containing a collection of highly specialized functional networks, which are able to flexibly interact with each other when distinct brain functions are performed. In such a scenario, activity in one brain region might need to recruit multiple neuroanatomical regions in a temporal sequence. Meanwhile, the same brain area might also participate in multiple functional processes simultaneously with different internal or external circumstances. As a consequence, the corresponding fMRI BOLD signal from a single voxel tends to be composed of various components corresponding to multiple functional sources. Therefore traditional subtraction-based tfMRI analysis methods (eg, GLM) might be insufficient to reconstruct all or most concurrent spatially and temporally overlapping brain functional networks. Indeed, ICA and its variations, such as spatial ICA and temporal ICA, perform better than GLM in the identification of concurrent functional components (eg, ICNs). Nevertheless, given the fact that we still have very limited understanding about the exact mechanism of the brain's functional operations, we might need to carefully consider the rationale and validity of applying any preset assumptions, like statistical independence in ICA.

In our opinion, an ideal solution to fMRI data analysis should have at least the following characteristics: (1) it should consider the complexity of fMRI BOLD signals which arise from different signal sources including the interactions of multiple activated functional networks; (2) it should conform to the neuroscience principle that multiple functional networks can be activated simultaneously and they might interact with each other; (3) it should impose as few assumptions as possible. As an attempt to better understand and interpret fMRI BOLD time series, in the following part of this chapter, we introduce a novel machine learning-based alternative methodology, called *holistic atlases of functional networks and interactions* (HAFNI), which employs a sparse representation of whole-brain fMRI signals for functional network identification in both tfMRI and rsfMRI data (Lv et al., 2015b). The basic idea is that we aggregate hundreds of thousands of tfMRI or rsfMRI signals within the whole brain from a single subject into a big data matrix (represented as S in Fig. 15.1A), and factorize it by an over-complete dictionary basis matrix (represented as D in Fig. 15.1B) and a sparse reference weight matrix (represented as α in Fig. 15.1C) via an effective online dictionary learning algorithm (Mairal et al., 2010). The derived time series in the learned basis dictionary represent different activities of concurrent brain functional networks (the white curves in Fig. 15.1B). Their corresponding weight vectors (each row of α) stand for the spatial maps of these concurrent functional networks (the 3D volume images in Fig. 15.1C). One important characteristic of this framework is that the decomposed weight matrix naturally reflects the spatial overlap/interaction patterns of reconstructed brain networks (Lv et al., 2015b).

The advantages of HAFNI include: (1) Sparse representation naturally accounts for the complexity of BOLD signals in that each single fMRI time series is represented as a linear combination of common "blocks," called dictionary atoms, which can correspond to different signal sources. Each dictionary atom also contributes to multiple fMRI signals simultaneously and the proportion of each contribution is

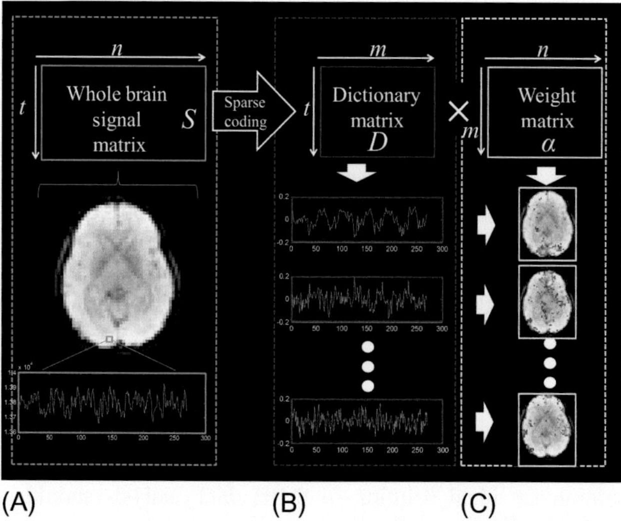

FIG. 15.1

The computational framework of sparse representation of whole-brain fMRI signals (from single brain) using online dictionary learning. (A) The whole-brain fMRI signals are integrated into a big data matrix, in which each row represents the whole-brain fMRI BOLD data at one time point and each column stands for the time series of a single voxel (green rectangle). (B) Illustration of the learned dictionary. Each column represents a latent functional network. Three exemplar dictionary atoms are shown in the bottom panels. (C) The decomposed sparse reference weight matrix. Each row contains the weight parameters of the corresponding functional network component.

encoded in the weight matrix. There is extensive literature suggesting that, through sparse learning, certain task-evoked and intrinsic connectivity networks (ICNs) can be successfully recovered (Varoquaux et al., 2013). (2) Based on the learned dictionary atoms, we are able to identify latent functional networks, especially those corresponding to external stimulus (designed tasks) or ICNs (eg, default mode network, DMN), which influence the behavior of the brain in a temporal sequence. (3) Unlike ICA, we do not enforce any predefined constraints except sparsity of the weight matrix. However, sparsity reflects a widely accepted precondition of how the brain works: only parts of the brain are involved in performing a task at one time. In the rest of this chapter, we will first discuss the results after applying HAFNI to the publicly released large-scale Human Connectome Project (HCP) high-quality fMRI data (Section 15.2), which is followed by three applications of HAFNI, including clinical studies, cerebral cortex structural/functional architecture exploration, and neuroimaging-informed multimedia analysis (Section 15.3). Then we will consider some new HAFNI-based machine learning methods (Section 15.4) and provide insights into the future directions of HAFNI (Section 15.5).

15.2 HAFNI FOR FUNCTIONAL BRAIN NETWORK IDENTIFICATION

We have applied HAFNI to the Q1 release of the HCP tfMRI dataset, which includes seven tasks (motor—M, emotion—E, gambling—G, language—L, relational—R, social—S, and working memory—WM) for 77 participants. More details of demographics, preprocessing, and data acquisition can be found in Smith et al. (2013). In total, we have identified and confirmed 23 group-wise consistent task-evoked networks, called task-evoked HAFNI components, for motor (M1–M5 in Fig. 15.2A), emotion (E1–E3 in Fig. 15.2A), gambling (G1 and G2 in Fig. 15.2A), language (L1 and L2 in Fig. 15.2A), relational (R1 and R2 in Fig. 15.2B), social (S1–S3 in Fig. 15.2B), and working memory (WM) (W1–W6 in Fig. 15.2B) networks. The details of the identification of these 23 HAFNI networks can be found in Lv et al. (2015b). In addition, these HAFNI networks correspond to some specific task stimuli (see Table 15.1).

In particular, these 23 task-evoked HAFNI components are consistent and can be reproduced across all of the HCP subjects of the Q1 release (Lv et al., 2015b). In Fig. 15.2A and B, the averaged spatial maps of each HAFNI network across all

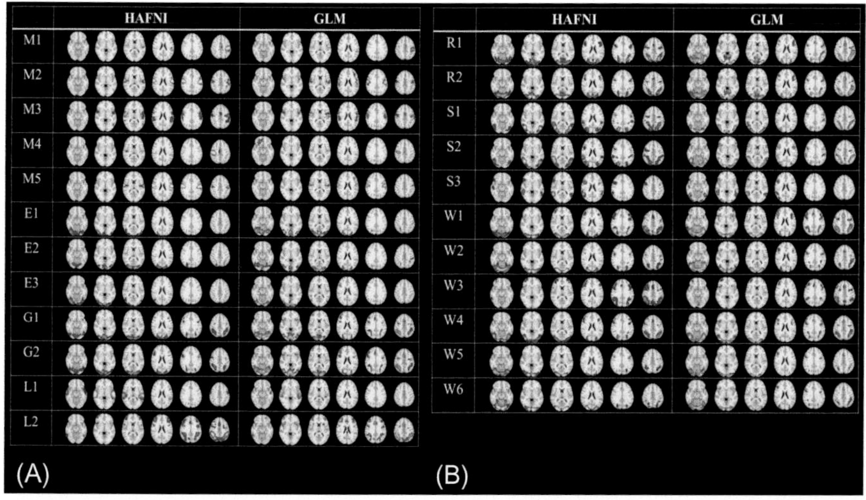

FIG. 15.2

The task-evoked HAFNI networks and the comparison with GLM-derived activation maps. The seven tasks are motor network (M), emotion network (E), gambling network (G), language network (L), relational network (R), social network (S), and working memory network (WM). (A, B) Group-wise averages of 23 identified HAFNI networks across all HCP subjects for the four tasks as well as the corresponding averaged GLM-derived activation maps (right column). Six representative volume slices are selected for visualization for each component.

Table 15.1 Task Names of Task-Evoked HAFNI Components

HAFNI Component	Task Name	HAFNI Component	Task Name
M1	Right hand movement	R1	Match task
M2	Tongue movement	R2	Relational task
M3	Global motion	S1	Interaction behavior
M4	Left hand movement (over average)	S2	Random behavior
M5	Tongue movement (over average)	S3	Interaction over random
E1	Emotional faces	W1	2-back memory task
E2	Simple shapes	W2	0-back memory task
E3	Emotional faces over simple shapes	W3	2-back over 0-back
G1	Punishment	W4	Body part memorization
G2	Reward	W5	Face recognition
L1	Story	W6	Place recognition
L2	Math over story		

subjects are shown and compared to the group-wise GLM-derived activation map. We can see that the averaged HAFNI networks are similar to the group-wise GLM-derived maps. Note that all HAFNI networks are learned simultaneously from the optimally decomposed fMRI time series by sparse representation of whole-brain data (see Fig. 15.1). The GLM maps, however, are obtained from individual fMRI time series with separate model-driven subtraction procedures. For example, the five motor networks (M1–M5 in Fig. 15.2A) can be robustly identified by characterizing the most relevant atoms from a collection of candidate dictionary atoms, which can maximally interpret the whole-brain fMRI time series. On the contrary, GLM examines fMRI signals whose compositions can only arise from predefined activation models. As a consequence, GLM has difficulty in identifying latent and concurrent functional networks. In this situation, some spatially overlapping networks with distinct temporal curves such as ICNs (Fig. 15.3) in task data, other than the task paradigm, will be essentially ignored (Krekelberg et al., 2006; Logothetis, 2008).

HAFNI is also effective and efficient in reconstructing concurrent ICNs based on either tfMRI or rsfMRI data. Once the HAFNI framework is applied to the whole-brain tfMRI or rsfMRI signals for each individual subject, both quantitative measurement and visual inspection of the spatial pattern of dictionary atoms (functional networks) are integrated to identify and characterize the ICNs using existing brain science knowledge (Lv et al., 2015b). Specifically, the well-defined ICN templates provided in the literature (eg, Smith et al., 2009) are adopted as the references. We define the spatial similarity as the spatial overlap rate R between a dictionary atom's spatial pattern (S) and an ICN template (T) (Jiang et al., 2015; Lv et al., 2015b):

$$R(S,T) = \frac{|S \cap T|}{|T|}. \tag{15.1}$$

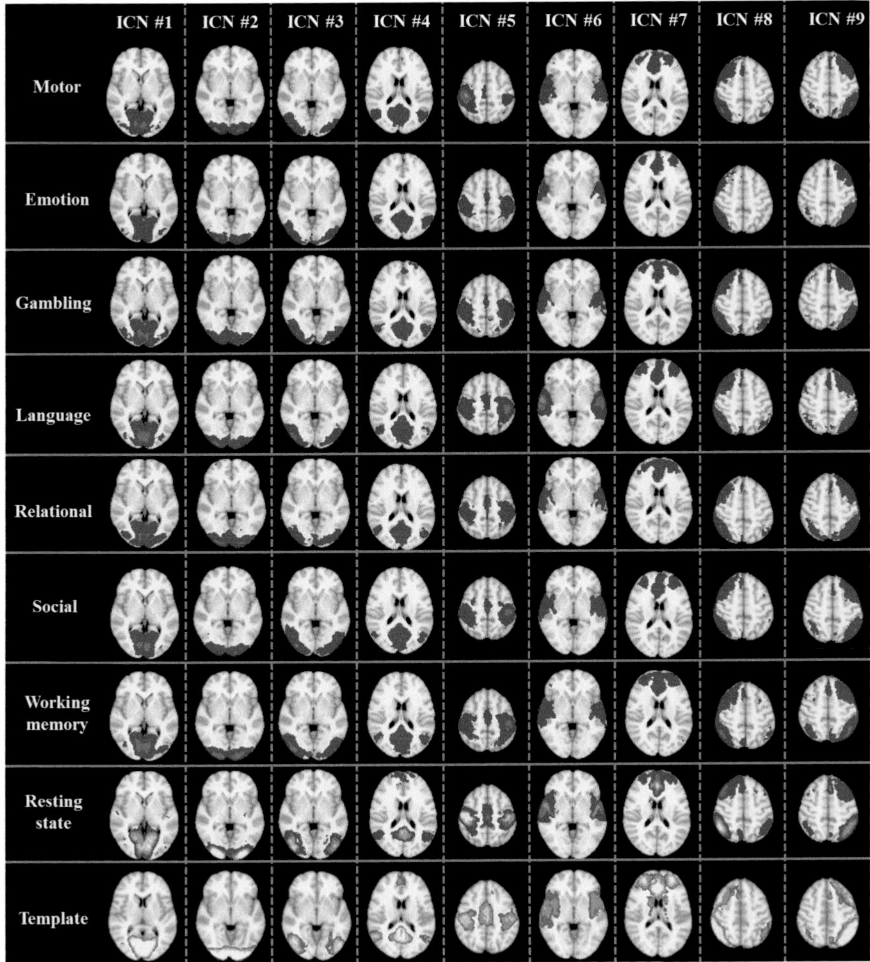

FIG. 15.3

The identified spatial patterns (averaged) of ICNs across all individual subjects in the seven tfMRI (motor, emotion, gambling, language, relational, social, and working memory) and one rsfMRI datasets in HCP Q1. The most informative slice superimposed on the MNI152 template image is visualized in each part. The color scale of the spatial pattern of ICNs ranges from 0.2 to 10 in the seven tfMRI data, and from 0.5 to 10 in rsfMRI data (Lv et al., 2015b).

Note that S and T are converted from continuous values to discrete labels (all values lass than or equal to 0 are labeled as 0, and others are labeled as 1). For the tfMRI or rsfMRI data of each subject, the dictionary atom with the highest spatial overlap rate to a specific ICN template (Eq. 15.1) is identified as the candidate of the

ICNs. A group of experts then quantitatively (spatial overlap rate) and qualitatively (visual inspection) examine the identified corresponding ICN candidates across a group of subjects. If all of the identified ICN candidates have high spatial overlap rate across all subjects, these dictionary atoms will be regarded as the identified ICNs for each subject. For example, in Lv et al. (2015a), the default mode network (DMN), which is a well known and interpreted ICN, was successfully identified in two tfMRI datasets (working memory and semantic decision making). In Lv et al. (2015b), nine ICNs were identified in the seven tfMRI and one rsfMRI datasets in the HCP Q1 release (Van Essen, 2013). Fig. 15.3 shows the averaged identified ICNs across all subjects in the seven tfMRI and one rsfMRI datasets. Specifically, ICNs 1–3 are located in the visual cortex. ICN 4 is in the DMN. ICNs 5–7 are the sensorimotor, auditory, and executive control networks respectively. ICNs 8–9 contain the frontal and parietal areas and have strong lateralization.

Based on the identified ICNs and task-evoked networks via HAFNI, we found significant spatial overlaps within task-evoked networks, within ICNs, and between task-evoked networks and ICNs (Lv et al., 2015a; Lv et al., 2015b), which may shed light on the exploration of the interaction among functional networks (both task-evoked networks and ICNs) to jointly fulfill brain function in the future (Lv et al., 2015b).

15.3 HAFNI APPLICATIONS

The HAFNI framework has gained increasing interest in a variety of applications. Here we briefly introduce three categories of applications. The first category is in clinical studies. For example, in our recent work (Jiang et al., 2014), the HAFNI framework was applied to the publicly available Alzheimer's Disease Neuroimaging Initiative (ADNI) rsfMRI datasets. Alzheimer's disease (AD) is one of the most common types of dementia and a major cause of death for elderly people (65 or older) (Thies and Bleiler, 2013). However, there is still no effective treatment for AD (Thies and Bleiler, 2013). Meanwhile, it has been demonstrated that the pre-stages of AD (including late mild cognitive impairment (LMCI), early mild cognitive impairment (EMCI), and significant memory concern (SMC)) have the potential to predict the conversion of AD or other neurodegenerative diseases (Petersen et al., 2001). Therefore effective and efficient classification of AD and its pre-stages as distinct from healthy people has received increasing interest. As illustrated in Fig. 15.4, the core idea in Jiang et al. (2014) is to identify the concurrent intrinsic functional networks including ICNs based on the HAFNI framework in five populations (normal control (NC), SMC, EMCI, LMCI, and AD groups), and to adopt meaningful features derived from those identified intrinsic functional networks to classify each of the four diseased groups from the NC group. Firstly, we identify 10 meaningful ICNs for each of the subjects based on the 10 ICN templates (Smith et al., 2009) via the HAFNI framework, as detailed in Section 15.2. Fig. 15.5 shows the 10 identified ICNs in one example subject brain of each of the five populations. ICNs 1–3 are located in

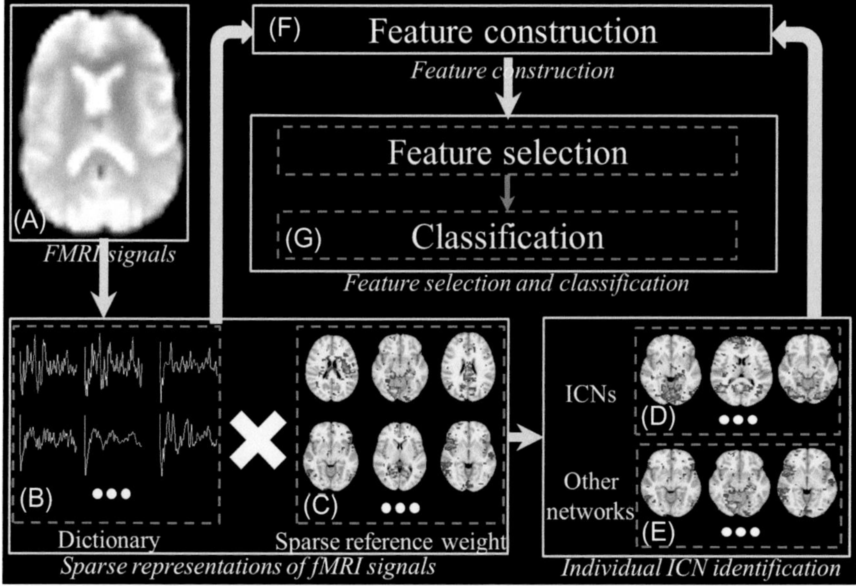

FIG. 15.4

Computational framework for discrimination of Alzheimer's disease (AD) and its pre-stages from healthy people based on the HAFNI framework. (A) Whole-brain rsfMRI data of an individual subject. (B) Dictionary matrix. (C) Sparse reference weight matrix. (D) Ten identified ICNs based on ICN templates. (E) Other dictionary atoms. (F) Six types of features constructed based on the dictionary matrix and identified ICNs. (G) Correlation-based feature selection (CFS) and support vector machine (SVM) classifier based classification between a diseased group and normal control group.

the visual cortex. ICNs 4–8 are in the DMN, cerebellum, sensorimotor, auditory, and executive control networks respectively. ICNs 9–10 are located in the frontal and parietal areas and have strong lateralization. Secondly, we construct a collection of meaningful features based on the dictionary matrix and identified intrinsic functional networks. Specifically, six types of features which can efficiently and effectively represent both spatial and functional characteristics of brains during the resting state are constructed for each subject (Jiang et al., 2014). The first type of feature is the spatial overlap rate R between an identified ICN and the corresponding ICN template as defined in Eq. (15.1). The second type is the functional connectivity within ICNs. We calculate the Pearson correlation value between any pair of temporal patterns of the 10 ICNs to obtain a 10 by 10 symmetric matrix, and adopt its 45 unique elements as the features. The third type is the functional connectivity within all dictionary components. Besides the connectivity within ICNs in the third type, we also examine the functional connectivity between the ICNs and the other functional

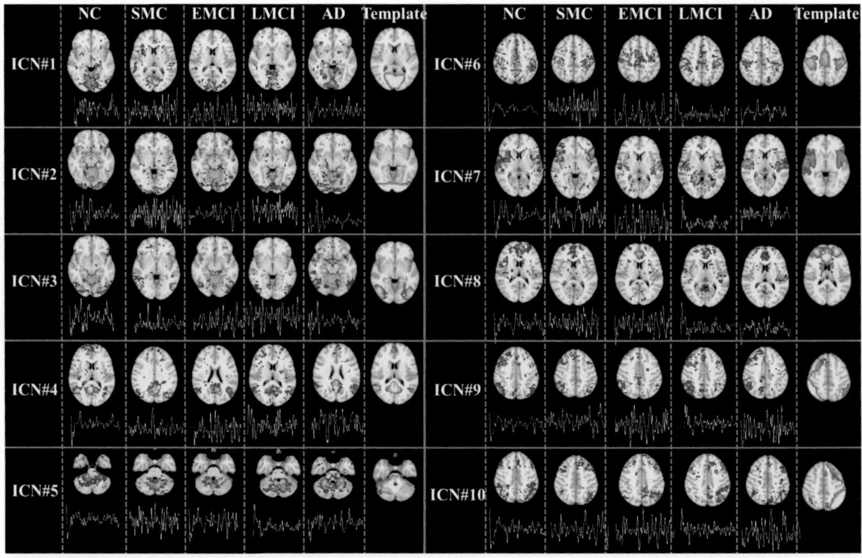

FIG. 15.5

Ten identified ICNs based on HAFNI in one example subject brain of each of the five population groups. Each part shows the most informative slice, which is superimposed on the MNI152 template image and the temporal pattern of the corresponding dictionary atom. The color scale of each part ranges from 0.1 to 10. The 10 ICN templates (thresholded at $z = 3$) are provided in Smith et al. (2009).

components. The second and third types of feature are complementary since they represent the functional connectivities between dictionary components from different perspectives. Instead of the traditional functional connectivity measurement based on raw fMRI signals, we measure the functional connectivity based on the temporal patterns of dictionary atoms, which reflect the intrinsic functional activities of brain networks. The fourth type is the entropy of functional connectivity. We construct a histogram which represents the functional connectivity distribution based on the third type of feature. Twenty equal-distance bins are adopted to cover $[-1, +1]$ and the Shannon entropy is calculated. The fifth type of feature is the entropy of component distribution within ICNs. For each voxel involved in an ICN component, we count the dictionary atoms which are involved when representing the original rsfMRI signals to obtain a distribution histogram of the number of voxels involved in each dictionary, and calculate the Shannon entropy of the distribution histogram as the feature. The sixth type of feature is the common dictionary distribution. Based on the dictionary matrix obtained for each subject, we perform second-round dictionary learning and sparse coding on the dictionaries in order to obtain the common dictionary atoms among all atoms. We then obtain the distribution of the number of individual dictionary atoms which can be represented by each common dictionary

Table 15.2 The Classification Accuracy for Each Group Pair

	NC-SMC (%)	NC-EMCI (%)	NC-LMCI (%)	NC-EMCI+LMCI (%)	NC-AD (%)
Accuracy	92.31	80.00	80.68	92.00	94.12
Specificity	96.15	76.00	79.55	90.00	94.12
Sensitivity	88.46	84.00	81.82	94.00	94.12

atom as the sixth type of feature. Thirdly, in order to preserve only those features with most differentiation power, we perform feature selection on all six types of features via correlation-based feature selection (CFS) (Hall and Smith, 1999). Then the support vector machine (SVM) classifier (Chang and Lin, 2011) is performed on the discriminative features for classification between each diseased group and the normal control group respectively. The experimental results indicate that the HAFNI-based computational framework achieves high discriminative accuracy for AD and its pre-stages from normal control, as detailed in Table 15.2.

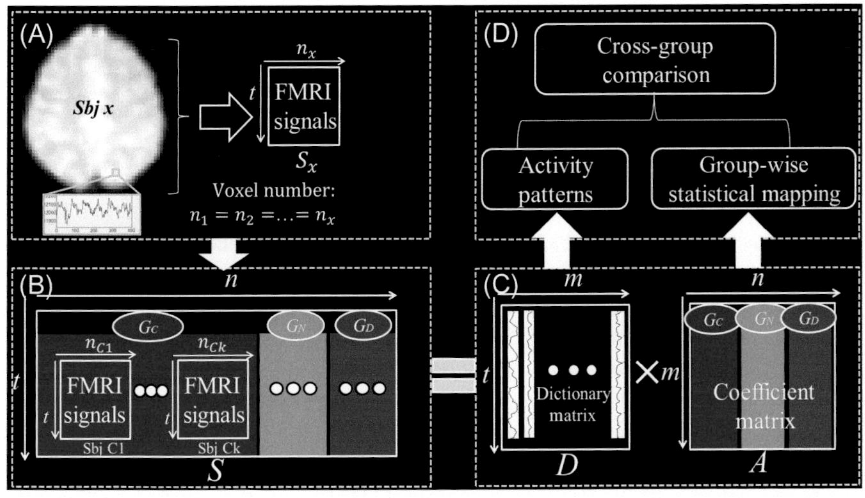

FIG. 15.6

The computational framework of group-wise sparse representation of fMRI signals for identification of affected functional networks among healthy control, nondysmorphic PAE, and dysmorphic PAE groups. (A) Aggregated signal matrix of one single subject. (B) Aggregated large signal matrix *S* from all signal matrices of the three groups of subjects. G_C, healthy control group; G_N, nondysmorphic PAE group; G_D, dysmorphic PAE group. *t* is the fMRI time point. (C) The learned dictionary matrix *D* and the sparse reference weight matrix *A*. *A* is decomposed into multiple submatrices corresponding to the sparse reference weight matrix of each subject. (D) Cross-group comparison is performed based on *D* and *A*.

Another example of HAFNI-based application in clinical studies is to identify possible functional abnormalities among healthy control, exposed nondysmorphic prenatal alcohol exposure (nondysmorphic PAE), and exposed dysmorphic prenatal alcohol exposure (dysmorphic PAE) groups using the HAFNI-based group-wise sparse coding framework (Lv et al., 2015c). As illustrated in Fig. 15.6, firstly, the whole-brain tfMRI signals of each subject are extracted and aggregated into a signal matrix. Secondly, the signal matrices of all three groups of subjects (healthy control, nondysmorphic PAE, and dysmorphic PAE) are arranged into a large signal matrix S. Thirdly, the dictionary learning and sparse coding framework is applied on S to learn a common dictionary D and a sparse reference weight matrix A. Since the sparse reference weight matrix A preserves the organization of subjects and groups in S, it is decomposed into multiple submatrices corresponding to the sparse reference weight matrix of each subject. Finally, cross-group comparisons are performed based on the multiple submatrices and those brain functional networks/regions which are affected by nondysmorphic PAE or dysmorphic PAE can be assessed. Experimental results show that the proposed approach effectively identifies a collection of brain networks/regions that are affected by nondysmorphic PAE or dysmorphic PAE group. Fig. 15.7 illustrates the six dominant networks of which the size (number of voxels involved in the network) decreases with the increment of severity of PAE. Specifically, networks 27, 126, and 180 (the indexes of networks are referred to in Lv et al., 2015c) belong to diverse dynamic networks, networks 73 and 390 are task-evoked networks, and network 354 is an antitask network (Lv et al., 2015c). We can see that for each of the six networks, the size is largest in the control group, smallest in the dysmorphic PAE group, and moderate in the nondysmorphic PAE group. The decreased regions include the visual cortex and default mode network for diverse dynamic networks, left superior and right inferior parietal regions and medial frontal gyrus for task-evoked networks, and subcortical regions and medial prefrontal cortex for the antitask network.

The second category of HAFNI applications is computational modeling of cerebral cortex structural/functional architecture. The human cerebral cortex is composed of highly convoluted cortical folding, including convex gyri and concave sulci (Rakic, 1988). To explore the possible functional difference between cortical gyri and sulci, Jiang et al. (2015) applied the HAFNI framework on the HCP gray-ordinate tfMRI data to identify both task-evoked networks and ICNs, to systematically characterize task-based heterogeneous functional regions (THFRs) on the cortical surface (the regions that are involved in multiple task conditions when performing a specific task), and to assess the spatial patterns of those task-based heterogeneous functional regions on cortical gyri and sulci. Specifically, both meaningful task-evoked networks and ICNs are firstly identified for each subject in each task based on HAFNI, as detailed in Section 15.2 and illustrated in Fig. 15.8. Secondly, we define THFR by assessing the number of involved functional networks (dictionary components) of each gray-ordinate g_i ($i = 1, \ldots, n$, n is the number of gray-ordinates):

$$\text{THFR} = \forall g_i \text{ s.t. } \|\boldsymbol{\alpha}_i\|_0 > q, \tag{15.2}$$

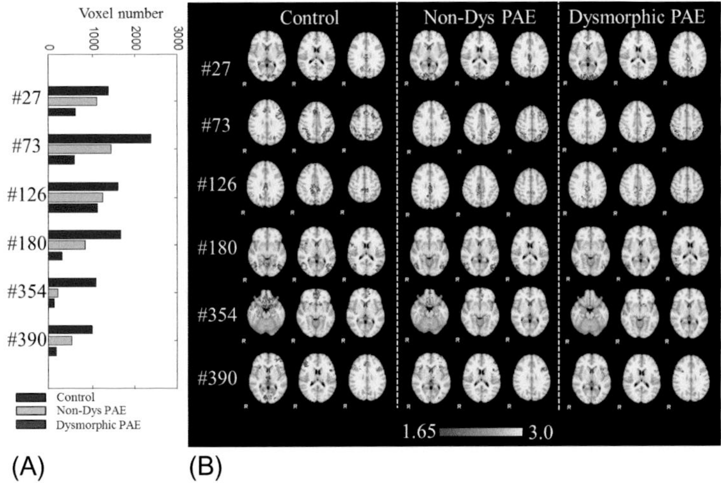

FIG. 15.7

Six dominant networks of which the size (number of voxels involved in the network) decreases with the increment of severity of PAE, that is, V(Control)>V(Non-Dys PAE)>V(Dys PAE). (A) Comparison of voxel numbers ($P < 0.05$, $Z > 1.65$) involved in the six networks among the three groups. (B) Comparison of the spatial pattern (z-score map) of the six networks among the three groups. The indexes of networks are referred to in Lv et al. (2015c).

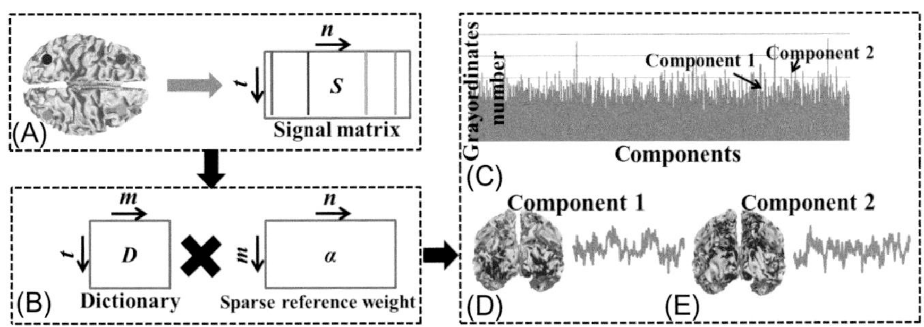

FIG. 15.8

Computational framework of sparse representation of gray-ordinate-based fMRI signals based on HAFNI. (A) Aggregated signal matrix S of one example subject. Four example cortical vertices (gray-ordinates) and associated signals are represented by four different colors (red, blue, green and orange) (four gray dots in print versions). t is the fMRI time point and n is the gray-ordinate. (B) Learned dictionary matrix D and sparse reference weight matrix α. m is the number of dictionary components. (C) The distribution histogram of the number of gray-ordinates involved in each dictionary component. (D, E) The spatial pattern on the cortical surface (highlighted by blue) (gray in print versions) and temporal pattern (gray curve) (gray in print versions) of two example dictionary components respectively.

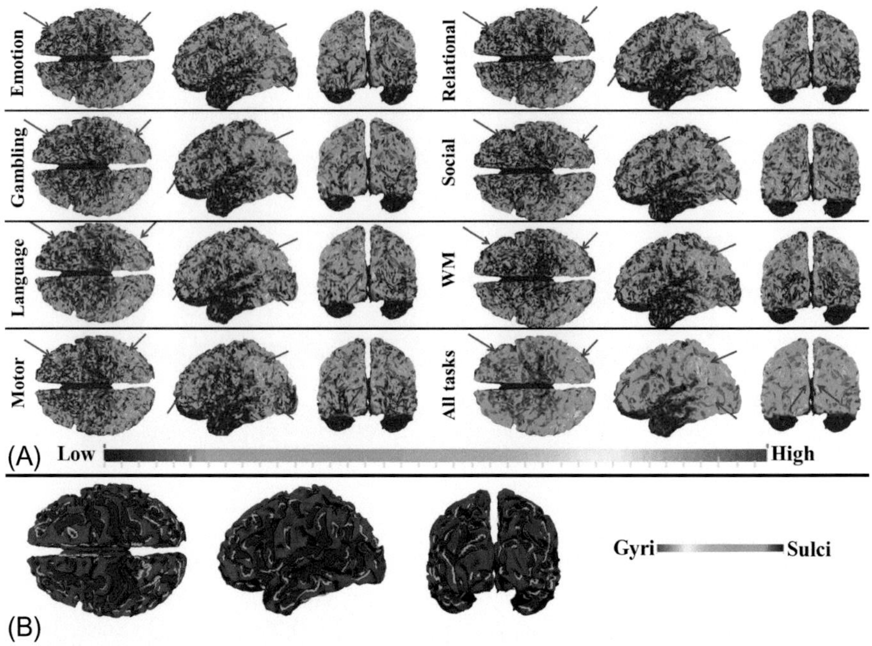

FIG. 15.9

(A) Spatial distribution density map of identified THFRs across all subjects in each of the seven tfMRI data sets (emotion, gambling, language, motor, relational, social, and working memory (WM)) and across all seven tfMRI data sets shown on one example subject. The THFRs which are relatively consistent across seven tasks and with higher distribution density are highlighted by red arrows (gray in print versions). (B) The gyri/sulci regions of the example subject.

where α_i is the ith column of α representing the functional network composition of g_i THFR is a collection of gray-ordinates of which the involved functional networks (number of nonzero elements in α_i) is larger than a threshold q (Jiang et al., 2015). Fig. 15.9 shows the spatial distribution density map of identified THFRs across all subjects in each of the seven tfMRI data sets. We can see that the THFRs are mainly located at the bilateral parietal lobe, frontal lobe, and visual association cortices. We further assess the distribution percentage of the THFRs on cortical gyral and sulcal regions respectively. Experimental results demonstrate that THFRs are located significantly more on gyri than on sulci across all seven tasks, as reported in Table 15.3. The proposed HAFNI-based framework for identification of THFRs indicates the functional difference between gyri and sulci during task performance, and might advance the understanding of the exact functional mechanisms of human cerebral cortex in the future (Jiang et al., 2015).

The third category of HAFNI applications is neuroimaging-informed multimedia analysis. For example, Hu et al. (2015) proposed a HAFNI-based fMRI decoding

Table 15.3 Ratio of Distribution Percentage of THFRs on Gyri vs. That on Sulci in All Subjects (the Ratio Is Represented as Mean ± Standard Deviation)

	Emotion	Gambling	Language	Motor	Rational	Social	WM
Ratio	2.22±1.09	3.14±3.87	2.63±2.07	2.95±2.47	3.38±4.29	2.50±1.32	2.76±1.59
p-Value	9.48E−22	9.60E−19	2.98E−20	1.56E−19	1.49E−17	8.25E−22	8.29E−22

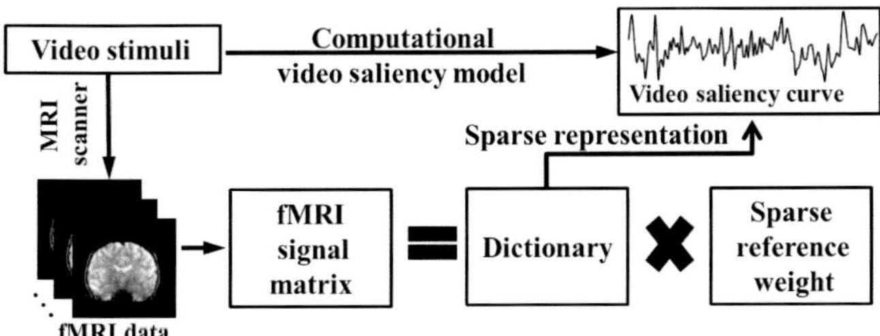

FIG. 15.10

The computational framework of the HAFNI-based fMRI decoding model to decode the bottom-up visual saliency in video streams based on whole-brain natural stimulus fMRI signals.

model to decode the bottom-up visual saliency in video streams using recorded whole-brain natural stimulus fMRI signals. As illustrated in Fig. 15.10, the video stimuli are analyzed based on a computational video saliency model to obtain the video saliency curves. Meanwhile, the whole-brain natural stimulus fMRI signals under the video stimuli are decomposed into the dictionary and sparse reference weight matrix via the HAFNI framework. The learned dictionary components are then adopted to sparsely represent the video saliency curves. The experimental results show that the learned dictionary atoms of whole-brain fMRI signals can well decode the temporal visual saliency information in a naturalistic video stream, indicating that HAFNI-based sparse representation of brain activities measured by fMRI may benefit the multimedia content analysis field (Hu et al., 2015).

15.4 HAFNI-BASED NEW METHODS

As mentioned above, HAFNI is a pure data-driven approach. Recently, there has been increasing interest in introducing novel methodology or integrating brain science domain knowledge into the HAFNI framework, and the results are promising. For example, Zhao et al. (2015) proposed a novel supervised dictionary learning and sparse representation framework of tfMRI data for concurrent functional brain network inference. As illustrated in Fig. 15.11, the basic idea is to predefine the task stimulus curves in tfMRI data as the fixed model-driven dictionary atoms, and merely optimize the other data-driven dictionary atoms. Specifically, the whole-brain tfMRI signals of one subject are firstly aggregated into a signal matrix. Then the supervised dictionary learning and sparse representation are performed to decompose the signal matrix into a dictionary matrix and a sparse reference weight matrix. In contrast to merely data-driven conventional dictionary learning, a constant part of dictionary atoms is defined as the model-driven task stimulus curves, and only the other dictionary atoms are optimized during dictionary learning (Fig. 15.11b). As a result, the part of sparse reference weight matrix which corresponds to the constant model-driven dictionary atoms represents the identified model-driven task-evoked functional networks. The other part of the sparse reference weight matrix represents the identified data-driven functional networks, including ICNs. This model-driven and data-driven integrated approach is effective and efficient in identifying both task-evoked networks and ICNs in all seven tfMRI datasets in HCP (Zhao et al., 2015).

A second example is the group-wise sparse representation of fMRI data of multiple groups of subjects (Lv et al., 2015c), which has already been illustrated in Fig. 15.6. Instead of the conventional sparse representation of fMRI signals for one single subject, the group-wise sparse representation learns a common dictionary matrix from the aggregated signal matrix of multiple groups of subjects. As illustrated in Fig. 15.7, this approach has the potential to be applied to clinical data

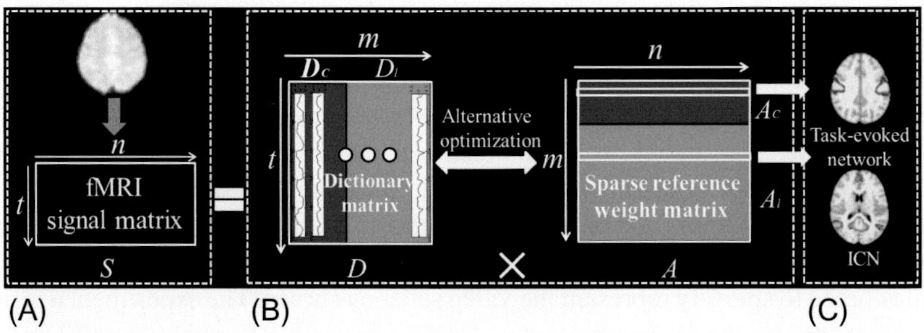

FIG. 15.11

The computational framework of supervised dictionary learning and sparse representation of tfMRI data for identification of concurrent functional brain networks. (A) FMRI signal matrix S aggregated from the whole-brain fMRI signals of one subject. t is the fMRI time point and n is the voxel number. (B) The learned dictionary D and sparse reference weight matrix A based on the supervised dictionary learning and sparse representation. m is the number of dictionary atoms. D_c refers to the predefined dictionary atoms based on model-driven task stimulus curves and is kept unchanged. D_l denotes the other part of dictionary atoms which is optimized during dictionary learning. A_c, which corresponds to D_c, represents the identified model-driven task-evoked functional networks. A_l, which corresponds to D_l, represents other identified data-driven functional networks including ICNs. (C) Two examples of identified model-driven task-evoked functional network and data-driven ICNs.

which includes multiple groups (eg, normal control and diseased groups) to identify a collection of functional brain networks/regions which are possibly affected by the disease (Lv et al., 2015c). A third example is a two-stage dictionary learning and sparse representation framework to differentiate the tfMRI and rsfMRI data (Zhang et al., 2016). The core idea is to perform dictionary learning and sparse representation of whole-brain fMRI signals twice. In the first stage, the HAFNI framework is applied to the whole-brain tfMRI or rsfMRI signals of each subject. In the second stage, all learned dictionary matrices of tfMRI and rsfMRI from the first stage are aggregated into one matrix and the matrix is further decomposed into a dictionary matrix and a sparse reference weight matrix (Zhang et al., 2016). This approach is capable of identifying the distinctive and descriptive dictionary atoms within the dictionary matrix in the second stage, and is able to effectively discriminate tfMRI and rsfMRI signals (Zhang et al., 2016).

15.5 FUTURE DIRECTIONS OF HAFNI APPLICATIONS

The HAFNI framework can be enhanced in the future in the following aspects. First, advanced algorithms such as signal sampling strategies (Ge et al., 2015) may be introduced and integrated to speed up the dictionary learning and sparse representation procedure. Second, our recently developed HAFNI-Enabled Largescale Platform for Neuroimaging Informatics (HELPNI) (Makkie et al., 2015) or Hadoop/SPARK-based systems can be implemented to handle fMRI "big data." Third, the HAFNI framework can be adopted to analyze brain dynamics and to establish functional dynamic HAFNI. Finally, HAFNI can also be adopted to identify network-based neuroimaging biomarkers to characterize, subtype, and diagnose other neurological/psychiatric disorders (eg, autism, schizophrenia, post-traumatic stress disorder, etc.).

ACKNOWLEDGMENTS

The HAFNI framework and its applications summarized in this chapter are based on the works of CAID lab members (http://caid.cs.uga.edu/) and their collaborators. We would like to thank all of them.

REFERENCES

Andersen, A.H., Gash, D.M., Avison, M.J., 1999. Principal component analysis of the dynamic response measured by fMRI: a generalized linear systems framework. Magn. Reson. Imaging 17 (6), 795–815.

Anderson, M.L., Kinnison, J., Pessoa, L., 2013. Describing functional diversity of brain regions and brain networks. NeuroImage 73, 50–58.

Bandettini, P.A., Jesmanowicz, A., Wong, E.C., Hyde, J.S., 1993. Processing strategies for time-course data sets in functional MRI of the human brain. Magn. Reson. Med. 30 (2), 161–173.

Biswal, B.B., Yetkin, F.Z., Haughton, V.M., Hyde, J.S., 1995. Functional connectivity in the motor cortex of resting human brain using echo-planar MRI. Magn. Reson. Med. 34, 537–541.

Bullmore, E., Fadili, J., Breakspear, M., Salvador, R., Suckling, J., Brammer, M., 2003. Wavelets and statistical analysis of functional magnetic resonance images of the human brain. Stat. Meth. Med. Res. 12 (5), 375–399.

Calhoun, V.D., Adali, T., 2006. 'Unmixing' functional magnetic resonance imaging with independent component analysis. IEEE Eng. Med. Biol. Mag. 25 (2), 79–90.

Chang, C.C., Lin, C.J., 2011. LIBSVM: a library for support vector machines. ACM Trans. Intell. Syst. Technol. 2 (3), 27.

Cohen, A.L., Fair, D.A., Dosenbach, N.U.F., Miezin, F.M., Dierker, D., Van Essen, D.C., Schlaggar, B.L., Petersen, S.E., 2008. Defining functional areas in individual human brains using resting functional connectivity MRI. NeuroImage 41, 45–57.

Deng, F., Zhu, D., Lv, J., Guo, L., Liu, T., 2012. fMRI signal analysis using empirical mean curve decomposition. IEEE Trans. Biomed. Eng. 60 (1), 42–54.

Descombes, X., Kruggel, F., von Cramon, D.Y., 1998. fMRI signal restoration using a spatio-temporal Markov random field preserving transitions. NeuroImage 8 (4), 340–349.

DuBois Bowman, F., Caffo, B., Bassett, S.S., Kilts, C., 2008. Bayesian hierarchical framework for spatial modeling of fMRI data. NeuroImage 39, 146–156.

Duncan, J., 2010. The multiple-demand (MD) system of the primate brain: mental programs for intelligent behaviour. Trends Cogn. Sci. 14 (4), 172–179.

Fedorenko, E., Duncan, J., Kanwisher, N., 2013. Broad domain generality in focal regions of frontal and parietal cortex. Proc. Natl. Acad. Sci. USA 110 (41), 16616–16621.

Fox, M.D., Raichle, M.E., 2007. Spontaneous fluctuations in brain activity observed with functional magnetic resonance imaging. Nat. Rev. Neurosci. 8 (9), 700–711.

Friston, K.J., 2009. Modalities, modes, and models in functional neuroimaging. Science 326, 399–403.

Friston, K.J., Holmes, A.P., Worsley, K.J., Poline, J.P., Frith, C.C., Frackowiak, R.S.J., 1994. Statistical parametric maps in functional imaging: a general linear approach. Hum. Brain Map. 2 (4), 189–210.

Ge, B., Makkie, M., Wang, J., Zhao, S., Jiang, X., Li, X., Lv, J., Zhang, S., Zhang, W., Han, J., Guo, L., Liu, T., 2015. Signal sampling for efficient sparse representation of resting state fMRI data. Brain Imaging Behav. In press. Epub ahead of print. http://dx.doi.org/10.1007/s11682-015-9487-0.

Hall, M.A., Smith, L.A., 1999. Feature selection for machine learning: comparing a correlation-based filter approach to the wrapper. In: In FLAIRS Conference 1999, pp. 235–239.

Hartvig, N.V., Jensen, J.L., 2000. Spatial mixture modeling of fMRI data. Hum. Brain Map. 11 (4), 233–248.

Hu, X., Lv, C., Cheng, G., Lv, J., Guo, L., Han, J., Liu, T., 2015. Sparsity constrained fMRI decoding of visual saliency in naturalistic video streams. IEEE Trans. Auton. Ment. Dev. 7 (2), 65–75.

Jiang, X., Zhang, X., Zhu, D., 2014. Intrinsic functional component analysis via sparse representation on Alzheimer's disease neuroimaging initiative database. Brain Connect. 4 (8), 575–86.

Jiang, X., Li, X., Lv, J., Zhang, T., Zhang, S., Guo, L., Liu, T., 2015. Sparse representation of HCP grayordinate data reveals novel functional architecture of cerebral cortex. Hum. Brain Map. 36 (12), 5301–5319.

Kanwisher, N., 2010. Functional specificity in the human brain: a window into the functional architecture of the mind. Proc. Natl. Acad. Sci. USA 107 (25), 11163–11170.

Krekelberg, B., Boynton, G.M., Van Wezel, R.J., 2006. Adaptation: from single cells to bold signals. Trends Neurosci. 29 (5), 250–256.

Logothetis, N.K., 2008. What we can do and what we cannot do with fMRI. Nature 453, 869–878.

Lv, J., Jiang, X., Li, X., Zhu, D., Chen, H., Zhang, T., Zhang, S., Hu, X., Han, J., Huang, H., Zhang, J., Guo, L., Liu, T., 2015a. Sparse representation of whole-brain fMRI signals for identification of functional networks. Med. Image Anal. 20 (1), 112–134.

Lv, J., Jiang, X., Li, X., Zhu, D., Zhang, S., Zhao, S., Chen, H., Zhang, T., Hu, X., Han, J., Ye, J., Guo, L., Liu, T., 2015b. Holistic atlases of functional networks and interactions reveal reciprocal organizational architecture of cortical function. IEEE Trans. Biomed. Eng. 62 (4), 1120–1131.

Lv, J., Jiang, X., Li, X., Zhu, D., Zhao, S., Zhang, T., Hu, X., Han, J., Guo, L., Li, Z., Coles, C., Hu, X., Liu, T., 2015c. Assessing effects of prenatal alcohol exposure using group-wise sparse representation of fMRI data. Psychiat. Res. Neuroimaging 233 (2), 254–68.

Mairal, J., Bach, F., Ponce, J., Sapiro, G., 2010. Online learning for matrix factorization and sparse coding. J. Mach. Learn. Res. 11, 19–60.

Makkie, M., Zhao, S., Jiang, X., Lv, J., Zhao, Y., Ge, B., Li, X., Han, J., Liu, T., 2015. HAFNI-enabled largescale platform for neuroimaging informatics (HELPNI). Brain Inform. 2 (4), 1–17.

McKeown, M.J., Jung, T.P., Makeig, S., Brown, G., Kindermann, S.S., Lee, T.W., Sejnowski, T.J., 1998. Spatially independent activity patterns in functional MRI data during the stroop color-naming task. Proc. Natl. Acad. Sci. USA 95 (3), 803.

McKeown, M.J., Hansen, L.K., Sejnowsk, T.J., 2003. Independent component analysis of functional MRI: what is signal and what is noise? Curr. Opin. Neurobiol. 13 (5), 620–629.

Ogawa, S., Lee, T.M., Kay, A.R., Tank, D.W., 1990a. Brain magnetic resonance imaging with contrast dependent on blood oxygenation. Proc. Natl. Acad. Sci. USA 87 (24), 9868–9872.

Ogawa, S., Lee, T.M., Nayak, A.S., Glynn, P., 1990b. Oxygenation-sensitive contrast in magnetic resonance image of rodent brain at high magnetic fields. Magn. Reson. Med. 14, 68–78.

Pessoa, L., 2012. Beyond brain regions: network perspective of cognition-emotion interactions. Behav. Brain Sci. 35 (03), 158–159.

Petersen, R.C., Stevens, J.C., Ganguli, M., Tangalos, E.G., Cummings, J.L., DeKosky, S.T., 2001. Practice parameter: early detection of dementia: mild cognitive impairment (an evidence-based review). report of the quality standards subcommittee of the American Academy of Neurology. Neurology 56, 1133–1142.

Rakic, P., 1988. Specification of cerebral cortical areas. Science 241, 170–176.

Shimizu, Y., Barth, M., Windischberger, C., Moser, E., Thurner, S., 2004. Wavelet-based multifractal analysis of fMRI time series. NeuroImage 22, 1195–1202.

Smith, S.M., Fox, P.T., Miller, K.L., Glahn, D.C., Fox, P.M., Mackay, C.E., Filippini, N., Watkins, K.E., Torod, R., Laird, A.R., Beckmann, C.F., 2009. Correspondence of the brain's functional architecture during activation and rest. Proc. Natl. Acad. Sci. USA 106 (31), 13040–13045.

Smith, S.M., Andersson, J., Auerbach, E.J., Beckmann, C.F., Bijsterbosch, J., Douaud, G., Duff, E., Feinberg, D.A., Griffanti, L., Harms, M.P., Kelly, M., Laumann, T., Miller, K.L., Moeller, S., Petersen, S., Power, J., Salimi-Khorshidi, G., Snyder, A.Z., Vu, A.T., Woolrich, M.W., Xu, J., Yacoub, E., Urbil, K., Van Essen, D.C., Glasser, M.F., 2013. Resting-state fMRI in the human connectome project. NeuroImage 80 (15), 144–168.

Thies, W., Bleiler, L., 2013. Alzheimer's disease facts and figures. Alzheimer's Dementia 9 (2), 208–45.

van den Heuvel, V., Pol, H.H., 2010. Exploring the brain network: a review on resting-state fMRI functional connectivity. Eur. Neuropsychopharmacol. 20 (8), 519–534.

Van Den Heuvel, M., Mandl, R., Luigjes, J., Hulshoff Pol, H., 2008. Microstructural organization of the cingulum tract and the level of default mode functional connectivity. J. Neurosci. 28, 10844–51.

Van Essen, D.C., Smith, S.M., Barch, D.M., Behrens, T.E., Yacoub, E., Ugurbil, K., WU-Minn HCP Consortium, 2013. The Wu-Minn human connectome project: an overview. NeuroImage 80, 62–79.

Varoquaux, G., Schwartz, Y., Pinel, P., Thirion, B., Wells, W.M., Joshi, S., Pohl, K., 2013. Cohort-level brain mapping: learning cognitive atoms to single out specialized regions. In: IPMI-Information Processing in Medical Imaging, LNCS vol. 7917. Springer, Heidelberg, pp. 438–449.

Wagenmakers, E.J., 2015. An Introduction to Model-Based Cognitive Neuroscience. Forstmann, B.U. (Ed.). Springer, New York.

Woolrich, M., Ripley, B., Brady, J., Smith, S., 2001. Temporal autocorrelation in univariate linear modelling of fMRI data. NeuroImage 14 (6), 1370–1386.

Worsley, K.J., 1997. An overview and some new developments in the statistical analysis of pet and fMRI data. Hum. Brain Map. 5 (4), 254–258.

Zhang, S., Li, X., Lv, J., Jiang, X., Guo, L., Liu, T., 2016. Characterizing and differentiating task-based and resting state fMRI signals via two-stage sparse representations. Brain Imaging Behav. 10 (1), 21–32.

Zhao, S., J., H., Lv, J., Jiang, X., Hu, X., Zhao, Y., Ge, B., Guo, L., Liu, T., 2015. Supervised dictionary learning for inferring concurrent brain networks. IEEE Trans. Med. Imaging 34 (10), 2036–2045.

CHAPTER

Neuronal network architecture and temporal lobe epilepsy: A connectome-based and machine learning study

16

B.C. Munsell[1], G. Wu[2], S. Keller[3], J. Fridriksson[4], B. Weber[5], M. Stoner[2], D. Shen[2], L. Bonilha[6]

College of Charleston, Charleston, SC, United States[1] University of North Carolina at Chapel Hill, Chapel Hill, NC, United States[2] University of Liverpool, Liverpool, United Kingdom[3] University of South Carolina, Columbia, SC, United States[4] University of Bonn, Bonn, Germany[5] Medical University of South Carolina, Charleston, SC, United States[6]

CHAPTER OUTLINE

Machine Learning and Medical Imaging. http://dx.doi.org/10.1016/B978-0-12-804076-8.00016-5

16.1 INTRODUCTION

Improvements in computational neuroimaging analyses now permit the assessment of whole brain maps of structural connectivity (Bartzokis, 2004; Sporns, 2011, 2013). By combining segmented gray matter tissue data from T1 weighted magnetic resonance imaging (MRI) with white matter fiber tractography from diffusion tensor imaging (DTI) MRI, it is possible to chart the organization of white matter connectivity across the entire brain, that is, the structural brain connectome (Sporns et al., 2005; Hagmann et al., 2008; Sporns, 2013). The brain connectome provides an unprecedented degree of information about the organization of neuronal network architecture, both at a regional level, as well as regarding the entire brain network. For this reason, the brain connectome has recently become instrumental in the investigation of neuronal systems organization and its relationship with health and disease, notably in the context of neurological conditions such as epilepsy (Bonilha, 2013, 2015; Besson, 2014; DeSalvo, 2014), schizophrenia (Rubinov and Bullmore, 2013; Crossley, 2014), and dementia (Xie and He, 2011; Daianu, 2013).

Epilepsy is a neurological disorder directly associated with pathological changes in brain network organization. Even though most forms of epilepsy are believed to arise from epileptogenic activity emerging from localized brain areas, there is a growing body of evidence suggesting that focal seizures are in reality the result of hyperexcitation of localized networks, rather than isolated cortical regions (Spencer, 2002; Richardson, 2012). Focal epilepsy is considered to be a disease of neuronal networks, where the aberrant structural and functional organization of neuronal systems leads to unabated neuronal hyper-excitability and seizures (Richardson, 2012). Even though histopathological changes were originally considered to be restricted to isolated brain regions, that is, hippocampal sclerosis in individuals with temporal lobe epilepsy (TLE) (Babb and Brown, 1987), there is a large and growing body of evidence from neuroimaging to suggest that more subtle abnormalities are pervasively distributed across multiple brain regions. In the case of TLE, regions that are functionally or structurally associated with the hippocampus demonstrate gray matter cell loss, white matter deafferentation, and pathological network architecture reorganization (Bernasconi, 2003, 2004; Bonilha, 2003, 2004; Bernhardt, 2008; Focke, 2008; Keller and Roberts, 2008). For this reason, epilepsy is now considered to be a condition that emanates from abnormal neuronal systems rather than isolated brain regions. Indeed, the concept of epilepsy as a neuronal network disorder features prominently in the newest epilepsy and seizures classification from the International League against epilepsy (Berg, 2010).

The clinical relevance of pervasive network abnormalities regarding seizure control is not yet completely understood. It stands to reason that more abnormally distributed networks may lead to different clinical endophenotypes regarding seizure control (Bonilha, 2012). Even though the advent of connectome-based research is still relatively recent, preliminary studies have not only demonstrated a consistent pattern of intra- and extratemporal connectivity abnormalities, but also individualized patterns of network abnormalities that may be directly associated with prognosis (Bonilha, 2015). Even though these are initial findings, they are a promising indication of the pathological significance of connectome-based abnormalities in epilepsy.

Here, two different clinical studies are performed that apply machine learning techniques to create computational models capable of identifying abnormal network connections in structural brain connectomes reconstructed using white matter fiber tracts from DTI data. The aim and goal of each study is provided below, and the specific details are provided in Sections 16.2 and 16.3.

16.1.1 TREATMENT OUTCOME PREDICTION OF PATIENTS WITH TLE

The objective of this study is to evaluate machine learning algorithms aimed at predicting surgical treatment outcomes in groups of patients with TLE using only the structural brain connectome. A two-stage connectome-based prediction framework was developed that gradually selects a small number of abnormal network connections that contribute to the surgical treatment outcome, and in each stage a linear kernel operation is used to further improve the accuracy of the learned classifier. Using a 10-fold cross-validation strategy, the connectome-based prediction framework is able to separate patients with TLE from healthy controls with 80% accuracy, and is able to correctly predict the surgical treatment outcome of patients with TLE with 70% accuracy. Compared to existing state-of-the-art methods that use voxel-based morphometry data, the proposed connectome-based prediction framework is a suitable alternative with comparable prediction performance. Our results additionally show that machine learning algorithms that exclusively use structural connectome data can predict treatment outcomes in epilepsy with similar accuracy compared with "expert-based" clinical decision.

16.1.2 NAMING IMPAIRMENT PERFORMANCE OF PATIENTS WITH TLE

The objective of this study to evaluate the neuronal networks that support naming in TLE by using a machine learning algorithm intended to predict naming performance in subjects with medication refractory TLE using only the structural brain connectome reconstructed from DTI. A connectome-based prediction framework was developed using network properties from anatomically defined brain regions across the entire brain, which were used in a multitask machine learning algorithm followed by support vector regression. Nodal eigenvector centrality, a measure of regional network integration, predicted approximately 60% of the variance in naming. The nodes with the highest regression weight were bilaterally distributed among perilimbic subnetworks involving mainly the medial and lateral temporal lobe regions. In the context of emerging evidence regarding the role of large structural networks that support language processing, our results suggest intact naming relies on the integration of subnetworks, as opposed to being dependent on isolated brain areas. In the case of TLE, these subnetworks may be disproportionately indicative naming processes that are dependent on semantic integration from memory and lexical retrieval, as opposed to multimodal perception or motor speech production.

16.2 TREATMENT OUTCOME PREDICTION OF PATIENTS WITH TLE

16.2.1 PARTICIPANTS

A cohort of 70 patients with refractory TLE with hippocampal sclerosis or with medical refractory lesional TLE and a cohort of 48 healthy controls (HCs) were included in this study. TLE was diagnosed according to the criteria defined by the International League Against Epilepsy (ILAE, 1989), including a comprehensive neurological evaluation, ictal electroencephalography (EEG) recordings, diagnostic MRI, and, when appropriate, nuclear medicine studies. All TLE cases exhibited unilateral temporal lobe seizure onset during ictal EEG monitoring, and had routine diagnostic MRI revealing unilateral hippocampal atrophy. All patients with TLE underwent anterior temporal lobectomy or amygdalohippocampectomy. The surgical outcome was assessed based on the Engel Surgical Outcome scale (Engel, 2003) defined at least one year after surgery. In general, patients with TLE were classified into two groups: (1) free of disabling of seizures (ie, seizure-free), equivalent to Engel Class I (including Class 1b patients with auras only); and (2) not seizure-free, equivalent to Engel Classes II, III, or IV.

16.2.2 PRESURGICAL IMAGE ACQUISITION AND PROCESSING

The same imaging protocol was applied to all study participants (patients and controls). Images were acquired on a Siemens 3T Verio MRI scanner and the imaging protocol yielded a high-resolution T1-weighted image, with an isotropic voxel size of 1 mm. Diffusion-weighted images were obtained using two diffusion weightings along either 30 or 60 diffusion-encoding directions. DICOM images were converted to NIfTI format using the MRIcron software toolbox.[1] The FMRIB software library diffusion toolkit[2] (FDT) was used for preprocessing diffusion-weighted images and also for diffusion tensor estimation (Heiervang, 2006; Behrens, 2007). The images underwent eddy current correction through affine transformation of each DWI to the base $b = 0$, T2-weighted image.

16.2.3 PRESURGICAL CONNECTOME RECONSTRUCTION

Probabilistic tractography was used to define the number of white matter streamlines connecting each pair of cortical regions according to the Lausanne anatomical atlas that defines 82 regions of interest (ROIs). This step was iteratively performed until the connectivity between all possible pairs of cortical regions was determined. Structural connectivity was obtained by applying FDT's probabilistic method for fiber tracking (Behrens, 2003, 2007; Ciccarelli, 2006) and was performed on diffusion data after

[1] http://www.mccauslandcenter.sc.edu/mricro/mricron/dcm2nii.html
[2] http://www.fmrib.ox.ac.uk/fsl

voxel-wise calculation of the diffusion tensor. In general, probabilistic tractography was chosen because it is theoretically capable of accommodating intravoxel fiber crossings (Behrens, 2007; Nucifora, 2007). Cortical seed regions for tractography were obtained from an automatic segmentation process on the T1-weighted images that subdivided the human cerebral cortex into sulco-gyral-based cortical and subcortical ROIs defined in the anatomical atlas. All processed images were visually inspected to ensure the cortical segmentation quality. The ROIs were then transformed into each subject's DTI space using an affine transformation obtained with FSL's FLIRT.

For each subject, a comprehensive presurgical neural connectivity map or connectome is calculated, where the connectivity is measured by the number of probabilistic white matter fiber tract streamlines arriving at ROI j when ROI i was seeded, averaged with the number of probabilistic white matter fiber tract streamlines arriving at ROI i when ROI j was seeded. The step is iteratively repeated to ensure all 82 cortical ROIs are treated as seed regions. Once all iterations are completed, a symmetric 82×82 density connectivity map D is constructed, where D_{ij} corresponds to the weighted network between ROIs i and j. Since the number of streamlines between two different ROIs is averaged, D is symmetric with respect to the main diagonal. Three different example connectomes using the described reconstruction procedure are shown in Fig. 16.1. In particular, three examples are shown: (1) normal control, (2) patient with TLE that is seizure-free after surgery is performed, and (3) patient with TLE who is not seizure-free (ie, continues to experience seizures) after surgery.

16.2.4 CONNECTOME PREDICTION FRAMEWORK

The block diagram shown in Fig. 16.2 illustrates the basic design and operation of the proposed connectome-based prediction framework that defines a *Stage-1* prediction pipeline that is able to separate patients with TLE from normal controls, and a *Stage-2* prediction pipeline that is able to predict the surgical treatment outcome of patients with TLE. It is important to point out that the Stage-2 prediction pipeline is dependent on the Stage-1 prediction pipeline. That is, the output of the Stage-1 connectome feature selection component is the input to the Stage-2 connectome feature selection component. Furthermore, each prediction pipeline defines three trained components, that is, connectome feature selection, linear kernel operation, and linear SVM classifier, which are sequentially applied one after the other. The rationale behind the two-stage design is directly related to the number of network connections (ie, features) defined in the connectome. Specifically, using only one stage to identify a small subset of features (less than a hundred) from thousands that contribute to the surgical treatment outcome is a very challenging feature selection problem for any machine learning algorithm. Instead, the proposed two-stage design takes a more controlled approach by gradually reducing a high-dimension connectome feature space to a lower-dimension one, thus making the problem more tractable.

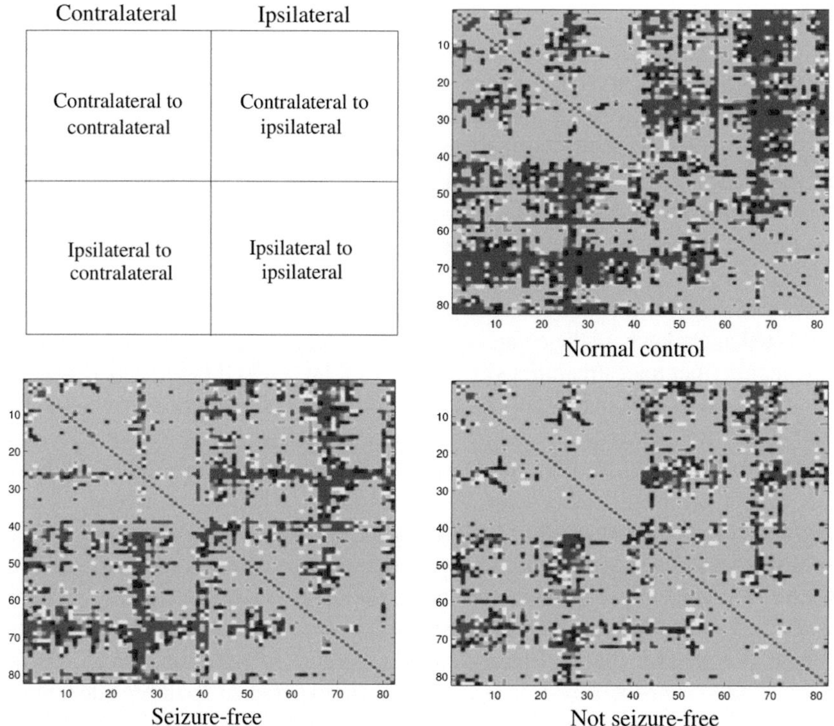

FIG. 16.1

Example connectomes are shown for a healthy control, a seizure-free patient, and a nonseizure-free patient. The brain structures are numbered from 1 to 82 in accordance with the chosen anatomical atlas. In particular, regions 1–42 represent the hemisphere contralateral to seizure onset, and 43–82 represent the hemisphere ipsilateral to seizure onset. Within each hemisphere, the regions are grouped as follows: frontal lobe, temporal lobe, basal nuclei, parietal lobe, and occipital lobe.

16.2.4.1 Connectome feature selection component

Assume an $n \times m$ training data matrix $A = [\boldsymbol{a}_1, \boldsymbol{a}_2, \ldots, \boldsymbol{a}_n]$ of n subjects where row vector $\boldsymbol{a}_i = (a_{i1}, a_{i2}, \ldots, a_{im})$ is an m dimension presurgical connectome feature vector for subject i whose element values are the upper diagonal of the connectivity matrix D, and $\boldsymbol{y} = (y_1, y_2, \ldots, y_n)$ is an n-dimension vector of binary element values that represent the clinical outcome of each subject in the training data set (ie, the clinical outcome for row vector $\boldsymbol{a}_i$ is y_i). Because $m \gg n$, the elastic net (Zou and Hastie, 2005) sparse learning technique is used to find a sparse m dimension weight vector $\boldsymbol{x}$ that minimizes

$$\min_x \tfrac{1}{2}||\tilde{A}\boldsymbol{x} - \boldsymbol{y}||_2^2 + \lambda||\boldsymbol{x}||_1 + \tfrac{\rho}{2}||\boldsymbol{x}||_2^2,$$

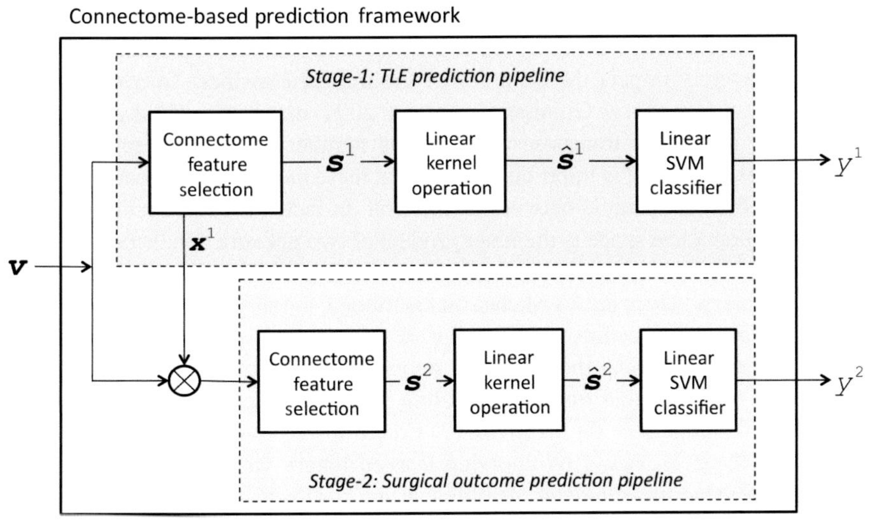

FIG. 16.2

Block diagram that illustrates the basic design and operation of the proposed connectome-based prediction framework. The framework defines two different prediction pipelines, specifically a Stage-1 prediction pipeline and a Stage-2 prediction pipeline. Each prediction pipeline has three trained components: (1) connectome feature selection, (2) linear kernel operation, and (3) linear SVM classifier. Note that the superscript value identifies the stage.

where $\lambda||\boldsymbol{x}||_1$ is the ℓ_1 regularization (sparsity) term, $\frac{\rho}{2}||\boldsymbol{x}||_2^2$ is the ℓ_2 regularization (smoothness) term, and $\tilde{A}$ is an $n \times m$ matrix with normalized training data. Specifically, $\tilde{A}\,(i,j) = (a_{ij} - a_{ij})/\sigma_j$, where a_{ij} is network connection j for subject i, a_{ij} is the mean value of column vector j in matrix A, and σ_j is the standard deviation of column vector j in matrix A. After optimization, $\boldsymbol{x}$ has weight values in [0 1], where weight values equal to zero indicate network connections that do not contribute to the clinical outcome, and weight values greater than zero indicate network connections that do contribute to the clinical outcome. In general, $\boldsymbol{x}$ is referred to as the sparse representation of the training data set. Lastly, each weight value in $\boldsymbol{x}$ greater than zero is set to one. Therefore the resulting sparse representation can be perceived as a binary mask, that is, the network connection is turned on (value of 1) or turned off (value of 0). A new sparse training data matrix is created, $S = [\boldsymbol{s}_1, \boldsymbol{s}_2, \ldots, \boldsymbol{s}_n]^t$, where row vector $\boldsymbol{s}_i = (a_{i1}x_1, a_{i2}x_2, \ldots, a_{im}x_m)$.

Even though the learned sparse representation can greatly reduce the dimension of the input connectome feature vector, the number of nonzero features in the newly created sparse training data matrix will most likely not be equal to the number of training data subjects. This condition may result in an overdetermined system of equations (with more subjects than features), or an underdetermined system of equations (with more features than subjects). In both cases, there may be an infinite number of solutions, or no solution, to this system of linear equations, which in turn may severely impact the accuracy of the trained classifier. To overcome this limitation, a square $\tilde{m} \times \tilde{m}$ Gramian (Lanckriet, 2004) matrix $\hat{S} = S^t S$ is constructed, where S^t is the matrix transpose and $\tilde{m}$ is the number of nonzero features in the sparse representation. It is important to note that these more compact features cannot be mapped back to a single network connection. In fact, each feature in this newly formed mathematical space is the inner product of two network connection vectors.

Finally, $\hat{S}$ and $\boldsymbol{y}$ are used to train a linear two-class SVM classifier based on the LIBSVM library.[3] Once the SVM classifier is trained, the surgical treatment outcome of a high-dimension feature vector, say $\boldsymbol{v} = (v_1, \ldots, v_m)$, not in the training data set can be predicted using the following steps: (1) Normalize each value in $\boldsymbol{v}$ using the learned centering and magnitude scaling values for the $j = 1, \ldots, m$ network connection features $v_i = (v_i - a_{ij})/\sigma_j$. (2) Create sparse connectome feature vector $\mathbf{s} = (v_1 x_1, v_2 x_2, \ldots, v_m x_m)$ by applying learned binary weights. All features that have a zero value are removed, resulting in an $\tilde{m}$ dimension connectome feature vector. (3) Apply learned linear transformation to obtain $\tilde{m}$-dimension feature vector $\hat{\mathbf{s}} = (\hat{s}_1, \hat{s}_2, \ldots, \hat{s}_{\tilde{m}})$, where $\hat{s}_i = \sum_{j=1}^{\tilde{m}} s_j \hat{S}(j, i)$ for $i = 1, \ldots, \tilde{m}$. (4) Calculate the predicted class label $y = \sum_{i=0}^{\tilde{m}} \alpha_i \kappa \left(\boldsymbol{\delta}_i, \hat{s}\right) + b$, where α are the weights, $\boldsymbol{\delta}$ are the support vectors, $\kappa\,(\cdot)$ is the inner product of the two vectors, and b is the bias that defines the linear hyperplane (decision boundary) learned by the SVM algorithm. The sign of the calculated prediction value (ie, $y \geq 0$ or $y < 0$) determines which of the two diagnosis labels the subject is assigned to.

16.2.5 RESULTS AND EVALUATION

The predictive power of the connectome-based framework is evaluated using a 10-fold cross-validation strategy. In particular, the Bonn and MUSC subjects are first combined into one data set, and then partitioned into 10 different folds, where each fold contains connectomes of randomly selected patients (ie, a mixture of seizure-free and not seizure-free) and/or randomly selected normal controls. The prediction framework is iteratively trained using the connectome data in nine of the 10 folds, and then tested using the connectome data in the remaining (or left-out) fold. This iterative process terminates when each fold has been selected as the test one. Using the combined confusion matrix (TP=true positive, FP=false positive, FN=false negative, and TN=true negative) results of each test fold, the prediction performance is reported using the specificity (SPE), sensitivity (SEN), positive predictive value (PPV), negative predictive value (NPV), and accuracy (ACC) measures.

[3] http://www.csie.ntu.edu.tw/$\sim$cjlin/libsvm/

16.2.5.1 Stage-1 TLE prediction pipeline

In this experiment, the total number of subjects in the connectome data set is 118, including 70 patients with TLE and 48 HCs. This data set was randomly partitioned into 10 folds, where 8 of the 10 folds have 12 subjects, and 2 of the 10 folds have 11 subjects. The number of subjects n in the training population is approximately 106, where each training subject is defined by $m = (81 \times 82)/2 = 3321$ dimension connectome feature vector. Lastly, the class labels used to train the prediction pipeline are 0=patient with TLE and 1=HC. The resulting performance of the TLE prediction pipeline was SEN=74%, SPE=88%, PPV=90%, NPV=70%, and ACC=80%. The total number of nonzero network connections $|w_t| = \cup_{f=1}^{10} w_f$ selected by the elastic net algorithm that can differentiate patients with TLE from HCs is 383, where w_t is the union of each learned sparse representation for each fold. Using a two-sample t-test with $\alpha = 0.05$ a paired p-value is calculated for each nonzero network connection in w_t and then sorted in ascending order, where the null hypothesis represents data that are independent random samples from normal distributions with equal means and equal but unknown variances. The top 15 nonzero network connections with the smallest p-values, that is, those with the greatest difference between the two groups, can be seen and visualized in Fig. 16.3.

16.2.5.2 Stage-2 surgical treatment outcome prediction pipeline

In this experiment, the total number of patients with TLE in this data set is 70 and it was also randomly partitioned into 10 folds, where each fold has 7 subjects. The

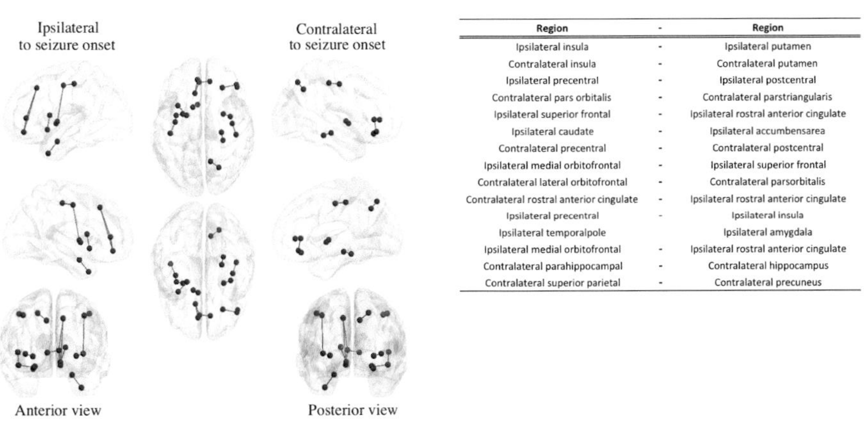

Region	-	Region
Ipsilateral insula	-	Ipsilateral putamen
Contralateral insula	-	Contralateral putamen
Ipsilateral precentral	-	Ipsilateral postcentral
Contralateral pars orbitalis	-	Contralateral parstriangularis
Ipsilateral superior frontal	-	Ipsilateral rostral anterior cingulate
Ipsilateral caudate	-	Ipsilateral accumbensarea
Contralateral precentral	-	Contralateral postcentral
Ipsilateral medial orbitofrontal	-	Ipsilateral superior frontal
Contralateral lateral orbitofrontal	-	Contralateral parsorbitalis
Contralateral rostral anterior cingulate	-	Ipsilateral rostral anterior cingulate
Ipsilateral precentral	-	Ipsilateral insula
Ipsilateral temporalpole		Ipsilateral amygdala
Ipsilateral medial orbitofrontal	-	Ipsilateral rostral anterior cingulate
Contralateral parahippocampal	-	Contralateral hippocampus
Contralateral superior parietal	-	Contralateral precuneus

FIG. 16.3

Stage-1 TLE results. The top 15 connected regions with the smallest p-value (ie, the network connections with the greatest difference between patients with TLE and HCs). The p-values are calculated using a two-sample t-test. The brain regions, defined using the Lausanne anatomical atlas, are represented by the red nodes (dark gray in print versions), and the edge connecting two brain regions represents a network connection in the connectome.

number of subjects n in the training population is 63, where each training subject is defined by an $m = 383$ dimension sparse connectome feature vector, found by the stage-1 connectome feature selection component. The class labels used to train the prediction pipeline are 0=not seizure-free and 1=seizure-free. The resulting performance of the TLE prediction pipeline was SEN=59%, SPE=76%, PPV=63%, NPV=74%, and ACC=70%. The total number of nonzero network connections $|w_t|$ selected by the elastic net algorithm (in the second stage of the proposed two-stage framework) that is able to differentiate the seizure-free postsurgery group from the not-seizure-free postsurgery group is 132. Using a two-sample t-test with $\alpha = 0.05$ a paired p-value is calculated for each nonzero network connection in w_t and then sorted in ascending order, where the null hypothesis represents data that are independent random samples from normal distributions with equal means and equal but unknown variances. The top 15 nonzero network connections with the smallest p-values, that is, those with the greatest difference between the two groups, can be seen and visualized in Fig. 16.4.

The results from this study also support the notion that epilepsy in general, and specifically TLE, are associated with temporal and extratemporal network architecture abnormalities (Bonilha, 2013; DeSalvo, 2013, 2014; Liu, 2014). They also indicate that a pattern of network abnormalities may be relevant on an individual basis to guide the estimation of clinical outcomes. While most studies to date have demonstrated the average effects on TLE on the structural connectome, the

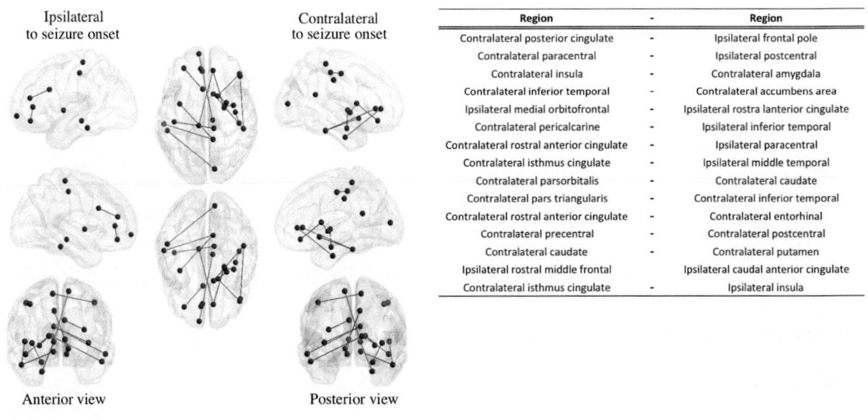

Region	-	Region
Contralateral posterior cingulate	-	Ipsilateral frontal pole
Contralateral paracentral	-	Ipsilateral postcentral
Contralateral insula	-	Contralateral amygdala
Contralateral inferior temporal	-	Contralateral accumbens area
Ipsilateral medial orbitofrontal	-	Ipsilateral rostra lanterior cingulate
Contralateral pericalcarine	-	Ipsilateral inferior temporal
Contralateral rostral anterior cingulate	-	Ipsilateral paracentral
Contralateral isthmus cingulate	-	Ipsilateral middle temporal
Contralateral parsorbitalis	-	Contralateral caudate
Contralateral pars triangularis	-	Contralateral inferior temporal
Contralateral rostral anterior cingulate	-	Contralateral entorhinal
Contralateral precentral	-	Contralateral postcentral
Contralateral caudate	-	Contralateral putamen
Ipsilateral rostral middle frontal		Ipsilateral caudal anterior cingulate
Contralateral isthmus cingulate	-	Ipsilateral insula

FIG. 16.4

Stage-1 TLE results. The top 15 connected regions with the smallest p-value (ie, the network connections with the greatest difference between the patients that are seizure-free after surgery and those that are not seizure-free after surgery). The p-values are calculated using a two-sample t-test. The brain regions, defined using the Lausanne anatomical atlas, are represented by the red nodes (dark gray in print versions), and the edge connecting two brain regions represents a network connection in the connectome.

application of machine learning to the connectome can disclose how the complexity of the connectome can be abridged to yield classifiers with clinical relevance. Importantly, the connectome is a rich and complex data set, and individuals with TLE may harbor abnormalities with inter-individual variability. Thus the use of machine learning can overcome some of these challenges, while incorporating the crucial parameters in the connectome that are relevant to epilepsy management. In this context, the connectome can be used not only to provide information about the neurobiology of the disease, but also to provide information about the personalized clinical trajectory. This trajectory cannot be accurately defined based on the existing clinical measures, and machine learning applied to the connectome may unveil a completely new avenue for additional clinical phenotyping and management planning. One very interesting observation is that the connectome-based prediction framework can achieve roughly the same accuracy as expert clinical opinion. Historically speaking, presurgical diagnostics using expert-based clinical information are approximately 70% accurate for patients that choose to have surgery. In this study, the prediction framework is also 70% accuracy. This level of accuracy is achieved based on the connectome alone, which is pretty remarkable, and is an important clinical finding that may advance outcome prediction for patients with epilepsy.

16.3 NAMING IMPAIRMENT PERFORMANCE OF PATIENTS WITH TLE

16.3.1 PARTICIPANTS

We retrospectively studied a group of 24 patients with medication refractory TLE. All patients had language dominance mapped to the left hemisphere through the WADA test. All patients had medically refractory TLE, diagnosed according to the criteria defined by the International League Against Epilepsy (ILAE, 1989), including a comprehensive neurological evaluation, electroencephalography (EEG) recordings, diagnostic MRI, and, when appropriate, nuclear medicine studies.

16.3.2 LANGUAGE ASSESSMENT

Table 16.1 outlines the language tests used in this study. Specifically, testing included six subtests from the 1983 edition of Boston Diagnostic Aphasia Examination (BDAE) that were designed to be understood by all patients, at all reading levels. For convenience the descriptions provided below are from the Assessment of Aphasia and Related Disorders (Goodglass et al., 2000). Even though the purpose of this study was to evaluate naming performance, measures of repetition and comprehension were included to provide a more comprehensive evaluation of language performance and to elucidate whether individual naming impairments were isolated deficits or part of global language impairments. All the test scores were normalized to a value in

Table 16.1 Name and Description of Each Test Used in the Aphasia Language Assessment

Test	Description
Basic word discrimination (BWD)	Stimulus items included five body parts and 31 picture cards (color, letter, or number). For body parts, patients were instructed to point to the corresponding location. For picture cards, patients were instructed to point to the picture corresponding to the spoken test word. Patients received 1 point per item if the response was correct within 5 s and 1/2 point if the response was correct in more than 5 s. Responses were scored out of 37. The choice of items samples many different categories, targeting categories that are frequently affected or unaffected in aphasic patients.
Verbal agility (VBA)	Patients repeated seven groups of test words as rapidly as possible and the number of repetitions completed within 5 s was recorded. Results were recorded out of a total possible score of 14. Performance on the verbal agility test is correlated with aphasic articulation difficulties and does not rely explicitly on memory or word retrieval, it just requires patients to produce continuous speech.
Repetition (REP)	Patients were instructed to repeat single words back to the examiner. Credit was only given if the word was intelligible, but records were made for articulation impairment. Results were scored out of a possible 10.
Boston naming test (BNT)	The BNT was administered using the Standard form of 60 picture items. Correct responses, latency in seconds, items requiring stimulus cues, and items requiring phonemic cues were recorded. Responses that were erroneous were also recorded. Items that could be answered correctly with the addition of multiple-choice options were also recorded. Credit was allowed for correct oral naming only.
Picture word matching (PWM)	Ten pictures were used for word identification examination. The examiner pointed to each picture and asked the patient to find its name among four word choices offered. Answer choices include the correct answer along with semantic and structural distracters of varying difficulty.
Sentence paragraph comprehension (SPC)	Patients were shown a primary sample sentence and four choices to complete it. Patients were given the option to have the examiner read the sentence and each of the choices aloud and select the correct completion. Patients were then instructed to read the 10 test sentences to him or her and point to the correct completion. Results were recorded out of a possible score of 10.

[0 1], where 0 indicates the patient did not correctly answer any test questions, and 1 indicates the patient correctly answered all the test questions correctly.

16.3.3 IMAGE ACQUISITION AND PROCESSING

The same imaging protocol was applied to all study participants. All subjects were scanned in a Siemens 3T Verio MRI scanner equipped with a 12-channel head coil. The imaging protocol yielded a high-resolution T1-weighted image with an isometric voxel size of 1 mm. Diffusion-weighted images were obtained using a twice-refocused echo planar sequence with three diffusion weightings. DICOM images were converted to NIfTI format using the MRIcron software toolbox. The FMRIB software library diffusion toolkit (FDT) was used for preprocessing diffusion-weighted images and also for diffusion tensor estimation. The images underwent eddy current correction through affine transformation of each DWI to the base $b = 0$, T2-weighted image.

16.3.4 CONNECTOME RECONSTRUCTION

Probabilistic tractography was used to define the number of white matter streamlines connecting each pair of cortical regions, which were defined according to the Lausanne anatomical atlas. This step was performed iteratively until the connectivity between all possible pairs of cortical regions was determined. The connectivity information was then compiled in an individual brain connectome (ie, symmetric two-dimensional connectivity matrix). Specifically, cortical seed regions for tractography were obtained through an automatic segmentation process employing FreeSurfer on the T1-weighted images, dividing the human cerebral cortex into cortical and subcortical ROIs, automatically assigning a neuroanatomical label to each location on a cortical surface model yielding 82 ROIs in the subjects' native T1-weighted space (41 regions in each hemisphere). All processed images were visually inspected to ensure the cortical segmentation quality. The ROIs were transformed into each subject's DTI space using an affine transformation obtained with FSL's FLIRT. Probabilistic tractography was performed using each of the 82 cortical ROIs in diffusion space as the seed region.

For each subject, a comprehensive neuronal connectivity map, or connectome, was calculated where the connectivity is measured by the number of probabilistic white matter fiber tract streamlines arriving at ROI j when ROI i was seeded, averaged with the number of probabilistic white matter fiber tract streamlines arriving at ROI i when ROI j was seeded. This step is iteratively repeated to ensure all 82 cortical ROIs are treated as seed regions, resulting in a symmetric 82×82 density connectivity map D, where D_{ij} is the weighted network connection between ROIs i and j. Since the numbers of streamlines are averaged between each ROI, D is symmetric with respect to the main diagonal. Example connectomes are shown in Fig. 16.5 that is created using the described reconstruction procedure.

16.3.5 CONNECTOME PREDICTION FRAMEWORK

Fig. 16.6 illustrates the basic design and operation of the proposed language prediction framework. In particular, the framework includes a connectome feature selection component that uses a multitask machine learning algorithm followed by a prediction component that consists of a bank of support vector regression (SVR) prediction models, one for each language assessment task used in this study. It is

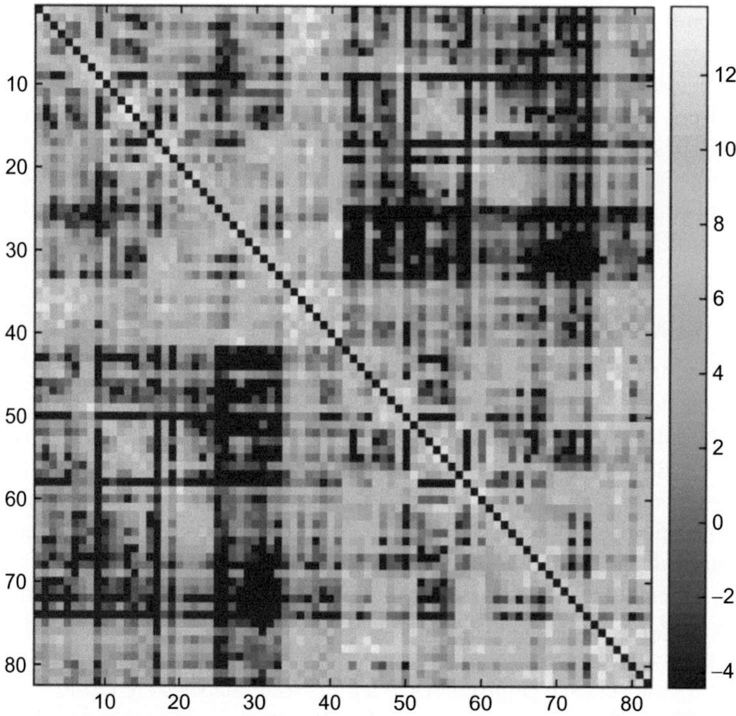

FIG. 16.5

Example of a symmetric 82 × 82 connectivity map for one representative subject. The brain structures are numbered from 1 to 82 in accordance with the atlas. The values in the scale bar illustrate logarithm of link strength.

important to note that each component in the prediction framework is trained with connectome features that are calculated using only one graph-theoretic measure (v). Therefore for the prediction framework to work correctly the graph-theoretic measure used to calculate the input connectome features represented by a^{υ} must be the same graph-theoretic measure used to train the entire framework. The multitask feature selection is used to choose nodes whose graph-theoretic measures are jointly predictive of all language tasks. Given that TLE is typically associated with poor performance on the BNT that is out of proportion to other language impairment, most of the variance in the multitask feature selection comes from the BNT. Nonetheless, by performing the multitask feature selection including all language tests, we ensured that nodes associated with language performance other than naming were also chosen for the subsequent steps, which then directly tests the subnetwork in relationship with the BNT. For this reason, the multitask feature selection controls for the influence of the other language measures (besides naming) on the subnetwork that is subsequently

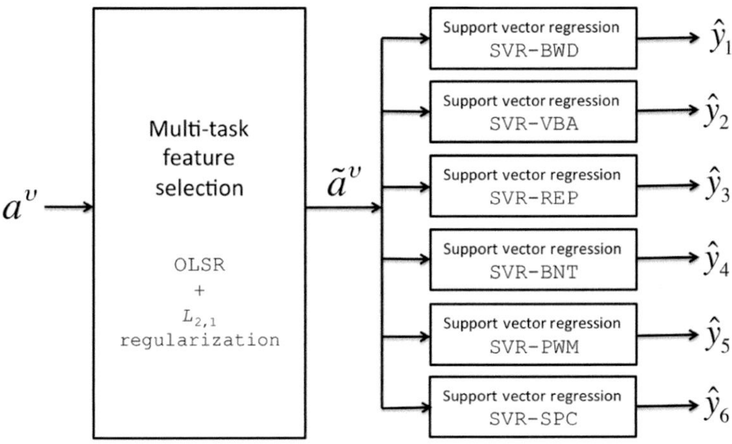

FIG. 16.6

Proposed language prediction framework that includes two trained components that are sequentially applied to connectome features *a* (found using graph-theoretic measure *v*) that is able to predict the test scores for the six language tasks: multitask feature selection using basic word discrimination (BWD), verbal agility (VBA), repetition (REP), Boston naming test (BNT), picture word matching (PWM), and sentence paragraph comprehension (SPC).

tested as predictive of BNT scores. Lastly, the output of the multitask feature selection component is input into a bank of six independent prediction models (one for each language task) that are able to predict the basic word discrimination (BWD), verbal agility (VBA), repetition (REP), Boston naming test (BNT), picture word matching (PWM), and sentence paragraph comprehension (SPC) language scores.

For each subject we created an $m = 82$ dimension connectome feature vector $\boldsymbol{a}^{\upsilon} = (a_1^{\upsilon}, \ldots, a_i^{\upsilon}, \ldots, a_m^{\upsilon})$ using the connectivity values in D as the eigenvector centrality graph-theoretic measure (Rubinov and Sporns, 2010) (denoted by ν). In general, the selected graph-theoretic measure allows us to quantify how a brain region influences the network. Specifically, this measure quantifies regional connectivity by identifying highly connected brain regions, which are connected to neighboring brain regions that are also highly connected. The eigenvector centrality measure for ROI i is $a_i = |\lambda_i|$, where $|\lambda_i|$ is the absolute value of the ith eigenvalue. More specifically, the m eigenvalues $(\lambda_1, \lambda_2, \ldots, \lambda_m)$ are found by performing the eigen decomposition on D.

Given a normalized $n \times m$ training data matrix $A^{\upsilon} = \left[\boldsymbol{a}_1^{\upsilon}, \boldsymbol{a}_2^{\upsilon}, \ldots, \boldsymbol{a}_n^{\upsilon}\right]$ of n subjects where row vector $\boldsymbol{a}_i^{\upsilon}$ is the connectome feature vector for subject i found using graph measure ν, and an $n \times k$ training label matrix $Y = \left[\boldsymbol{y}_1, \boldsymbol{y}_2, \ldots, \boldsymbol{y}_n\right]$ where the elements in row vector $\boldsymbol{y}_i = (y_{i1}, y_{i2}, \ldots, y_{ik})$ represent the known $k = 6$ language test scores for subject i (y_{i1} is the BWD test score, y_{i2} is the VBA test score, etc.), an ordinary least squares linear regression (OLSR) algorithm, which includes an $\ell_{2,1}$

regularization term (Gong, 2013), is used to find a sparse $m \times k$ dimension weight matrix X^{υ} that minimizes

$$\min_{X} \frac{1}{2}||A^{\upsilon}X^{\upsilon} - Y||_2^2 + \lambda||X^{\upsilon}||_{2,1},$$

where $\lambda||X^{\upsilon}||_{2,1}$ is the regularization term. The above equation is optimized using the *mcLeastR* function in the Sparse Learning with Efficient Projections software package.[4] After optimization, X^{υ} has weight values in [0 1] where values greater than zero indicate brain regions that contribute to the language tasks. The weight matrix X^{υ} is then used to find an m dimension vector:

$$\boldsymbol{x}^{\upsilon} = \left(\sqrt{\sum\nolimits_{j=1}^{k}(X_{1j}^{\upsilon}X_{1j}^{\upsilon})}, \ldots, \sqrt{\sum\nolimits_{j=1}^{k}(X_{mj}^{\upsilon}X_{mj}^{\upsilon})}\right),$$

that is commonly referred to as the sparse representation of the training data set. In our approach weight values greater than zero are set to 1, therefore the resulting sparse representation becomes a binary mask. That is, a brain region is turned on (value of 1) or turned off (value of 0). A new $n \times m$ sparse training data matrix is created $\tilde{A}^{\upsilon} = \left[\tilde{\boldsymbol{a}}_1^{\upsilon}, \tilde{\boldsymbol{a}}_2^{\upsilon}, \ldots, \tilde{\boldsymbol{a}}_n^{\upsilon}\right]$, where row vector $\tilde{\boldsymbol{a}}_i^{\upsilon} = \left(a_{i1}^{\upsilon}x_1^{\upsilon}, a_{i2}^{\upsilon}x_2^{\upsilon}, \ldots, a_{im}^{\upsilon}x_m^{\upsilon}\right)$.

The sparse connectome feature matrix $\tilde{A}^{\upsilon}$ and the known language test score matrix Y are then used to train a bank of support vector regression (SVR) prediction models, specifically one prediction model for each language task as shown in Fig. 16.2. In particular, a linear ε-SVR prediction model (Vapnik, 1998), which is based on the LIBSVM library (Chang and Lin, 2011), is independently trained for each language task. Once each ε-SVR prediction model is trained, the six language scores can be predicted using the three steps: (1) Calculate the normalized $\boldsymbol{a}^{\upsilon}$, the connectome feature vector using graph measure υ; (2) create sparse feature vector $\tilde{\boldsymbol{a}}^{\upsilon}$ by applying learned binary weights in x^{υ} to $\boldsymbol{a}^{\upsilon}$; and (3) calculate the predicted language score $\hat{y}_k = \sum_{i=1}^{m}\left(\alpha_i - \alpha_i^*\right)\langle\boldsymbol{\phi}_i, \tilde{\boldsymbol{a}}^{\upsilon}\rangle + b$ for tasks $k = 1$–6, where $\langle\cdot\rangle$ is the inner product of two vectors, $\alpha_i^{(*)}$ is the regression weight, ϕ_i is the support vector, and b is the bias that defines the linear regression line learned by the ε-SVR algorithm.

16.3.6 RESULTS AND EVALUATION

For each graph-theoretic measure (v) a prediction framework is constructed and the performance is assessed using a leave-one-out cross-validation (LOOCV) strategy that iteratively removes one subject from the population as the test subject, and the remaining subjects are used to train the prediction framework. For each leave-one-out test subject $l = 1, 2, \ldots, n$, the predicted language scores $\hat{\boldsymbol{y}}_l = \left(\hat{y}_{l1}, \hat{y}_{l2}, \ldots, \hat{y}_{lk}\right)$ are calculated. Finally, the mean square error $\text{MSE}_k = \frac{1}{n}\sum_{l=1}^{n}\left|y_{lk} - \hat{y}_{lk}\right|$ for the $k = 1, 2, \ldots, 6$ language tests are calculated where $\boldsymbol{y}_l = (y_{l1}, y_{l2}, \ldots, y_{lk})$ is the

[4] *http://www.public.asu.edu/~jye02/Software/SLEP*

Table 16.2 Prediction Framework Results for the Six Language Tests, and for the Eigenvector Graph-Theoretic Measure

	Eigenvector Centrality	
Language Tests	**MSE**	R^2
Basic word discrimination (BWD)	0.002	0.99
Verbal agility (VBA)	0.010	0.44
Repetition (REP)	0.008	0.99
Boston naming test (BNT)	0.035	0.60
Picture word matching (PWM)	0.013	0.72
Sent paragraph comprehension (SPC)	0.041	0.75

The average regression correlation coefficient (R^2) measures how well the linear regression line found by the ε-SVR algorithm fits the connectome features to the known language scores.

known language scores for test k, and $|\cdot|$ is the absolute value. If the MSE value is zero then the predicted language test score is identical to the known language test score.

The language prediction results are summarized in Table 16.2, trained and tested using connectome features found using the eigenvector graph-theoretic measure. Additionally, Table 16.2 includes the average regression correlation coefficient (R^2) value, which discloses the linear regression found by the e-SVR algorithm in relationship with the language scores, that is, how well the predicted scores correlated with the known scores. Given that in the LOOCV training procedure $n = 24$ prediction frameworks are created, 24 prediction models are created for each of the six language tests, therefore the reported R^2 language test k is $\frac{1}{n}\sum_{i=1}^{n} R_k^2$.

The results in Fig. 16.7 and Table 16.3 show the brain regions that have the greatest frequency in the learned binary sparse representation found by the multitask feature selection component when connectome features are calculated using the eigenvector centrality graph-theoretic measure. The results in Table 16.3 also show the average ROI regression weight calculated using each of the $k = 6\varepsilon$-SVR regression models in each of the $n = 24$ prediction frameworks. Specifically, the average regression weight for ROI i is $\bar{r}_i^{\upsilon} = \frac{1}{nk}\sum_{j\in n, t\in k} r_{tj}^{\upsilon}$ where r_{tj}^{υ} is the ε-SVR regression weight for prediction framework j and language task t when graph measure υ is used. In general, an average regression weight that makes a value of zero indicates the brain region does not contribute to any of the language tasks and a very large average regression weight indicates the brain region has a significant contribution to one or more of the language tasks. In summary, if a brain region in Table 16.3 has a large frequency value and a large average regression weight value, then this brain region makes a significant contribution to one or more of the language tasks.

From nodal regression weights within the eigenvector centrality model, it is possible to observe that this subnetwork is strongly dependent on medial temporal structures, including the entorhinal cortex and basal nuclei (pallidum and putamen),

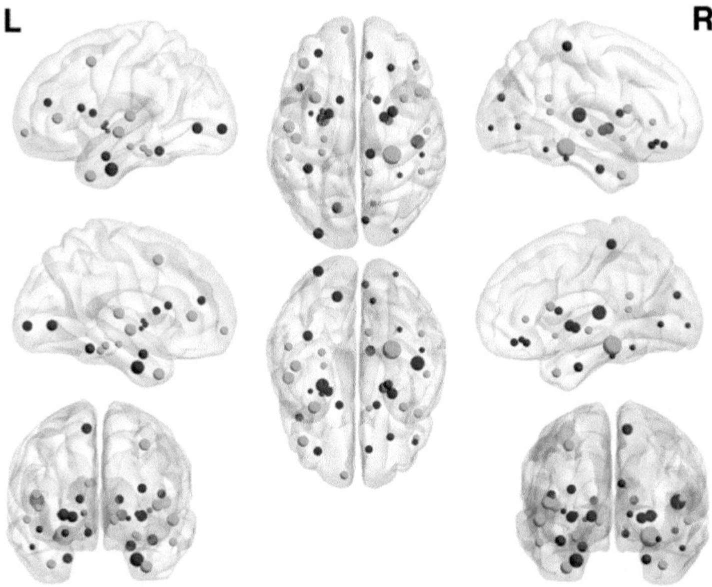

FIG. 16.7

Brain regions that have the greatest frequency among the $n = 25$ binary sparse representations found by the multitask feature selection component and connectome features calculated using the eigenvector graph-theoretic measure. Node color is based on frequency in sparse representation (gray = 100%, light gray 99–90%, dark gray 89–80%). These features are described in detail in the accompanying Table 16.3. Furthermore, the size of the node is related to the average regression weight shown in Table 16.3.

Table 16.3 Accompanying Legend to Fig. 16.7

Rank	Frequency in Sparse Representation	Average Regression Weight (Across the Six SVR Models)	Node Color in Visualization	Region Description
1	100%	4.12	Red	Right transverse temporal
2	100%	3.81	Red	Left entorhinal
3	100%	2.83	Red	Right putamen
4	100%	2.81	Red	Left lingual
5	100%	2.57	Red	Right pallidum
6	100%	2.48	Red	Right paracentral
7	100%	2.46	Red	Left fusiform
8	100%	2.30	Red	Left amygdala
9	100%	2.14	Red	Right entorhinal

Continued

Table 16.3 Accompanying Legend to Fig. 16.7—cont'd

Rank	Frequency in Sparse Representation	Average Regression Weight (Across the Six SVR Models)	Node Color in Visualization	Region Description
10	100%	1.69	Red	Right caudate
11	100%	1.67	Red	Left parsopercularis
12	100%	1.54	Red	Right cuneus
13	100%	1.47	Red	Right lateral orbitofrontal
14	100%	0.98	Red	Right parsorbitalis
15	100%	0.89	Red	Right lingual
16	100%	0.85	Red	Left putamen
17	100%	0.66	Red	Right inferior temporal
18	100%	0.65	Red	Right lateral occipital
19	100%	0.64	Red	Right fusiform
20	100%	0.55	Red	Left insula
21	100%	0.46	Red	Left pallidum
22	96%	5.76	Green	Right parahippocampal
23	96%	2.96	Green	Left temporal pole
24	96%	2.62	Green	Left caudal middle frontal
25	96%	2.00	Green	Right parsopercularis
26	96%	1.45	Green	Left inferior temporal
27	96%	1.41	Green	Right isthmuscingulate
28	96%	1.37	Green	Left parahippocampal
29	96%	1.12	Green	Right bankssts
30	96%	1.11	Green	Left hippocampus
31	96%	1.05	Green	Right parstriangularis
32	96%	0.72	Green	Right accumbens area
33	92%	3.10	Green	Left superior temporal
34	92%	2.85	Green	Left transverse temporal
35	92%	2.66	Green	Left parstriangularis
36	92%	1.84	Green	Right temporal pole
37	92%	1.55	Green	Left frontal pole
38	92%	1.39	Green	Right superior temporal

Continued

Table 16.3 Accompanying Legend to Fig. 16.7—cont'd

Rank	Frequency in Sparse Representation	Average Regression Weight (Across the Six SVR Models)	Node Color in Visualization	Region Description
39	88%	1.84	Blue	Left rostral middle frontal
40	88%	1.74	Blue	Left caudate
41	83%	2.57	Blue	Left lateraloccipital
42	83%	1.44	Blue	Right medial orbitofrontal

Within the multitask feature selection, the largest frequency in sparse representation is 100%, and the smallest frequency is 0. A frequency of 100% indicates that this brain region was selected by the group LASSO algorithm in all 24 LOOCV folds, and a frequency of 0% means the group LASSO algorithm, in any LOOCV fold, did not select the brain region. Within all six ε-SVR regression models and all 24 LOOCV folds an average regression weight that has a value of zero indicates that this brain region does not contribute to any of the language tasks and a very large average regression weight indicates that this brain region makes a significant contribution to one or more of the language tasks.

followed by other medial temporal structures (eg, parahippocampal gyri, hippocampi and amydgala) and frontal regions. Given the high frequency of memory problems in subjects with difficult to control TLE (Elger, 2004), these results may indicate that naming performance in TLE is mostly associated with the structures supporting memory function, notably the circuitry involving the connections between *medial temporal to lateral temporal to basal nuclei to frontal regions.* As such, among all necessary components involved in naming may disproportionately emphasize the semantic integration and symbol retrieval processes, as opposed to multimodal recognition and speech production.

Naming impairments in TLE are associated with the structural preservation of a subnetwork, which includes gray matter regions located mostly in the temporal and frontal areas. Approximately 60% of naming performance on the BNT can be predicted by a measure of integration of the nodes in this subnetwork termed eigenvector centrality. These results confirm the assumption that language and, more specifically, naming, which are higher cognitive functions, rely on the concerted action among multiple regions, as opposed to being dependent on isolated brain areas. Nonetheless, the generalization of these results to other neurological conditions associated with naming impairment, or to normal brain function, should take into account that the disproportionate participation of right hemisphere nodes in the predictive model may be related to chronic and gradual network rearrangements in epilepsy and the chronic compensatory role of the right hemisphere. Furthermore, this distributed network may also be unequally influenced by memory problems associated with TLE, being more sensitive to memory-dependent aspects of naming such as semantic integration and communicative symbol retrieval, as opposed to multimodal perception or speech production.

REFERENCES

Babb, T.L., Brown, W.J., 1987. Pathological Findings in Epilepsy. Raven Press, New York, pp. 511–540 .

Bartzokis, G., 2004. Age-related myelin breakdown: a developmental model of cognitive decline and Alzheimer's disease. Neurobiol. Aging 25 (1), 5–18.

Behrens, T., 2003. Characterization and propagation of uncertainty in diffusion-weighted MR imaging. Magn. Reson. Med. 50 (5), 1077–1088.

Behrens, T., 2007. Probabilistic diffusion tractography with multiple fibre orientations: what can we gain? NeuroImage 34 (1), 144–155.

Berg, A.T., 2010. Revised terminology and concepts for organization of seizures and epilepsies: report of the ILAE commission on classification and terminology, 2005–2009. Epilepsia 51 (4), 676–685.

Bernasconi, N., 2003. Mesial temporal damage in temporal lobe epilepsy: a volumetric MRI study of the hippocampus, amygdala and parahippocampal region. Brain 126 (Pt 2), 462–469.

Bernasconi, N., 2004. Whole-brain voxel-based statistical analysis of gray matter and white matter in temporal lobe epilepsy. NeuroImage 23 (2), 717–723.

Bernhardt, B.C., 2008. Mapping limbic network organization in temporal lobe epilepsy using morphometric correlations: insights on the relation between mesiotemporal connectivity and cortical atrophy. NeuroImage 42 (2), 515–524.

Besson, P., 2014. Structural connectivity differences in left and right temporal lobe epilepsy. NeuroImage 100, 135–144.

Bonilha, L., 2003. Medial temporal lobe atrophy in patients with refractory temporal lobe epilepsy. J. Neurol. Neurosurg. Psychiat. 74 (12), 1627–1630.

Bonilha, L., 2004. Voxel-based morphometry reveals gray matter network atrophy in refractory medial temporal lobe epilepsy. Arch. Neurol. 61 (9), 1379–1384.

Bonilha, L., 2012. Subtypes of medial temporal lobe epilepsy: influence on temporal lobectomy outcomes? Epilepsia 53 (1), 1–6.

Bonilha, L., 2013. Presurgical connectome and postsurgical seizure control in temporal lobe epilepsy. Neurology 81 (19), 1704–1710.

Bonilha, L., 2015. The brain connectome as a personalized biomarker of seizure outcomes after temporal lobectomy. Neurology 84 (18), 1846–1853.

Chang, C.C., Lin, C.J., 2011. LIBSVM: a library for support vector machines. ACM Trans. Interact. Intell. Syst. 2 (3), 1–27.

Ciccarelli, O., 2006. Probabilistic diffusion tractography: a potential tool to assess the rate of disease progression in amyotrophic lateral sclerosis. Brain 129 (7), 1859–1871.

Crossley, N.A., 2014. The hubs of the human connectome are generally implicated in the anatomy of brain disorders. Brain 137 (Pt 8), 2382–2395.

Daianu, M., 2013. Breakdown of brain connectivity between normal aging and alzheimer's disease: a structural k-core network analysis. Brain Connect. 3 (4), 407–422.

DeSalvo, M.N., 2013. Altered structural connectome in temporal lobe epilepsy. Radiology 270 (3), 842–848.

DeSalvo, M.N., 2014. Altered structural connectome in temporal lobe epilepsy. Radiology 270 (3), 842–848.

Elger, C.E., 2004. Chronic epilepsy and cognition. Lancet Neurol. 3 (11), 663–672.

Engel, J., 2003. Practice parameter: temporal lobe and localized neocortical resections for epilepsy report of the quality standards subcommittee of the American Academy of Neurology, in association with the American Epilepsy Society and the American Association of Neurological Surgeons. Neurology 60 (4), 538–547.

Focke, N.K., 2008. Voxel-based diffusion tensor imaging in patients with mesial temporal lobe epilepsy and hippocampal sclerosis. NeuroImage 40 (2), 728–737.

Gong, P., 2013. Multi-stage multi-task feature learning. Adv. Neural Inform. Process. Syst. 14, 2979–3010.

Goodglass, H., et al., 2000. The Assessment of Aphasia and Related Disorders, third ed. Lippincott Williams & Wilkins, Philadelphia.

Hagmann, P., et al., 2008. Mapping the structural core of human cerebral cortex. PLoS Biol. 6 (7), e159.

Heiervang, E., 2006. Between session reproducibility and between subject variability of diffusion MR and tractography measures. NeuroImage 33 (3), 867–877.

ILAE, 1989. Proposal for revised classification of epilepsies and epileptic syndromes. Epilepsia 30 (4), 389–399.

Keller, S.S., Roberts, N., 2008. Voxel-based morphometry of temporal lobe epilepsy: an introduction and review of the literature. Epilepsia 49 (5), 741–757.

Lanckriet, G.R., 2004. Learning the kernel matrix with semidefinite programming. J. Mach. Learn. Res. 5, 27–72.

Liu, M., 2014. Disrupted anatomic white matter network in left mesial temporal lobe epilepsy. Epilepsia 55 (5), 674–682.

Nucifora, P.G., 2007. Diffusion-tensor MR imaging and tractography: exploring brain microstructure and connectivity 1. Radiology 245 (2), 367–384.

Richardson, M.P., 2012. Large scale brain models of epilepsy: dynamics meets connectomics. J. Neurol. Neurosurg. Psychiat. 83 (12), 1238–1248.

Rubinov, M., Sporns, O., 2010. Complex network measures of brain connectivity: uses and interpretations. NeuroImage 52 (3), 1059–1069.

Rubinov, M., Bullmore, E., 2013. Schizophrenia and abnormal brain network hubs. Dial. Clin. Neurosc. 15 (3), 339–349.

Spencer, S.S., 2002. Neural networks in human epilepsy: evidence of and implications for treatment. Epilepsia 43 (3), 219–227.

Sporns, O., 2011. The human connectome: a complex network. Ann. NY Acad. Sci. 1224, 109–125.

Sporns, O., 2013. The human connectome: origins and challenges. NeuroImage 80, 53–61.

Sporns, O., et al., 2005. The human connectome: a structural description of the human brain. PLoS Comput. Biol. 1 (4), e42.

Vapnik, V.N., 1998. Statistical Learning Theory. Wiley, New York.

Xie, T., He, Y., 2011. Mapping the Alzheimer's brain with connectomics. Front. Psychiat. 2, 77.

Zou, H., Hastie, T., 2005. Regularization and variable selection via the elastic net. J. R. Stat. Soc. B 67 (2), 301–320.

Index

Note: Page numbers followed by *f* indicate figures and *t* indicate tables.

D

N

Printed in the United States
By Bookmasters